T0251299

An Introduction to Stochastic Processes with Applications to Biology

Second Edition

An Introduction to Stochastic Processes with Applications to Biology

Second Edition

Linda J. S. Allen

Texas Tech University
Lubbock, Texas, USA

CRC Press
Taylor & Francis Group
Boca Raton London New York

CRC Press is an imprint of the
Taylor & Francis Group an **informa** business

A CHAPMAN & HALL BOOK

CRC Press
Taylor & Francis Group
6000 Broken Sound Parkway NW, Suite 300
Boca Raton, FL 33487-2742

© 2011 by Taylor & Francis Group, LLC
CRC Press is an imprint of Taylor & Francis Group, an Informa business

No claim to original U.S. Government works

International Standard Book Number: 978-1-4398-1882-4 (Hardback)

Library of Congress Cataloging-in-Publication Data

Allen, Linda J. S.
    An introduction to stochastic processes with applications to biology / Linda J. S. Allen. -- 2nd ed.
        p. cm.
    Includes bibliographical references and index.
    ISBN 978-1-4398-1882-4 (hardback)
    1. Stochastic processes. 2. Biomathematics. I. Title.

QA274.A63 2010
519.2'3--dc22   2010043676

Visit the Taylor & Francis Web site at
http://www.taylorandfrancis.com

and the CRC Press Web site at
http://www.crcpress.com

Dedication

This book is dedicated to my husband Edward, my daughter Anna, and in memory of my parents Vivian and Jerry Svoboda.

Preface to the First Edition

My goal in writing this book is to provide an introduction to the basic theory of stochastic processes and to some of the many biological applications of stochastic processes. The mathematical and biological background required is kept to a minimum so that the topics are accessible to students and scientists in biology, mathematics, and engineering. Many of the biological applications are from the areas of population dynamics and epidemiology due to personal preference and expertise and because these applications can be readily understood.

Interdisciplinary areas such as mathematical biology, biostatistics, and bioengineering are growing rapidly. Modeling and analysis of biological phenomena require techniques and tools from various disciplines. Many recent books have addressed the need for a better understanding of the underlying theory of modeling in biology. However, much more attention has been paid to the area of deterministic modeling in biology than to stochastic modeling, and this book helps to fill the gap.

The topics in this book are covered in a one-semester graduate course offered by the Department of Mathematics and Statistics at Texas Tech University. This book is intended for an introductory course in stochastic processes. The targeted audiences for this book are advanced undergraduate students and beginning graduate students in mathematics, statistics, biology, or engineering. The level of material in this book requires only a basic background in probability theory, linear algebra, and analysis. Measure theory is not required. Exercises at the end of each chapter help reinforce concepts discussed in each chapter. To better visualize and understand the dynamics of various stochastic processes, students are encouraged to use the MATLAB® programs provided in the appendices. These programs can be modified for other types of processes or adapted to other programming languages. In addition, research on current stochastic biological models in the literature can be assigned as individual or group research projects.

The book is organized according to the following three types of stochastic processes: discrete-time Markov chains, continuous-time Markov chains, and continuous-time and continuous-state Markov processes. Because many biological phenomena can be modeled by one or more of these three modeling approaches, there may be different stochastic models for the same biological phenomena (e.g., logistic growth and epidemics). Biological applications are presented in each chapter. Some chapters and sections are devoted entirely to the discussion of biological applications and their analysis (e.g., Chapter 7).

In Chapter 1, topics from probability theory are briefly reviewed that are

vii

particularly relevant to stochastic processes. In Chapters 2 and 3, the theory and biological applications of discrete-time Markov chains are discussed, including the classical gambler's ruin problem, birth and death processes, and epidemic processes. In Chapter 4, the topic of branching process is discussed, a discrete-time Markov chain important to applications in biology and medicine. An application to an age-structured population is discussed in Chapter 4. Chapters 5, 6, and 7 present the theory and biological applications of continuous-time Markov chains. Chapter 6 concentrates on birth and death processes, and in Chapter 7 there are applications to epidemic, competition, predation, and population genetics processes. The last chapter, Chapter 8, is a brief introduction to continuous-time and continuous-state Markov processes; that is, diffusion processes and stochastic differential equations. Chapter 8 is a nonmeasure theoretic introduction to stochastic differential equations. These eight chapters can be covered in a one-semester course. One may be selective about the particular applications covered, particularly in Chapters 3, 7, and 8. In addition, Section 1.6 on the simple birth process and Section 2.10 on the random walk in two and three dimensions are optional.

Numerous applications of stochastic processes important in areas outside of biology, including finance, economics, physics, chemistry, and engineering, can be found in the references. This book stresses biological applications and, therefore, some topics in stochastic processes important to these other areas are omitted or discussed very briefly. For example, martingales are not discussed, and queueing theory is discussed only briefly in Chapter 6.

Throughout this book, the emphasis is on Markov processes due to their rich structure and the numerous biological models satisfying the Markov property. However, there are also many biological applications where the Markov restriction does not apply. A stochastic process is discussed in Section 7.2, which is a non-Markovian, age-dependent process belonging to a class of stochastic processes known as regenerative processes. It is important to note that in some applications the Markov restriction is not necessary (e.g., first passage time in Chapter 2 and the waiting time distribution in Chapter 5). This latter theory can be discussed in the more general context of renewal theory.

In writing this book, I received much help and advice from colleagues and friends and I would like to acknowledge their contributions. First, I thank my husband, Edward Allen, for his careful proofreading of many drafts of this book, especially for his help with Chapter 8, and for his continuous encouragement throughout the long process of writing and rewriting. I thank Robert Paige, Texas Tech University, for reviewing Chapter 1 and Thomas Gard, University of Georgia, for reviewing Chapter 8; I thank both of them for their numerous suggestions on these chapters. I am grateful to my graduate students, Nadarajah Kirupaharan and Keith Emmert, for their help in checking for errors. Also, I am grateful to the students in the biomathematics classes at Texas Tech University during spring 2000, 2002, and 2003 for their feedback on the exercises. I thank Texas Tech University for granting me a leave of absence during the spring semester of 2001. During that time I organized my notes and wrote a preliminary draft of this book. In addition, I thank Prentice Hall editor George Lobell for his advice, Adam Lewenberg for his assistance in formatting and setting the final page layout for this book, and the following Prentice Hall reviewers for their many helpful comments and suggestions: Wei-Min Huang, Lehigh University; John Borkowski, Montana State University; Michael Neubauer, California State University at Northridge and Aparna Huzurbazar, The University of New Mexico (Chapters 1–3); Andrea Brose, UCLA (Chapters 1–5); Xiuli Chao, North Carolina State University; Bozenna Pasik-Duncan, University of Kansas; Magda Peligrad, University of Cincinnati; and Andre Adler, Illinois Institute of Technology (Chapters 1–8). Many books, monographs, and articles were sources of reference in writing this book. These sources of reference are too numerous to mention here but are acknowledged in the list of references at the end of each chapter. Finally, and most important, I thank God for His constant support and guidance. After countless revisions based on suggestions from knowledgeable colleagues and friends, I assume full responsibility for any omissions and errors in this final draft.

Preface to the Second Edition

Two of the most significant changes in the second edition are the expansion of Chapter 8 to two chapters and the inclusion of examples and exercises from cellular and molecular biology. Chapter 8 concentrates on the basic theory of diffusion processes and stochastic differential equations, Sections 8.1-8.9 and 8.13 from the first edition. The new Chapter 9 extends the basic theory to multivariate processes, including multivariate forward and backward Kolmogorov differential equations and the multivariate Itô's formula. Systems of Itô stochastic differential equations are derived based on the possible changes that occur during the biological process. Several examples illustrate the derivation procedure. Biological examples from cellular and molecular biology are included in the examples and exercises throughout Chapters 2-9.

Additional changes and revisions in the second edition include a new section on Monte Carlo simulation in Chapter 2. This technique is applied to Markov chain models in later chapters. The number of exercises at the end of each chapter and number of MATLAB® programs has increased by over 50%. Answers and hints to selected exercises are included in Appendix A. The discussion on the n-step transition matrix, formerly Section 2.8, has been moved to the Appendix for Chapter 2. Errors in the first edition were corrected, sections on first passage time were expanded, and several sections in each chapter were revised to clarify some of the topics and to provide additional references from the literature. Certainly, new errors were introduced during the revision but, hopefully, many of the errors were caught before the second edition went to press.

I would like to acknowledge the assistance of family, friends, and colleagues for their help with the second edition: Edward Allen, Texas Tech University, for proofreading Chapters 8 and 9 and for many helpful suggestions; Harvey Qu, Oakland University and George Fegan, Santa Clara University, for help in correcting some of the errors in the first edition; Lih-Ing Roeger, Texas Tech University, for proofreading Chapters 5 and 6; Sukhitha Vidurupola, Yuan Yuan, and Chelsea Lewis, graduate students at Texas Tech University, for reading several chapters and checking the solutions; Louis Gross, University of Tennessee, for suggesting Monte Carlo simulations and examples from systems biology be included in the second edition; and Frédéric Hamelin, Université de Rennes, France, for help on the diffusion approximation of the quasistationary distribution. In addition, I acknowledge the financial support of the National Science Foundation under the grants DMS-0201105 and DMS-0718302. Finally, I would like to thank the editorial staff at Taylor and Francis, Bob Stern, Executive Editor, Marsha Pronin, Project Coordinator, Amy Rodriguez, Project Editor, and Shashi Kumar, LATEX help desk, for their support during the final preparation of this book. I take full responsibility for omissions and errors in this second edition.

For product information on MATLAB, please contact:
The MathWorks, Inc.
3 Apple Hill Drive
Natick, MA 01760-2098 USA
Tel: 508-647-7000
Fax: 508-647-7001
E-mail: info@mathworks.com
Web: www.mathworks.com

Linda J. S. Allen
Texas Tech University

Contents

List of Tables

List of Figures

Chapter 1

Review of Probability Theory and an Introduction to Stochastic Processes

1.1 Introduction

The underlying mathematical theory of stochastic modeling lies within *stochastic processes*. The theory of stochastic processes is based on probability theory. Therefore, this first chapter is a brief review of some basic principles from probability theory. An excellent reference for stochastic processes with applications to biology is the classic textbook by Bailey, *The Elements of Stochastic Processes with Applications to the Natural Sciences*, which has been referenced frequently since its initial publication in 1964 and is cited frequently in this book as well. John Wiley & Sons republished this classic textbook in 1990 and it is this reference that is cited. Other references for various types of stochastic processes are listed at the end of each of the chapters.

It is the goal of this book to introduce the basic theory of stochastic processes necessary in understanding and applying these methods to biological problems, such as population extinction, enzyme kinetics, two-species competition and predation, spread of epidemics, and the genetics of inbreeding. Some of the classical stochastic models from biology have a deterministic counterpart. For these models, the dynamics of the deterministic model will be discussed and compared to that of the corresponding stochastic model. One of the most important differences between deterministic and stochastic models is that deterministic models predict an outcome with absolute certainty, whereas stochastic models provide only the probability of an outcome. Stochastic models take into account the variability in the process due to births, deaths, transitions, emigration, immigration, and environmental variability that impact the biological process. For example, a deterministic model for a population of cells, animals, plants, or genes is often expressed in terms of difference equations or differential equations which follow the growth and decline in the size of the populations over time. The solution of these equations follows a prescribed path or trajectory over time. Given a fixed initial population size, a unique deterministic solution will follow the same path over time. A stochastic model of a population process may also be expressed

1

in terms of difference equations (transition matrix) or differential equations (Kolmogorov differential equations or stochastic differential equations). However, unlike a unique deterministic solution, for a fixed initial population size, a stochastic solution does not follow the same path over time because each solution represents only a single realization (a sample path) from a unique probability distribution. Thus, to predict the stochastic behavior of a biological process it is necessary to know the unique probability distribution of the process. When this is not possible, which is the case for most realistic stochastic models, other analytical and computational methods are applied–analysis of the moments such as mean and variance of the distribution, estimation of the probability of extinction, numerical simulation of sample paths, and computation of probability histograms. All of these methods are applied to various biological processes in the following chapters.

This first chapter begins with a review of probability theory. In Sections 1.3 and 1.4, generating functions are introduced and the Central Limit Theorem is stated. A variety of examples illustrate and reinforce these important concepts from probability theory. In the last two sections of this chapter, stochastic processes are defined and an example is given of a simple birth process. This chapter is not intended to be a comprehensive review of probability theory but only to be a brief review of some concepts from probability theory that are important to the theory of stochastic processes.

1.2 Brief Review of Probability Theory

Many books may be consulted for a more extensive review of probability. A classical reference for probability theory is the book by Feller (1968). Another good reference for probability theory and probability models is the book by Ross (2006). Other references on the basic theory of probability and statistics include Hogg and Craig (1995), Hogg and Tanis (2001), and Hsu (1997).

1.2.1 Basic Probability Concepts

Let S be a set, any collection of elements, which shall be referred to as the *outcome space* or *sample space*. For example, the sample space could be $S = \{H, T\}$, $S = \{0, 1, 2, \ldots\}$, $S = \{s : s \in [0, \infty)\}$, or any collection of objects or elements. Each element of S is called a *sample point* and each subset of S is referred to as an *event*. For example, suppose a coin is tossed and whether the coin lands with heads or tails showing is recorded. Then the sample space is $\{H, T\}$, a sample point is H or T, and an event may be $\{H\}$, $\{T\}$, $\{H, T\}$, or $\emptyset$. If the coin is fair, then the probability of a head appearing is $1/2$, the probability of a tail is $1/2$, and the probability of any other event is zero. In general, for any experiment, there is an underlying probability space, an ordered triple $(S, \mathcal{A}, P)$, where S is the sample space, $\mathcal{A}$ is the collection of subsets or *events* in S, and P is a *probability measure* defined on $\mathcal{A}$. The set $\mathcal{A}$ must satisfy three properties.

DEFINITION 1.1   *Let $\mathcal{A}$ be a collection of subsets of S. Then $\mathcal{A}$ is called a σ-algebra and the ordered pair $(S, \mathcal{A})$ is called a* measurable space *if $\mathcal{A}$ has the following properties:*

(i) $S \in \mathcal{A}$.

(ii) If $B \in \mathcal{A}$, then the complement of B, denoted B^c, is in $\mathcal{A}$, i.e.,

$$B^c = \{s : s \in S, s \notin B\} \in \mathcal{A}.$$

(iii) For any sequence $\{B_n\}_{n=1}^{\infty}$, the union $\cup_{n=1}^{\infty} B_n \in \mathcal{A}$.

The probability measure P is defined on $\mathcal{A}$.

DEFINITION 1.2   *Let $(S, \mathcal{A})$ be a measurable space. Let P be a real-valued set function defined on the σ-algebra $\mathcal{A}$. The set function $P : \mathcal{A} \to [0, 1]$ is called a* probability measure *if it has the following properties:*

(1) $P(B) \geq 0$ for all $B \in \mathcal{A}$.

(2) $P(S) = 1$.

(3) If $B_i \cap B_j = \emptyset$ for $i, j = 1, 2, \ldots, i \neq j$ (pairwise disjoint), then

$$P\left(\cup_{i=1}^{\infty} B_i\right) = \sum_{i=1}^{\infty} P(B_i),$$

where $B_i \in \mathcal{A}$, $i = 1, 2, \ldots$.

The probability measure P along with the measurable space $(S, \mathcal{A})$ define a *probability space*, the ordered triple $(S, \mathcal{A}, P)$. Often we do not define the underlying measurable space $(S, \mathcal{A}, P)$ but concentrate on the probability measure P. For example, if S is a finite set, $\mathcal{A}$ can be taken to be all subsets of S or if S is the set of real numbers $\mathbb{R}$, the set $\mathcal{A}$ can be taken as the Borel measurable sets. For more theoretical introductions to measurable spaces, please consult the references. Next, the concepts of conditional probability and independence are defined.

DEFINITION 1.3   *Let $(S, \mathcal{A}, P)$ be a probability space. Let B_1 and B_2 be two events (elements of $\mathcal{A}$). The* conditional probability of event B_1 *given* event B_2, *denoted* $P(B_1|B_2)$, *is defined as*

$$P(B_1|B_2) = \frac{P(B_1 \cap B_2)}{P(B_2)},$$

provided that $P(B_2) > 0$. *Similarly, the* conditional probability of event B_2 given event B_1 *is*

$$P(B_2|B_1) = \frac{P(B_1 \cap B_2)}{P(B_1)},$$

provided that $P(B_1) > 0$.

The events B_1 and B_2 are independent if the occurrence of either one of the events does not affect the probability of occurrence of the other. More formally,

DEFINITION 1.4   *Let* $(S, \mathcal{A}, P)$ *be a probability space. Let* B_1 *and* B_2 *be two events. Events* B_1 *and* B_2 *are said to be* independent *if and only if*

$$P(B_1 \cap B_2) = P(B_1)P(B_2).$$

If the events B_1 *and* B_2 *are not independent, they are said to be* dependent.

Therefore, B_1 and B_2 are independent if and only if $P(B_2|B_1) = P(B_2)$ or $P(B_1|B_2) = P(B_1)$. In other words, the events B_1 and B_2 are independent if the probability of B_2 does not depend on whether B_1 has occurred or the probability of B_1 does not depend on the occurrence of B_2.

The concept of a random variable is central to probability theory.

DEFINITION 1.5   *Let* $(S, \mathcal{A})$ *be a measurable space. A* random variable X *is a real-valued function defined on the sample space* S,

$$X : S \rightarrow \mathbb{R} = (-\infty, \infty),$$

such that

$$X^{-1}(-\infty, a] = \{s : X(s) \leq a\} \in \mathcal{A}.$$

Let A *denote the range of* X,

$$A = \{x : X(s) = x, s \in S\},$$

also known as the space of X *or* state space *of* X.

If the range of X is finite or countably infinite, then X is said to be a *discrete random variable*, whereas if the range is an interval (finite or infinite in length), then X is said to be a *continuous random variable*. However, the random variable could be of *mixed type*, having properties of both a discrete and a continuous random variable. The distinction between discrete and continuous random variables can be seen in the examples. We shall be primarily concerned with discrete or continuous random variables, rather than the mixed type.

Example 1.1. Suppose two fair coins are tossed sequentially and the outcomes are HH, HT, TH, and TT (i.e., $S = \{HH, HT, TH, TT\}$). Let X be the discrete random variable associated with this experiment having state space $A = \{1, 2, 3, 4\}$, where $X(HH) = 1$, $X(HT) = 2$, and so on. Assume each of the outcomes has an equal probability of $1/4$. $P(\{HH\}) = 1/4$, $P(\{HT\}) = 1/4$, etc. Let B_1 be the event that the first coin is a head, $B_1 = \{HH, HT\}$, and B_2 be the event that the second coin is a head, $B_2 = \{HH, TH\}$. Then $P(B_1) = 1/2$, $P(B_2) = 1/2$, and $P(B_1 \cap B_2) = P(\{HH\}) = 1/4$. Then

$$P(B_2|B_1) = \frac{P(B_1 \cap B_2)}{P(B_1)} = \frac{1/4}{1/2} = \frac{1}{2}.$$

Since $P(B_2) = 1/2$, B_1 and B_2 are independent events. ∎

According to Definition 1.5, events associated with a random variable X can be related to the subsets of $\mathbb{R}$. For example, we shall use the shorthand notation $X = x$ to mean the event $\{s : X(s) = x, s \in S\}$ and the notation $X \leq x$ or $(-\infty, x]$ for the event $\{s : X(s) \leq x, s \in S\}$. In Example 1.1, $B_1 = \{HH, HT\}$ expressed in terms of the state space of X is represented by the set $\{1, 2\}$ and $B_2 = \{HH, TH\}$ by the set $\{1, 3\}$. With this convention, the probability measure P associated with the random variable X can be defined on $\mathbb{R}$. Sometimes this measure is denoted as $P_X : \mathbb{R} \to [0, 1]$ and referred to as the *induced probability measure* (Hogg and Craig, 1995). The subscript X is often omitted, but it will be clear from the context that the induced probability measure is implied. This notation is used in defining the cumulative distribution function of a random variable X that is defined on the set of real numbers.

DEFINITION 1.6 *The* cumulative distribution function *(c.d.f.) of the random variable X is the function $F : \mathbb{R} \to [0, 1]$, with domain $\mathbb{R}$ and range $[0, 1]$, defined by*

$$F(x) = P_X((-\infty, x]).$$

It can be shown that F is nondecreasing, right continuous, and satisfies

$$\lim_{x \to -\infty} F(x) = F(-\infty) = 0 \quad \text{and} \quad \lim_{x \to \infty} F(x) = 1.$$

The cumulative distribution describes how the probabilities accumulate (Examples 1.2 and 1.3).

Functions that define the probability measure for discrete and continuous random variables are the probability mass function and the probability density function.

DEFINITION 1.7   *Suppose X is a discrete random variable. Then the function f defined as $f(x) = P_X(X = x)$ for each x in the range of X is called the* probability mass function *(p.m.f.) of X.*

It follows from Definition 1.2 that f has the following two properties:

$$\sum_{x \in A} f(x) = 1 \text{ and } P_X(X \in B) = \sum_{x \in B} f(x), \tag{1.1}$$

for any $B \subset A$, where A is the range of X. In addition, the c.d.f. F of a discrete random variable satisfies

$$F(x) = \sum_{a_i \leq x} f(a_i),$$

where A is the space of X, a collection of elements $\{a_i\}_i$ and $F(x) = 0$ if $x < \inf_i\{a_i\}$.

Example 1.2. Let the space of the discrete random variable X be $A = \{1, 2, 3, 4, 5\}$ and $f(x) = 1/5$ for $x \in A$. The c.d.f. F of X is

$$F(x) = \begin{cases} 0, & x < 1, \\ 1/5, & 1 \leq x < 2, \\ 2/5, & 2 \leq x < 3, \\ \vdots & \vdots \\ 1, & 5 \leq x. \end{cases}$$

The c.d.f. F, graphed in Figure 1.1, is known as a *discrete uniform distribution*. ∎

DEFINITION 1.8   *Suppose X is a continuous random variable with c.d.f. F. If there exists a nonnegative, integrable function f, $f : \mathbb{R} \to [0, \infty)$, such that*

$$F(x) = \int_{-\infty}^{x} f(y) \, dy,$$

then the function f is called the probability density function *(p.d.f.) of X.*

The p.d.f. of a continuous random variable can be used to compute the probability associated with an outcome or an event. Suppose A is the space of X and $B \subset A$ is an event. Then

$$P_X(X \in A) = \int_A f(x) \, dx = 1 \text{ and } P_X(X \in B) = \int_B f(x) \, dx. \tag{1.2}$$

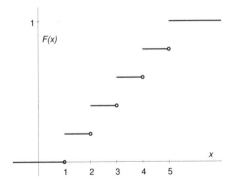

FIGURE 1.1: Discrete uniform c.d.f.

In particular,

$$P_X(a \leq X \leq b) = \int_a^b f(x)\, dx = F(b) - F(a).$$

For a continuous random variable X,

$$P_X(a < X < b) = P_X(a \leq X < b) = P_X(a < X \leq b) = P_X(a \leq X \leq b),$$

which follows from (1.2). In addition, if the cumulative distribution function F is differentiable for $x \in B$, then

$$\frac{dF(x)}{dx} = f(x) \ \text{ for } \ x \in B.$$

Example 1.3. Let the space of a continuous random variable X be $A = [0, 1]$ and the p.d.f. be $f(x) = 1$ for $x \in (0, 1)$. The c.d.f. F is

$$F(x) = \begin{cases} 0, \ x < 0, \\ x, \ 0 \leq x < 1, \\ 1, \ 1 \leq x. \end{cases}$$

The c.d.f. F, graphed in Figure 1.2, is known as a *continuous uniform distribution*. ∎

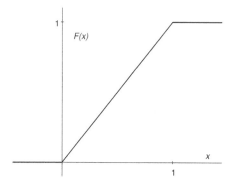

FIGURE 1.2: Continuous uniform c.d.f.

Sometimes we shall use the term *probability density function* to include both the p.d.f. of a continuous random variable and the p.m.f. of a discrete random variable. In addition, sometimes the notation Prob{·} is used in place of $P(\cdot)$ or $P_X(\cdot)$ to emphasize the fact that a probability is being computed. For example, for a discrete random variable X,

$$P_X(X = x) = \text{Prob}\{X = x\} = f(x)$$

and for either a continuous or discrete random variable X,

$$P_X(X \le x) = \text{Prob}\{X \le x\} = F(x).$$

1.2.2 Probability Distributions

Some well-known discrete distributions include the uniform, binomial, geometric, negative binomial, and Poisson. The binomial distribution will be seen in many applications of discrete-time and continuous-time Markov chains. For example, the binomial distribution appears as the solution in the simple death process, whereas the negative binomial distribution appears in the simple birth process. The Poisson distribution is especially important in the study of continuous-time Markov chain models and will be discussed in more detail in Chapter 5. The p.m.f.s for some discrete distributions are summarized below.

Discrete Uniform:

$$f(x) = \begin{cases} \dfrac{1}{n}, & x = 1, 2, \ldots, n, \\ 0, & \text{otherwise.} \end{cases}$$

Geometric:

$$f(x) = p(1 - p)^x, \quad x = 0, 1, 2, \ldots, \quad 0 < p < 1.$$

The value of $f(x)$ can be thought of as the probability of one success in $x+1$ trials, where p is the probability of success.

Binomial:

$$f(x) = \begin{cases} \dbinom{n}{x} p^x(1-p)^{n-x}, & x = 0, 1, 2, \ldots, n, \\ 0, & \text{otherwise,} \end{cases}$$

where n is a positive integer and $0 < p < 1$. The notation $\dbinom{n}{x}$ for the binomial coefficient is defined as

$$\binom{n}{x} = \frac{n!}{x!(n-x)!}.$$

For example,

$$\binom{5}{1} = 5 \text{ and } \binom{5}{3} = 10.$$

It is assumed that $0! = 1$. The binomial probability distribution is denoted as $b(n, p)$. The value of $f(x)$ can be thought of as the probability of x successes in n trials, where p is the probability of success.

Negative Binomial:

$$f(x) = \begin{cases} \dbinom{x+n-1}{n-1} p^n(1-p)^x, & x = 0, 1, 2, \ldots, \\ 0, & \text{otherwise,} \end{cases} \tag{1.3}$$

where n is a positive integer and $0 < p < 1$. The value of $f(x)$ can be thought of as the probability of n successes in $n + x$ trials, where p is the probability of success. For $n = 1$ the negative binomial distribution simplifies to the geometric distribution.

Poisson:

$$f(x) = \begin{cases} \dfrac{\lambda^x e^{-\lambda}}{x!}, & x = 0, 1, 2, \ldots, \\ 0, & \text{otherwise,} \end{cases}$$

where λ is a positive constant.

Some well-known continuous distributions include the uniform, normal, gamma, and exponential. The uniform distribution is the basis for a random number generator, which is used extensively in numerical simulations of stochastic models. The normal distribution is the underlying distribution for Brownian motion, a diffusion process studied in Chapters 8 and 9. The gamma and exponential distributions are associated with waiting time distributions, the time until one or more than one event occurs. The exponential distribution will be seen extensively in the continuous-time Markov chain models discussed in Chapters 5, 6, and 7. The p.d.f.s for each of these distributions are defined below.

Uniform:

$$f(x) = \begin{cases} \dfrac{1}{b-a}, & a \le x \le b, \\ 0, & \text{otherwise,} \end{cases}$$

where $a < b$ are constants. The uniform distribution is denoted as $U(a,b)$.

Gamma:

$$f(x) = \begin{cases} \dfrac{1}{\Gamma(\alpha)\beta^\alpha} x^{\alpha-1} e^{-x/\beta}, & x \ge 0, \\ 0, & x < 0, \end{cases}$$

where α and β are positive constants and

$$\Gamma(\alpha) = \int_0^\infty \frac{1}{\beta^\alpha} x^{\alpha-1} e^{-x/\beta} \, dx.$$

For a positive integer n, $\Gamma(n) = (n-1)!$.

Exponential:

$$f(x) = \begin{cases} \lambda e^{-\lambda x}, & x \ge 0, \\ 0, & x < 0, \end{cases}$$

where λ is a positive constant.

Normal:

$$f(x) = \frac{1}{\sigma\sqrt{2\pi}} \exp\left(-\frac{(x-\mu)^2}{2\sigma^2}\right), \quad -\infty < x < \infty,$$

where μ and σ are constants. A random variable X that is normally distributed with mean μ and variance σ^2 is denoted as $X \sim N(\mu, \sigma^2)$. The standard normal distribution is denoted as $N(0,1)$, where $\mu = 0$ and $\sigma^2 = 1$.

Note that the exponential distribution is a special case of the gamma distribution, where $\alpha = 1$ and $\beta = 1/\lambda$. Graphs of a Poisson mass function and the standard normal density are given in Figure 1.3. The MATLAB® program which generated the Poisson mass function is given in the Appendix for Chapter 1.

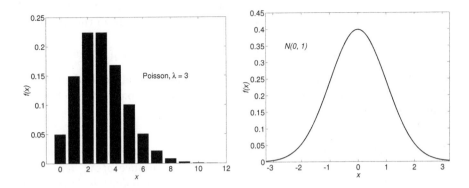

FIGURE 1.3: Poisson mass function with parameter $\lambda = 3$ and the standard normal density.

1.2.3 Expectation

An important concept that helps characterize the p.d.f. of a random variable is the expectation.

DEFINITION 1.9 *Suppose X is a continuous random variable with p.d.f. f. Then the* expectation *of X, denoted as $E(X)$, is defined as*

$$E(X) = \int_{\mathbb{R}} x f(x)\, dx.$$

Suppose X is a discrete random variable with probability function f defined on the space $A = \{a_i\}_{i=1}^{\infty}$. Then the expectation *of X is defined as*

$$E(X) = \sum_{i=1}^{\infty} a_i f(a_i).$$

The expectation of X is a weighted average. In particular, the p.d.f. f is weighted by the values of the random variable X.

The definition of the expectation of a random variable can be extended to the expectation of a function of a random variable. Suppose X is a continuous random variable. Then the *expectation of $u(X)$* is

$$E(u(X)) = \int_{\mathbb{R}} u(x) f(x)\, dx.$$

If X is a discrete random variable, then the *expectation of $u(X)$* is

$$E(u(X)) = \sum_{i=1}^{\infty} u(a_i) f(a_i).$$

It can be seen that the expectation is a linear operator defined on the set of functions $u(X)$: If a_1 and a_2 are constants, then it follows from the definition that

$$E(a_1u_1(X) + a_2u_2(X)) = a_1E(u_1(X)) + a_2E(u_2(X)).$$

In addition, for $b \equiv$ constant, $E(b) = b$. The mean, variance, and moments of X, which measure the center and the spread of the p.d.f., are defined in terms of the expectation.

DEFINITION 1.10   *The* mean *of the random variable X, denoted as μ or μ_X, is the expectation of X, $\mu_X = E(X)$. The* variance *of X, denoted as σ^2, σ_X^2, or $\mathrm{Var}(X)$, is $\mathrm{Var}(X) = E([X - \mu_X]^2)$. The* standard deviation *of X is $\sigma = \sqrt{\mathrm{Var}(X)}$. The* nth moment *of X about the point a is $E([X - a]^n)$.*

The subscript X on the mean and variance is used to avoid confusion, especially if more than one random variable is being discussed. The first moment about the origin is the mean, and the second moment about the mean is the variance. An important identity for the variance can be derived from the linearity property of the expectation,

$$\sigma_X^2 = E([X - \mu_X]^2) = E(X^2) - 2\mu_X E(X) + \mu_X^2 = E(X^2) - \mu_X^2.$$

Example 1.4. Suppose X is a random variable with a discrete uniform distribution and Y is a random variable with a continuous uniform distribution [i.e., Y is distributed as $U(0,1)$]. The mean and variance for each of these two random variables are computed. The mean of X is

$$\mu_X = E(X) = \sum_{x=1}^{n}\left(x\frac{1}{n}\right) = \frac{1}{n}\sum_{x=1}^{n}x = \frac{n+1}{2},$$

because $\sum_{x=1}^{n} x = n(n+1)/2$. Then

$$E(X^2) = \sum_{x=1}^{n}\left(x^2\frac{1}{n}\right) = \frac{1}{n}\sum_{x=1}^{n}x^2 = \frac{(n+1)(2n+1)}{6},$$

because $\sum_{x=1}^{n} x^2 = n(n+1)(2n+1)/6$. The variance of X is

$$\sigma_X^2 = E(X^2) - \mu_X^2 = \frac{(n+1)(2n+1)}{6} - \frac{(n+1)^2}{4} = \frac{n^2-1}{12}.$$

The mean of Y is

$$\mu_Y = E(Y) = \int_0^1 y\,dy = \frac{1}{2}$$

and $E(Y^2) = \int_0^1 y^2 \, dy = 1/3$ so that the variance of Y is

$$\sigma_Y^2 = E(Y^2) - \mu_Y^2 = \frac{1}{3} - \frac{1}{4} = \frac{1}{12}.$$

∎

Example 1.5. Consider the normal distribution $N(0, \sigma^2)$. The first, second, third, and fourth moments about the origin of this normal distribution are calculated. The normal p.d.f. corresponding to $N(0, \sigma^2)$ is

$$f(x) = \frac{\exp\left(-x^2/2\sigma^2\right)}{\sigma\sqrt{2\pi}}.$$

It is an even function over the interval $(-\infty, \infty)$. Therefore, $xf(x)$ and $x^3 f(x)$ are odd functions on $(-\infty, \infty)$. It follows that the first and third moments about the origin are zero,

$$E(X) = \int_{-\infty}^{\infty} xf(x) \, dx = 0 = \int_{-\infty}^{\infty} x^3 f(x) \, dx = E(X^3).$$

In addition,

$$\begin{aligned}
E(X^2) &= \int_{-\infty}^{\infty} x^2 f(x) \, dx = \int_{-\infty}^{\infty} \frac{x^2 \exp\left(-x^2/2\sigma^2\right)}{\sigma\sqrt{2\pi}} \, dx \\
&= -\left. \frac{\sigma x \exp(-x^2/2\sigma^2)}{\sqrt{2\pi}} \right|_{-\infty}^{\infty} + \int_{-\infty}^{\infty} \frac{\sigma \exp(-x^2/2\sigma^2)}{\sqrt{2\pi}} \, dx \\
&= \sigma^2 \int_{-\infty}^{\infty} f(x) \, dx = \sigma^2,
\end{aligned}$$

where integration by parts is used in the first integral, $u = x$, and $dv = xf(x) \, dx$. Therefore, the normal distribution $N(0, \sigma^2)$ has mean and variance, $\mu_X = 0$ and $\sigma_X^2 = \sigma^2$. In a similar manner, it can be shown that the normal distribution $N(\mu, \sigma^2)$ has mean and variance, $\mu_X = \mu$ and $\sigma_X^2 = \sigma^2$. The fourth moment is computed using the same technique,

$$\begin{aligned}
E(X^4) &= \int_{-\infty}^{\infty} x^4 f(x) \, dx = \int_{-\infty}^{\infty} \frac{x^4 \exp\left(-x^2/2\sigma^2\right)}{\sigma\sqrt{2\pi}} \, dx \\
&= -\left. \frac{\sigma x^3 \exp(-x^2/2\sigma^2)}{\sqrt{2\pi}} \right|_{-\infty}^{\infty} + \int_{-\infty}^{\infty} \frac{3\sigma x^2 \exp(-x^2/2\sigma^2)}{\sqrt{2\pi}} \, dx \\
&= 3\sigma^2 \int_{-\infty}^{\infty} x^2 f(x) \, dx = 3\sigma^2 E(X^2) = 3(\sigma^2)^2.
\end{aligned}$$

∎

The standard normal random variable Z, where $Z \sim N(0,1)$, can be obtained from any normally distributed random variable X. If X is distributed as $N(\mu, \sigma^2)$, then

$$Z = \frac{X - \mu}{\sigma}$$

is distributed as $N(0,1)$. This relationship can be verified by showing that the c.d.f. of Z corresponds to a standard normal:

$$\text{Prob}\{Z \leq z\} = \text{Prob}\left\{\frac{X - \mu}{\sigma} \leq z\right\} = \text{Prob}\{X \leq z\sigma + \mu\}$$

$$= \int_{-\infty}^{z\sigma + \mu} \frac{1}{\sigma\sqrt{2\pi}} \exp\left(-\frac{(x - \mu)^2}{2\sigma^2}\right) dx.$$

A change of variables in the integral, $y = (x - \mu)/\sigma$ and $dy = dx/\sigma$, results in

$$\text{Prob}\{Z \leq z\} = \int_{-\infty}^{z} \frac{1}{\sqrt{2\pi}} e^{-y^2/2} dy.$$

The latter integral is the c.d.f. of the standard normal distribution.

Example 1.6. The standard normal distribution $N(0,1)$ satisfies $\text{Prob}\{Z \leq 0\} = 0.5$ and thus, for an arbitrary normal distribution, where X is distributed as $N(\mu, \sigma^2)$, it follows that

$$\text{Prob}\left\{\frac{X - \mu}{\sigma} \leq 0\right\} = \text{Prob}\{X \leq \mu\} = 0.5.$$

Values of the c.d.f $F(z)$, $z \in [0,3]$, for the standard normal distribution can be found in tabular form in many textbooks. In addition, these values can be numerically approximated directly from the integral. For example,

$$\text{Prob}\{-2.1 \leq Z < 1\} = \int_{-2.1}^{1} \frac{1}{\sqrt{2\pi}} e^{-z^2/2} dz \approx 0.8235.$$

If X is distributed as $N(1,4)$, then

$$\text{Prob}\{X \leq 3\} = \text{Prob}\left\{\frac{X - 1}{2} \leq \frac{3 - 1}{2}\right\} = \text{Prob}\{Z \leq 1\} \approx 0.8413.$$

∎

An interesting property of the normal distribution is that approximately two-thirds of the values lie within one standard deviation of the mean, approximately 95% are within two standard deviations, and more than 99% of the values are within three standard deviations. In particular,

$$\text{Prob}\{-k\sigma \leq X - \mu \leq k\sigma\} = \text{Prob}\{|Z| \leq k\} \approx \begin{cases} 0.6827, & k = 1, \\ 0.9545, & k = 2, \\ 0.9973, & k = 3. \end{cases}$$

(Refer to Figure 1.3.)

1.2.4 Multivariate Distributions

When several random variables, $X_1, X_2, \ldots$, and X_n, are associated with the same sample space, a multivariate probability density function or probability mass function $f(x_1, x_2, \ldots, x_n)$ can be defined. Definitions associated with two random variables, X_1 and X_2, are given, that is, for a random vector (X_1, X_2), where $(X_1, X_2) : S \to \mathbb{R}^2$. These definitions can be easily extended to more than two random variables. For each element s in the sample space S, there is associated a unique ordered pair $(X_1(s), X_2(s))$. The set of all ordered pairs

$$A = \{(X_1(s), X_2(s)) | s \in S\} \subset \mathbb{R}^2, \tag{1.4}$$

is known as the *state space* or *space* of the random vector (X_1, X_2).

DEFINITION 1.11 *Suppose X_1 and X_2 are two continuous random variables defined on the common sample space S, having probability measure $P : S \to [0, 1]$. If there exists a function $f : \mathbb{R}^2 \to [0, \infty)$ such that*

$$\iint\limits_{\mathbb{R}^2} f(x_1, x_2) \, dx_1 \, dx_2 = \iint\limits_{A} f(x_1, x_2) \, dx_1 \, dx_2 = 1$$

and if for $B \subset A$,

$$P_{(X_1, X_2)}(B) = \text{Prob}\,\{(X_1, X_2) \in B\} = \iint\limits_{B} f(x_1, x_2) \, dx_1 \, dx_2,$$

then f is called the joint probability density function (p.d.f.) *or* joint density function *of the random variables X_1 and X_2. The* marginal p.d.f. *of X_1 is defined as*

$$f_1(x_1) = \int\limits_{\mathbb{R}} f(x_1, x_2) \, dx_2.$$

The marginal p.d.f. of X_2, $f_2(x_2)$, can be defined in a similar manner. The set A in Definition 1.11 is defined by equation (1.4), and the function $P_{(X_1, X_2)}$ refers to the induced probability measure on $\mathbb{R}^2$.

DEFINITION 1.12 *Suppose X_1 and X_2 are two discrete random variables defined on a common sample space S, having probability measure $P : S \to [0, 1]$. If there exists a function $f : A \to [0, 1]$ such that*

$$\sum_{A} f(x_1, x_2) = 1$$

and if for $B \subset A$,

$$P_{(X_1, X_2)}(B) = \text{Prob}\,\{(X_1, X_2) \in B\} = \sum_{B} f(x_1, x_2),$$

then f is called the joint probability mass function (p.m.f.) *or* joint mass function *of the random variables* X_1 *and* X_2. *The* marginal probability mass function (p.m.f.) *of* X_1 *is defined as*

$$f_1(x_1) = \sum_{x_2} f(x_1, x_2).$$

The marginal p.m.f. of X_2, $f_2(x_2)$, can be defined in a similar manner.

If the set A can be written as the *product space*, $A = A_1 \times A_2$, then the integrals in Definition 1.11 can be expressed as double integrals and the sums in Definition 1.12 as double sums:

$$\iint_A = \int_{A_1} \int_{A_2} \quad \text{and} \quad \sum_A = \sum_{A_1} \sum_{A_2}.$$

In addition, the sum in Definition 1.12 can be expressed as

$$\sum_{x_2} = \sum_{x_2 \in A_2}.$$

The marginal p.d.f.s or p.m.f.s $f_1(x_1)$ and $f_2(x_2)$ are indeed p.d.f.s or p.m.f.s in their own right, satisfying either (1.2) or (1.1), respectively. The definitions of conditional probability and independence of events are extended to random variables, important concepts in stochastic processes.

DEFINITION 1.13 *Let the random variables* X_1 *and* X_2 *have the joint p.d.f.* $f(x_1, x_2)$ *and marginal p.d.f.s* $f_1(x_1)$ *and* $f_2(x_2)$, *respectively. Let* $X_1|x_2$ *denote the random variable* X_1, *given that the random variable* $X_2 = x_2$, *and* $X_2|x_1$ *denote the random variable* X_2, *given that the random variable* $X_1 = x_1$. *The* conditional p.d.f. *of the random variable* $X_1|x_2$ *is defined as*

$$f(x_1|x_2) = \frac{f(x_1, x_2)}{f_2(x_2)}, \quad f_2(x_2) > 0.$$

The conditional p.d.f. *of* $X_2|x_1$ *is*

$$f(x_2|x_1) = \frac{f(x_1, x_2)}{f_1(x_1)}, \quad f_1(x_1) > 0.$$

DEFINITION 1.14 *Let the random variables* X_1 *and* X_2 *have the joint p.d.f.* $f(x_1, x_2)$ *and marginal p.d.f.s* $f_1(x_1)$ *and* $f_2(x_2)$, *respectively. Assume* $A = A_1 \times A_2$. *The random variables* X_1 *and* X_2 *are said to be* independent *if and only if*

$$f(x_1, x_2) = f_1(x_1)f_2(x_2),$$

for all $x_1 \in A_1$ *and* $x_2 \in A_2$. *Otherwise, they are said to be* dependent.

Definitions 1.13 and 1.14 apply to discrete and continuous random variables. It follows from these definitions that if X_1 and X_2 are independent, then

$$f(x_1|x_2) = f_1(x_1) \text{ and } f(x_2|x_1) = f_2(x_2).$$

DEFINITION 1.15 *The covariance of X_1 and X_2, two jointly distributed random variables, denoted $cov(X_1, X_2)$, is defined as*

$$cov(X_1, X_2) = E[(X_1 - E(X_1))(X_2 - E(X_2))]$$
$$= E(X_1 X_2) - E(X_1)E(X_2).$$

If $cov(X_1, X_2) = 0$, then X_1 and X_2 are said to be uncorrelated.

If the random variables X_1 and X_2 are independent, then they are uncorrelated, i.e., $E(X_1 X_2) = E(X_1)E(X_2)$. The converse of this statement is not true (Exercise 10).

Example 1.7. Suppose the joint p.d.f of the random variables X_1 and X_2 is $f(x_1, x_2) = 8x_1 x_2$ for $0 < x_1 < x_2 < 1$ and 0 otherwise. The marginal p.d.f.s of X_1 and X_2 are

$$f_1(x_1) = \int_{x_1}^{1} 8x_1 x_2 \, dx_2 = 4x_1(1 - x_1^2), \quad 0 < x_1 < 1$$

and

$$f_2(x_2) = \int_{0}^{x_2} 8x_1 x_2 \, dx_1 = 4x_2^3, \quad 0 < x_2 < 1.$$

The random variables X_1 and X_2 are dependent. One reason $f_1(x_1)f_2(x_2) \neq f(x_1, x_2)$ is that $A = \{(x_1, x_2) : 0 < x_1 < x_2 < 1\}$ is not a product space, $A \neq A_1 \times A_2$. A necessary condition for independence of the random variables is that the state space A be a product space. ∎

Example 1.8. Suppose the joint p.d.f. of the random variables X_1 and X_2 is $f(x_1, x_2) = 4x_1 x_2$ for $0 < x_1 < 1$ and $0 < x_2 < 1$ and 0 otherwise. It is easy to see that the random variables X_1 and X_2 are independent and therefore, uncorrelated. The marginal p.d.f.s of X_1 and X_2, respectively, are $f_1(x_1) = 2x_1$ for $0 < x_1 < 1$ and $f_2(x_2) = 2x_2$ for $0 < x_2 < 1$. Hence, to compute $E(X_1 X_2)$ we need only compute

$$E(X_1) = \int_{0}^{1} x f_1(x) \, dx = \int_{0}^{1} 2x^2 \, dx = \frac{2}{3} = E(X_2).$$

Then $E(X_1 X_2) = E(X_1)E(X_2) = 4/9$. ∎

Linear combinations of random variables often occur in applications. Suppose $\{X_i\}_{i=1}^{n}$ is a collection of n random variables, where with the mean of

X_i is equal to μ_i and the variance is equal to σ_i^2, $i = 1, 2, \ldots, n$. Then it is easy to show that the random variable

$$Y = \sum_{i=1}^{n} a_i X_i$$

has mean and variance equal to

$$\mu_Y = \sum_{i=1}^{n} a_i \mu_i \quad \text{and} \quad \sigma_Y^2 = \sum_{i=1}^{n} a_i^2 \sigma_i^2 + \sum_{i<j} a_i a_j cov(X_i, X_j). \qquad (1.5)$$

If the collection of n random variables $\{X_i\}_{i=1}^{n}$ are mutually independent, then the covariance term is zero and the formula for the variance σ_Y^2 in (1.5) simplifies to

$$\sigma_Y^2 = \sum_{i=1}^{n} a_i^2 \sigma_i^2.$$

The notation $\bar{X}$ is used to denote the *average* of a sum of n independent random variables,

$$\bar{X} = \sum_{i=1}^{n} X_i / n.$$

Suppose $\mu_i = \mu$ and $\sigma_i^2 = \sigma^2$ so that the independent random variables have the same mean and variance. Then, according to (1.5) and the independent assumption,

$$\mu_{\bar{X}} = \sum_{i=1}^{n} \frac{1}{n}\mu = \mu \quad \text{and} \quad \sigma_{\bar{X}}^2 = \sum_{i=1}^{n} \left(\frac{1}{n}\right)^2 \sigma^2 = \frac{\sigma^2}{n}.$$

In addition to these basic concepts from probability theory, the concept of generating functions is helpful when analyzing stochastic models. A short review of some important facts concerning generating functions is given in the next section.

1.3   Generating Functions

Generating functions are defined in terms of a discrete random variable, then the definitions are extended to a continuous random variable.

Assume X is a discrete random variable and, for convenience, assume the state space is $\{0, 1, 2, \ldots\}$. Let f denote the p.m.f. of X, defined as

$$f(j) = \text{Prob}\{X = j\} = p_j, \ j = 0, 1, 2, \ldots, \quad \text{where} \quad \sum_{j=0}^{\infty} p_j = 1.$$

The mean and variance of X are

$$\mu_X = E(X) = \sum_{j=0}^{\infty} jp_j$$

and

$$\sigma_X^2 = E[(X - \mu_X)^2] = E(X^2) - \mu_X^2 = \sum_{j=0}^{\infty} j^2 p_j - \mu_X^2.$$

DEFINITION 1.16 *The* probability generating function (p.g.f.) *of the discrete random variable X is a function defined on a subset of the reals, denoted as $\mathcal{P}_X$, and defined by*

$$\mathcal{P}_X(t) = E(t^X) = \sum_{j=0}^{\infty} p_j t^j, \tag{1.6}$$

for some $t \in \mathbb{R}$.

The script notation $\mathcal{P}$ is used for the p.g.f. to distinguish this function from the probability measure P and the subscript X is used to denote its association with the random variable X. This subscript is often omitted when it is clear from the context that the associated random variable is X or some other random variable. Because $\sum_{j=0}^{\infty} p_j = 1$, the sum in (1.6) converges absolutely for $|t| \leq 1$. Thus, $\mathcal{P}(t)$ is well defined for $|t| \leq 1$. As the name implies, the p.g.f. generates the probabilities associated with the distribution

$$\mathcal{P}_X(0) = p_0, \quad \mathcal{P}_X'(0) = p_1, \quad \mathcal{P}_X''(0) = 2!p_2.$$

In general, the kth derivative of the p.g.f. of X is

$$\mathcal{P}_X^{(k)}(0) = k!p_k.$$

Since the series for the p.g.f. converges absolutely on $|t| < 1$, it is infinitely differentiable inside the interval of convergence.

The p.g.f. can be used to calculate the mean and variance of a random variable X. Note that $\mathcal{P}_X'(t) = \sum_{j=1}^{\infty} jp_j t^{j-1}$ for $-1 < t < 1$. Letting t approach one from the left, $t \to 1^-$, yields

$$\mathcal{P}_X'(1) = \sum_{j=1}^{\infty} jp_j = E(X) = \mu_X.$$

The second derivative of $\mathcal{P}_X$ is

$$\mathcal{P}_X''(t) = \sum_{j=1}^{\infty} j(j-1)p_j t^{j-2},$$

so that as $t \to 1^-$,

$$P_X''(1) = \sum_{j=1}^{\infty} j(j-1)p_j = E(X^2 - X).$$

Suppose the mean is finite. Then the variance of X is

$$\begin{aligned} \sigma_X^2 &= \mathrm{Var}(X) = E(X^2) - E(X) + E(X) - [E(X)]^2 \\ &= P_X''(1) + P_X'(1) - [P_X'(1)]^2. \end{aligned}$$

There are several other generating functions useful to the study of stochastic processes: the moment generating function, the characteristic function, and the cumulant generating function.

DEFINITION 1.17   *The* moment generating function *(m.g.f.) of the discrete random variable X with state space $\{0, 1, 2, \ldots\}$ and probability function $f(j) = p_j$, $j = 0, 1, 2, \ldots$, denoted $M_X(t)$ is defined as*

$$M_X(t) = E(e^{tX}) = \sum_{j=0}^{\infty} p_j e^{jt}$$

for some $t \in \mathbb{R}$.

For the state space $\{0, 1, 2, \ldots\}$, $M_X(t)$ is defined for $t \leq 0$, but since the series may not converge for $t > 0$, it may not be defined for $t > 0$. The values of t for which the series converges depend on the particular values of the probabilities. The m.g.f. generates the moments $E(X^k)$ of the distribution of the random variable X provided the summation converges in some interval about the origin:

$$M_X(0) = 1, \quad M_X'(0) = \mu_X = E(X), \quad M_X''(0) = E(X^2),$$

and, in general,

$$M_X^{(k)}(0) = E(X^k).$$

DEFINITION 1.18   *The* characteristic function *(ch.f.) of the discrete random variable X is*

$$\phi_X(t) = E(e^{itX}) = \sum_{j=0}^{\infty} p_j e^{ijt}, \quad \text{where} \quad i = \sqrt{-1}.$$

The characteristic function is defined for all real t because the summation converges for all t.

DEFINITION 1.19 *The* cumulant generating function (c.g.f) *of the discrete random variable* X *is the natural logarithm of the moment generating function, denoted* $K_X(t)$,

$$K_X(t) = \ln[M_X(t)].$$

The generating functions for continuous random variables are defined in a similar manner.

DEFINITION 1.20 *Assume* X *is a continuous random variable with p.d.f.* f. *The* probability generating function (p.g.f.) *of* X *is*

$$\mathcal{P}_X(t) = E(t^X) = \int_{\mathbb{R}} f(x)t^x \, dx.$$

The moment generating function (m.g.f.) *of* X *is*

$$M_X(t) = E(e^{tX}) = \int_{\mathbb{R}} f(x)e^{tx} \, dx$$

and the characteristic function (ch.f.) *of* X *is*

$$\phi_X(t) = E(e^{itX}) = \int_{\mathbb{R}} f(x)e^{itx} \, dx.$$

Finally, the cumulant generating function (c.g.f.) *of* X *is* $K_X(t) = \ln[M_X(t)]$.

The p.g.f. is defined for $|t| < 1$, the ch.f. for all real t, and the m.g.f. and c.g.f. for $t \leq 0$. One generating function can be transformed into another by applying the following identities:

$$\mathcal{P}_X(e^t) = M_X(t) \quad \text{and} \quad M_X(it) = \phi_X(t). \tag{1.7}$$

The same relationships established between the generating functions and the mean and the variance that were shown for discrete random variables hold for continuous random variables as well. Formulas for the mean and the variance, computed from the c.g.f., are verified in the exercises. These formulas are summarized below.

$$\boxed{\mu_X = \mathcal{P}'_X(1) = M'_X(0) = K'_X(0)}$$

and

$$\boxed{\sigma_X^2 = \begin{cases} \mathcal{P}''_X(1) + \mathcal{P}'_X(1) - [\mathcal{P}'_X(1)]^2 \\ M''_X(0) - [M'_X(0)]^2 \\ K''_X(0) \end{cases}}$$

Generating functions for linear combinations of independent random variables can be defined in terms of the generating functions of the individual random variables. Suppose $X_1, X_2, \ldots, X_n$ are n independent random variables and Y is a linear combination of these random variables,

$$Y = \sum_{i=1}^{n} a_i X_i.$$

The m.g.f. of Y has a simple form; it is the product of the individual moment generating functions:

$$\begin{aligned} M_Y(t) &= E(e^{tY}) = E(e^{t(\sum a_i X_i)}) \\ &= E(e^{a_1 t X_1}) E(e^{a_2 t X_2}) \cdots E(e^{a_n t X_n}) \\ &= \prod_{i=1}^{n} M_{X_i}(a_i t). \end{aligned}$$

Often information about the stochastic process is obtained by studying the generating functions of the process. In the next example, the p.g.f. and m.g.f. of a binomial distribution are calculated.

Example 1.9. Let X be a binomial random variable, $b(n,p)$, with p.d.f.

$$f(j) = \binom{n}{j} p^j (1-p)^{n-j}$$

for $j = 0, 1, 2, \ldots, n$. The p.g.f. of X can be computed directly from the binomial expansion formula, e.g., $\sum_{j=0}^{n} f(j) = (p + 1 - p)^n = 1$. The p.g.f. is

$$\begin{aligned} \mathcal{P}_X(t) &= \sum_{j=0}^{n} \binom{n}{j} (pt)^j (1-p)^{n-j} \\ &= (pt + 1 - p)^n. \end{aligned}$$

The m.g.f. can be obtained from the identity (1.7), so that

$$M_X(t) = (pe^t + 1 - p)^n.$$

Calculation of the derivatives,

$$\mathcal{P}_X'(t) = np(pt + 1 - p)^{n-1} \quad \text{and} \quad \mathcal{P}_X''(t) = n(n-1)p^2(pt + 1 - p)^{n-2},$$

leads to $\mu_X = \mathcal{P}_X'(1) = np$ and

$$\begin{aligned} \sigma_X^2 &= \mathcal{P}_X''(1) + \mathcal{P}_X'(1) - [\mathcal{P}_X'(1)]^2 \\ &= n(n-1)p^2 + np - n^2 p^2 \\ &= np(1-p). \end{aligned}$$

An important result concerning generating functions states that the m.g.f. *uniquely* defines the probability distribution (provided the m.g.f. exists in an open interval about zero). For example, if the m.g.f. of X equals $M_X(t) = [0.75 + 0.25e^t]^{20}$, then the distribution is binomial with $n = 20$ and $p = 0.25$ [i.e., $b(n, p) = b(20, 0.25)$].

For reference purposes, the m.g.f. for each of the discrete and continuous random variables that were discussed in Section 1.2.2 are put in the Appendix for Chapter 1.

1.4 Central Limit Theorem

An important theorem in probability theory relates the sum of independent random variables to the normal distribution. The Central Limit Theorem states that the mean of n independent and identically distributed random variables, $\bar{X}$, has an approximate normal distribution if n is large. Generally, the expressions *random sample of size n* or *n independent and identically distributed (i.i.d.)* random variables are used to denote a collection of n independent random variables with the same distribution. We shall use this standard terminology.

The Central Limit Theorem is an amazing result when one realizes this normal approximation applies to a collection of random variables from any distribution with a finite mean and variance, discrete or continuous. If the distribution is skewed and discrete, the size of the random sample may need to be large to ensure a good approximation.

Recall that the mean and variance of $\bar{X} = \sum_{i=1}^{n} X_i/n$, where $\{X_i\}_{i=1}^n$ is a random sample with $\mu_{X_i} = \mu$ and $\sigma^2_{X_i} = \sigma^2$, $i = 1, 2, \ldots, n$, is

$$\mu_{\bar{X}} = \mu \quad \text{and} \quad \sigma^2_{\bar{X}} = \sigma^2/n.$$

The Central Limit Theorem is stated in terms of the random variable W_n, where

$$W_n = \frac{\bar{X} - \mu}{\sigma/\sqrt{n}}.$$

THEOREM 1.1 Central Limit Theorem
Let $X_1, X_2, \ldots, X_n, \ldots$, be a sequence of i.i.d. random variables with finite mean, $|\mu| < \infty$, and positive standard deviation, $0 < \sigma < \infty$. Then, as $n \to \infty$, the limiting distribution of

$$W_n = \frac{\sum_{i=1}^{n} X_i/n - \mu}{\sigma/\sqrt{n}},$$

is a standard normal distribution.

The sequence $\{W_n\}_{n=1}^{\infty}$ *converges in distribution* to a standard normal distribution. That is, for each $z \in (-\infty, \infty)$,

$$\lim_{n\to\infty} \text{Prob}\{W_n \leq z\} = F(z) = \int_{-\infty}^{z} \frac{1}{\sqrt{2\pi}} e^{-y^2/2}\, dy,$$

where $F(z)$ is the c.d.f. of the standard normal distribution. Based on the assumption that the m.g.f. exists, it can be shown that the m.g.f. of W_n approaches the m.g.f. of a standard normal distribution, $\lim_{n\to\infty} M_{W_n}(t) = e^{t^2/2}$ (Exercise 27). For a more general proof based on characteristic functions, see Cramér (1945) or Schervish (1995).

Example 1.10. Suppose $X_1, X_2, \ldots, X_{15}$ are i.i.d. random variables from the binomial distribution $b(6, 1/3)$. The mean and variance for each of the X_i are $\mu = 2$ and $\sigma^2 = 4/3$. A graph of the approximate probability density (histogram) of

$$W_{15} = \frac{\sum_{i=1}^{15} X_i/15 - 2}{\sqrt{4/3}/\sqrt{15}}$$

$(n = 15)$ is compared to the standard normal density on the interval $[-3.5, 3.5]$ (Figure 1.4). The histogram is generated from many random samples of size $n = 15$. For each random sample, the value of w_{15} is calculated. It can be seen that the probability histogram is in close agreement with the standard normal p.d.f. ∎

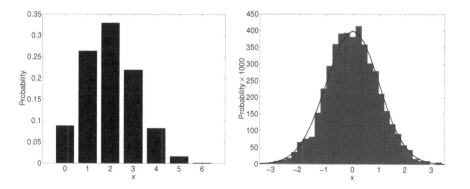

FIGURE 1.4: Binomial mass function $b(6, 1/3)$, and the standard normal density. Approximate probability histogram of W_{15} (defined in the Central Limit Theorem) is graphed against the standard normal density.

A practical application of the Central Limit Theorem is to approximate $\text{Prob}\{a < \bar{X} < b\}$, where $\bar{X} = \sum_{i=1}^{n} X_i/n$ is the average of a random sample of size n.

Example 1.11. Let $X_1, \ldots, X_n$ be a random sample of size $n = 25$ from the uniform distribution $U(0, 1)$. The mean $E(X_i) = 1/2$ and the variance $Var(X_i) = 1/12$ (Example 1.4). Thus, by the Central Limit Theorem,

$$W_{25} = \frac{\bar{X} - 1/2}{1/(5\sqrt{12})}$$

has a distribution close to $N(0, 1)$. Then

$$\text{Prob}\{\bar{X} < 1/2\} = \text{Prob}\{W_{25} < 0\} \approx 0.5$$

and

$$\text{Prob}\{\bar{X} < 1/3\} = \text{Prob}\left\{ \frac{\bar{X} - 1/2}{1/(5\sqrt{12})} < \frac{1/3 - 1/2}{1/(5\sqrt{12})} \right\}$$
$$\approx \text{Prob}\{W_{25} < -2.887\} \approx 0.0019.$$

$\blacksquare$

The Law of Large Numbers is closely related to the Central Limit Theorem. The Law of Large Numbers states that the average of a collection of n i.i.d. random variables converges to the mean, $\bar{X} \to \mu$, as $n \to \infty$ (also known as the *strong* Law of Large Numbers). Convergence is in the mean square sense or with probability one as defined in the next theorem.

THEOREM 1.2 Law of Large Numbers
Let $X_1, X_2, \ldots, X_n, \ldots$, be a sequence of i.i.d. random variables with finite mean, $|\mu| < \infty$, and positive standard deviation, $0 < \sigma < \infty$. Then the mean $\bar{X} \equiv \bar{X}(n)$ satisfies

$$\lim_{n \to \infty} E(|\bar{X}(n) - \mu|^2) = 0 \tag{1.8}$$

and

$$\text{Prob}\left\{ \lim_{n \to \infty} |\bar{X}(n) - \mu| = 0 \right\} = 1. \tag{1.9}$$

The convergence in (1.8) is known as *mean square convergence* and the convergence in (1.9) is known as *convergence with probability one* or as *convergence almost everywhere* (*a.e.*). The probability in (1.9) is often expressed as

$$\lim_{n \to \infty} \bar{X}(n) = \mu$$

with probability one. Notice that mean square convergence for the random sample $\bar{X}(n)$ is equivalent to convergence of the variance to zero,

$$E(|\bar{X}(n) - \mu|^2) = Var(\bar{X}(n)) = \frac{\sigma^2}{n} \to 0$$

as $n \to \infty$.

1.5 Introduction to Stochastic Processes

The history of the development of the theory of stochastic process began with the study of biological as well as physical problems. According to Guttorp (1995), one of the first occurrences of a Markov chain may have been in explaining rainfall patterns in Brussels by Quetelet in 1852. The simple branching process was invented by Bienaymé in 1845 to compute the probability of extinction of a family surname. In 1910, Rutherford and Geiger and the mathematician Bateman described the disintegration of radioactive substances using a Poisson process. In 1905, Einstein described Brownian motion of gold particles in solution, and in 1900, Bachelier used this same process to describe bond prices (Guttorp, 1995). The simple birth and death process was introduced by McKendrick in 1914 to describe epidemics, and Gibbs in 1902 used nearest-neighbor models to describe the interactions among large systems of molecules (Guttorp, 1995). Stochastic processes are now used to model many different types of phenomena from a variety of different areas, including biology, physics, chemistry, finance, economics, and engineering.

DEFINITION 1.21   *A stochastic process* is a collection of random variables $\{X_t(s) : t \in T, s \in S\}$, where T is some index set and S is the common sample space *of the random variables. For each fixed t, $X_t(s)$ denotes a single random variable defined on S. For each fixed $s \in S$, $X_t(s)$ corresponds to a function defined on T that is called a* sample path *or a* stochastic realization *of the process.*

A stochastic process may consist of a collection of random vectors. For example, for two random variables, a stochastic process is a collection of random vectors $\{(X_t^1(s), X_t^2(s)) : t \in T, s \in S\}$. When speaking of a stochastic process, sometimes the variable s is omitted; the random variables are denoted simply as X_t or $X(t)$. We follow this practice. A stochastic model is based on a stochastic process in which specific relationships among the set of random variables $\{X_t\}$ are assumed to hold; they are dependent. In the probability space $(S, \mathcal{A}, P)$, on which the collection of random variables are defined, the probability measure P is the joint distribution of this collection.

Methods and techniques for formulating, analyzing, and numerically computing solutions of stochastic processes depend on whether the random variables or the index set are discrete or continuous. These distinctions between discrete versus continuous random variables and discrete versus continuous index set determine the type of stochastic model and the techniques that can be used to study the process. The set T is often referred to as *time*, and in our stochastic models it will frequently represent time. The first type of stochastic models, discussed in Chapters 1-4, is a type where both time and the random variables are discrete-valued: discrete-time Markov chain models and discrete-time branching processes. In Chapters 5-7, the stochastic models are continuous in time but the random variables are discrete-valued: continuous-time Markov chain models. These types of models have received the most attention in terms of biological applications: cellular, molecular, genetic, competition, predation, and epidemic processes. It will be easy to see the connection between deterministic differential equation models and the stochastic models of this type. In Chapter 4, a brief discussion of stochastic population models with environmental variation leads to equations for the expectation of a structured population that are stochastic difference equations. In this case, the random variables for the expected population size are continuous-valued but the time is discrete-valued. Finally, the last type of model is discussed in Chapters 8 and 9, where both time and the random variables are continuous-valued. These models are referred to as diffusion processes, where the stochastic realization $X(t)$ is a solution of a stochastic differential equation. Inclusion of the variability due to births, deaths, transitions, immigration, emigration, and the environment in biological processes leads to the formulation of a stochastic process, which may take one of the forms discussed in Chapters 2-9 (discrete-time Markov chain, continuous-time Markov chain, stochastic difference equation, or stochastic differential equation).

Four examples of stochastic processes from population biology are described, where the state space or the index set are discrete or continuous:

1. X_t is the position of an object at time t, during a 24 hour period. The object's directional distance from a particular point 0 is measured in integer units. In this case, $T = \{0, 1, 2, \ldots, 24\}$ and the state space is $\{0, \pm 1, \pm 2, \ldots\}$. Both time and state space are *discrete*.

2. X_t is the number of births in a given population during the time period $[0, t]$. In this case, $T = \mathbb{R}_+ = [0, \infty)$ and the state space is $\{0, 1, 2, \ldots\}$. Time is *continuous* and the state space is *discrete*.

3. X_t is the population density at time $t \in T = \mathbb{R}_+ = [0, \infty)$. The state space is also $\mathbb{R}_+$; both time and state are *continuous*.

4. X_t is the expected density of an annual plant species in year t, subject to environmental variation, where $T = \{0, 1, 2, \ldots\}$ and the state space is $\mathbb{R}_+$. Time is *discrete* but the state space is *continuous*.

Examples such as these will be studied in more detail in Chapters 2 through 9. In the case where time is discrete, assumptions will be made regarding the relationship between the state of the system at time t to the state at time $t + 1$. The first example is a random walk model, a discrete-time Markov chain, that will be discussed in Chapter 2. The second example is a simple birth process, a continuous-time Markov chain, discussed in Chapter 6. The third example is a birth and death process with a continuous state space, a diffusion process, which can be modeled by a stochastic differential equation (Chapter 9). The fourth example is a birth and death process, modeled by a stochastic difference equation (Chapter 4).

One of the simplest examples of a stochastic process is a simple birth process (the analogue of exponential growth in deterministic theory). Recall that exponential growth can be modeled by the equation $y(t) = ae^{bt}$, $a, b > 0$, $t \in [0, \infty)$. Three stochastic realizations of the simple birth process when $a = 1 = b$ are graphed in Figure 1.5, and compared to the exponential growth model, $y(t) = e^t$. The corresponding stochastic model has an infinite number of stochastic realizations, only three of them are graphed. For a fixed time t, the random variable X_t has an associated p.m.f. f_t. At a fixed value of t, the stochastic realization equals n, $X_t = n$, for some $n = 0, 1, 2, \ldots$, with probability

$$f_t(n) = \text{Prob}\{X_t = n\}.$$

If $f_t(n) > 0$, then there is a positive probability that $X_t = n$. It will be shown in Chapter 6, for this example, that the mean of X_t equals the deterministic solution,

$$\mu_t = E(X_t) = \sum_{n=0}^{\infty} n f_t(n) = e^t.$$

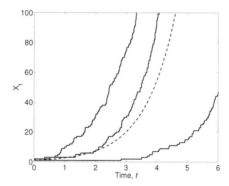

FIGURE 1.5: Three stochastic realizations of the simple birth process and corresponding deterministic exponential growth model $y = e^t$ (dashed curve).

In the simple birth process, time is continuous and the state space is discrete. The simple birth process is discussed briefly in the next section. Some techniques useful to the study of stochastic processes are introduced and some of the differences between deterministic and stochastic models are illustrated. The simple birth process will be studied in more detail in Chapter 6.

1.6 An Introductory Example: A Simple Birth Process

Deterministic and stochastic exponential growth models are derived from first principles. The stochastic model is continuous in time but discrete in the state space.

Three assumptions are made in developing this simple birth model:

(i)   No individuals die.
(ii) There are no interactions between individuals.
(iii) The birth rate b is the same for all individuals.

See also Renshaw (1993). For example, the term population could be a population of animals, plants, cells, or genes.

First, a deterministic model is derived. Let $n(t)$ denote the population size at time t. In a small time period Δt, the increase in the total population size due to a single individual is $b \times \Delta t$ and the increase in the total population size due to all individuals is $b \Delta t \times n(t)$. Thus,

$$n(t + \Delta t) = n(t) + b \Delta t \, n(t).$$

Rewriting this expression leads to

$$\frac{n(t + \Delta t) - n(t)}{\Delta t} = bn(t).$$

Letting $\Delta t \to 0$, we arrive at the differential equation for exponential growth,

$$\frac{dn\,(t)}{dt} = bn(t).$$

If the initial population size is $n(0) = a$, then the solution to this ordinary differential equation (ODE) is

$$n(t) = ae^{bt}.$$

The population size is predicted at time t with absolute certainty, once the initial size a and the birth rate b are known.

Next, a stochastic model is formulated. In this case, the population size is not known with certainty but with some probability; the population size will be n at time t. It is assumed that the population size is discrete-valued but time is continuous. Since the state space is discrete and time is continuous, the stochastic process satisfies $X_t \in \{0, 1, 2, \ldots\}$, $t \in [0, \infty)$, where X_t is the discrete random variable for the size of the population at time t. Let the p.m.f. associated with the random variable X_t be denoted $\{p_n(t)\}_{n=0}^{\infty}$, where

$$p_n(t) = \text{Prob}\{X_t = n\}.$$

Note that the notation is different from the previous sections. This notation for the p.m.f. $p_n(t)$ is consistent with the notation used in later chapters.

The random variables $\{X_t\}_{t=0}^{\infty}$ are related (and dependent) through the following assumptions. Assume that in a sufficiently small period of time Δt,

1. The probability that a birth occurs is approximately $b\,\Delta t$.
2. The probability of more than one birth in time Δt is negligible.
3. At $t = 0$, $\text{Prob}\{X_0 = a\} = 1$.

The assumption that the probability is negligible means it is of order Δt or $o(\Delta t)$. That is, $\lim_{\Delta t \to 0} o(\Delta t)/\Delta t = 0$ or $o(\Delta t)$ approaches zero faster than Δt. The first assumption can be stated more precisely as the probability that a birth occurs is $b\Delta t + o(\Delta t)$.

Based on assumptions 1 and 2, for the population to be of size n at time $t + \Delta t$, either it is of size n at time t and no birth occurs in $(t, t + \Delta t)$, or else it is of size $n - 1$ at time t and one birth occurs in $(t, t + \Delta t)$. The probability that a population of size n increases to $n + 1$ in $(t, t + \Delta t)$ is approximately $b\,\Delta t \times n$, and the probability that the population fails to increase in that time period is then approximately $1 - b\,\Delta t \times n$.

The assumptions relate the state of the process at time $t + \Delta t$, $X_{t+\Delta t}$ to the state at time t, X_t. In terms of the probabilities, the probability that the state equals n at time $t + \Delta t$ depends on whether at time t the population size was $n - 1$ and there was a birth or the size was n and there was no birth; that is,

$$p_n(t + \Delta t) = p_{n-1}(t)b(n - 1)\,\Delta t + p_n(t)(1 - bn\,\Delta t).$$

There is an inherent assumption about independence in deriving this equation. That is, the state of the process at time $t + \Delta t$ only depends on the state at time t and not on the times prior to t (this is known as the *Markov property*). Subtracting $p_n(t)$ from both sides of the equation above, dividing by Δt, and letting $\Delta t \to 0$, a system of differential equations is obtained that is known as the *forward Kolmogorov differential equations*:

$$\frac{dp_n(t)}{dt} = b(n - 1)p_{n-1}(t) - bnp_n(t),$$

where $n = 1, 2, \ldots$. For example,

$$\frac{dp_1(t)}{dt} = -bp_1(t) \quad \text{and} \quad \frac{dp_2(t)}{dt} = bp_1(t) - 2bp_2(t).$$

If the initial population size is zero, then no births occur and $p_0(t) = 1$ for all time.

The process often begins with a known value for X_0. That is, initially, the population size is fixed at a, $X_0 = a$. Written in terms of the probabilities, $p_a(0) = 1$ and $p_n(0) = 0$, $n \neq a$. The forward Kolmogorov equations can be solved iteratively or by using generating function techniques. Methods of solution will be discussed in Chapter 5. It will be shown in Chapter 6 that the solution $p_n(t)$ has the form of a negative binomial distribution for each fixed time t:

$$p_n(t) = \binom{n-1}{a-1} e^{-abt}(1 - e^{-bt})^{n-a}, \quad n = a, a+1, a+2, \ldots.$$

This distribution is a shift of a units to the right of the negative binomial distribution defined in (1.3) (see Exercise 14). The parameter p in the negative binomial distribution is $p = e^{-bt}$. The probability distribution $\{p_n(t)\}_{n=a}^{\infty}$ is graphed in Figure 1.6 when $a = 1$ and $b = 1$ at $t = 0, 1, 2, 3$.

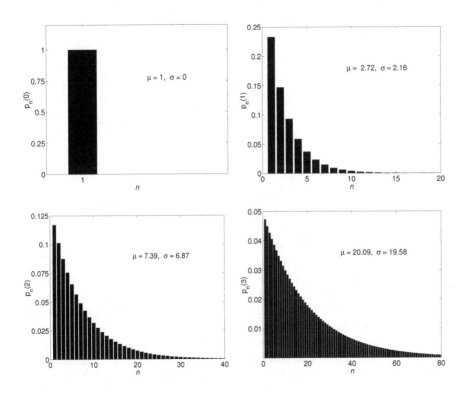

FIGURE 1.6: Graphs of the p.m.f. $p_n(t)$, $n = 0, 1, 2, \ldots$, at $t = 0, 1, 2, 3$ when $X_0 = a = 1$ and $b = 1$ equals the birth rate.

Finding the probability distribution solves the stochastic modeling problem, just as finding the solution $n(t)$ solves the deterministic modeling problem. Other information about the stochastic process can be assessed from the probability distribution (e.g., mean and variance). Therefore, the goal in stochastic modeling is to determine the probability distribution associated with the stochastic process. If this is not possible, then information about the mean or variance or other properties of the distribution is obtained through application of other techniques.

In the simple birth process X_t, the mean and variance can be obtained directly from the negative binomial probability distribution:

$$\mu_t = ae^{bt}$$
$$\sigma_t^2 = ae^{bt}(e^{bt} - 1).$$

Note for this example that the mean equals the solution of the deterministic model. However, the variance increases as time increases. Specific values for the mean and variance when $a = 1$ and $b = 1$ are given in Table 1.1.

Table 1.1: Mean μ_t, variance σ_t^2, standard deviation σ_t, and ODE solution $n(t)$ for the simple birth process at $t = 0, 1, 2, 3$ when $a = 1$ and $b = 1$

t	μ_t	σ_t^2	σ_t	$n(t)$
0	1	0	0	1
1	2.72	4.67	2.16	2.72
2	7.39	47.21	6.87	7.39
3	20.09	383.34	19.58	20.09

An important part of modeling is numerical simulation. Many different programming languages can be used to simulate the dynamics of a stochastic model. The output from a MATLAB program for the simple birth process is graphed in Figure 1.7. MATLAB and FORTRAN programs for the simple birth process are given in the Appendix for Chapter 1. Three stochastic realizations of the simple birth process when $b = 1$, $a = 1$, and $p_1(0) = 1$ are graphed in Figure 1.7. The corresponding deterministic exponential growth model, $n(t) = e^t$, is also graphed. Table 1.2 lists the times at which a birth occurs (up to a population size of 50) for two different realizations or sample paths for the simple birth process. Notice that it requires times of 2.89 and 4.68, respectively, for the two realizations to reach a population size of 50. Recall that the time to reach a population size of 50 in the deterministic model with $b = 1 = a = n(0)$ is found by solving $50 = \exp(t)$, or $t = \ln 50 \approx 3.91$. In general, the time between births is much longer when the population size is small. As the population size builds up, then births occur more frequently and the interevent time decreases. See the Appendix for Chapter 1 for a brief discussion of interevent time.

The values of the stochastic realizations at a particular time t, $X_t = n$,

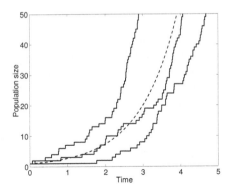

FIGURE 1.7: Three stochastic realizations of the simple birth process when $b = 1$, $a = 1$, and $X_0 = 1$. The deterministic exponential growth model, $n(t) = e^t$, is the dashed curve.

Table 1.2: For two stochastic realizations, the times at which a birth occurs for a simple birth process with $b = 1$ and $a = 1$

Realization 1		Realization 2	
Size $n(t)$	Event Time t	Size $n(t)$	Event Time t
1	0	1	0
2	0.138	2	1.764
3	0.407	3	2.174
4	0.575	4	2.269
5	0.755	5	2.390
6	0.766	6	2.664
7	0.943	7	2.839
8	1.173	8	2.909
9	1.463	9	3.044
10	1.511	10	3.095
⋮	⋮	⋮	⋮
50	2.890	50	4.678

depend on the probability $p_n(t)$,

$$p_n(t) = \mathrm{Prob}\{X_t = n\}.$$

If $p_n(t) > 0$, then it is possible for a stochastic realization to have the value n at time t. For example, for $t = 0, 1, 2$, and 3, the probability distributions, $p_n(1)$, $p_n(2)$ and $p_n(3)$ show that it is possible for the stochastic realization to reach any positive integer value n at $t = 1, 2$, or 3 (Figure 1.6). However, the probability of a very large value of n is small.

1.7 Exercises for Chapter 1

1. Calculate the mean and variance of X, if X has one of the following probability distributions:

   (a) binomial p.m.f.

   (b) uniform p.d.f.

   (c) exponential p.d.f.

2. Suppose the random variable X has a Cauchy probability density:

   $$f(x) = \frac{1}{\pi(1 + x^2)}, \quad -\infty < x < \infty.$$

   Show that $\int_{-\infty}^{\infty} f(x)\, dx = 1$ but the improper integral $\int_{-\infty}^{\infty} x f(x)\, dx$ does not converge. Hence, the mean of X and all higher-order moments are not finite.

3. Let X be a discrete random variable with a geometric distribution.

   (a) Show that the probability generating function (p.g.f.) of X is $\mathcal{P}_X(t) = p(1 - (1 - p)t)^{-1}$.

   (b) Use the p.g.f. to find the mean μ_X and variance σ_X^2.

4. Show that the cumulative distribution function for a random variable X with a geometric distribution is $F(x) = 0$ for $x < 0$, $F(x) = p$ for $0 \le x < 1$, and, in general, $F(x) = 1 - (1 - p)^n$ for $n - 1 \le x < n$ for $n = 2, 3, \ldots$.

5. Show that the cumulative distribution function for a random variable X with an exponential distribution is $F(x) = 1 - e^{-\lambda x}$ for $x \ge 0$ and that $\mathcal{P}_X([x, \infty)) = \mathrm{Prob}\{X \ge x\} = e^{-\lambda x} = 1 - F(x)$ for $x \ge 0$.

6. A special case of the gamma distribution in which $\beta = 2$ and $\alpha = r/2$, r a positive integer, is known as the *chi-square distribution* with r degrees of freedom. The notation for a chi-square random variable is $X \sim \chi^2(r)$. Calculate the mean and the variance of $X \sim \chi^2(r)$.

7. Let $X \sim N(\mu, \sigma^2)$. Show that $Y = (X - \mu)^2/\sigma^2$ is a chi-square random variable with one degree of freedom, $Y \sim \chi^2(1)$.

8. The continuous random variables X_1 and X_2 have the joint probability density function (p.d.f.),

$$f(x_1, x_2) = e^{-x_1 - x_2}, \quad 0 < x_1 < \infty, \quad 0 < x_2 < \infty \qquad (1.10)$$

and zero otherwise.

   (a) Show that X_1 and X_2 are independent and have exponential distributions.

   (b) Write the moment generating function (m.g.f.) of X_1 and X_2.

   (c) Compute $E\left(e^{t(X_1 + X_2)}\right)$.

9. Suppose X_1 and X_2 are independent, continuous random variables with joint p.d.f. equal to $f(x_1, x_2) = f_1(x_1)f_2(x_2)$.

   (a) Show that the joint distribution function $F(x_1, x_2)$ also has this property, $F(x_1, x_2) = F_1(x_1)F_2(x_2)$, where $F_i(x_i) = \text{Prob}\{X_i \leq x_i\}$. The cumulative distribution of X_i is independent of X_j, $i \neq j$, and $F(x_1, x_2) = \text{Prob}\{X_1 \leq x_1, X_2 \leq x_2\}$.

   (b) For the joint p.d.f. defined in equation (1.10), calculate $E(X_1 X_2)$ and $E(X_1^2 X_2^2)$.

10. Let $X_1 \sim N(0, \sigma^2)$ and $X_2 = (X_1)^2$.

   (a) Use the preceding exercise (part (a)) to show that X_1 and X_2 are dependent random variables, i.e., $F(x_1, x_2) \neq F_1(x_1)F_2(x_2)$.

   (b) Show that X_1 and X_2 are uncorrelated, $cov(X_1, X_2) = 0$.

11. Assume that the m.g.f. of X, $M_X(t)$, converges in some open interval about the origin. Use the fact that the cumulant generating function (c.g.f) is

$$K_X(t) = \ln M_X(t)$$

and the properties of the m.g.f. to show that the c.g.f. satisfies

$$K_X(0) = 0, \quad K_X'(0) = \mu_X, \quad K_X''(0) = \sigma_X^2.$$

12. Show that the p.g.f. of the Poisson distribution is $\mathcal{P}_X(t) = e^{\lambda(t-1)}$. Then use the p.g.f. to show that the mean and variance are $\mu_X = \lambda = \sigma_X^2$.

13. Show that the p.g.f. of the negative binomial distribution defined in equation (1.3) is

$$P_X(t) = \frac{p^n}{[1 - (1 - p)t]^n}.$$

Then use the p.g.f. to find the mean and variance of X.

14. The following p.m.f. is associated with a *negative binomial distribution* Y:

$$f(y) = \begin{cases} \binom{y - 1}{n - 1} p^n (1 - p)^{y-n}, & y = n, n + 1, n + 2, \ldots, \\ 0, & \text{otherwise}, \end{cases}$$

where n is a positive integer and $0 < p < 1$. This function of y is a shift of n units to the right of the function $f(x)$ given in (1.3). Show that the p.g.f. of Y satisfies $P_Y(t) = t^n P_X(t)$, $\mu_Y = \mu_X + n$, and $\sigma_Y^2 = \sigma_X^2$, where X is the random variable defined by (1.3) (Exercise 13).

15. For the gamma distribution,

   (a) Show that the m.g.f. is $M_X(t) = (1 - \beta t)^{-\alpha}$ for $t < 1/\beta$.

   (b) Use the m.g.f. to find the mean and variance of the gamma distribution.

   (c) For the special case of the exponential distribution, $\alpha = 1$ and $\beta = 1/\lambda$, find the mean and variance. Show that the *coefficient of variation (CV)* equals one for the exponential distribution, $CV = \sigma/\mu = 1$.

16. Suppose that the p.g.f. of a continuous random variable X is $P_X(t) = 1/(1 - \theta \ln(t))$, $\theta > 0$.

   (a) Find the m.g.f. $M_X(t)$ and the c.g.f. $K_X(t)$.

   (b) Compute the mean and variance of X.

17. Suppose the m.g.f. of a continuous random variable X is $M_X(t) = (1 - \theta)^{-n}$. Find the p.g.f. of X. Use the p.g.f. or the m.g.f. to find the mean μ_X and variance σ_X^2 of X.

18. Let X_1 and X_2 be independent binomial random variables such that $X_1 \sim b(N_1, p)$ and $X_2 \sim b(N_2, p)$. Use the m.g.f. of X_i, $M_{X_i}(t) = E(e^{tX_i}) = (1 - p + pe^t)^{N_i}$, $i = 1, 2$, to show that the random variable $Z = X_1 + X_2$ has a binomial distribution, that is, $Z \sim b(N_1 + N_2, p)$.

19. Suppose the continuous random variable X is exponentially distributed. Show that X has the following property:

$$\text{Prob}\{X \geq t + \Delta t | X \geq t\} = \text{Prob}\{X \geq \Delta t\}.$$

This property of the exponential distribution is known as the *memory-less property.*

20. Suppose the discrete random variable X has a geometric distribution (Exercise 4). Show that X has the memoryless property (Exercise 19):

$$\text{Prob}\{X \geq j | X \geq i\} = \text{Prob}\{X \geq j - i\}, \quad i \leq j.$$

21. This exercise will illustrate the memoryless property of the exponential and geometric distributions as defined in Exercises 19 and 20.

   (a) Compute $\text{Prob}\{X \geq 4 | X \geq 1\}$ and $\text{Prob}\{X \geq 3\}$ for the geometric distribution.

   (b) Compute $\text{Prob}\{X \geq 4 | X \geq 1\}$ and $\text{Prob}\{X \geq 3\}$ for the exponential distribution.

22. Suppose the continuous random variable X has an exponential distribution with mean $\mu = 10$. Compute

   (a) $\text{Prob}\{5 < X < 15\}$
   (b) $\text{Prob}\{X > 15\}$
   (c) $\text{Prob}\{X > 20 | X > 5\}$

23. Suppose the random variable X has a distribution that is $N(2, 1)$. Find

   (a) $\text{Prob}\{X \leq 2\}$
   (b) $\text{Prob}\{-1 \leq X - 2 \leq 1\}$
   (c) $\text{Prob}\{-3 \leq X - 2 \leq 3\}$
   (d) $\text{Prob}\{-0.9 \leq X \leq 1.5\}$

24. Show that the m.g.f. of the normal distribution is $M_X(t) = e^{\mu t + \sigma^2 t^2 / 2}$ and the c.g.f. is $K_X(t) = \mu t + \sigma^2 t^2 / 2$. Then use the c.g.f. to show that μ and σ^2 are the mean and variance of X, respectively.

25. A random variable X has a *lognormal distribution* if $Y = \ln X$ has a normal distribution, i.e., $Y = \ln X \sim N(\mu, \sigma^2)$. Suppose X is lognormally distributed. Use Exercise 24 to show that $E(X^t) = M_Y(t) = e^{\mu t + \sigma^2 t^2 / 2}$. Use the preceding relationship to calculate the mean and variance of X.

26. If the cumulant generating function of the random variable X is expressed as $K_X(t) = \sum_{i=1}^{n} \kappa_n t^n / n!$, then κ_n is called the *nth cumulant.*

   (a) Show that the first cumulant is the mean and the second cumulant is the variance.

   (b) Find all of the cumulants, κ_n, $n = 1, 2, \ldots$ for a normal distribution $X \sim N(\mu, \sigma^2)$.

27. Let $\{X_i\}_{i=1}^{\infty}$ be a collection of independent and identically distributed random variables with mean $\mu_i = \mu$ and variance $\sigma_i^2 = \sigma^2$. Define $W_n = \frac{\sum_{i=1}^{n} X_i - n\mu}{\sqrt{n}\sigma}$. Then consider the m.g.f. of W_n, $M_{W_n}(t) = E(e^{tW_n})$, where $-h < t < h$, $h > 0$. The following steps verify the Central Limit Theorem, that is, $M_{W_n}(t) \to M_Z(t)$, as $n \to \infty$, where $Z \sim N(0,1)$ (Hogg and Craig, 1995).

   (a) Show that

   $$E(e^{tW_n}) = \prod_{i=1}^{n} E\left(\exp\left[\frac{t}{\sqrt{n}}Y_i\right]\right) = \left[f\left(\frac{t}{\sqrt{n}}\right)\right]^n,$$

   where $Y_i = \dfrac{X_i - \mu}{\sigma}$, $f(t) = E(e^{tY_i})$, $E(Y_i) = 0$, and $E(Y_i^2) = 1$.

   (b) Use the fact that $f(t)$ is the m.g.f. of Y_i to show that $f(0) = 1$, $f'(0) = 0$, and $f''(0) = 1$.

   (c) Consider a Taylor's series with remainder for f, expanded about zero. Show that there exists $\xi \in [0, t]$ so that

   $$f(t) = 1 + f''(\xi)\frac{t^2}{2} = 1 + \frac{t^2}{2} + [f''(\xi) - 1]\frac{t^2}{2}.$$

   (d) Now use the following facts for $f(t/\sqrt{n})$: $\xi \in [0, t/\sqrt{n}]$, $\xi \to 0$ as $n \to \infty$, and f'' is continuous at $t = 0$, to show

   $$\lim_{n\to\infty} E(e^{tW_n}) = \lim_{n\to\infty}\left[f\left(\frac{t}{\sqrt{n}}\right)\right]^n = \lim_{n\to\infty}\left(1 + \frac{t^2}{2n}\right)^n = e^{t^2/2},$$

   the m.g.f. for $Z \sim N(0,1)$.

28. Suppose that the random variable X has a m.g.f. $M_X(t)$.

   (a) Verify that the m.g.f. of $Y = X - \mu_X$ is $M_Y(t) = e^{-\mu_X t}M_X(t)$ and that $M_Y^{(k)}(0) = E(Y^k) = E[(X - \mu_X)^k]$ gives the kth moment of X about its mean (Bailey, 1990).

   (b) Suppose that X is a binomial random variable. Find $M_Y(t)$; then find the first three moments of X about its mean value.

29. Suppose $X_1, \ldots, X_n$ is a random sample of size $n = 25$ from a Poisson distribution with parameter λ. Apply the Central Limit Theorem to approximate $\text{Prob}\{\bar{X} \leq \lambda\}$ and $\text{Prob}\{\bar{X} \leq \lambda + \sqrt{\lambda}/5\}$.

30. A well-known inequality in probability theory is *Markov's inequality*. The Markov inequality states that for a nonnegative random variable X with expectation μ,

$$\text{Prob}\{X \geq c\} \leq \frac{\mu}{c}, \quad c > 0.$$

31. Another well-known inequality in probability theory is *Chebyshev's inequality*. The Chebyshev inequality states that for any random variable X with finite expectation μ and positive variance σ^2,

$$\text{Prob}\{|X - \mu| < k\sigma\} \geq 1 - \frac{1}{k^2}, \quad k > 0.$$

Use the Markov inequality in Exercise 30 with $Y^2 = (X - \mu)^2$ and $E(Y^2) = \sigma^2$ to verify the Chebyshev inequality.

32. A weaker form of convergence than mean square convergence or convergence with probability one is convergence in probability. Let $\{X_n\}_{n=1}^{\infty}$ be a sequence of i.i.d. random variables with finite expectation μ and positive variance σ^2. Then the mean $\bar{X}(n) = \sum_{i=1}^{n} X_i$ is said to *converge in probability* to the expectation μ if

$$\lim_{n \to \infty} \text{Prob}\{|\bar{X}(n) - \mu| \geq \epsilon\} = 0 \qquad (1.11)$$

for all $\epsilon > 0$. The limit in (1.11) is also known as the *weak* law of large numbers; it can also be expressed as follows:

$$\lim_{n \to \infty} \text{Prob}\{|\bar{X}(n) - \mu| < \epsilon\} = 1. \qquad (1.12)$$

Use the Chebyshev inequality in Exercise 31 to verify the inequality (1.12).

33. Consider the exponential growth model, $dn/dt = bn$, $n(0) = a$. Compute the doubling time, $T_{2a,a}$, the first time the population size is $2a$. Then find $T_{ka,a}$.

34. Modify the MATLAB program in the Appendix and graph Poisson probability mass functions for $\lambda = 5$ and $\lambda = 10$. On each of these graphs, superimpose the graph of the normal density, $N(\lambda, \lambda)$. How do these two distributions compare? It can be shown that if X has a Poisson distribution, then the random variable $Z = (X - \lambda)/\sqrt{\lambda}$ approaches a standard normal distribution as $\lambda \to \infty$.

1.8 References for Chapter 1

Anderson, S. L. 1990. Random number generators on vector supercomputers and other advanced architectures. *SIAM Review*. 32: 221–251.

Bailey, N. T. J. 1990. *The Elements of Stochastic Processes with Applications to the Natural Sciences*. John Wiley & Sons, New York.

Cramér, H. 1945. *Mathematical Methods of Statistics.* Princeton Univ. Press, Princeton, N. J.

Feller, W. 1968. *An Introduction to Probability Theory and Its Applications.* Vol. 1. 3rd ed. John Wiley & Sons, New York.

Guttorp, P. 1995. *Stochastic Modeling of Scientific Data.* Chapman & Hall, London.

Hogg, R. V. and A. T. Craig. 1995. *Introduction to Mathematical Statistics.* 5th ed. Prentice Hall, Upper Saddle River, N. J.

Hogg, R. V. and E. A. Tanis. 2001. *Probability and Statistical Inference.* 6th ed. Prentice Hall, Upper Saddle River, N. J.

Hsu, H. P. 1997. *Schaum's Outline of Theory and Problems of Probability, Random Variables, and Random Processes.* McGraw-Hill, New York.

Renshaw, E. 1993. *Modelling Biological Populations in Space and Time.* Cambridge Studies in Mathematical Biology. Cambridge Univ. Press, Cambridge, U. K.

Ross, S. M. 2006. *A First Course in Probability.* 7th ed. Pearson Prentice Hall, Upper Saddle River, N. J.

Schervish, M. J. 1995. *Theory of Statistics.* Springer-Verlag, New York, Berlin, Heidelberg.

1.9 Appendix for Chapter 1

1.9.1 Probability Distributions

The mean, μ, variance, σ^2, and m.g.f., $M(t)$, for each of the probability distributions discussed in Section 1.2.2 are given here for reference purposes.

Discrete Distributions

Discrete Uniform: $\mu = \dfrac{1+n}{2}$, $\sigma^2 = \dfrac{n^2-1}{12}$, $M(t) = \dfrac{e^{(n+1)t} - e^t}{n(e^t - 1)}$, $t \neq 0$.

Geometric: $\mu = \dfrac{1-p}{p}$, $\sigma^2 = \dfrac{1-p}{p^2}$, $M(t) = \dfrac{p}{1 - (1-p)e^t}$.

Binomial, $b(n, p)$: $\mu = np$, $\sigma^2 = np(1 - p)$, $M(t) = (1 - p + pe^t)^n$.

Neg. Binomial: $\mu = \dfrac{n(1-p)}{p}$, $\sigma^2 = \dfrac{n(1-p)}{p^2}$, $M(t) = \dfrac{p^n}{[1-(1-p)e^t]^n}$.

Poisson: $\mu = \lambda$, $\sigma^2 = \lambda$, $M(t) = e^{\lambda(e^t-1)}$.

Continuous Distributions

Uniform, $U(a,b)$: $\mu = \dfrac{a+b}{2}$, $\sigma^2 = \dfrac{(b-a)^2}{12}$, $M(t) = \dfrac{e^{bt}-e^{at}}{t(b-a)}$, $t \neq 0$.

Gamma: $\mu = \alpha\beta$, $\sigma^2 = \alpha\beta^2$, $M(t) = \dfrac{1}{(1-\beta t)^\alpha}$, $t < 1/\beta$.

Exponential: $\mu = \dfrac{1}{\lambda}$, $\sigma^2 = \dfrac{1}{\lambda^2}$, $M(t) = \dfrac{\lambda}{\lambda - t}$, $t < \lambda$.

Normal, $X \sim N(\mu, \sigma^2)$: $E(X) = \mu$, $Var(X) = \sigma^2$, $M(t) = e^{\mu t + \sigma^2 t^2/2}$.

1.9.2 MATLAB® and FORTRAN Programs

The following MATLAB program can be used to graph the Poisson mass function given in Figure 1.3.

```
clear all %Clears variables and functions from memory.
set(0,'DefaultAxesFontSize',18); %Increases axes labels.
lastx=25; %Truncates the Poisson function at x=25.
x=linspace(0,lastx,lastx+1);
w(1)=1; w(2)=1; lambda=3;
for i=2:lastx
     w(i+1)=i*w(i);
end
y=lambda.^x*exp(-lambda)./w;
bar(x,y,'k'); %Graphs a histogram of y.
axis([-1,12,0,0.25])
xlabel(x); ylabel(f(x));
```

The following FORTRAN and MATLAB programs can be used to generate sample paths for the simple birth process.

```
REAL*8 N(50), T(50),Y,B,XX
PRINT *, 'SEED (POSITIVE NUMBER < M)'
READ *, XX
T(1)=0.0
N(1)=1.0
B=1.0
Y=RAND(XX)
DO I=1,49
      Y=RAND(XX)
```

```
                    T(I+1)=-DLOG(Y)/(B*N(I))+T(I)
                    N(I+1)=N(I)+1
                    PRINT *, 'T', T(I+1), 'N', N(I+1)
            ENDDO
            STOP
            END
C SUBROUTINE for the uniform random number on [0,1]
            FUNCTION RAND(XX)
            REAL*8 XX,A,M,D
            A=16807.0
            M=2147483647.0
            ID=A*XX/M
            D=ID
            XX=A*XX-D*M
            RAND=XX/M
            RETURN
            END
```

```
clear all
set(0,'DefaultAxesFontSize',18);
b=1;
x=linspace(0,50,51); %Defines vector [0,1,2,...,50].
y=exp(x);
n=linspace(1,50,50); %Defines population vector [1,2,...,50].
for j=1:3; %Three sample paths.
      t(1)=0;
      for i=1:49;
            t(i+1)=t(i)-log(rand)/(b*n(i));
      end
      s= stairs(t,n); %Draws stairstep graph of n.
      set(s,'LineWidth',2);
      hold on
end
plot(x,y,'k--','LineWidth',2); %Plots the exponential solution.
axis([0,5,0,50]);
xlabel('Time'); ylabel('Population Size');
hold off
```

Note: A statement following % is a comment statement; these statements are not executable.

1.9.3 Interevent Time

To simulate the simple birth process, it is necessary to know the random variable for the time between births or *interevent time*. It is shown in Chapter 5 that the random variable for the interevent time is exponentially distributed; if the population is of size N, then the time H to the next event (or interevent time) has a distribution equal to

$$P(H \geq h) = \exp(-bNh).$$

To simulate a value $h \in H$, a uniformly distributed random number Y is selected in the range $0 \leq Y \leq 1$ [i.e., from the uniform distribution $U(0,1)$]. Then

$$Y = \exp(-bNh),$$

which yields

$$h = -\frac{\ln(Y)}{bN}$$

(Renshaw, 1993). Notice in the simple birth process that as N increases, $h = -\ln(Y)/bN$ decreases; the interevent time decreases as the population size increases. To simulate an interevent time, it is necessary to apply a random number generator that generates uniformly distributed numbers in $[0,1]$. The function subroutine RAND in the FORTRAN program is a pseudo–random number generator and is based on the recursion relation $y_{n+1} = (Ay_n) \bmod M$, where $RAND = y_{n+1}/M \in [0,1)$ and the modulus $M = 2^{31} - 1$ is a Mersenne prime (Anderson, 1990). The command "rand" in the MATLAB program is a built-in MATLAB function for a uniform random number generator on $[0,1]$.

Chapter 2

Discrete-Time Markov Chains

2.1 Introduction

Discrete-time Markov chains are introduced in this chapter, where the time and state variables are discrete-valued. Some basic notation and theory for discrete-time Markov chains are presented in the first section. Then discrete-time Markov chains are classified into one of several types, irreducible or reducible, periodic or aperiodic, and recurrent or transient. This classification scheme determines the asymptotic behavior of the Markov chain. For example, an aperiodic, irreducible, and recurrent Markov chain is shown to have a stationary limiting distribution. Some well-known examples of discrete-time Markov chains are discussed, including the random walk model in one, two, and three dimensions and a random walk model on a finite domain, often referred to as the gambler's ruin problem. Another biological example that is discussed is the genetics of inbreeding.

The theory and application of Markov chains is probably one of the most well-developed in stochastic processes. A classic textbook on finite Markov chains (where the state space is finite) is by Kemeny and Snell (1960). Some additional references on the theory and numerical methods for discrete-time Markov chains include *Finite Markov Processes and Their Applications*, by Iosifescu (1980); *A First Course in Stochastic Processes*, by Karlin and Taylor (1975); *An Introduction to Stochastic Modeling*, by Taylor and Karlin (1998); *Classical and Spatial Stochastic Processes*, by Schinazi (1999); *Markov Chains*, by Norris (1997); and *Introduction to the Numerical Solution of Markov Chains*, by Stewart (1994).

2.2 Definitions and Notation

Consider a discrete-time stochastic process $\{X_n\}_{n=0}^{\infty}$, where the random variable X_n is a discrete random variable defined on a finite or countably infinite *state space*. For convenience, denote the state space as $\{1, 2, \ldots\}$. However, this set could be finite and could include nonpositive integer values.

The variable n is used for the index set instead of t because it is common notation in discrete-time processes. The index set is defined as $\{0, 1, 2, \ldots\}$, since it often represents the progression of time, which is also the reason for the terminology, discrete-time processes. Therefore, the index n shall be referred to as "time" n.

A Markov stochastic process is a stochastic process in which the future behavior of the system depends only on the present and not on its past history.

DEFINITION 2.1 *A discrete-time stochastic process $\{X_n\}_{n=0}^{\infty}$ is said to have the* Markov property *if*

$$\mathrm{Prob}\{X_n = i_n | X_0 = i_0, \ldots, X_{n-1} = i_{n-1}\} = \mathrm{Prob}\{X_n = i_n | X_{n-1} = i_{n-1}\},$$

where the values of $i_k \in \{1, 2, \ldots\}$ for $k = 0, 1, 2, \ldots, n$. The stochastic process is then called a Markov chain *or, more specifically, a* discrete-time Markov chain (DTMC). *It is called a* finite state Markov chain *if the state space is finite.*

The stochastic process is referred to as a *chain* when the state space is discrete. The name *Markov* refers to Andrei A. Markov, a Russian probabilist (1856–1922), whose work in this area contributed much to the theory of stochastic processes.

The notation Prob is used to denote the induced probability measure, $\mathrm{Prob}\{\cdot\} = P_{X_n}(\cdot)$, because P will refer to the transition matrix that is defined below. Denote the p.m.f. associated with the random variable X_n by $\{p_i(n)\}_{i=0}^{\infty}$, where

$$p_i(n) = \mathrm{Prob}\{X_n = i\}. \tag{2.1}$$

The state of the process at time n, X_n, is related to the process at time $n+1$ through what is known as the transition probabilities. If the process is in state i at time n, at the next time step $n + 1$, it will either stay in state i or move or transfer to another state j. The probabilities for these changes in state are defined by the one-step transition probabilities.

DEFINITION 2.2 *The* one-step transition probability, *denoted as $p_{ji}(n)$, is defined as the following conditional probability:*

$$p_{ji}(n) = \mathrm{Prob}\{X_{n+1} = j | X_n = i\},$$

the probability that the process is in state j at time $n+1$ given that the process was in state i at the previous time n, for $i, j = 1, 2, \ldots$.

DEFINITION 2.3 *If the transition probabilities $p_{ji}(n)$ in a Markov chain do not depend on time n, they are said to be* stationary *or* time-homogeneous *or simply* homogeneous. *In this case, the notation p_{ji} is used. If the transition* *probabilities are time dependent, $p_{ji}(n)$, then they are said to be* nonstationary *or* nonhomogeneous.

A Markov chain may have either stationary or nonstationary transition probabilities. Unless stated otherwise, it shall be assumed that the transition probabilities of the Markov chain are stationary. For each state, the one-step transition probabilities satisfy

$$\boxed{\sum_{j=1}^{\infty} p_{ji} = 1 \text{ and } p_{ji} \geq 0} \tag{2.2}$$

for $i, j = 1, 2, \ldots,$ The first summation means, with probability one, the process in any state i must transfer to some other state j, $j \neq i$ or stay in state i during the next time interval. This identity also implies for a fixed i, $\{p_{ji}\}_{j=1}^{\infty}$ is a probability distribution. The one-step transition probabilities can be expressed in matrix form.

DEFINITION 2.4 *The* transition matrix *of the DTMC $\{X_n\}_{n=0}^{\infty}$ with state space $\{1, 2, \ldots\}$ and one-step transition probabilities, $\{p_{ij}\}_{i,j=1}^{\infty}$, is denoted as $P = (p_{ij})$, where*

$$P = \begin{pmatrix} p_{11} & p_{12} & p_{13} & \cdots \\ p_{21} & p_{22} & p_{23} & \cdots \\ p_{31} & p_{32} & p_{33} & \cdots \\ \vdots & \vdots & \vdots & \end{pmatrix}.$$

If the set of states is finite, $\{1, 2, \ldots, N\}$, then P is an $N \times N$ matrix. The column elements sum to one because of the identity in (2.2). A nonnegative matrix with the property that each column sum equals one is called a *stochastic matrix*. The transition matrix P is a stochastic matrix. It is left as an exercise to show that if P is a stochastic matrix, then P^n is a stochastic matrix, for n any positive integer (for a finite or infinite number of states). If the row sums also equal one, then the matrix is called *doubly stochastic*.

The notation used here differs from that used in some textbooks in two respects. First, the transition matrix is sometimes defined as the transpose of P, P^{tr}. Then the definition of a stochastic matrix is defined as a nonnegative matrix whose row sums equal one (rather than column sums equal one) (Bailey, 1990; Karlin and Taylor, 1975; Kemeny and Snell, 1960; Norris, 1997; Schinazi, 1999; Stewart, 1994; Taylor and Karlin, 1998). Second, generally, the one-step transition probability p_{ij} is defined as the probability of a transition from state i to state j rather than a transition from j to i as in our notation. The notation used in this textbook closely resembles the notation used for deterministic matrix models, where matrix multiplication is on the left of the vector $Y_n : PY_n = Y_{n+1}$ (Caswell, 2001; Tuljapurkar, 1997). As an aid in setting up and understanding how the elements of P are related, note that the nonzero elements in the ith row of P represent all those states j (column j) that can transfer into state i in one time step.

DEFINITION 2.5   *The n-step transition probability, denoted $p_{ji}^{(n)}$, is the probability of transferring from state i to state j in n time steps,*

$$p_{ji}^{(n)} = \text{Prob}\{X_n = j | X_0 = i\}.$$

The n-step transition matrix is denoted as $P^{(n)} = \left(p_{ji}^{(n)} \right)$. For the cases $n = 0$ and $n = 1$, $p_{ji}^{(1)} = p_{ji}$ and

$$p_{ji}^{(0)} = \delta_{ji} = \begin{cases} 1, & j = i, \\ 0, & j \neq i, \end{cases}$$

where δ_{ji} represents the Kronecker delta symbol. Then $P^{(1)} = P$ and $P^{(0)} = I$, where I is the identity matrix.

Relationships exist between the n-step transition probabilities and s-step and $(n - s)$-step transition probabilities. These relationships are known as the *Chapman-Kolmogorov equations*:

$$p_{ji}^{(n)} = \sum_{k=1}^{\infty} p_{jk}^{(n-s)} p_{ki}^{(s)}, \quad 0 < s < n.$$

Verification of the Chapman-Kolmogorov equations can be shown as follows (Stewart, 1994):

$$p_{ji}^{(n)} = \text{Prob}\{X_n = j | X_0 = i\},$$

$$= \sum_{k=1}^{\infty} \text{Prob}\{X_n = j, X_s = k | X_0 = i\}, \tag{2.3}$$

$$= \sum_{k=1}^{\infty} \text{Prob}\{X_n = j | X_s = k, X_0 = i\}\text{Prob}\{X_s = k | X_0 = i\}, \tag{2.4}$$

$$= \sum_{k=1}^{\infty} \text{Prob}\{X_n = j | X_s = k\}\text{Prob}\{X_s = k | X_0 = i\}, \tag{2.5}$$

$$= \sum_{k=1}^{\infty} p_{jk}^{(n-s)} p_{ki}^{(s)}, \tag{2.6}$$

where equations (2.3)–(2.6) hold for $0 < s < n$. The relationship (2.4) follows from conditional probabilities (Exercise 4). The relationship (2.5) follows from the Markov property. The preceding identity written in terms of matrix notation yields

$$P^{(n)} = P^{(n-s)} P^{(s)}. \tag{2.7}$$

However, because $P^{(1)} = P$, it follows from the Chapman-Kolmogorov equations (2.7) that $P^{(2)} = P^2$ and, in general, $P^{(n)} = P^n$. The n-step transition matrix $P^{(n)}$ is just the nth power of P. The elements of P^n are the n-step transition probabilities, $p_{ij}^{(n)}$. Be careful not to confuse the notation p_{ij}^n with $p_{ij}^{(n)}$, $p_{ij}^{(n)} \neq p_{ij}^n$. The notation p_{ij}^n is the nth power of the element p_{ij}, whereas $p_{ij}^{(n)}$ is the ij element in the nth power of P.

Let $p(n)$ denote the vector form of the p.m.f. associated with X_n; that is, $p(n) = (p_1(n), p_2(n), \ldots)^{tr}$, where $p_i(n)$ is defined in (2.1) and the states are arranged in increasing order in the column vector $p(n)$. The probabilities satisfy

$$\sum_{i=1}^{\infty} p_i(n) = 1.$$

Given the probability distribution associated with X_n, the probability distribution associated with X_{n+1} can be found by multiplying the transition matrix P by $p(n)$; that is,

$$p_i(n+1) = \sum_{j=1}^{\infty} p_{ij} p_j(n)$$

or

$$p(n+1) = Pp(n).$$

This latter equation projects the process "forward" in time. In general,

$$p(n+m) = P^{n+m} p(0) = P^n \left(P^m p(0) \right) = P^n p(m).$$

Example 2.1. A simple model for molecular evolution is modeled by a finite Markov chain: the probability of a base substitution occurring in a segment of DNA (deoxyribonucleic acid) in one generation. The four DNA bases are adenine (A), guanine (G), cytosine (C), and thymine (T), where $p(0) = (p_A(0), p_G(0), p_C(0), p_T(0))^{tr}$ is the initial proportion of each type of base in the DNA segment (expected fraction of sites occupied by each base). The mutation process is modeled by assuming only one base substitution can occur per generation, $n \to n+1$ (Allman and Rhodes, 2004). Suppose $p_{A,G} = p_{12} = \text{Prob}\{X_{n+1} = A | X_n = G\} = a$, $0 < a < 1/3$ and $p_{ij} = a$ for $i \neq j$. The transition matrix in this case is doubly stochastic:

$$P = \begin{pmatrix} 1 - 3a & a & a & a \\ a & 1 - 3a & a & a \\ a & a & 1 - 3a & a \\ a & a & a & 1 - 3a \end{pmatrix}.$$

This model is known as the Jukes-Cantor model for molecular evolution (Allman and Rhodes, 2004, Jukes and Cantor, 1969). ∎

2.3 Classification of States

Relationships between the states of a Markov chain lead to a classification scheme for the states and ultimately classification for Markov chains.

DEFINITION 2.6 *The state j can be* reached *from the state i (or state j is* accessible *from state i) if there is a nonzero probability, $p_{ji}^{(n)} > 0$, for some $n \geq 0$. This relationship is denoted as $i \to j$. If i can be reached from j, we use $j \to i$, and if j can be reached from i, we use $i \to j$. In addition if $i \to j$ and $j \to i$, i and j are said to* communicate, *or to be in the* same class, *denoted $i \leftrightarrow j$; that is, there exists n and n' such that*

$$p_{ji}^{(n)} > 0 \quad \text{and} \quad p_{ij}^{(n')} > 0.$$

The relation $i \to j$ can be represented in graph theory as a directed graph (Figure 2.1).

FIGURE 2.1: Directed graph with $i \to j$ $(p_{ji} > 0)$ and $i \to k$ $(p_{ki}^{(2)} > 0)$, but it is not the case that $k \to i$.

The relation $i \leftrightarrow j$ is an *equivalence relation* on the state space $\{1, 2, \ldots\}$ (Karlin and Taylor, 1975):

(1) reflexivity: $i \leftrightarrow i$, because $p_{ii}^{(0)} = 1$. Beginning in state i, the system stays in state i if there is no time change.

(2) symmetry: $i \leftrightarrow j$ implies $j \leftrightarrow i$ follows from the definition.

(3) transitivity: $i \leftrightarrow j$, $j \leftrightarrow k$ implies $i \leftrightarrow k$. To verify this last property, note that the the the first two properties imply there exist nonnegative integers n and m such that $p_{ji}^{(n)} > 0$ and $p_{kj}^{(m)} > 0$. Thus,

$$
\begin{aligned}
p_{ki}^{(n+m)} &= \text{Prob}\{X_{n+m} = k | X_0 = i\}, \\
&\geq \text{Prob}\{X_{n+m} = k, X_n = j | X_0 = i\}, \\
&= \text{Prob}\{X_{n+m} = k | X_n = j\}\text{Prob}\{X_n = j | X_0 = i\}, \quad (2.8) \\
&= p_{kj}^{(m)} p_{ji}^{(n)},
\end{aligned}
$$

where probability (2.8) follows from conditional probabilities and the Markov property. Thus, $p_{ki}^{(n+m)} > 0$ and $i \to k$. Similarly, it can be shown that $p_{ik}^{(n+m)} > 0$, which implies $k \to i$.

The equivalence relation on the states of the Markov chain define a set of equivalence classes. These equivalence classes are known as classes of the Markov chain.

DEFINITION 2.7 *The set of equivalence classes in a DTMC are called the* communication classes *or, more simply, the* classes *of the Markov chain.*

If every state in the Markov chain can be reached from every other state, then there is only one communication class (all the states are in the same class).

DEFINITION 2.8 *If there is only one communication class, then the Markov chain is said to be* irreducible, *but if there is more than one communication class, then the Markov chain is said to be* reducible.

A communication class may have the additional property that it is closed.

DEFINITION 2.9 *A set of states C is said to be* closed *if it is impossible to reach any state outside of C from any state in C by one-step transitions; $p_{ji} = 0$ if $i \in C$ and $j \notin C$.*

A sufficient condition to show that a Markov chain is irreducible is the existence of a positive integer n such that $p_{ji}^{(n)} > 0$ for all i and j; that is, every element in P^n is positive, $P^n > 0$, for some positive integer n. If a matrix has this latter property it is said to be *regular*. For a finite Markov chain, irreducibility can be checked from the directed graph for that chain. A finite Markov chain with states $\{1, 2, \ldots, N\}$ is irreducible if there is a directed path from i to j for every $i, j \in \{1, 2, \ldots, N\}$.

The definitions of *irreducible* and *reducible* apply more generally to $N \times N$ matrices, $A = (a_{ij})$. A *directed graph* or *digraph* with N nodes can be constructed from an $N \times N$ matrix. There is a single directed path from node i to node j if $a_{ji} \neq 0$. Then node j can be reached from node i in one step. A more general signed digraph can be constructed, where the sign of a_{ji} is associated with each directed path. Node j can be reached from node i in n steps if $a_{ji}^{(n)} \neq 0$, where $a_{ji}^{(n)}$ is the element in the jth row and ith column of A^n. A directed graph with N nodes constructed from a matrix A is said to be *strongly connected* if there exists a series of directed paths from i to j for every $i, j \in \{1, 2 \ldots, N\}$ ($i \leftrightarrow j$). Then a directed graph is strongly connected if it is possible to start from any node i and reach any other node

j in a finite number of steps. Matrix irreducibility is defined as a strongly connected digraph (Ortega, 1987). If matrix A is a transition matrix of a finite Markov chain, then this definition applies to the chain as well.

DEFINITION 2.10 *A transition matrix P of a finite Markov chain said to be* irreducible *if its directed graph is strongly connected. Alternately, a transition matrix P of a finite Markov chain is is said to be* reducible *if its directed graph is not strongly connected.*

The simple model of molecular evolution in Example 2.1 has an irreducible transition matrix P and hence, the DTMC has a single communication class.

Example 2.2. Suppose a DTMC with four states $\{1, 2, 3, 4\}$ has the following transition matrix:

$$P = \begin{pmatrix} 0 & 0 & p_{13} & 0 \\ p_{21} & 0 & p_{23} & p_{24} \\ 0 & 0 & 0 & 0 \\ 0 & p_{42} & 0 & 0 \end{pmatrix},$$

where p_{ij} is positive (a transition occurs from j to i). Then it is easy to see that $4 \leftrightarrow 2 \leftarrow 1 \leftarrow 3$ and $4 \leftrightarrow 2 \leftarrow 3$ (Figure 2.2). Since it is impossible to return to states 1 or 3 after having left them, each of these states forms a single communication class, $\{1\}$, $\{3\}$. The set $\{2, 4\}$ is a third communication class. The Markov chain is reducible. In addition, the set $\{2, 4\}$ is closed, but the sets $\{1\}$ and $\{3\}$ are not closed. If one of the elements, either p_{12} or p_{14}, is positive, then the communication classes consist of $\{1, 2, 4\}$ and $\{3\}$. Instead, if one of the elements, either p_{32} or p_{34}, was positive, then there would be a single communication class $\{1, 2, 3, 4\}$ so that the Markov chain would be irreducible. ∎

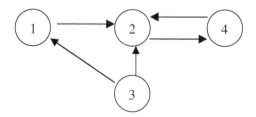

FIGURE 2.2: Directed graph for Example 2.2.

The following example illustrates the random walk model with absorbing barriers or the classical gambler's ruin problem.

Example 2.3. The state space is the set $\{0, 1, 2 \ldots, N\}$. The states represent the amount of money of one of the players (gambler). The gambler bets \$1 per game and either wins or loses each game. The gambler is ruined if he/she reaches state 0 and the gambler stops playing if he/she reaches state N, a winner. The probability of winning (moving to the right) is $p > 0$ and the probability of losing (moving to the left) is $q > 0$, $p + q = 1$ (i.e., $p_{i,i+1} = q$ and $p_{i+1,i} = p$, $i = 1, \ldots, N - 1$). In addition, $p_{00} = 1$ and $p_{NN} = 1$, which are referred to as *absorbing boundaries*. All other elements of the transition matrix are zero. In general, a state i is called *absorbing* if $p_{ii} = 1$. See the directed graph in Figure 2.3 and the corresponding $(N+1) \times (N+1)$ transition matrix:

$$P = \begin{pmatrix} 1 & q & 0 & \cdots & 0 & 0 & 0 \\ 0 & 0 & q & \cdots & 0 & 0 & 0 \\ 0 & p & 0 & \cdots & 0 & 0 & 0 \\ 0 & 0 & p & \cdots & 0 & 0 & 0 \\ \vdots & \vdots & \vdots & \cdots & \vdots & \vdots & \vdots \\ 0 & 0 & 0 & \cdots & 0 & q & 0 \\ 0 & 0 & 0 & \cdots & p & 0 & 0 \\ 0 & 0 & 0 & \cdots & 0 & p & 1 \end{pmatrix}.$$

The Markov chain, graphed in Figure 2.3, has three communication classes: $\{0\}$, $\{1, 2, \ldots, N - 1\}$, and $\{N\}$. The Markov chain is reducible. The sets $\{0\}$ and $\{N\}$ are closed, but the set $\{1, 2, \ldots, N - 1\}$ is not closed. Also, states 0 and N are absorbing; the remaining states are *transient*. A transient state is defined more formally later. ∎

FIGURE 2.3: Probability of winning is p and losing is q. Boundaries, 0 and N, are absorbing, $p_{00} = 1 = p_{NN}$.

Example 2.4. In an infinite-dimensional random walk or unrestricted random walk, the states are the integers, $0, \pm 1, \pm 2, \ldots$. Let $p > 0$ be the probability of moving to the right and $q > 0$ be the probability of moving to the left, $p + q = 1$. There are no absorbing boundaries, $p_{i,i+1} = q$ and $p_{i+1,i} = p$ for $i \in \{0, \pm 1, \pm 2, \ldots\}$. From the directed graph in Figure 2.4 it is easy to see that the Markov chain is irreducible. Every state in the system communicates with every other state. The set of states forms a closed set. In this case, the transition matrix P is infinite dimensional. If the states are ordered such that

54

..., $-1, 0, 1, \ldots$, then matrix P is an extension of the matrix in Example 2.3 with q along the superdiagonal and p along the subdiagonal. ∎

FIGURE 2.4: Unrestricted random walk with the probability of moving right equal to p and left equal to q.

Example 2.5. Suppose the states of the system are $\{1, 2, 3, 4, 5\}$ with directed graph in Figure 2.5 and transition matrix

$$P = \begin{pmatrix} 1/2 & 1/3 & 0 & 0 & 0 \\ 1/2 & 2/3 & 0 & 0 & 0 \\ 0 & 0 & 0 & 1/4 & 0 \\ 0 & 0 & 1 & 1/2 & 1 \\ 0 & 0 & 0 & 1/4 & 0 \end{pmatrix}.$$

There are two communication classes: $\{1, 2\}$ and $\{3, 4, 5\}$. Both classes are closed. The Markov chain is reducible. ∎

FIGURE 2.5: Directed graph for Example 2.5.

Example 2.6. Suppose the states of the system are $\{1, 2, \ldots, N\}$ with directed graph given in Figure 2.6 and transition matrix

$$P = \begin{pmatrix} 0 & 0 & \cdots & 0 & 1 \\ 1 & 0 & \cdots & 0 & 0 \\ 0 & 1 & \cdots & 0 & 0 \\ \vdots & \vdots & \cdots & \vdots & \vdots \\ 0 & 0 & \cdots & 1 & 0 \end{pmatrix}.$$

The Markov chain is irreducible. The set $\{1, 2, \ldots, N\}$ is closed. ∎

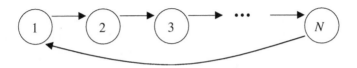

FIGURE 2.6: Directed graph for Example 2.6.

The chain in Example 2.6 has the property that beginning in state i it takes exactly N time steps to return to state i. That is, $P^N = I$. The chain is periodic with period equal to N.

DEFINITION 2.11   *The* period of state i, *denoted as* $d(i)$, *is the greatest common divisor of all integers* $n \geq 1$ *for which* $p_{ii}^{(n)} > 0$. *That is,*

$$d(i) = \text{g.c.d}\{n | p_{ii}^{(n)} > 0 \text{ and } n \geq 1\}.$$

If a state i *has period* $d(i) > 1$, *it is said to be* periodic *of period* $d(i)$. *If the period of a state equals one, it is said to be* aperiodic. *If* $p_{ii}^{(n)} = 0$ *for all* $n \geq 1$, *define* $d(i) = 0$.

It follows from the definition that $d(i)$ is a nonnegative integer.

Example 2.7. The directed graph of a Markov chain with three states $\{1, 2, 3\}$ is given in Figure 2.7 with transition matrix

$$P = \begin{pmatrix} 0 & 0 & 0 \\ 1 & 0 & 0 \\ 0 & 1 & 1 \end{pmatrix}.$$

It is easy to see that there are three communication classes, $\{1\}$, $\{2\}$, and $\{3\}$. The value of $d(i) = 0$ for $i = 1, 2$ because $p_{ii}^{(n)} = 0$ for $i = 1, 2$ and $n = 1, 2, \ldots$. Also, $d(3) = 1$; state 3 is aperiodic. ∎

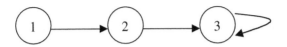

FIGURE 2.7: Directed graph for Example 2.7.

In the case that $d(i) = 0$, it can be shown that the set $\{i\}$ is a communication class (Exercise 7). In the case that $p_{ii} > 0$, then $d(i) = 1$. The term periodic is reserved for the case $d(i) > 1$. In Example 2.2, the classes are $\{1\}$, $\{3\}$ and $\{2, 4\}$. States 1 and 3 satisfy $d(1) = 0 = d(3)$. States 2 and 4 are periodic with period 2, $d(2) = 2 = d(4)$; $p_{ii}^{(2n)} = 1$ and $p_{ii}^{(2n+1)} = 0$ for $i = 2, 4$ and $n = 1, 2, \ldots$.

Periodicity is a class property; that is, if $i \leftrightarrow j$, then $d(i) = d(j)$. All states in one class have the same period. Thus, we can speak of a periodic class or a periodic chain. This result is verified in the next theorem.

THEOREM 2.1
If $i \leftrightarrow j$, then $d(i) = d(j)$.

Proof. The case $d(i) = 0$ is trivial. Suppose $d(i) \geq 1$ and $p_{ii}^{(s)} > 0$ for some $s > 0$. Then $d(i)$ divides s. Since $i \leftrightarrow j$, there exists m and n such that $p_{ij}^{(m)} > 0$ and $p_{ji}^{(n)} > 0$. Then

$$p_{jj}^{(n+s+m)} \geq p_{ji}^{(n)} p_{ii}^{(s)} p_{ij}^{(m)} > 0.$$

Also, since $p_{ii}^{(2s)} > 0$, $p_{jj}^{(n+2s+m)} > 0$. Thus, $d(j)$ divides $n + s + m$ and $n + 2s + m$ and must divide $(n + 2s + m) - (n + s + m) = s$. Since s is arbitrary, $d(j) \leq d(i)$.

Reverse the argument by assuming $p_{jj}^{(r)} > 0$. Then it can be shown that $d(i) \leq d(j)$. Combining these two inequalities gives the desired result, $d(i) = d(j)$. $\qquad\square$

In the random walk model with absorbing boundaries, Example 2.3, the classes $\{0\}$ and $\{N\}$ are aperiodic. The class $\{1, 2, \ldots, N - 1\}$ has period 2. In the unrestricted random walk model, Example 2.4, the entire chain is periodic of period 2. In this case, the notation $d = 2$ will be used rather than state that $d(i) = 2$ for each of the i states. The two classes in Example 2.5 are both aperiodic.

Some additional definitions and notation are needed to define a transient state. This is done in the next section.

2.4 First Passage Time

Assume the process begins in state i, $X_0 = i$. We will define a first return to state i and a first passage to state j for $j \neq i$.

DEFINITION 2.12 Let $f_{ii}^{(n)}$ denote the probability that, starting from *state i, $X_0 = i$, the* first return *to state i is at the nth time step,*

$$f_{ii}^{(n)} = \text{Prob}\{X_n = i, X_m \neq i, m = 1, 2, \ldots, n-1 | X_0 = i\}, \ n \geq 1.$$

The probabilities $f_{ii}^{(n)}$ are known as first return probabilities. *Define $f_{ii}^{(0)} = 0$.*

Note that $f_{ii}^{(1)} = p_{ii}$ but, in general, $f_{ii}^{(n)}$ is not equal to $p_{ii}^{(n)}$. The first return probabilities represent the *first time* the chain returns to state i; thus,

$$0 \leq \sum_{n=1}^{\infty} f_{ii}^{(n)} \leq 1.$$

A transient state is defined in terms of these first return probabilities.

DEFINITION 2.13 *State i is said to be* transient *if $\sum_{n=1}^{\infty} f_{ii}^{(n)} < 1$. State i is said to be* recurrent *if $\sum_{n=1}^{\infty} f_{ii}^{(n)} = 1$.*

The term *persistent* is sometimes used instead of recurrent (Bailey, 1990). If state i is recurrent, then the set $\{f_{ii}^{(n)}\}_{n=0}^{\infty}$ defines a probability distribution for a random variable T_{ii} which defines the first return time. That is, $T_{ii} = n$ with probability $f_{ii}^{(n)}$, $n = 0, 1, 2, \ldots$. When state i is transient, the set $\{f_{ii}^{(n)}\}_{n=0}^{\infty}$ does not define a complete set of probabilities necessary to define a probability distribution. However, for a transient state i, if we let $f_{ii} = \sum_{n=0}^{\infty} f_{ii}^{(n)} < 1$, we can define $1 - f_{ii}$ as the probability of never returning to state i. It is precisely for this reason that state i is transient: there is a positive probability of never returning to state i. The random variable T_{ii} may be thought of as a "waiting time" until the chain returns to state i.

DEFINITION 2.14 *The mean of the distribution of T_{ii} is referred to as the* mean recurrence time *to state i, denoted $\mu_{ii} = E(T_{ii})$. For a recurrent state i,*

$$\mu_{ii} = \sum_{n=1}^{\infty} n f_{ii}^{(n)}. \tag{2.9}$$

The term "recurrence" is often used when the return is to the same state. The definition of μ_{ii} in equation (2.9) does not apply to a transient state. The mean recurrence time for a transient state is always infinite ($T_{ii} = \infty$ with probability $1 - f_{ii}$). For a recurrent state, the the mean recurrence time can be finite or infinite.

DEFINITION 2.15 *If a recurrent state i satisfies $\mu_{ii} < \infty$, then it is said to be* positive recurrent, *and if it satisfies $\mu_{ii} = \infty$, then it is said to be* null recurrent.

Sometimes the term *nonnull recurrent* is used instead of positive recurrent (Bailey, 1990). An absorbing state is a simple example of a positive recurrent state with mean recurrence time $\mu_{ii} = 1$. This is a straightforward consequence of the fact that if i is an absorbing state, then $p_{ii} = 1$ which implies $f_{ii}^{(1)} = 1$ and $f_{ii}^{(n)} = 0$ for $n \neq 1$.

Example 2.8. Suppose the transition matrix of a two-state Markov chain is

$$P = \begin{pmatrix} p_{11} & p_{12} \\ p_{21} & p_{22} \end{pmatrix},$$

where $0 < p_{ii} < 1$ for $i = 1, 2$. Then all of the elements of matrix P are positive, $p_{ij} > 0$, $i, j = 1, 2$. It is shown that both states are positive recurrent. First $f_{11}^{(1)} = p_{11}$, $f_{11}^{(2)} = p_{12}p_{21}$, $f_{11}^{(3)} = p_{12}p_{22}p_{21}$, and, in general,

$$f_{11}^{(n)} = p_{12}p_{22}^{n-2}p_{21}, \quad n \geq 3.$$

Because $p_{22} < 1$,

$$\sum_{n=1}^{\infty} f_{11}^{(n)} = p_{11} + p_{12}p_{21} \sum_{n=0}^{\infty} p_{22}^n = p_{11} + \frac{p_{12}p_{21}}{1 - p_{22}}.$$

From the definition of a stochastic matrix, $p_{11} + p_{21} = 1$ and $p_{12} + p_{22} = 1$, it follows that

$$\sum_{n=1}^{\infty} f_{11}^{(n)} = p_{11} + p_{21} = 1,$$

which implies that state 1 is recurrent. Similarly, it can be shown that state 2 is recurrent. Computing the mean recurrence times,

$$\mu_{11} = p_{11} + p_{12}p_{21} \sum_{n=0}^{\infty} (n + 2)p_{22}^n < \infty$$

(Exercise 8). Similarly, $0 < \mu_{22} < \infty$. ∎

Note that in the definitions of first return probabilities and mean recurrence time, the Markov property was not assumed. These concepts do not require the Markov assumption and are sometimes discussed in the context of renewal processes. These definitions are extended to the case $i \neq j$.

Define the probability $f_{ji}^{(n)}$ for $j \neq i$ in a manner analogous to $f_{ii}^{(n)}$.

DEFINITION 2.16 *Let $f_{ji}^{(n)}$ denote the probability that, starting from state i, $X_0 = i$, the first return to state j is at the nth time step,*

$$f_{ji}^{(n)} = \text{Prob}\{X_n = j, X_m \neq j, m = 1, 2, \ldots, n - 1 | X_0 = i\}, \quad j \neq i, \ n \geq 1,$$

The probabilities $f_{ji}^{(n)}$ are known as first passage time probabilities. *Define*
$f_{ji}^{(0)} = 0.$

It follows from the definition that $0 \leq \sum_{n=0}^{\infty} f_{ji}^{(n)} \leq 1$. If $\sum_{n=0}^{\infty} f_{ji}^{(n)} = 1$, $\{f_{ji}^{(n)}\}_{n=0}^{\infty}$ defines a probability distribution for a random variable T_{ji}, known as the first passage to state j from state i. If $i = j$, then Definition 2.16 is the same as Definition 2.12.

DEFINITION 2.17   *If $X_0 = i$, then the* mean first passage time *to state j, denoted $\mu_{ji} = E(T_{ji})$, is defined as*

$$\mu_{ji} = \sum_{n=1}^{\infty} n f_{ji}^{(n)}, \quad j \neq i.$$

This definition can be extended to include the case $f_{ji} = \sum_{n=0}^{\infty} f_{ji}^{(n)} < 1$ by defining the probability of never reaching state j from state i as $1 - f_{ji}$. If $f_{ji} < 1$, then the mean first passage time is infinite.

There exist relationships between the n-step transition probabilities of a Markov chain and the first return probabilities. The transition from state i to i at the nth step, $p_{ii}^{(n)}$, may have its first return to state i at any of the steps $j = 1, 2, \ldots, n$. It is easy to see that

$$
\begin{aligned}
p_{ii}^{(n)} &= f_{ii}^{(0)} p_{ii}^{(n)} + f_{ii}^{(1)} p_{ii}^{(n-1)} + \cdots + f_{ii}^{(n)} p_{ii}^{(0)} \\
&= \sum_{k=1}^{n} f_{ii}^{(k)} p_{ii}^{(n-k)},
\end{aligned}
\tag{2.10}
$$

since $f_{ii}^{(0)} = 0$ and $p_{ii}^{(0)} = 1$. A similar relationship exists for $f_{ji}^{(n)}$ and $p_{ji}^{(n)}$:

$$p_{ji}^{(n)} = \sum_{k=1}^{n} f_{ji}^{(k)} p_{jj}^{(n-k)}, \quad j \neq i. \tag{2.11}$$

Let the generating function for the sequence $\{f_{ji}^{(n)}\}$ be

$$F_{ji}(s) = \sum_{n=0}^{\infty} f_{ji}^{(n)} s^n, \quad |s| < 1$$

and the generating function for the sequence $\{p_{ji}^{(n)}\}$ be

$$P_{ji}(s) = \sum_{n=0}^{\infty} p_{ji}^{(n)} s^n, \quad |s| < 1$$

for all states i, j of the Markov chain. Note that these functions may not be probability generating functions since the sets of probabilities $\{f_{ji}^{(n)}\}_{n=0}^{\infty}$ and $\{p_{ji}^{(n)}\}_{n=0}^{\infty}$ may not represent probability distributions (the sums may be less than one). Some relationships between these two generating functions are used to prove results about Markov chains.

Multiply $F_{ii}(s)$ by $P_{ii}(s)$ using the definition for the product of two series. The product $C(s)$ of two series $A(s)$ and $B(s)$, where $A(s) = \sum_0^{\infty} a_k s^k$ and $B(s) = \sum_0^{\infty} b_l s^l$, is

$$C(s) = A(s)B(s) = \sum_{r=0}^{\infty} c_r s^r,$$

where

$$c_r = a_0 b_r + a_1 b_{r-1} + \cdots + a_r b_0 = \sum_{k=0}^{r} a_k b_{r-k}.$$

If $A(s)$ and $B(s)$ converge on the interval $(-1, 1)$, then $C(s)$ also converges on $(-1, 1)$ (Wade, 2000). Identify the coefficient a_k of $A(s)$ with $f_{ii}^{(k)}$ of $F_{ii}(s)$ and the coefficient b_l of $B(s)$ with $p_{ii}^{(l)}$ of $P_{ii}(s)$ and apply equation (2.10) so that $c_r = p_{ii}^{(r)}$. The following relationship between the generating functions is obtained:

$$F_{ii}(s)P_{ii}(s) = \sum_{r=1}^{\infty} p_{ii}^{(r)} s^r = P_{ii}(s) - 1,$$

where $p_{ii}^{(r)} = \sum_{k=1}^{r} f_{ii}^{(k)} p_{ii}^{(r-k)}$ and $|s| < 1$. Note that the number one is subtracted from $P_{ii}(s)$ since the first term $c_0 = f_{ii}^{(0)} p_{ii}^{(0)}$ in the product of $F_{ii}(s)P_{ii}(s)$ is zero but the first term in $P_{ii}(s)$ is $p_{ii}^{(0)} = 1$. Hence,

$$P_{ii}(s) = \frac{1}{1 - F_{ii}(s)}. \tag{2.12}$$

A similar relationship exists between $P_{ji}(s)$ and $F_{ji}(s)$ that follows from (2.11):

$$F_{ji}(s)P_{jj}(s) = P_{ji}(s), \quad i \neq j. \tag{2.13}$$

For equation (2.13), the number one is not subtracted from $P_{ji}(s)$ since the first term in its series representation is $p_{ji}^{(0)} = 0$, $i \neq j$, which equals the first term in the series representation of $F_{ji}(s)P_{jj}(s)$; that is, $c_0 = f_{ji}^{(0)} p_{ji}^{(0)} = 0$.

2.5   Basic Theorems for Markov Chains

The relationships between the generating functions are used to relate a recurrent state i to the n-step transition probability $p_{ii}^{(n)}$. First, a lemma is needed.

LEMMA 2.1 Abel's Convergence Theorem

(i) If $\sum\limits_{k=0}^{\infty} a_k$ converges, then $\lim\limits_{s \to 1^-} \sum\limits_{k=0}^{\infty} a_k s^k = \sum\limits_{k=0}^{\infty} a_k = a$.

(ii) If $a_k \geq 0$ and $\lim\limits_{s \to 1^-} \sum\limits_{k=0}^{\infty} a_k s^k = a \leq \infty$, then $\sum\limits_{k=0}^{\infty} a_k = a$.

For a proof of Abel's convergence theorem, consult Karlin and Taylor (1975, pp. 64–65). The lemma is straightforward if the series converges absolutely. Lemma 2.1 is used to verify the following theorem which provides a practical method to check for recurrent or transient states.

THEOREM 2.2

A state i is recurrent (transient) if and only if (iff) $\sum\limits_{n=0}^{\infty} p_{ii}^{(n)}$ diverges (converges), i.e.,

$$\sum_{n=0}^{\infty} p_{ii}^{(n)} = \infty \ (< \infty).$$

Proof. The theorem is verified in the case of a recurrent state. The proof in the case of a transient state follows as a direct consequence because if a state i is not recurrent it is transient. Assume state i is recurrent which means

$$\sum_{n=1}^{\infty} f_{ii}^{(n)} = 1.$$

By part (i) of Lemma 2.1,

$$\lim_{s \to 1^-} \sum_{n=1}^{\infty} f_{ii}^{(n)} s^n = \lim_{s \to 1^-} F_{ii}(s) = 1.$$

From the identity (2.12), it follows that

$$\lim_{s \to 1^-} P_{ii}(s) = \lim_{s \to 1^-} \frac{1}{1 - F_{ii}(s)} = \infty.$$

Because $P_{ii}(s) = \sum_{n=0}^{\infty} p_{ii}^{(n)} s^n$, it follows from Lemma 2.1 part (ii) that

$$\sum_{n=0}^{\infty} p_{ii}^{(n)} = \infty.$$

The converse of the theorem is proved by contradiction. Assume that $\sum_{n=0}^{\infty} p_{ii}^{(n)} = \infty$ and state i is transient which means

$$\sum_{n=1}^{\infty} f_{ii}^{(n)} < 1.$$

Applying Lemma 2.1 part (i),

$$\lim_{s \to 1^-} F_{ii}(s) = \lim_{s \to 1^-} \sum_{n=0}^{\infty} f_{ii}^{(n)} s^n = \sum_{n=0}^{\infty} f_{ii}^{(n)} = \sum_{n=1}^{\infty} f_{ii}^{(n)} < 1.$$

Now, applying the identity (2.12), it follows that

$$\lim_{s \to 1^-} P_{ii}(s) = \lim_{s \to 1^-} \frac{1}{1 - F_{ii}(s)} < \infty.$$

Finally, Lemma 2.1 part (ii) and the preceding inequality yield

$$\sum_{n=0}^{\infty} p_{ii}^{(n)} < \infty,$$

which contradicts the original assumption. The theorem is verified. $\square$

Next it is shown that if state i is recurrent, then all states in the same communicating class are recurrent. Thus, recurrence and transience are class properties. If the chain is irreducible, then the chain is either recurrent or transient.

COROLLARY 2.1

Assume $i \leftrightarrow j$. State i is recurrent (transient) if and only if state j is recurrent (transient).

Proof. Suppose $i \leftrightarrow j$ and state i is recurrent. Then there exist $n, m \geq 1$ such that

$$p_{ij}^{(n)} > 0 \quad \text{and} \quad p_{ji}^{(m)} > 0.$$

Let k be a nonnegative integer,

$$p_{jj}^{(m+n+k)} \geq p_{ji}^{(m)} p_{ii}^{(k)} p_{ij}^{(n)}.$$

Summing on k,

$$\sum_{k=0}^{\infty} p_{jj}^{(k)} \geq \sum_{k=0}^{\infty} p_{jj}^{(n+m+k)} \geq \sum_{k=0}^{\infty} p_{ji}^{(m)} p_{ii}^{(k)} p_{ij}^{(n)} = p_{ji}^{(m)} p_{ij}^{(n)} \sum_{k=0}^{\infty} p_{ii}^{(k)}.$$

The right-hand side is infinite by Theorem 2.2 because state i is recurrent. Thus, $\sum_{k=0}^{\infty} p_{jj}^{(k)}$ is divergent and state j is recurrent. The theorem also holds for transient states because if a state is not recurrent it is transient. $\square$

An important property about recurrent classes follows from the definition of a recurrent state. The next corollary shows that a recurrent class forms a closed set.

COROLLARY 2.2

Every recurrent class in a DTMC is a closed set.

Proof. Let C be a recurrent class. Suppose C is not closed. Then for some $i \in C$ and $j \notin C$, $p_{ji} > 0$. Because $j \notin C$, it is impossible to return to the set C from state j (otherwise $i \leftrightarrow j$). Thus, beginning from state i, the probability of never returning to C is at least p_{ji} or $\sum_n f_{ii}^{(n)} \leq 1 - p_{ji} < 1$, a contradiction to the fact that i is a recurrent state. Hence, C must be closed. $\qquad\square$

Example 2.9. Suppose the transition matrix of a Markov chain with states $\{1, 2, 3, \ldots\}$ is

$$P = \begin{pmatrix} a_1 & 0 & 0 & \cdots \\ a_2 & a_1 & 0 & \cdots \\ a_3 & a_2 & a_1 & \cdots \\ \vdots & \vdots & \vdots & \end{pmatrix},$$

where $a_i > 0$ and $\sum_{i=1}^{\infty} a_i = 1$. The communication classes consist of $\{1\}$, $\{2\}$, $\{3\}$, and so on. Each state represents a communication class. In addition, none of the classes are closed. Hence, by Corollary 2.2, it follows that none of the classes are recurrent, they must all be transient. In fact, each class is aperiodic and transient. $\qquad\blacksquare$

Example 2.4, the one-dimensional unrestricted random walk, illustrates Theorem 2.2. It will be shown that the Markov chain for this process is recurrent if and only if (iff) the probabilities of moving right or left are equal, $p = 1/2 = q$, which means it is a symmetric random walk.

Example 2.10. Consider the one-dimensional, unrestricted random walk in Example 2.4. The chain is irreducible and periodic of period 2. Let p be the probability of moving to the right and q be the probability of moving left, $p + q = 1$. We verify that the state 0 or the origin is recurrent iff $p = 1/2 = q$. However, if the origin is recurrent, then all states are recurrent because the chain is irreducible. Notice that starting from the origin, it is impossible to return to the origin in an odd number of steps,

$$p_{00}^{(2n+1)} = 0 \quad \text{for} \quad n = 0, 1, 2, \ldots.$$

The chain has period 2 because only in an even numbers of steps is the transition probability positive. In $2n$ steps, there are a total of n steps to the right and a total of n steps to the left, and the n steps to the left must be the reverse of those steps taken to the right in order to return to the origin. In particular, in $2n$ steps, there are

$$\binom{2n}{n} = \frac{(2n)!}{n!n!}$$

different paths (combinations) that begin and end at the origin. Also, the probability of occurrence of each one of these paths is $p^n q^n$. Thus,

$$\sum_{n=0}^{\infty} p_{00}^{(n)} = \sum_{n=0}^{\infty} p_{00}^{(2n)} = \sum_{n=0}^{\infty} \binom{2n}{n} p^n q^n.$$

An asymptotic formula for $n!$, known as Stirling's formula, is needed to verify recurrence:

$$n! \sim n^n e^{-n} \sqrt{2\pi n}.$$

The notation $f(n) \sim g(n)$ means $f(n)$ and $g(n)$ grow at the same rate as $n \to \infty$:

$$\lim_{n \to \infty} \frac{f(n)}{g(n)} = 1.$$

Verification of Stirling's formula can be found in Feller (1968) or Norris (1997).

Application of Stirling's formula yields the following approximation:

$$\begin{aligned}
p_{00}^{(2n)} &= \frac{(2n)!}{n!n!} p^n q^n \\
&\sim \frac{\sqrt{4\pi n}(2n)^{2n} e^{-2n}}{2\pi n^{2n+1} e^{-2n}} p^n q^n \\
&= \frac{(4pq)^n}{\sqrt{\pi n}}.
\end{aligned} \tag{2.14}$$

Thus, there exists a positive integer N such that for $n \geq N$,

$$\frac{(4pq)^n}{2\sqrt{\pi n}} < p_{00}^{(2n)} < \frac{2(4pq)^n}{\sqrt{\pi n}}.$$

Considered as a function of p, the expression $4pq = 4p(1-p)$ has a maximum at $p = 1/2$. If $p = 1/2$, then $4pq = 1$ and if $p \neq 1/2$, then $4pq < 1$. When $pq \neq 1/4$, then

$$\sum_{n=0}^{\infty} p_{00}^{(2n)} < N + \sum_{n=N}^{\infty} \frac{2(4pq)^n}{\sqrt{\pi n}} < \infty.$$

The latter series converges by the ratio test. When $pq = 1/4$,

$$\sum_{n=0}^{\infty} p_{00}^{(2n)} > \sum_{n=N}^{\infty} \frac{(4pq)^n}{2\sqrt{\pi n}} = \frac{1}{2\sqrt{\pi}} \sum_{n=N}^{\infty} \frac{1}{\sqrt{n}} = \infty.$$

The latter series diverges because it is just a multiple of a divergent p-series. Therefore, the series $\sum_{n=0}^{\infty} p_{00}^{(2n)}$ diverges iff $p = 1/2 = q$, which means the one-dimensional random walk is recurrent iff it is a symmetric random walk. If $p \neq q$, then all states are transient; there is a positive probability that an object starting from the origin will never return to the origin. What happens if an object never returns to the origin? It can be shown that either the object tends to $+\infty$ or to $-\infty$. ∎

Before giving additional examples, some important results concerning irreducible and recurrent Markov chains are stated. Verification of these results is quite lengthy and the proofs are not given. They depend on a result from renewal theory known as the discrete renewal theorem. A statement and proof of the discrete renewal theorem and proofs of the following theorems can be found in Karlin and Taylor (1975; Theorem 1.2 and Remarks 1.3 and 1.4). The first result is known as the Basic Limit Theorem for aperiodic Markov chains, which gives conditions for a recurrent, irreducible, and aperiodic Markov chain to have a limiting probability distribution. The second result applies to periodic Markov chains.

THEOREM 2.3 Basic Limit Theorem, aperiodic Markov chains
Let $\{X_n\}_{n=0}^{\infty}$ be a recurrent, irreducible, and aperiodic DTMC with transition matrix $P = (p_{ij})$. Then

$$\lim_{n \to \infty} p_{ij}^{(n)} = \frac{1}{\mu_{ii}},$$

where μ_{ii} is the mean recurrence time for state i defined by (2.9) and i and j are any states of the chain. [If $\mu_{ii} = \infty$, then $\lim_{n \to \infty} p_{ij}^{(n)} = 0$.]

THEOREM 2.4 Basic Limit Theorem, periodic Markov chains
Let $\{X_n\}_{n=0}^{\infty}$ be a recurrent, irreducible, and d-periodic DTMC, $d > 1$, with transition matrix $P = (p_{ij})$. Then

$$\lim_{n \to \infty} p_{ii}^{(nd)} = \frac{d}{\mu_{ii}}$$

and $p_{ii}^{(m)} = 0$ if m is not a multiple of d, where μ_{ii} is the mean recurrence time for state i defined by (2.9). [If $\mu_{ii} = \infty$, then $\lim_{n \to \infty} p_{ii}^{(nd)} = 0$.]

Example 2.11. The Markov chain in Example 2.6 is irreducible, recurrent, and periodic with period $d = N$. Applying Theorem 2.4, $\lim_{n \to \infty} p_{ii}^{(nN)} = N/\mu_{ii}$. Since $P^N = I$, $p_{ii}^{(nN)} = 1$ so that $1 = N/\mu_{ii}$, which implies the mean recurrence time is $\mu_{ii} = N$, an obvious result for this example. ∎

The basic theorems apply to recurrent classes as well as to recurrent chains. Suppose C is a recurrent communication class. Then since C is closed, $p_{ki}^{(n)} = 0$ for $i \in C$ and $k \notin C$ for $n \geq 1$. Therefore, the submatrix P_C of P given by $P_C = (p_{ij})_{i,j \in C}$ is a transition matrix for C. By Corollary 2.1, the associated Markov chain for C is irreducible and recurrent. The extension of Theorem 2.3 to aperiodic recurrent classes is stated as a corollary.

COROLLARY 2.3
If i and j are any states in a recurrent and aperiodic class of a Markov chain, then

$$\lim_{n \to \infty} p_{ij}^{(n)} = \frac{1}{\mu_{ii}},$$

where μ_{ii} is defined in (2.9).

Example 2.12. The Markov chain in Example 2.5 has two aperiodic, recurrent classes, $\{1, 2\}$ and $\{3, 4, 5\}$. Then $\lim_{n \to \infty} p_{ij}^{(n)} = 1/\mu_{ii}$ for $i, j = 1, 2$ and for $i, j = 3, 4, 5$. Note that $\lim_{n \to \infty} p_{ij}^{(n)} = 0$ when $i \in \{1, 2\}$ and $j \in \{3, 4, 5\}$ or when $i \in \{3, 4, 5\}$ and $j \in \{1, 2\}$. Instead of calculating the limit to find μ_{ii}, an alternate method will be given in the next section. ■

If $\mu_{ii} = \infty$, then state i is null recurrent and if $0 < \mu_{ii} < \infty$, then state i is positive recurrent. It can be shown that if one state is positive recurrent in a communication class, then all states in that class are positive recurrent. In this case, the entire class is positive recurrent. In addition, it follows that if one state is null recurrent in a communication class, then all states are null recurrent. Verification of these results is left as an exercise (Exercise 15). Hence, null recurrence and positive recurrence are class properties. Therefore, it follows from the previous results that every irreducible Markov chain can be classified as either periodic or aperiodic and as either transient, null recurrent, or positive recurrent:

(1) periodic or (2) aperiodic.

(i) transient or (ii) null recurrent or (iii) positive recurrent.

The classifications (1) and (2) are disjoint, and the three classifications (i), (ii), and (iii) are disjoint. This classification scheme can be applied to communication classes as well, provided the period $d \geq 1$. The special case where $d = 0$ consists of a class with only a single element and the class must be transient (Example 2.7 and Exercise 7). The term ergodic is used to classify states or irreducible chains that are aperiodic and positive recurrent.

DEFINITION 2.18 *A state is* strongly ergodic *if it is both aperiodic and positive recurrent. An* ergodic chain *is a Markov chain that is irreducible, aperiodic, and positive recurrent.*

When the entire class or chain is ergodic, it is also referred to as *ergodic* (Karlin and Taylor, 1975). If the ergodic class or chain is null recurrent rather than positive recurrent, then it is said to be *weakly ergodic* (Karlin and Taylor, 1975). In the case of a finite Markov chain, it will be shown that an ergodic chain means a strongly ergodic chain. The term strongly ergodic will be used if there is a possibility of a weak ergodic chain; otherwise, the term ergodic chain will be used.

The next example reconsiders the unrestricted random walk model. It has already been shown that this Markov chain is irreducible and periodic. In the case of a symmetric random walk, it is shown that the chain is null recurrent.

Example 2.13. The unrestricted random walk model is irreducible and periodic with period $d = 2$. The chain is recurrent iff it is a symmetric random walk, $p = 1/2 = q$ (Example 2.10). Recall that the $2n$-step transition probability is

$$p_{00}^{(2n)} \sim \frac{1}{\sqrt{\pi n}}$$

[equation (2.14)] and hence, $\lim_{n \to \infty} p_{00}^{(2n)} = 0$. It follows from the Basic Limit Theorem for periodic Markov chains that $d/\mu_{00} = 0$. Thus, $\mu_{00} = \infty$; the zero state is null recurrent. Since the chain is irreducible, all states are null recurrent. Thus, when $p = 1/2 = q$, the chain is periodic and null recurrent and when $p \neq 1/2$, the chain is periodic and transient. ∎

2.6 Stationary Probability Distribution

A stationary probability distribution represents an "equilibrium" of the Markov chain; that is, a probability distribution that remains fixed in time. For instance, if the chain is initially at a stationary probability distribution, $p(0) = \pi$, then $p(n) = P^n \pi = \pi$ for all time n.

DEFINITION 2.19 *A stationary probability distribution of a DTMC with states* $\{1, 2, \ldots\}$ *is a nonnegative vector* $\pi = (\pi_1, \pi_2, \ldots)^{tr}$ *that satisfies* $P\pi = \pi$ *and whose elements sum to one. That is,* $\sum_{i=1}^{\infty} \pi_i = 1$.

Definition 2.19 also applies to a finite Markov chain, where the vector $\pi = (\pi_1, \pi_2, \ldots, \pi_N)^{tr}$ and $\sum_{i=1}^{N} \pi_i = 1$. In the finite case, π is a right eigenvector of P corresponding to the eigenvalue $\lambda = 1$, $P\pi = \lambda\pi$. There may be one or more than one linearly independent eigenvector corresponding to the eigenvalue $\lambda = 1$. If there is more than one nonnegative eigenvector, then the stationary probability distribution is not unique.

Example 2.14. In the finite Markov chain model for molecular evolution, Example 2.1, the stationary probability distribution $\pi = (1/4, 1/4, 1/4, 1/4)^{tr}$ is the unique normalized eigenvector corresponding to $\lambda = 1$ (Exercise 18). ∎

Example 2.15. If the transition matrix P is the $N \times N$ identity matrix, then there exists N linearly independent eigenvectors, $e_1 = (1, 0, \ldots, 0)^{tr}, \ldots, e_N = (0, 0, \ldots, 1)^{tr}$, corresponding to the eigenvalue $\lambda = 1$. Hence, any nonzero vector $\pi = (\pi_1, \pi_2, \ldots, \pi_N)^{tr}$ with $\pi_i \geq 0$ for $i = 1, 2, \ldots, N$, and $\sum_{i=1}^{N} \pi_i = 1$ is a stationary probability distribution. The Markov chain with this transition matrix is not very interesting; it is stationary for every initial state. ∎

In Section 2.7, it will be shown that if the finite Markov chain is irreducible, the stationary probability distribution is unique. The finite Markov chain in Example 2.14 is irreducible but in Example 2.15 it is reducible. For an infinite Markov chain, a stationary probability distribution may not exist, as illustrated in the next example.

Example 2.16. Consider the transition matrix in Example 2.9,

$$P = \begin{pmatrix} a_1 & 0 & 0 & \cdots \\ a_2 & a_1 & 0 & \cdots \\ a_3 & a_2 & a_1 & \cdots \\ \vdots & \vdots & \vdots & \end{pmatrix},$$

where $0 < a_i < 1$ and $\sum_{i=1}^{\infty} a_i = 1$. There exists no stationary probability distribution because $P\pi = \pi$ implies $\pi = \mathbf{0}$, the zero vector. It is impossible for the sum of the elements of π to equal one. ∎

The next theorem shows that a positive recurrent, irreducible, and aperiodic Markov chain has a unique stationary probability distribution and this stationary distribution is the limiting distribution as stated in the Basic Limit Theorem for aperiodic Markov chains (Theorem 2.3). The proof is given in the Appendix for Chapter 2.

THEOREM 2.5
Let $\{X_n\}_{n=0}^{\infty}$ be a positive recurrent, irreducible, and aperiodic DTMC (strongly ergodic) with transition matrix $P = (p_{ij})$. Then there exists a unique positive stationary probability distribution π, $P\pi = \pi$, such that

$$\lim_{n \to \infty} p_{ij}^{(n)} = \pi_i, \quad i, j = 1, 2, \ldots.$$

Theorem 2.5 applies to finite and infinite Markov chains. The hypotheses required for finite Markov chains can be simplified, as will be seen in Section 2.7. All that is needed for convergence to a stationary probability distribution in a finite Markov chain is irreducibility and aperiodicity. It follows that a transition matrix of a strongly ergodic Markov chain satisfies

$$\lim_{n \to \infty} P^n = \begin{pmatrix} \pi_1 & \pi_1 & \pi_1 & \cdots \\ \pi_2 & \pi_2 & \pi_2 & \cdots \\ \vdots & \vdots & \vdots & \end{pmatrix}$$

which implies

$$\lim_{n \to \infty} P^n p(0) = \pi. \tag{2.15}$$

The limit is independent of the initial distribution and equals the stationary probability distribution. According to the Basic Limit Theorem for aperiodic

Markov chains, Theorem 2.3, it follows that

$$\boxed{\pi_i = \frac{1}{\mu_{ii}} > 0,}$$

where μ_{ii} is the mean recurrence time for state i. Thus, the mean recurrence time of a positive recurrent, irreducible, and aperiodic chain can be computed from the stationary probability distribution, $\mu_{ii} = 1/\pi_i$.

Example 2.17. Consider the two recurrent classes in the Markov chain discussed in Examples 2.5 and 2.12. The two transition matrices are

$$P_1 = \begin{pmatrix} 1/2 & 1/3 \\ 1/2 & 2/3 \end{pmatrix} \quad \text{and} \quad P_2 = \begin{pmatrix} 0 & 1/4 & 0 \\ 1 & 1/2 & 1 \\ 0 & 1/4 & 0 \end{pmatrix}.$$

There exist unique limiting stationary probability distributions. The stationary distribution for transition matrix P_1 is $\pi = (2/5, 3/5)^{tr}$ so that the mean recurrence times are $\mu_{11} = 5/2$ and $\mu_{22} = 5/3$. After leaving state 1, it takes on average 2.5 time units to return to state 1, and after leaving state 2, it takes 1.67 time units to return to state 2. For matrix P_2, the stationary probability distribution is $\pi = (1/6, 2/3, 1/6)^{tr}$ so that $\mu_{11} = 6$, $\mu_{22} = 3/2$, and $\mu_{33} = 6$. The columns of P_i^n approach the stationary probability distribution,

$$\lim_{n \to \infty} P_1^n = \begin{pmatrix} 2/5 & 2/5 \\ 3/5 & 3/5 \end{pmatrix} \quad \text{and} \quad \lim_{n \to \infty} P_2^n = \begin{pmatrix} 1/6 & 1/6 & 1/6 \\ 2/3 & 2/3 & 2/3 \\ 1/6 & 1/6 & 1/6 \end{pmatrix}.$$

∎

If the Markov chain is periodic, then the limit exists when the time steps are taken in increments of the period d (Basic Limit Theorem for periodic Markov chains, Theorem 2.4). For an irreducible, periodic, and finite Markov chain, the limit $\lim_{n \to \infty} P^{nd}$ depends on all of the eigenvalues λ_i such that $|\lambda_i| = 1$ and their associated eigenvectors. See the Appendix for Chapter 2 and Exercise 22.

2.7 Finite Markov Chains

An important property of finite Markov chains is that there are no null recurrent states and not all states can be transient. Therefore, an irreducible, finite Markov chain is positive recurrent. The assumption of recurrence is not needed when the Basic Limit Theorems are applied to finite Markov chains. A lemma is needed to verify these results.

LEMMA 2.2

Suppose j is a transient state of a Markov chain and i is any state in the Markov chain. Then $\lim_{n\to\infty} p_{ji}^{(n)} = 0$. In addition, a submatrix $T = (t_{jk})$ of the transition matrix P, where indices j, k are from the set of transient states has the following property:

$$\boxed{\lim_{n\to\infty} T^n = \mathbf{0}.}$$ (2.16)

Actually, Lemma 2.2 applies to finite and infinite Markov chains. The proof is left as an exercise.

Example 2.18. Consider Example 2.3, the gambler's problem with two absorbing states. The $N-1 \times N-1$ submatrix of P, consisting of the transient states is

$$T = \begin{pmatrix} 0 & q & \cdots & 0 & 0 \\ p & 0 & \cdots & 0 & 0 \\ 0 & p & \cdots & 0 & 0 \\ \vdots & \vdots & \cdots & \vdots & \vdots \\ 0 & 0 & \cdots & 0 & q \\ 0 & 0 & \cdot & p & 0 \end{pmatrix}.$$

It is clear that T satisfies property (2.16). ∎

THEOREM 2.6

In a finite DTMC, not all states can be transient and no states can be null recurrent. In particular, an irreducible and finite Markov chain is positive recurrent.

The proof of this theorem can be found in the Appendix for Chapter 2. Since there are no null recurrent states in finite Markov chains, there are only four different types of classification schemes based on periodicity and recurrence. The states of a finite Markov chain can be classified as either periodic or aperiodic and either transient or positive recurrent. Recurrence in a finite Markov chain will always mean positive recurrence.

Example 2.19. Let P be the transition matrix of a Markov chain:

$$P = \begin{pmatrix} 1/2 & 0 & 0 & 1/2 \\ 1/2 & 1 & 0 & 0 \\ 0 & 0 & 0 & 1/2 \\ 0 & 0 & 1 & 0 \end{pmatrix}.$$

There are three communication classes, $\{1\}$, $\{2\}$, and $\{3,4\}$. Class $\{1\}$ is aperiodic and transient. Class $\{2\}$ is aperiodic and recurrent. Class $\{3,4\}$ is periodic and transient. State 2 is an absorbing state. Reordering the states,

2, 1, 3, 4, matrix P can be partitioned according to the recurrent class $\{2\}$ and the transient class $\{1, 3, 4\}$.

$$P = \begin{pmatrix} 1 & | & 1/2 & 0 & 0 \\ \hline 0 & | & 1/2 & 0 & 1/2 \\ 0 & | & 0 & 0 & 1/2 \\ 0 & | & 0 & 1 & 0 \end{pmatrix} = \begin{pmatrix} 1 & A \\ \mathbf{0} & T \end{pmatrix}.$$

Matrix T corresponds to the transient class and thus, satisfies property (2.16):

$$\lim_{n \to \infty} T^n = \mathbf{0}.$$

In addition, rows 2, 3, and 4 all tend to zero as $n \to \infty$ (Lemma 2.2). Eventually, all transient classes are absorbed into the recurrent classes. In this example, there is eventual absorption into state 2. ∎

An additional property of finite Markov chains is that a communication class that is closed is recurrent. This is verified in the next theorem. See Schinazi (1999).

THEOREM 2.7
In a finite DTMC, a class is recurrent iff it is closed.

Proof. It has already been shown that if a class is recurrent, it is closed. The reverse implication is verified by contradiction. Assume a class of states C is closed, but C is not recurrent. Then C is transient. By Lemma 2.2, if j is a transient state, then $\lim_{n \to \infty} p_{ji}^{(n)} = 0$ for all states i. In particular, for state $i \in C$,

$$\sum_{j \in C} \lim_{n \to \infty} p_{ji}^{(n)} = 0. \tag{2.17}$$

But because C is closed, the submatrix P_C consisting of all states in C is a stochastic matrix. In addition, $P_C^{(n)}$ is a stochastic matrix (Exercise 2). The column sums of $P_C^{(n)}$ equal one and must equal one in the limit as $n \to \infty$, a contradiction to (2.17). Hence, C cannot be transient; C must be recurrent. □

Example 2.20. The finite Markov chain in Examples 2.5 and 2.12 has two recurrent classes, $\{1, 2\}$ and $\{3, 4, 5\}$. The transition matrix can be partitioned according to these classes,

$$P = \begin{pmatrix} 1/2 & 1/3 & | & 0 & 0 & 0 \\ 1/2 & 2/3 & | & 0 & 0 & 0 \\ \hline 0 & 0 & | & 0 & 1/4 & 0 \\ 0 & 0 & | & 1 & 1/2 & 1 \\ 0 & 0 & | & 0 & 1/4 & 0 \end{pmatrix}.$$

Since both classes are closed, they are both recurrent. They are also aperiodic. In addition, from the partition, it is easy to see that each of the diagonal submatrices forms a stochastic matrix. ∎

The following result is a corollary of Theorems 2.5 and 2.6.

COROLLARY 2.4

Suppose a finite DTMC is irreducible and aperiodic. Then there exists a unique stationary probability distribution $\pi = (\pi_1, \ldots, \pi_N)^{tr}$ such that

$$\lim_{n \to \infty} p_{ij}^{(n)} = \pi_i, \quad i, j = 1, 2, \ldots, N.$$

The unique stationary probability distribution also follows from linear algebra and the theorems of Perron and Frobenius (Appendix for Chapter 2). The value $\lambda = 1$ is an eigenvalue of P because $P^{tr}\mathbf{1} = \mathbf{1}$, where $\mathbf{1}$ is a column vector of ones and P and P^{tr} have the same eigenvalues. The Frobenius theorem ensures that the right eigenvector of P is positive. In finite Markov chain theory, a stochastic matrix with the property $p_{ji}^{(n)} > 0$, for some $n > 0$ and all $i, j = 1, 2, \ldots, N$ ($P^n > 0$), is referred to as a *regular* matrix. If the transition matrix is regular, then the Markov chain is irreducible and aperiodic (Why?). The converse is also true; an irreducible, aperiodic finite Markov chain is regular. Therefore, a regular Markov chain is strongly ergodic. The Perron theorem guarantees that a regular matrix P has a positive dominant eigenvalue λ which is simple and satisfies $\lambda > |\lambda_i|$, where λ_i is any other eigenvalue of P (Gantmacher, 1964; Ortega, 1987). For a stochastic matrix, this dominant eigenvalue is $\lambda = 1$. The eigenvector $\pi > 0$ associated with $\lambda = 1$ and satisfying $\sum \pi_j = 1$ defines a unique stationary probability distribution, $P\pi = \pi$ for a strongly ergodic Markov chain. If the assumptions are weakened, so that the stochastic matrix P is irreducible and periodic, then the Frobenius theorem still implies that $\lambda = 1$ is a simple eigenvalue satisfying $\lambda \geq |\lambda_i|$ with associated positive eigenvector π (Ortega, 1987). Therefore, all that is required for existence of a unique stationary probability distribution is that P be irreducible. The additional property of aperiodicity is needed to show convergence to the stationary probability distribution.

The next biological example applies finite Markov chain theory to study the dynamics of an introduced species.

Example 2.21. The introduction of a new or alien species into an environment will often disrupt the dynamics of native species (Hengeveld, 1989; Shigesada and Kawasaki, 1997; Williamson, 1996). For example, the gray squirrel, *Sciurus carolinensis*, was introduced into Great Britain in the late nineteenth century and it quickly invaded many regions previously occupied by the native red squirrel, *Sciurus vulgaris* (Reynolds, 1985). Data were collected in various regions in Great Britain as to the presence of red squirrels only (R), gray squirrels only (G), both squirrels (B), or absence of both squirrels (A).

The data were summarized and a Markov chain model was developed with the four states, $\{R, G, B, A\}$. The model is reported in Mooney and Swift (1999). Each region was classified as being in one of these states, and the transitions between states over a period of one year were estimated (e.g., $p_{RR} = 0.8797$, $p_{RG} = 0.0382$). If the states 1, 2, 3, and 4 are ordered as R, G, B, and A, respectively, then the transition matrix has the form

$$P = \begin{pmatrix} 0.8797 & 0.0382 & 0.0527 & 0.0008 \\ 0.0212 & 0.8002 & 0.0041 & 0.0143 \\ 0.0981 & 0.0273 & 0.8802 & 0.0527 \\ 0.0010 & 0.1343 & 0.0630 & 0.9322 \end{pmatrix}.$$

It is easy to see that the corresponding Markov chain is irreducible and aperiodic (P is regular). There exists a unique, limiting stationary distribution π which can be found by calculating the eigenvector of P corresponding to the eigenvalue $\lambda = 1$:

$$\pi = (0.1705, 0.0560, 0.3421, 0.4314)^{tr}.$$

Over the long term, the model predicts that 17.05% of the region will be populated by red squirrels, 5.6% by gray squirrels, 34.21% by both species, and 43.14% by neither species. The mean recurrence times, $\mu_{ii} = 1/\pi_i$, $i = 1, 2, 3, 4$, are given by the vector

$$\mu = (5.865, 17.857, 2.923, 2.318)^{tr}.$$

For example, a region populated by red squirrels (R) may change to other states (G, B, or A) but, on the average, it will again be populated by red squirrels after about six years. ∎

2.7.1 Mean First Passage Time

Methods are derived for calculating the mean first passage times for irreducible finite Markov chains and the mean time until absorption for absorbing Markov chains. The methods are primarily useful for numerical computation.

First, assume that the Markov chain is irreducible. Denote the matrix of mean first passage times by

$$M = (\mu_{ij}) = \begin{pmatrix} \mu_{11} & \mu_{12} & \cdots & \mu_{1N} \\ \mu_{21} & \mu_{22} & \cdots & \mu_{2N} \\ \vdots & \vdots & \cdots & \vdots \\ \mu_{N1} & \mu_{N2} & \cdots & \mu_{NN} \end{pmatrix}.$$

Instead of calculating the matrix elements via their definitions, using $\{f_{ii}^{(n)}\}$ and $\{f_{ji}^{(n)}\}$, an alternate method is applied.

Consider what happens at the first time step. Either state j can be reached in one time step with probability p_{ji} or it takes more than one time step. If it takes more than one time step to reach j, then in one step another state k is reached, $k \neq j$, with probability p_{ki}. Then the time it takes to reach state j is $1 + \mu_{jk}$, one time step plus the mean time it takes to reach state j from state k. This relationship is given by

$$\mu_{ji} = p_{ji} + \sum_{k=1, k \neq j}^{N} p_{ki}(1 + \mu_{jk}) = 1 + \sum_{k=1, k \neq j}^{N} p_{ki}\mu_{jk}. \qquad (2.18)$$

The relationship (2.18) assumes matrix P is irreducible; every state j can be reached from any other state i.

The equations in (2.18) can be expressed in matrix form,

$$M = E + (M - \text{diag}(M))P, \qquad (2.19)$$

where E is an $N \times N$ matrix of ones. Since P is irreducible, the Markov chain is irreducible, which means it is positive recurrent, $1 \leq \mu_{ii} < \infty$, $i = 1, 2, \ldots, N$. It follows that $1 \leq \mu_{ji} < \infty$ for $j \neq i$. The system (2.19) can be written as a linear system of equations, N^2 equations and N^2 unknowns (the μ_{ji}'s). It can be shown that the linear system has a unique solution given by the μ_{ji}'s. Stewart (1994) presents an iterative method based on equation (2.19) to estimate the mean first passage times.

Example 2.22. Suppose

$$P = \begin{pmatrix} 0 & 1 \\ 1 & 0 \end{pmatrix}.$$

Then equation (2.19) can be expressed as

$$\begin{pmatrix} \mu_{11} & \mu_{12} \\ \mu_{21} & \mu_{22} \end{pmatrix} = \begin{pmatrix} 1 + \mu_{12} & 1 \\ 1 & 1 + \mu_{21} \end{pmatrix}.$$

Hence, $\mu_{12} = 1 = \mu_{21}$ and $\mu_{11} = 2$ and $\mu_{22} = 2$. This result is obvious once one recognizes that the chain is periodic of period 2. It takes two time steps to return to states 1 or 2 and only one time step to go from state 1 to state 2 or from state 2 to state 1. ∎

Now suppose the Markov chain has k absorbing states. Partition the matrix as in Example 2.19 into k absorbing states and $m - k$ transient states:

$$P = \begin{pmatrix} I & A \\ \mathbf{0} & T \end{pmatrix}.$$

Matrix I is the $k \times k$ identity matrix corresponding to the k absorbing states and matrix T is an $m - k \times m - k$ matrix corresponding to the $m - k$ transient states. It follows from Lemma 2.2, property (2.16),

$$\lim_{n \to \infty} T^n = \mathbf{0}. \qquad (2.20)$$

Let ν_{ij} be the random variable for the number of visits (prior to absorption) to transient state i beginning from transient state j. Then the expected number of visits to state i from j before absorption is

$$(E[\nu_{ij}]) = I + T + T^2 + T^3 + \cdots = (I - T)^{-1},$$

where the last equality follows from property (2.20) (Iosifescu, 1980; Ortega, 1987). Alternately, the inverse of the matrix $I - T$ exists because the matrix is irreducibly diagonally dominant.

DEFINITION 2.20 *An $N \times N$ matrix $M = (m_{ij})$ is said to be* irreducibly diagonally dominant *if it is irreducible and the following inequality holds for the elements m_{ij} with strict inequality for at least one i:*

$$|m_{ii}| \geq \sum_{j=1, j \neq i}^{N} |m_{ij}|, \quad i = 1, \ldots, N.$$

That is, the absolute value of each of the diagonal elements of M dominates the sum of the absolute values of all of the off-diagonal elements in that row. An irreducibly diagonally dominant matrix is invertible (Ortega, 1987). To show that a matrix is invertible, the preceding definition can be applied to the columns or the rows. If the columns of M satisfy the preceding definition and M is irreducible, then M^{tr} is irreducibly diagonally dominant which means M^{tr} is invertible. But if M^{tr} is invertible, M is also invertible.

Matrix $(I - T)^{-1}$ is known as the *fundamental matrix* in DTMC theory and is denoted

$$\boxed{F = (I - T)^{-1}.} \tag{2.21}$$

To calculate the expected time until absorption beginning from transient state j, it is necessary to sum the columns of F, since the expected time until absorption is the time spent in each of the transient states. Let $\mathbf{1}$ be a column vector of ones. Then the *expected time until absorption* m can be calculated as follows:

$$\boxed{m = \mathbf{1}^{tr} F.}$$

Example 2.23. Consider Example 2.19. Matrix P is partitioned into a single absorbing state, 2, and three transient states, 1, 3, and 4. The matrix corresponding to the transient class $\{1, 3, 4\}$ is

$$T = \begin{pmatrix} 1/2 & 0 & 1/2 \\ 0 & 0 & 1/2 \\ 0 & 1 & 0 \end{pmatrix}.$$

The fundamental matrix is

$$F = (I - T)^{-1} = \begin{pmatrix} 2 & 2 & 2 \\ 0 & 2 & 1 \\ 0 & 2 & 2 \end{pmatrix}.$$

Thus, the mean time until absorption is $m = \mathbf{1}^{tr} F = (2, 6, 5)$. Beginning from transient states 1, 3, or 4, it takes, on average, 2, 6 or 5 time intervals to reach the absorbing state 2. ∎

2.8 An Example: Genetics Inbreeding Problem

Inheritance depends on the information contained in the chromosomes that are passed down from generation to generation. Humans have two sets of chromosomes (diploid), one obtained from each parent. Certain locations along the chromosomes contain the instructions for some physical characteristic. The collections of chemicals at these locations are called *genes* and their locations are called *loci* (see, e.g., Hoppensteadt, 1975). At each locus, the gene may take one of several forms referred to as an *allele*.

Suppose there are only two types of alleles for a given gene, denoted a and A. A diploid individual could then have one of three different combinations of alleles: AA, Aa, or aa, known as the *genotypes* of the locus. The combinations AA and aa are called *homozygous*, whereas Aa is called *heterozygous*.

Bailey (1990) and Feller (1968) formulate a DTMC model for the genetics of inbreeding which is discussed below. Assume two individuals are randomly mated. Then, in the next generation, two of their offspring of opposite sex are randomly mated. The process of brother and sister mating or inbreeding continues each year. This process can be formulated as a finite DTMC whose states consist of the six mating types,

1. $AA \times AA$, 2. $AA \times Aa$, 3. $Aa \times Aa$, 4. $Aa \times aa$, 5. $AA \times aa$,
6. $aa \times aa$.

Suppose the parents are of type 1, $AA \times AA$. Then the next generation of offspring from these parents will be AA individuals, so that crossing of brother and sister will give only type 1, $p_{11} = 1$. Now, suppose the parents are of type 2, $AA \times Aa$. Offspring of type $AA \times Aa$ will occur in the following proportions, 1/2 AA and 1/2 Aa, so that crossing of brother and sister will give 1/4 type 1 ($AA \times AA$), 1/2 type 2 ($AA \times Aa$), and 1/4 type 3 ($Aa \times Aa$). If the parents are of type 3, $Aa \times Aa$, offspring are in the proportions 1/4 AA, 1/2 Aa, and 1/4 aa, so that brother and sister mating will give 1/16 type 1, 1/4 type 2, 1/4 type 3, 1/4 type 4, 1/8 type 5, and 1/16 type 6. Continuing in this manner, the complete transition matrix P is

$$P = \begin{pmatrix} 1 & 1/4 & 1/16 & 0 & 0 & 0 \\ 0 & 1/2 & 1/4 & 0 & 0 & 0 \\ 0 & 1/4 & 1/4 & 1/4 & 1 & 0 \\ 0 & 0 & 1/4 & 1/2 & 0 & 0 \\ 0 & 0 & 1/8 & 0 & 0 & 0 \\ 0 & 0 & 1/16 & 1/4 & 0 & 1 \end{pmatrix}$$

$$= \left(\begin{array}{c|cccc|c} 1 & 1/4 & 1/16 & 0 & 0 & 0 \\ \hline 0 & 1/2 & 1/4 & 0 & 0 & 0 \\ 0 & 1/4 & 1/4 & 1/4 & 1 & 0 \\ 0 & 0 & 1/4 & 1/2 & 0 & 0 \\ 0 & 0 & 1/8 & 0 & 0 & 0 \\ \hline 0 & 0 & 1/16 & 1/4 & 0 & 1 \end{array} \right)$$

$$= \begin{pmatrix} 1 & A & 0 \\ \mathbf{0} & T & \mathbf{0} \\ 0 & B & 1 \end{pmatrix}.$$

The Markov chain is reducible and has three communicating classes: $\{1\}$, $\{6\}$, and $\{2, 3, 4, 5\}$. The first two classes are positive recurrent and the third class is transient. States 1 and 6 are absorbing states, $p_{ii} = 1$, $i = 1, 6$.

Note that

$$P^n = \begin{pmatrix} 1 & A_n & 0 \\ \mathbf{0} & T^n & \mathbf{0} \\ 0 & B_n & 1 \end{pmatrix},$$

where A_n and B_n are functions of T, A, and B, $A_n = A \sum_{i=0}^{n-1} T^i$, and $B_n = B \sum_{i=0}^{n-1} T^i$. Using the theory from the previous section, the expected time until absorption or fixation from transient states, 2, 3, 4, or 5, is

$$m = \mathbf{1}^{tr} F = \mathbf{1}^{tr} \frac{1}{3} \begin{pmatrix} 8 & 4 & 2 & 4 \\ 4 & 8 & 4 & 8 \\ 2 & 4 & 8 & 4 \\ 1/2 & 1 & 1/2 & 4 \end{pmatrix},$$

where $F = (I - T)^{-1}$ is the fundamental matrix. Thus,

$$m = (29/6, 17/3, 29/6, 20/3) \approx (4.83, 5.67, 4.83, 6.67).$$

Unfortunately, the expected times do not tell us which one of the two absorbing states is reached. This problem can be solved by returning to the matrix P^n and calculating $\lim_{n \to \infty} A_n = AF$ and $\lim_{n \to \infty} B_n = BF$. The vector $AF = (3/4, 1/2, 1/4, 1/2)$ and $BF = (1/4, 1/2, 3/4, 1/2)$ are the (eventual) probabilities of absorption into states 1 or 6, respectively, beginning from the transient states 2, 3, 4, or 5.

A general formula can be found for T^n using the method described in the Appendix for Chapter 2. Once T^n is calculated, various questions can be addressed about the dynamics of the model at the nth time step or nth generation. For example, what is the probability of absorption and the proportion of heterozygotes in the population in the nth generation? Absorption at step n into state 1 implies that at the $(n-1)$st step state 2 or 3 is entered. Then state 1 is entered at the next step. Thus, absorption into state 1 at the nth step is

$$p_{12}p_{2i}^{(n-1)} + p_{13}p_{3i}^{(n-1)} = \frac{1}{4}p_{2i}^{(n-1)} + \frac{1}{16}p_{3i}^{(n-1)}.$$

The values of $p_{2i}^{(n-1)}$ and $p_{3i}^{(n-1)}$ can be calculated from T^{n-1}. Absorption into state 6 at the nth step is

$$p_{63}p_{3i}^{(n-1)} + p_{64}p_{4i}^{(n-1)} = \frac{1}{16}p_{3i}^{(n-1)} + \frac{1}{4}p_{4i}^{(n-1)}.$$

The proportion of heterozygous individuals, Aa, at the nth time step is

$$h_n = \frac{1}{2}p_2(n) + p_3(n) + \frac{1}{2}p_4(n),$$

where $p_i(n)$ is the proportion of the population in state i at time n (Appendix for Chapter 2). Because states 2, 3, and 4 are transient, $\lim_{n\to\infty} h_n = 0$.

2.9 Monte Carlo Simulation

The name "Monte Carlo" was first used by Nicholas Metropolis about 1947 in connection with a new statistical sampling procedure that was used to test the computational power of the first electronic computer (Metropolis, 1987). The name comes from a town in Monaco on the southern border of France, known for its casinos.

The Monte Carlo method is a statistical sampling procedure to obtain information about the expectation. A collection of m i.i.d. samples are used to calculate the expectation. For example, this method can be used to estimate the value of a definite integral,

$$\int_A g(x)\, dx.$$

By selecting an appropriate p.d.f. f, defined on A, where $\int_A f(x)\, dx = 1$, the definite integral can be viewed as an expectation:

$$\mu = E\left(\frac{g(x)}{f(x)}\right) = \int_A \frac{g(x)}{f(x)} f(x)\, dx.$$

In the Monte Carlo integration procedure, the p.d.f f is used to select m independent sample points $\{x_i\}_{i=1}^m$, then the integrand is evaluated at x_i, $g(x_i)/f(x_i)$. For a sufficient number of sample points, $i = 1, \ldots, m$, the integral is approximated as follows:

$$\mu \approx \frac{1}{m} \sum_{i=1}^m \frac{g(x_i)}{f(x_i)} = \hat{\mu}.$$

From the Law of Large Numbers (Theorem 1.2), it follows that $\hat{\mu}$ converges in the mean square sense and with probability one to μ.

Example 2.24. The following integral is evaluated using Monte Carlo integration:

$$\int_0^1 \exp(x^2)\, dx,$$

where f is the p.d.f. of a uniform random variable on $[0, 1]$. The following MATLAB® commands can be used to approximate this integral (which gives a value of 1.463, correct to three decimal places):

```
m=10^6;
x=rand(m,1); % generates m random numbers from U(0,1).
ans=mean(exp(x.^2))
```

■

Monte Carlo methods are frequently applied to Markov chain models. These methods will be applied in later chapters. The methods are particularly useful for large complex problems, with many states, for which it is difficult to compute expectation or probability distributions from a transition matrix or by analytical methods. Many algorithms for Markov chain models have been developed for Markov chain models, referred to as Markov Chain Monte Carlo methods–MCMC methods, which speed up the calculations, e.g., Metropolis-Hastings algorithm (Robert and Casella, 2004). A simple example is given for the unrestricted random walk model.

Example 2.25. Consider a one-dimensional, unrestricted random walk on the integers $\{0, \pm1, \pm2, \ldots\}$. Given $X_0 = 0$, let's calculate $E(X_{100}|X_0)$, where $p = 0.45$ and $q = 0.55$. A random value r is selected from the uniform distribution $U(0,1)$: if $r < p$, then $x_{i+1} = x_i - 1$ and if $r \geq p$, then $x_{i+1} = x_i + 1$. This method generates sample values $\{x_1, x_2, \ldots, x_{100}\}$. It is repeated m times, giving m values for x_{100}: $\{x_{100}^i\}_{i=1}^m$. Then

$$E(X_{100}|X_0) \approx \frac{1}{m} \sum_{i=1}^m x_{100}^i.$$

The expected value $E(X_{100}|X_0 = 0) = 10$. A MATLAB program for the Monte Carlo method is given below.

```
m=10^6; p=0.45;
x=zeros(m,1);
for i=1:100
    r=rand(m,1);
    y1=(r<p);
    x=x-y1;
    y2=(r>=p);
    x=x+y2;
end
ans=mean(x)
```

■

2.10   Unrestricted Random Walk in Higher Dimensions

The random walk model can be extended to two and three dimensions. It was shown for the unrestricted random walk in one dimension that the chain is null recurrent if and only if $p = 1/2 = q$. For two and three dimensions, it is assumed that the probability of moving in any one direction is the same. Thus, for two dimensions, the probability is $1/4$ of moving in any of the four directions: up, down, right, or left. For three dimensions, the probability is $1/6$ of moving in any of the six directions: up, down, right, left, forward, or backward. It is shown for two dimensions that the chain is null recurrent but for three dimension it is transient. These examples illustrate the distinctly different behavior between one and two dimensions and dimensions greater than two. These examples were first studied by Polya and are discussed in many textbooks (see, e.g., Bailey 1990; Karlin and Taylor, 1975; Norris, 1997; Schinazi, 1999). The verifications are quite lengthy.

The Markov chain represented by this unrestricted random walk is irreducible and periodic of period 2. Because of the symmetry, transience or recurrence can be checked at the origin. Let the origin be denoted as 0 and $p_{00}^{(n)}$ be the probability of returning to the origin after n steps. Note that $p_{00}^{(2n)} > 0$, but $p_{00}^{(2n+1)} = 0$ for $n = 0, 1, 2, \ldots$. It is impossible to begin at the origin and return to the origin in an odd number of steps.

2.10.1   Two Dimensions

In two dimensions, for a path length of $2n$ beginning and ending at 0, if k steps are taken to the right, then k steps must be also taken to the left, and if $n - k$ steps are taken in the upward direction, then $n - k$ steps must be taken downward, $k + k + n - k + n - k = 2n$. There are

$$\sum_{k=0}^{n} \frac{(2n)!}{k!k!(n-k)!(n-k)!}$$

different paths of length $2n$ that begin and end at the origin. Each of these paths is equally likely and has a probability of occurring equal to $(1/4)^{2n}$. Thus,

$$
\begin{aligned}
p_{00}^{(2n)} &= \sum_{k=0}^{n} \frac{(2n)!}{k!k!(n-k)!(n-k)!} \left(\frac{1}{4}\right)^{2n} \\
&= \frac{(2n)!}{(n!)^2} \sum_{k=0}^{n} \left(\frac{n!}{k!(n-k)!}\right)^2 \left(\frac{1}{4}\right)^{2n} \\
&= \frac{(2n)!}{(n!)^2} \sum_{k=0}^{n} \binom{n}{k}^2 \left(\frac{1}{4}\right)^{2n}.
\end{aligned}
$$

It can be shown that

$$\sum_{k=0}^{n} \binom{n}{k}^2 = \binom{2n}{n}$$

(Bailey, 1990). Hence, $p_{00}^{(2n)}$ can be simplified to

$$p_{00}^{(2n)} = \frac{(2n)!}{(n!)^2} \binom{2n}{n} \left(\frac{1}{4}\right)^{2n} = \left[\frac{(2n)!}{n!n!}\right]^2 \frac{1}{4^{2n}}.$$

Stirling's formula can be applied to the right side of the preceding expression $(n! \sim n^n \sqrt{2\pi n} e^{-n})$ so that

$$
\begin{aligned}
p_{00}^{(2n)} &\sim \left[\frac{(2n)^{2n}\sqrt{4\pi n}\, e^{-2n}}{n^{2n} 2\pi n e^{-2n}}\right]^2 \frac{1}{4^{2n}} \\
&= \left[\frac{4^n}{\sqrt{\pi n}}\right]^2 \frac{1}{4^{2n}} = \frac{1}{\pi n}.
\end{aligned}
$$

By comparing $\sum p_{00}^{(2n)}$ with the divergent harmonic series $\sum 1/[\pi n]$, it follows that $\sum p_{00}^{(2n)}$ also diverges. Thus, by Theorem 2.2, the origin is recurrent and all states must be recurrent. In addition, by applying the Basic Limit Theorem for periodic Markov chains, $\lim_{n \to \infty} p_{00}^{(2n)} = 2/\mu_{00}$. But this limit is zero; thus, $\mu_{00} = \infty$. The zero state is null recurrent and, hence, the Markov chain for the symmetric, two-dimensional random walk is null recurrent.

2.10.2 Three Dimensions

In three dimensions, in a path of length $2n$ beginning and ending at the origin, if k steps are taken to the right, then k steps must be taken to the left; if j steps are taken upward, then j steps must be taken downward; and if $n-k-j$ steps are taken forward, then $n-k-j$ steps must be taken backward, $k+k+j+j+n-k-j+n-k-j = 2n$. The total number of paths of length $2n$ is

$$\sum_{j+k\leq n} \frac{(2n)!}{(k!)^2(j!)^2[(n-k-j)!]^2},$$

where the sum is over all of the j and k, $j+k \leq n$. Because each path has probability $(1/6)^{2n}$, it follows that

$$p_{00}^{(2n)} = \sum_{j+k\leq n} \frac{(2n)!}{(k!)^2(j!)^2[(n-k-j)!]^2} \left(\frac{1}{6}\right)^{2n}$$

$$= \frac{(2n)!}{2^{2n}(n!)^2} \sum_{j+k\leq n} \left(\frac{n!}{j!k!(n-j-k)!}\right)^2 \left(\frac{1}{3}\right)^{2n}.$$

Use the fact that the trinomial distribution satisfies

$$\sum_{j+k\leq n} \frac{n!}{j!k!(n-j-k)!}\frac{1}{3^n} = 1.$$

For convenience, denote the trinomial coefficient as

$$\frac{n!}{j!k!(n-j-k)!} = \binom{n}{j\ k}.$$

The maximum value of the trinomial distribution can be shown to occur when $j \approx n/3$ and $k \approx n/3$ and is approximately equal to $M_n \approx n!(1/3)^n/[(n/3)!]^3$ when n is large. This can be seen as follows. Suppose the maximum value occurs at j' and k'. Then

$$\binom{n}{j'\ (k'-1)} \leq \binom{n}{j'\ k'}$$

$$\binom{n}{j'\ (k'+1)} \leq \binom{n}{j'\ k'}$$

$$\binom{n}{(j'-1)\ k'} \leq \binom{n}{j'\ k'}$$

$$\binom{n}{(j'+1)\ k'} \leq \binom{n}{j'\ k'}$$

so that

$$n-k'-1 \leq 2j' \leq n-k'+1, \quad n-j'-1 \leq 2k' \leq n-j'+1$$

or

$$\frac{n-1}{n} \leq \frac{2j'+k'}{n} \leq \frac{n+1}{n}, \quad \frac{n-1}{n} \leq \frac{2k'+j'}{n} \leq \frac{n-1}{n}.$$

Letting $n \to \infty$, then $2j' + k' \sim n$ and $2k' + j' \sim n$, from which it follows that $j' \sim n/3$ and $k' \sim n/3$.

These facts lead to an upper bound on $p_{00}^{(2n)}$. First,

$$p_{00}^{(2n)} \leq \frac{1}{2^{2n}} \frac{(2n)!}{(n!)^2} M_n \left[\sum_{j+k \leq n} \frac{n!}{j!k!(n-j-k)!} \frac{1}{3^n} \right]$$

$$= \frac{1}{2^{2n}} \frac{(2n)!}{(n!)^2} \frac{n!}{[(n/3)!]^3} \frac{1}{3^n},$$

because the expression in the square brackets is a trinomial distribution whose sum equals one. Next, Stirling's formula is used to approximate the right-hand side of the preceding inequality for large n:

$$\frac{1}{2^{2n}} \frac{(2n)!}{(n!)^2} \frac{n!}{[(n/3)!]^3} \frac{1}{3^n} \sim \frac{1}{2^{2n}} \frac{(2n)^{2n}\sqrt{4\pi n}e^{-2n}}{n^n\sqrt{2\pi n}e^{-n}(n/3)^n(\sqrt{2\pi n/3})^3 e^{-n}} \frac{1}{3^n}$$

$$= \frac{c}{n^{3/2}},$$

where $c = (1/2)(3/\pi)^{3/2}$. Thus, for large n, $p_{00}^{(2n)} \leq c/n^{3/2}$. Because $\sum_n c/n^{3/2}$ is a convergent p-series, it follows by comparison and Theorem 2.2 that the origin is a transient state. Hence, because the Markov chain is irreducible, all states are transient. The DTMC for the symmetric, three-dimensional random walk is transient.

The distinctly different behavior of a DTMC in three dimensions as opposed to one or two dimensions is not unusual. A path along a line or in a plane is much more restricted than a path in space. This difference in behavior is demonstrated in other models as well (e.g., systems of autonomous differential equations), where the behavior in three or higher dimensions is much more complicated and harder to predict than in one or two dimensions.

2.11   Exercises for Chapter 2

1. For the directed graph in Figure 2.8, write the corresponding stochastic matrix.

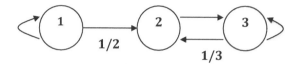

FIGURE 2.8: Directed graph of a stochastic matrix.

2. Suppose P is an $N \times N$ stochastic matrix (column sums equal one),

$$P = \begin{pmatrix} p_{11} & p_{12} & \cdots & p_{1N} \\ p_{21} & p_{22} & \cdots & p_{2N} \\ \vdots & \vdots & \cdots & \vdots \\ p_{N1} & p_{N2} & \cdots & p_{NN} \end{pmatrix}.$$

(a) Show that P^2 is a stochastic matrix. Then show that P^n is a stochastic matrix for all positive integers n.

(b) Suppose P is a doubly stochastic matrix (row and column sums equal one). Show that P^n is a doubly stochastic matrix for all positive integers n.

3. Let $P = (p_{ij})$ be transition matrix with an infinite number of states $i, j \in \{1, 2, \ldots\}$. Fubini's theorem (discrete form) states that if $a_{ij} \geq 0$, then the following sums can be interchanged (Norris, 1997, p. 219):

$$\sum_{i=1}^{\infty} \sum_{j=1}^{\infty} a_{ij} = \sum_{j=1}^{\infty} \sum_{i=1}^{\infty} a_{ij}.$$

Apply Fubini's theorem to show that if P is a stochastic matrix, then P^2 is a stochastic matrix. Then show that P^n is a stochastic matrix for all positive integers n.

4. Show that the relation (2.4) follows from conditional probabilities. In particular, show that

$$\text{Prob}\{A \cap B | C\} = \text{Prob}\{A | B \cap C\}\text{Prob}\{B | C\}.$$

5. If j is a transient state of a Markov chain with states $\{1, 2, \ldots\}$, prove that for all states $i = 1, 2, \ldots$,

$$\sum_{n=1}^{\infty} p_{ji}^{(n)} < \infty.$$

Since the series is convergent, it follows that $\lim_{n \to \infty} p_{ji}^{(n)} = 0$.

6. Suppose a finite Markov chain has N states. State 1 is absorbing and the remaining states are transient. Use Exercises 2 and 5 to show that

$$\lim_{n \to \infty} P^n = \begin{pmatrix} 1 & 1 & \cdots & 1 \\ 0 & 0 & \cdots & 0 \\ \vdots & \vdots & \cdots & \vdots \\ 0 & 0 & \cdots & 0 \end{pmatrix}.$$

Then for any initial probability distribution corresponding to X_0, $p(0) = (p_1(0), p_2(0), \ldots, p_N(0))^{tr}$, it follows that

$$\lim_{n \to \infty} P^n p(0) = (1, 0, \ldots, 0)^{tr}.$$

7. Verify the following two statements.

   (a) Assume the period of state i in a DTMC model is $d(i) = 0$. Then the set $\{i\}$ is a communication class in the Markov chain.

   (b) In an irreducible DTMC, the period $d \geq 1$.

8. Refer to Example 2.8. Show that the mean recurrence times for this example are finite, $\mu_{ii} < \infty$ for $i = 1, 2$.

9. A Markov chain has the following transition matrix:

$$P = \begin{pmatrix} 0 & 1/2 & 0 \\ 1 & 0 & 1 \\ 0 & 1/2 & 0 \end{pmatrix}.$$

   (a) Draw a directed graph for the chain.

   (b) Identify the communicating classes and classify them as periodic or aperiodic, transient or recurrent.

   (c) Calculate the probability of the first return to state i at the nth step, $f_{ii}^{(n)}$, for each state $i = 1, 2, 3$ and for each time step $n = 1, 2, \ldots$.

   (d) Use (c) to calculate the mean recurrence times for each state, μ_{ii}, $i = 1, 2, 3$.

10. Three different Markov chains are defined by the following transition matrices:

(i) $\begin{pmatrix} 1 & 0 & 1/2 \\ 0 & 0 & 1/2 \\ 0 & 1 & 0 \end{pmatrix}$, (ii) $\begin{pmatrix} 1 & 0 & 1/3 \\ 0 & 0 & 1/3 \\ 0 & 1 & 1/3 \end{pmatrix}$, (iii) $\begin{pmatrix} 0 & 1 & 0 & 0 \\ 1 & 0 & 0 & 0 \\ 0 & 0 & 0 & 1 \\ 0 & 0 & 1 & 0 \end{pmatrix}$.

   (a) Draw a directed graph for each chain. Is the Markov chain irreducible?

(b) Identify the communicating classes and classify them as periodic or aperiodic, transient or recurrent.

11. Three different Markov chains are defined by the following transition matrices:

(i) $\begin{pmatrix} 1/3 & 1/4 & 0 & 1/2 \\ 1/3 & 1/4 & 0 & 0 \\ 0 & 1/4 & 1 & 0 \\ 1/3 & 1/4 & 0 & 1/2 \end{pmatrix}$, (ii) $\begin{pmatrix} 1/2 & 1/3 & 0 & 0 & 1 \\ 0 & 0 & 1/3 & 0 & 0 \\ 0 & 1/3 & 0 & 0 & 0 \\ 0 & 1/3 & 1/3 & 1 & 0 \\ 1/2 & 0 & 1/3 & 0 & 0 \end{pmatrix}$, (iii) $\begin{pmatrix} 1/3 & 1/3 & 1/3 \\ 1/3 & 2/3 & 0 \\ 1/3 & 0 & 2/3 \end{pmatrix}$.

(a) Draw a digraph for each chain. Is the Markov chain irreducible?

(b) Identify the communicating classes and classify them as periodic or aperiodic, transient or recurrent.

12. Suppose the transition matrices of two finite Markov chains are

$$P_1 = \begin{pmatrix} 0 & a_1 & 0 \\ 1 & 0 & 1 \\ 0 & a_2 & 0 \end{pmatrix}, \quad P_2 = \begin{pmatrix} a_1 & 0 & 0 \\ 0 & 0 & 1 \\ a_2 & 1 & 0 \end{pmatrix}$$

where $0 < a_i < 1$, $i = 1, 2$ and $a_1 + a_2 = 1$.

(a) For the two Markov chains determine if the chains are reducible or irreducible. Then determine whether each of the classes in the Markov chains is transient or positive recurrent, periodic or aperiodic.

(b) The Markov chain with transition matrix P_1 is periodic with period $d = 2$. Compute P_1^2. Then show that $P_1^{2n} = P_1^2$ for $n = 2, 3, \ldots$. Apply the Basic Limit Theorem for periodic Markov chains to find the mean recurrence times μ_{ii} for $i = 1, 2, 3$.

13. Suppose the states of three different Markov chains are $\{1, 2, \ldots\}$ and their corresponding transition matrices are

$$P_1 = \begin{pmatrix} a_1 & 0 & 0 & 0 & 0 & 0 & \cdots \\ a_2 & 0 & a_1 & 0 & 0 & 0 & \cdots \\ 0 & 1 & 0 & 0 & 0 & 0 & \cdots \\ 0 & 0 & a_2 & 0 & 0 & a_1 & \cdots \\ 0 & 0 & 0 & 1 & 0 & 0 & \cdots \\ 0 & 0 & 0 & 0 & 1 & 0 & \cdots \\ 0 & 0 & 0 & 0 & 0 & a_2 & \cdots \\ \vdots & \vdots & \vdots & \vdots & \vdots & \vdots & \end{pmatrix}, \quad P_2 = \begin{pmatrix} 0 & 0 & 0 & \cdots \\ a_1 & 0 & 0 & \cdots \\ a_2 & a_1 & 0 & \cdots \\ a_3 & a_2 & a_1 & \cdots \\ \vdots & \vdots & \vdots & \end{pmatrix},$$

and

$$P_3 = \begin{pmatrix} 1 & 1/2 & 1/3 & \cdots \\ 0 & 1/2 & 1/3 & \cdots \\ 0 & 0 & 1/3 & \cdots \\ \vdots & \vdots & \vdots & \end{pmatrix}.$$

The elements a_i of P_1 and P_2 are positive and $\sum a_i = 1$.

(a) Draw a directed graph for each chain. Is the Markov chain irreducible?

(b) Identify the communicating classes, find their period, then classify them as transient, null recurrent, or positive recurrent.

14. Suppose the transition matrix of an infinite DTMC is given by

$$P = \begin{pmatrix} a_1 & a_1 & a_1 & \cdots \\ a_2 & 0 & 0 & \cdots \\ 0 & a_2 & 0 & \cdots \\ 0 & 0 & a_2 & \cdots \\ \vdots & \vdots & \vdots & \end{pmatrix},$$

where $0 < a_i < 1$, $i = 1, 2$ and $a_1 + a_2 = 1$.

(a) Show that the DTMC is irreducible and aperiodic.

(b) Show that the first two rows of P^2 are constant, i.e., $p_{1j}^{(2)} = a_1$ and $p_{2j}^{(2)} = a_1 a_2$. Then show that the first three rows of P^4 are constant. What is $p_{3j}^{(4)}$?

(c) The Markov chain is recurrent and the following limit exists, $\lim_{n\to\infty} P^n$. What is this limit?

(d) Use the Basic Limit Theorem for aperiodic Markov chains $\lim_{n\to\infty} p_{ij}^{(n)} = 1/\mu_{ii}$ to find μ_{ii} for $i = 1, 2, \ldots$.

(e) Is the Markov chain positive recurrent or null recurrent?

15. Assume $i \leftrightarrow j$ for states i and j in a Markov chain. Prove the following: State i is positive recurrent iff state j is positive recurrent. Show that the same result holds if positive recurrent is replaced by null recurrent.

16. The transition matrix for a three-state Markov chain is

$$P = \begin{pmatrix} 1 & q & 0 \\ 0 & r & q \\ 0 & p & p+r \end{pmatrix},$$

$p, q > 0$, $r \geq 0$, and $p + q + r = 1$.

(a) Draw the directed graph of the chain.

(b) Is the set $\{2, 3\}$ closed? Why or why not?

(c) Find an expression for $p_{11}^{(n)}$. Then verify that state 1 is positive recurrent.

(d) Show that the process has a unique stationary probability distribution, $\pi = (\pi_1, \pi_2, \pi_3)^{tr}$.

17. The transition matrix for a four-state Markov chain is

$$P = \begin{pmatrix} 0 & 1/4 & 0 & 1/2 \\ 1/2 & 0 & 3/4 & 0 \\ 0 & 3/4 & 0 & 1/2 \\ 1/2 & 0 & 1/4 & 0 \end{pmatrix}.$$

(a) Show that the chain is irreducible, positive recurrent, and periodic. What is the period?

(b) Find the unique stationary probability distribution.

18. Suppose the transition matrix P of a finite Markov chain is doubly stochastic; that is, row and column sums equal one, $p_{ij} \geq 0$,

$$\sum_{i=1}^{N} p_{ij} = 1, \quad \text{and} \quad \sum_{j=1}^{N} p_{ij} = 1.$$

Prove the following: If an irreducible, aperiodic finite Markov chain (ergodic chain) has a doubly stochastic transition matrix, then all stationary probabilities are equal, $\pi_1 = \pi_2 = \cdots = \pi_N$. In particular $\pi_i = 1/N$.

19. The transition matrix for a three-state Markov chain is

$$P = \begin{pmatrix} 0 & 0 & 1/2 \\ 0 & 0 & 1/2 \\ 1 & 1 & 0 \end{pmatrix}.$$

(a) Draw a directed graph of the chain and show that P is irreducible.

(b) Show that P is periodic of period 2 and find P^{2n}, $n = 1, 2, \ldots$.

(c) Use the identity (2.19) to find the mean recurrence times. Show that they agree with the formula given in the Basic Limit Theorem for periodic Markov chains.

20. Suppose that the transition matrix of a two-state Markov chain is

$$P = \begin{pmatrix} 1 - a & b \\ a & 1 - b \end{pmatrix}, \tag{2.22}$$

where $0 < a < 1$ and $0 < b < 1$. Use the identity (2.19) to find a general formula for the mean recurrence times and mean first passage times.

21. Suppose that the transition matrix of a two-state Markov chain is given by equation (2.22) in Exercise 20. Use the identity $P^n = UD^nU^{-1}$ (Appendix for Chapter 2) to show that P^n can be expressed as follows:

$$P^n = \frac{1}{a+b}\begin{pmatrix} b & b \\ a & a \end{pmatrix} + \frac{(1-a-b)^n}{a+b}\begin{pmatrix} a & -b \\ -a & b \end{pmatrix}.$$

22. Suppose an irreducible, aperiodic finite Markov chain of period 2 has a transition matrix P with two eigenvalues $\lambda_1 = 1$ and $\lambda_2 = -1$. Suppose the remaining eigenvalues of P are distinct but smaller in magnitude than λ_i, $i = 1, 2$. Use formula (2.29) in the Appendix for Chapter 2 to show that

$$\lim_{n\to\infty} P^{2n} = \pi(1, 1, \ldots, 1)^{tr} + x_2 y_2^{tr},$$

where π is the unique stationary distribution of the Markov chain and x_2 and y_2 are the right and left eigenvectors of P, respectively, corresponding to $\lambda_2 = -1$ that satisfy the orthonormal conditions (2.28).

23. Suppose $P = \begin{pmatrix} I & A \\ 0 & T \end{pmatrix}$ is a transition matrix of a finite Markov chain, where submatrix I corresponds to k absorbing states and submatrix T corresponds to $m - k$ transient states. Let

$$T = \begin{pmatrix} 1/2 & 1/2 & 1/2 & 0 \\ 0 & 1/4 & 0 & 1/2 \\ 1/4 & 0 & 0 & 0 \\ 0 & 0 & 1/2 & 1/4 \end{pmatrix}$$

and

$$A = \begin{pmatrix} 1/4 & 1/8 & 0 & 0 \\ 0 & 1/8 & 0 & 1/4 \end{pmatrix} = \begin{pmatrix} A_1 \\ A_2 \end{pmatrix}.$$

There are two absorbing states $\{1, 2\}$ and four transient states $\{3, 4, 5, 6\}$.

(a) Calculate the fundamental matrix $F = (I - T)^{-1}$. Then use F to find the expected time until absorption, $m = 1^{tr}F$, and the probability of absorption into state 1 or state 2, A_1F and A_2F, respectively (genetics inbreeding example).

(b) The variance of the time until absorption can be calculated from the fundamental matrix (Exercise 24). Use the following computations to calculate the variance until absorption. The second moment until absorption is

$$m_2 = 1^{tr}(2F^2 - F) = 1^{tr}F(2F - I)$$

so that the variance until absorption is

$$V = m_2 - m \circ m,$$

where $m \circ m$ is the entrywise product. That is, if $m = (a, b, c, d)$, then $m \circ m = (a^2, b^2, c^2, d^2)$.

24. Higher-order moments for the time until absorption for an absorbing Markov chain can be calculated using the fundamental matrix (Iosifescu, 1980). Let ν_i be the random variable for the number of time steps until absorption beginning in transient state i. Let $\mathcal{A}$ be the set of absorbing states, $\mathcal{T}$ be the set of transient states, and T be the submatrix of the transition matrix corresponding to the transient class. Then the kth moment for the time until absorption $E(\nu_i^k) = \sum n^k \mathrm{Prob}\{\nu_i = n | X_0 = i\}$ can be written as follows (Iosifescu, 1980, p. 105):

$$E(\nu_i^k) = \sum_{j \in \mathcal{A}} p_{ji} + \sum_{j \in \mathcal{T}} p_{ji} E((\nu_j + 1)^k).$$

(a) Expand $(\nu_j + 1)^k$ using the binomial theorem, take expectations, and substitute into the preceding expression to show that

$$E(\nu_i^k) = 1 + \sum_{j \in \mathcal{T}} p_{ji} \left[E(\nu_j^k) + \sum_{r=1}^{k-1} \binom{k}{r} E(\nu_j^{k-r}) \right].$$

(b) Let $m_k = (E(\nu_1^k), E(\nu_2^k), \ldots, E(\nu_m^k))$ be the kth moment, expressed as a row vector. Rewrite part (a) in matrix notation as follows:

$$m_k(I - T) = \mathbf{1}^{tr} + \sum_{r=1}^{k-1} \binom{k}{r} m_{k-r} T,$$

where $\mathbf{1}^{tr}$ is a row vector of ones. Then show that $m_1 = \mathbf{1}^{tr} F$ and $m_2 = m_1(2F - I)$, where F is the fundamental matrix..

25. Apply the Monte Carlo code for integration in Example 2.24 to approximate the value of π from the following integral:

$$\int_0^1 \frac{4}{1 + x^2} \, dx.$$

26. (a) For the one-dimensional, unrestricted random walk model, calculate $p_{i,0}^{(2)}$, $i = 0, \pm 1, \pm 2$, if $p = 0.45$ and $q = 0.55$. Then use these probabilities to calculate $E(X_2 | X_0 = 0)$ and compare your answer to the answer computed using the MATLAB code given in Example 2.25.

(b) For the one-dimensional, unrestricted random walk model in Example 2.25, show that if $p = 0.45$ and $q = 0.55$, then the exact value for the expectation is

$$E(X_{100} | X_0 = 0) = 10.$$

In particular, show that $X_{100} \in \{0, \pm 2, \pm 4, \ldots, \pm 100\}$ because the number of steps from 0 to 100 is even. Verify that the probabilities are binomially distributed $b(100, 0.45)$, i.e., $p_{-100,0}^{(100)} = (0.45)^{100}$, $p_{-98,0}^{(100)} = 100(0.45)^{99}(0.55)$, etc.

27. Let $P = \begin{pmatrix} 1 & 1/4 & 1/2 \\ 0 & 1/2 & 0 \\ 0 & 1/4 & 1/2 \end{pmatrix}$. A general formula for P^n can be derived using the method of Example 2.27 (Appendix for Chapter 2).

   (a) Show that the characteristic polynomial of P is $\lambda^3 - 2\lambda^2 + (5/4)\lambda - 1/4 = (\lambda - 1)(\lambda - 1/2)^2 = 0$. Therefore, three linearly independent solutions of the third order linear difference equation, $x(n + 3) - 2x(n + 2) + (5/4)x(n + 1) - (1/4)x(n) = 0$, are $y_1(n) = 1$, $y_2(n) = 1/2^n$, and $y_3(n) = n/2^n$.

   (b) Use the three linearly independent solutions, $y_i(n)$, $i = 1, 2, 3$, to find three solutions $x_i(n)$, $i = 1, 2, 3$ that satisfy the initial conditions.

   (c) Use the identity (2.32) to find a general expression for P^n.

28. Suppose that two unbiased coins are tossed repeatedly and after each toss the accumulated number of heads and tails that have appeared on each coin is recorded. Let the random variable X_n denote the difference in the accumulated number of heads on coin 1 and coin 2 after the nth toss [e.g, (Total # Heads Coin 1) − (Total # Heads Coin 2)]. Thus, the state space is $\{0, \pm 1, \pm 2, \ldots\}$. Show that the zero state (cumulative number of heads corresponding to each coin is the same) is null recurrent (Bailey, 1990).

29. Consider the genetics inbreeding problem. Let

$$p(0) = (0, 1/4, 1/4, 1/4, 1/4, 0)^{tr}.$$

   (a) Find a general formula for the proportion of heterozygotes h_n in terms of the eigenvalues:

$$h_n = a\lambda_1^n + b\lambda_2^n + c\lambda_3^n + d\lambda_4^n.$$

   (b) Find h_{20} and h_{40}. See the Appendix for Chapter 2.

30. A Markov chain model for the growth and replacement of trees assumes that there are three stages of growth based on the size of the tree: young tree, mature tree, and old tree. When an old tree dies, it is replaced by a young tree with probability $1 - p$. Order the states numerically, 1, 2, and 3. State 1 is a young tree, state 2 is a mature tree, and state 3 is an old tree. A Markov chain model for the transitions between each of the states over a period of eight years has the following transition matrix:

$$P = \begin{pmatrix} 1/4 & 0 & 1-p \\ 3/4 & 1/2 & 0 \\ 0 & 1/2 & p \end{pmatrix}.$$

Transitions occur over an eight-year period. For example, after a period of eight years the probability that a young tree becomes a mature tree is 3/4 and the probability it remains a young tree is 1/4.

(a) Suppose $0 \le p < 1$. Show that the Markov chain is irreducible and aperiodic. Find the unique limiting stationary distribution.

(b) Suppose $p = 7/10$. Compute the mean recurrence times for $i = 1, 2, 3$ (i.e., the mean number of eight-year periods it will take a tree in stage i to be replaced by another tree of stage i).

31. A Markov chain model for the growth and replacement of trees assumes that there are four stages of growth based on the size of the tree: seedling, young tree, mature tree, and old tree. When an old tree dies, it is replaced by a seedling. Order the states numerically, $1, 2, 3$, and 4. State 1 is a seedling, state 2 is a young tree, and so on. A Markov chain model for the transitions between each state over a period of five years has the following transition matrix:

$$P = \begin{pmatrix} p_{11} & 0 & 0 & 1 - p_{44} \\ 1 - p_{11} & p_{22} & 0 & 0 \\ 0 & 1 - p_{22} & p_{33} & 0 \\ 0 & 0 & 1 - p_{33} & p_{44} \end{pmatrix}.$$

Transition p_{ii} is the probability that a tree remains in the same state for five years and $1 - p_{ii}$ is the probability a tree is at the next stage after five years of growth.

(a) Suppose $0 < p_{ii} < 1$ for $i = 1, 2, 3, 4$. Show that the Markov chain is irreducible and aperiodic. Find the unique limiting stationary distribution.

(b) Suppose $p_{44} = 1$ and $0 < p_{ii} < 1$ for $i = 1, 2, 3$. What do these assumptions imply about the growth and replacement of trees? Show that $p_{ii}^{(n)} = p_{ii}^n$ and $\lim_{n \to \infty} p_{ii}^{(n)} = 0$. Identify the communicating classes and determine whether they are transient or recurrent.

2.12 References for Chapter 2

Allman, E. S. and J. A. Rhodes. 2004. *Mathematical Models in Biology: An Introduction*. Cambridge Univ. Press, Cambridge, U. K.

Bailey, N. T. J. 1990. *The Elements of Stochastic Processes with Applications to the Natural Sciences*. John Wiley & Sons, New York.

Caswell, H. 2001. *Matrix Population Models.* Sinauer Assoc., Inc. Pub., Sunderland, Mass.

Feller, W. 1968. *An Introduction to Probability Theory and Its Applications.* Vol. 1. 3rd ed. John Wiley & Sons, New York.

Gantmacher, F. R. 1964. *The Theory of Matrices.* Vol. II. Chelsea Pub. Co., New York.

Hengeveld, R. 1989. *Dynamics of Biological Invasions.* Chapman and Hall, London and New York.

Hoppensteadt, F. 1975. *Mathematical Methods of Population Biology.* Cambridge Univ. Press, Cambridge, U. K.

Iosifescu, M. 1980. *Finite Markov Processes and Their Applications.* John Wiley & Sons, Chichester, New York.

Jukes, T. H. and C. R. Cantor. 1969. Evolution of protein molecules. In: *Mammalian Protein Metabolism.* Monro, H. N. (ed.), pp. 21–132, Academic Press, N. Y.

Karlin, S. and H. Taylor. 1975. *A First Course in Stochastic Processes.* 2nd ed. Academic Press, New York.

Kemeny, J. G. and J. L. Snell. 1960. *Finite Markov Chains.* Van Nostrand, Princeton, N. J.

Kwapisz, M. 1998. The power of a matrix. *SIAM Review* 40: 703–705.

Leonard, I. E. 1996. The matrix exponential. *SIAM Review* 38: 507–512.

Metropolis, N. 1987. The beginning of the Monte Carlo method. *Los Alamos Science* (Special Issue 1987 dedicated to Stanislaw Ulam) 15: 125-130.

Mooney, D. and R. Swift. 1999. *A Course in Mathematical Modeling.* The Mathematical Association of America, Washington, D. C.

Norris, J. R. 1997. *Markov Chains.* Cambridge Series in Statistical and Probabilistic Mathematics, Cambridge Univ. Press, Cambridge, U. K.

Ortega, J. M. 1987. *Matrix Theory A Second Course.* Plenum Press, New York.

Reynolds, J. C. 1985. Details of the geographic replacement of the red squirrel (*Sciurus vulgaris*) by the grey squirrel (*Sciurus carolinensis*) in eastern England. *J. Anim. Ecol.* 54: 149–162.

Robert, C. P. and G. Casella. 2004. *Monte Carlo Statistical Methods*, 2nd ed. Springer-Verlag, New York, Berlin, Heidelberg.

Schinazi, R. B. 1999. *Classical and Spatial Stochastic Processes*. Birkhäuser, Boston.

Shigesada N. and K. Kawasaki. 1997. *Biological Invasions: Theory and Practice*. Oxford Univ. Press, Oxford, New York, and Tokyo.

Stewart, W. J. 1994. *Introduction to the Numerical Solution of Markov Chains*. Princeton Univ. Press, Princeton, N. J.

Taylor, H. M. and S. Karlin. 1998. *An Introduction to Stochastic Modeling*. 3rd ed. Academic Press, New York.

Tuljapurkar, S. 1997. Stochastic matrix models. In: *Structured-Population Models in Marine, Terrestrial, and Freshwater Systems*. Tuljapurkar, S. and H. Caswell (eds.), pp. 59–87, Chapman & Hall, New York.

Wade, W. R. 2000. *An Introduction to Analysis*. 2nd ed. Prentice Hall, Upper Saddle River, N. J.

Williamson, M. 1996. *Biological Invasions*. Chapman & Hall, London.

2.13   Appendix for Chapter 2

2.13.1   Proofs of Theorems 2.5 and 2.6

THEOREM 2.5 *Let* $\{X_n\}_{n=0}^{\infty}$ *be a positive recurrent, irreducible, and aperiodic DTMC (strongly ergodic) with transition matrix* $P = (p_{ij})$. *Then there exists a unique positive stationary probability distribution* π, $P\pi = \pi$, *such that*

$$\lim_{n \to \infty} p_{ij}^{(n)} = \pi_i, \quad i, j = 1, 2, \ldots.$$

Proof. The proof follows Karlin and Taylor (1975). Theorem 2.3, the fact that $\sum_{i=1}^{\infty} p_{ij}^{(n)} = 1$ for all n, and Fubini's theorem (Exercise 3) are used in the proof. Let $\pi_i = 1/\mu_{ii}$, $i = 1, 2, \ldots$; $\pi_i > 0$ because state i is positive recurrent, $i = 1, 2, \ldots$.

The proof proceeds in four steps. First, it is shown that $\sum_{i=1}^{\infty} \pi_i \leq 1$. For any positive integer M

$$\sum_{i=1}^{M} p_{ij}^{(n)} \leq \sum_{i=1}^{\infty} p_{ij}^{(n)} = 1.$$

Letting $n \to \infty$ in the sum on the left and applying Theorem 2.3 leads to $\sum_{i=1}^{M} \pi_i \leq 1$. Since the inequality holds for all positive integers M, it follows that $\sum_{i=1}^{\infty} \pi_i \leq 1$.

The second step is to show $P^n \pi = \pi$ for any positive integer n. Note that

$$p_{ij}^{(n+1)} \geq \sum_{k=1}^{M} p_{ik} p_{kj}^{(n)}.$$

Letting $n \to \infty$ and applying Theorem 2.3 leads to

$$\pi_i \geq \sum_{k=1}^{M} p_{ik} \pi_k.$$

Since the inequality holds for all positive integers M, it follows that

$$\pi_i \geq \sum_{k=1}^{\infty} p_{ik} \pi_k. \tag{2.23}$$

Multiplying the preceding inequality by p_{ji} and summing over i yields

$$\sum_{i=1}^{\infty} p_{ji} \pi_i \geq \sum_{i=1}^{\infty} \sum_{k=1}^{\infty} p_{ji} p_{ik} \pi_k = \sum_{k=1}^{\infty} \left(\sum_{i=1}^{\infty} p_{ji} p_{ik} \right) \pi_k.$$

The left side of the preceding identity is bounded above by π_j (from the inequality (2.23)) and the right side equals $\sum_{k=1}^{\infty} p_{jk}^{(2)} \pi_k$. Thus,

$$\pi_j \geq \sum_{k=1}^{\infty} p_{jk}^{(2)} \pi_k.$$

This can be continued to show that for any positive integer n,

$$\pi_j \geq \sum_{k=1}^{\infty} p_{jk}^{(n)} \pi_k.$$

We want to show that the last inequality is actually an equality. Suppose the inequality is strict for some j. Then

$$\sum_{i=1}^{\infty} \pi_i > \sum_{i=1}^{\infty} \sum_{k=1}^{\infty} p_{ik}^{(n)} \pi_k = \sum_{k=1}^{\infty} \left(\sum_{i=1}^{\infty} p_{ik}^{(n)} \right) \pi_k = \sum_{k=1}^{\infty} \pi_k.$$

But this is a contradiction. It follows that

$$\pi_i = \sum_{k=1}^{\infty} p_{ik}^{(n)} \pi_k \tag{2.24}$$

for all n. Since i is arbitrary $P^n \pi = \pi$.

The third step is to show that $\sum_{i=1}^{\infty} \pi_i = 1$. From the first step, it follows that the summation in (2.24) is uniformly bounded by a convergent series $\sum_{i=1}^{\infty} \pi_i \leq 1$. Thus, the limit as $n \to \infty$ of the right side of (2.24) can be interchanged with the summation to yield

$$\pi_i = \sum_{k=1}^{\infty} \pi_i \pi_k = \pi_i \sum_{k=1}^{\infty} \pi_k.$$

It follows that $\sum_{k=1}^{\infty} \pi_k = 1$.

The fourth and final step is to show uniqueness of the stationary distribution. Assume that $a = (a_1, a_2, \ldots)^{tr}$ is another stationary distribution. Then, by definition of a stationary distribution,

$$a_i = \sum_{k=1}^{\infty} p_{ik} a_k = \sum_{k=1}^{\infty} p_{ik}^{(n)} a_k.$$

Letting $n \to \infty$, as in the third step, verifies uniqueness:

$$a_i = \pi_i \sum_{k=1}^{\infty} a_k = \pi_i.$$

for all i. □

THEOREM 2.6 *In a finite DTMC, not all states can be transient and no states can be null recurrent. In particular, an irreducible and finite Markov chain is positive recurrent.*

Proof. From Lemma 2.2, it follows that if j is transient, then

$$\lim_{n \to \infty} p_{ji}^{(n)} = 0, \quad \text{for} \quad i = 1, 2, \ldots, N, \tag{2.25}$$

where N is the number of states. Suppose all states are transient. Then the identity (2.25) holds for all i and j, $i, j = 1, 2, \ldots, N$ and

$$\lim_{n \to \infty} P^n = \mathbf{0},$$

the zero matrix. Matrix P^n is a stochastic matrix (Exercise 2). Hence, $\sum_{j=1}^{N} p_{ji}^{(n)} = 1$; the column sums of P^n are one. Taking the limit as $n \to \infty$ and interchanging the limit and summation (possible because of the finite sum) leads to $\sum_{j=1}^{N} \lim_{n \to \infty} p_{ji}^{(n)} = 1$, a contradiction to the preceding limit. Thus, not all states can be transient.

Suppose there exists a null recurrent state i and $i \in C$, where C is a class of states. The class C is closed by Corollary 2.2 and all states in C are null recurrent. (See the remarks following Corollary 2.3 and Exercise 15.)

Suppose the class C is aperiodic and null recurrent. Then according to the Basic Limit Theorem for aperiodic Markov chains, $\lim_{n\to\infty} p_{ij}^{(n)} = 0$ for all $i, j \in C$. The submatrix P_C of P consisting of all states in C is a stochastic matrix ($p_{kj}^{(n)} = 0$ for $k \notin C$). But $\lim_{n\to\infty} P_C^n = \mathbf{0}$, an impossibility. Thus, all states are positive recurrent.

Suppose the class C is periodic and null recurrent. Then according to the Basic Limit Theorem for periodic Markov chains, $\lim_{n\to\infty} p_{ii}^{(n)} = 0$ for any $i \in C$. Furthermore, for any state $j \in C$, since $i \leftrightarrow j$, there exists positive integers m and n such that $p_{ij}^{(m)} > 0$ and $p_{ji}^{(n)} > 0$. Therefore,

$$p_{ii}^{(m+n)} \geq p_{ij}^{(n)} p_{ji}^{(m)} > 0.$$

Fix n and let $m \to \infty$. Then it follows that $\lim_{m\to\infty} p_{ji}^{(m)} = 0$. Also, fix m and let $n \to \infty$. Then $\lim_{n\to\infty} p_{ij}^{(n)} = 0$. Thus, $\lim_{n\to\infty} P_C^n = \mathbf{0}$, where P_C is the submatrix of P consisting of states in C. This is an impossibility since P_C^n is stochastic. Thus, all states are positive recurrent.

In the case that the finite Markov chain is irreducible, there is only one class and all states in that class must be either positive recurrent, null recurrent, or transient. Since they cannot all be transient and there are no null recurrent states, they all must be positive recurrent. □

2.13.2 Perron and Frobenius Theorems

Frobenius Theorem *An irreducible, nonnegative matrix P has a positive eigenvalue λ that is a simple root of the characteristic equation and is greater than or equal to the magnitude of all of the other eigenvalues. An eigenvector corresponding to λ has all positive entries.*

Perron Theorem *A positive matrix P has a positive eigenvalue λ that is a simple root of the characteristic equation and is strictly greater than the magnitude of all of the other eigenvalues. An eigenvector corresponding to λ has all positive entries.*

Proofs of these two theorems can be found in Gantmacher (1964). The *spectral radius* of an $N \times N$ matrix P is the largest magnitude of all of the eigenvalues of P : $\rho(P) = \max_i\{|\lambda_i|\}$. It is also well-known that $\rho(P) \leq \|P\|_1$, where $\|\cdot\|_1$ is the L^1 matrix norm (maximum of the absolute column sum) (Ortega, 1987). If P is a transition matrix of an irreducible DTMC, so that $\|P\|_1 = 1$, it follows from the Frobenius theorem that the largest eigenvalue of P is $\lambda = 1$ and the corresponding normalized eigenvector π, $\sum_{j=1}^{N} \pi_j = \|\pi\|_1 = 1$, is the unique stationary distribution of the DTMC model. In addition, if the finite DTMC model is irreducible and aperiodic, then the Perron theorem implies $\lim_{n\to\infty} P^n p(0) = \pi$. See Corollary 2.4.

2.13.3 The n-Step Transition Matrix

In the case of a finite Markov chain, several methods are presented for calculating a general form for the n-step transition matrix. A simple form for P^n can be generated if P can be expressed as

$$P = UDU^{-1},$$

where D is a diagonal matrix and U is a nonsingular matrix. In this case, matrix P^n can be expressed as

$$P^n = UD^nU^{-1}.$$

An important theorem in linear algebra states that P can be expressed as $P = UDU^{-1}$ iff P is diagonalizable iff P has n linearly independent eigenvectors (Ortega, 1987). Hence, it shall be assumed that P has n linearly independent eigenvectors.

We show how the matrices U and D can be formed. Assume P is an $N \times N$ matrix with N eigenvalues, $\lambda_1, \lambda_2, \cdots, \lambda_N$. Let x_j be a right eigenvector (column vector) corresponding to λ_j and y_j be a left eigenvector (column vector):

$$Px_j = \lambda_j x_j \quad \text{and} \quad y_j^{tr} P = \lambda_j y_j^{tr}. \tag{2.26}$$

Define $N \times N$ matrices

$$H = (x_1, x_2, \ldots, x_N) \quad \text{and} \quad K = (y_1, y_2, \ldots, y_N),$$

where the columns of H are the right eigenvectors and the columns of K are the left eigenvectors. These matrices are nonsingular because the vectors are linearly independent. Because of the identities in (2.26),

$$PH = HD \quad \text{and} \quad K^{tr}P = DK^{tr},$$

where $D = \text{diag}(\lambda_1, \lambda_2, \ldots, \lambda_N)$. Thus,

$$P = HDH^{-1} \quad \text{or} \quad P = (K^{tr})^{-1}DK^{tr};$$

$U = H$ or $U = (K^{tr})^{-1}$. The n-step transition matrix can be expressed as either

$$P^n = HD^nH^{-1} \quad \text{or} \quad P^n = (K^{tr})^{-1}D^nK^{tr}. \tag{2.27}$$

The identities in (2.27) demonstrate one method that can be used to calculate P^n. Another method is also demonstrated below (Bailey, 1990).

Note that

$$y_j^{tr} P x_i = y_j^{tr} \lambda_i x_i = y_j^{tr} \lambda_j x_i.$$

If $\lambda_i \neq \lambda_j$, then $y_j^{tr} x_i = 0$; the left and right eigenvectors are orthogonal. Thus, in this method, it is assumed that the eigenvalues are distinct (distinct

eigenvalues imply the corresponding eigenvectors are linearly independent).
Suppose y_j and x_j are chosen to be orthonormal:

$$y_j^{tr} x_i = \begin{cases} 0 & i \neq j \\ 1 & i = j. \end{cases} \tag{2.28}$$

Then $K^{tr} H = I$ (identity matrix) or $K^{tr} = H^{-1}$ and

$$\begin{aligned} P &= HDH^{-1} = HDK^{tr} \\ &= (\lambda_1 x_1, \lambda_2 x_2, \ldots, \lambda_N x_N)(y_1, y_2, \ldots, y_N)^{tr} \\ &= \lambda_1 x_1 y_1^{tr} + \lambda_2 x_2 y_2^{tr} + \cdots + \lambda_N x_N y_N^{tr} \end{aligned}$$

Therefore,

$$P = \sum_{i=1}^{N} \lambda_i x_i y_i^{tr}.$$

Because the matrix $x_i y_i^{tr} x_j y_j^{tr}$ is the zero matrix for $i \neq j$ and the sum $\sum_{i=1}^{N} x_i y_i^{tr} = HK^{tr} = I$, it follows that P^2 can be expressed in terms of the matrices $x_i y_i^{tr}$:

$$P^2 = \left(\sum_{i=1}^{N} \lambda_i x_i y_i^{tr} \right) \left(\sum_{k=1}^{N} \lambda_k x_k y_k^{tr} \right) = \sum_{i=1}^{N} \lambda_i^2 x_i y_i^{tr}.$$

In general, the n-step transition matrix satisfies

$$P^n = \sum_{i=1}^{N} \lambda_i^n x_i y_i^{tr}. \tag{2.29}$$

In the case that the Markov chain is regular (or ergodic), which means it is irreducible and aperiodic, then P^n has a limiting distribution. The limiting distribution is the stationary distribution corresponding to the eigenvalue $\lambda_1 = 1$. In this case,

$$\lim_{n \to \infty} P^n = x_1 y_1^{tr} = \begin{pmatrix} \pi_1 & \pi_1 & \cdots & \pi_1 \\ \pi_2 & \pi_2 & \cdots & \pi_2 \\ \vdots & \vdots & \ldots & \vdots \\ \pi_N & \pi_N & \cdots & \pi_N \end{pmatrix},$$

where $x_1 = \pi$ and $y_1^{tr} = (1, 1, \ldots, 1)$.

Note that both methods apply to any finite matrix with distinct eigenvalues. The two methods of computing P^n, given in (2.27) and (2.29), are illustrated in the next example.

Example 2.26. Consider a Markov chain with two states $\{1, 2\}$ and transition matrix

$$P = \begin{pmatrix} 1-p & p \\ p & 1-p \end{pmatrix},$$

where $0 < p < 1$. Note that this is a doubly stochastic matrix and all states are positive recurrent. Thus, $\lim_{n \to \infty} P^n p(0) = \pi = (\pi_1, \pi_1)^{tr}$, where $\pi_2 = \pi_1$ (Exercise 18).

The eigenvalues and corresponding eigenvectors of P are $\lambda_1 = 1$, $\lambda_2 = 1 - 2p$, $x_1^{tr} = (1, 1)$, $x_2^{tr} = (1, -1)$, $y_1^{tr} = (1, 1)$, and $y_2^{tr} = (1, -1)$. Note that $x_1/2 = \pi$ is the stationary probability distribution and $|\lambda_2| = |1 - 2p| < 1$. Using the first identity in (2.27), an expression for P^n is given by

$$P^n = H\Lambda^n H^{-1} = \begin{pmatrix} 1 & 1 \\ 1 & -1 \end{pmatrix} \begin{pmatrix} 1 & 0 \\ 0 & (1 - 2p)^n \end{pmatrix} \begin{pmatrix} 1 & 1 \\ 1 & -1 \end{pmatrix} \frac{1}{2}.$$

Multiplication of the three matrices yields

$$P^n = \frac{1}{2} \begin{pmatrix} 1 + (1 - 2p)^n & 1 - (1 - 2p)^n \\ 1 - (1 - 2p)^n & 1 + (1 - 2p)^n \end{pmatrix}$$
$$= \frac{1}{2} \begin{pmatrix} 1 & 1 \\ 1 & 1 \end{pmatrix} + \frac{(1 - 2p)^n}{2} \begin{pmatrix} 1 & -1 \\ -1 & 1 \end{pmatrix}.$$

For example, the probability $p_{11}^{(n)}$ is $1/2 + (1 - 2p)^n/2$. Note that $P^n p(0)$ approaches the stationary probability distribution given by $\pi = (1/2, 1/2)^{tr}$.

For the second method (2.29), the eigenvectors are normalized. Since $y_1^{tr} x_1 = y_2^{tr} x_2 = 2$, divide by 2. Thus,

$$P^n = \lambda_1^n \frac{x_1 y_1^{tr}}{2} + \lambda_2^n \frac{x_2 y_2^{tr}}{2}$$
$$= \frac{1}{2} \begin{pmatrix} 1 & 1 \\ 1 & 1 \end{pmatrix} + \frac{(1 - 2p)^n}{2} \begin{pmatrix} 1 & -1 \\ -1 & 1 \end{pmatrix}.$$

The two methods give the same expression for P^n. ∎

Another for computing P^n is based on the Cayley-Hamilton theorem from linear algebra. In this case, it is not necessary that P be diagonalizable. Let P is an $N \times N$ matrix with characteristic polynomial

$$c(\lambda) = \det(\lambda I - P) = \lambda^N + a_{N-1}\lambda^{N-1} + \cdots + a_0 = 0.$$

Note that matrix $Z(n) = P^n$ is the unique solution to the matrix difference equation,

$$Z(N + n) + a_{N-1}Z(N + n - 1) + \cdots + a_0 Z(n) = \mathbf{0}, \qquad (2.30)$$

with initial conditions $Z(0) = I$, $Z(1) = P, \ldots$, and $Z(N - 1) = P^{N-1}$. This follows from the Cayley-Hamilton theorem from linear algebra that P is a solution of the characteristic equation, $c(P) = \mathbf{0}$. The following theorem is due to Kwapisz (1998). It is based on a similar theorem for matrix exponentials by Leonard (1996).

THEOREM 2.8

Suppose $x_1(n), x_2(n), \ldots, x_N(n)$ are solutions of the Nth-order scalar difference equation,

$$x(N+n) + a_{N-1}x(N+n-1) + \cdots + a_0 x(n) = 0,$$

with initial conditions

$$
\left.\begin{aligned}
x_1(0) &= 1 \\
x_1(1) &= 0 \\
&\vdots \\
x_1(N-1) &= 0
\end{aligned}\right\}, \quad
\left.\begin{aligned}
x_2(0) &= 0 \\
x_2(1) &= 1 \\
&\vdots \\
x_2(N-1) &= 0
\end{aligned}\right\}, \quad \cdots, \quad
\left.\begin{aligned}
x_N(0) &= 0 \\
x_N(1) &= 0 \\
&\vdots \\
x_N(N-1) &= 1
\end{aligned}\right\}. \quad (2.31)
$$

Then

$$P^n = x_1(n)I + x_2(n)P + \cdots + x_N(n)P^{N-1}, \quad n = 0, 1, 2, \ldots. \quad (2.32)$$

Proof. Let $Z(n) = x_1(n)I + x_2(n)P + \cdots + x_N(n)P^{N-1}$ for $n = 0, 1, 2, \ldots$. Then substitution of $Z(n)$ into the difference equation (2.30) shows that $Z(n)$ is a solution of the equation. In addition, $Z(0) = I$, $Z(1) = P, \ldots$, and $Z(N-1) = P^{N-1}$. Because the solution of (2.30) is unique it follows that $Z(n) = P^n$ for $n = 0, 1, 2, \ldots$. $\qquad\square$

Example 2.27. The nth power of matrix $P = \begin{pmatrix} 1-p & p \\ p & 1-p \end{pmatrix}$, given in Example 2.26, is computed using the characteristic polynomial of P:

$$\lambda^2 - (2-2p)\lambda + 1 - 2p = (\lambda - 1)(\lambda - 1 + 2p) = 0.$$

The second-order linear difference equation,

$$x(n+2) - (2-2p)x(n+1) + (1-2p)x(n) = 0,$$

has two linearly independent solutions, 1 and $(1-2p)^n$, for $p \neq 0$. The general solution is a linear combination of these two solutions, $x(n) = c_1 + c_2(1-2p)^n$. Next, find the constants c_1 and c_2 so that the two solutions x_1 and x_2 satisfy the required initial conditions (2.31). For the first solution, $x_1(0) = c_1 + c_2 = 1$ and $x_1(1) = c_1 + c_2(1-2p) = 0$. Solving for c_1 and c_2 the first solution is

$$x_1(n) = \frac{2p-1}{2p} + \frac{(1-2p)^n}{2p}.$$

For the second solution, $x_2(0) = c_1 + c_2 = 0$ and $x_2(1) = c_1 + c_2(1-2p) = 1$. Solving for c_1 and c_2 the second solution is

$$x_2(n) = \frac{1}{2p} - \frac{(1-2p)^n}{2p}.$$

Then applying the identity (2.32),

$$P^n = x_1(n)I + x_2(n)P = \frac{1}{2}\begin{pmatrix} 1 + (1-2p)^n & 1 - (1-2p)^n \\ 1 - (1-2p)^n & 1 + (1-2p)^n \end{pmatrix}.$$

This latter formula agrees with the one given in Example 2.26. ∎

2.13.4 Genetics Inbreeding Problem

In the genetics inbreeding problem, h_n is the expected proportion of heterozygotes at time n,

$$h_n = \frac{1}{2}p_2(n) + p_3(n) + \frac{1}{2}p_4(n),$$

where $p_i(n)$ is the proportion of the population in state i at time n (Bailey, 1990). The dynamics of the three states depend on the matrix T^n. Let $\tilde{p}(0) = (p_2(0), p_3(0), p_4(0), p_5(0))^{tr}$ be the truncated vector $p(0)$. Then $T^n\tilde{p}(0) = \tilde{p}(n) = \sum_{i=1}^{4}\lambda_i^n x_i y_i^{tr}\tilde{p}(0)$. It follows that the $p_i(n)$ satisfy

$$p_i(n) = c_{i1}\lambda_1^n + c_{i2}\lambda_2^n + c_{i3}\lambda_3^n + c_{i4}\lambda_4^n, \quad i = 2, 3, 4, 5.$$

Hence,

$$h_n = a\lambda_1^n + b\lambda_2^n + c\lambda_3^n + d\lambda_4^n,$$

where a, b, c, d are combinations of the c_{ij}. The coefficients a, b, c, d can be found by solving the following four linear equations (linear in a, b, c, d):

$$h_i = a\lambda_1^i + b\lambda_2^i + c\lambda_3^i + d\lambda_4^i, \quad i = 0, 1, 2, 3.$$

Suppose, initially, the entire population is of type 2, $AA \times Aa$, $p(0) = (0, 1, 0, 0, 0, 0)^{tr}$. Then $h_0 = 1/2$ and $Pp(0) = (1/4, 1/2, 1/4, 0, 0, 0)^{tr}$ so that $h_1 = (1/2)(1/2) + (1/2)(1/2) + (1/2)(0) = 1/2$. By computing $P^2p(0)$ and $P^3p(0)$, values for h_2 and h_3 can be calculated, $h_2 = 3/8$ and $h_3 = 5/16$. The following Maple^{TM} program was used to calculate h_{20} and h_{30} and a general formula was obtained for h_n

$$h_{20} = 0.008445, \quad h_{30} = 0.001014,$$

and

$$h_n = (1/4 + 3\sqrt{5}/20)\lambda_3^n + (1/4 - 3\sqrt{5}/20)\lambda_4^n.$$

```
with(linalg):
P:=matrix(6,6,[1,1/4,1/16,0,0,0,0,1/2,1/4,0,0,0,0,1/4,1/4,1/4,1,
0,0,0,1/4,1/2,0,0,0,0,1/8,0,0,0, 0,0,1/16,1/4,0,1]):
T:=matrix(4,4,[1/2,1/4,0,0,1/4,1/4,1,0,0,1/4,1/2,0,0,1/8,0,0]):
p0:=vector([0,1,0,0,0,0]):
h0:=1/2*p0[2]+p0[3]+1/2*p0[4]:
p:=n->evalm(P^n&*p0):
h:=n->1/2*p(n)[2]+p(n)[3]+1/2*p(n)[4]:
evalm(p(3)); h(3);
```

$$\left[\frac{123}{256}, \frac{13}{64}, \frac{11}{64}, \frac{5}{64}, \frac{3}{128}, \frac{11}{256} \right]$$

$$\frac{5}{16}$$

```
ll:=[eigenvals(P)];
```

$$ll := \left[\frac{1}{2}, \frac{1}{4}, \frac{1}{4} + \frac{1}{4}\sqrt{5}, \frac{1}{4} - \frac{1}{4}\sqrt{5}, 1, 1 \right]$$

```
f:=n->a*ll[1]^n+b*ll[2]^n+c*ll[3]^n+d*ll[4]^n:
solve({f(0)=q0,f(1)=q(1),f(2)=q(2),f(3)=q(3)},{a,b,c,d});
```

$$\left\{ a = 0, b = 0, c = \frac{1}{4} + \frac{3}{20}\sqrt{5}, d = \frac{1}{4} - \frac{3}{20}\sqrt{5} \right\}$$

```
evalf(subs({a=0,b=0,c=1/4+3/20*sqrt(5),d=1/4-3/20*sqrt(5)},f(20)));
```

$$0.008445262873$$

```
evalf(subs({a=0,b=0,c=1/4+3/20*sqrt(5),d=1/4-3/20*sqrt(5)},f(30)));
```

$$0.001014354166$$

Chapter 3

Biological Applications of Discrete-Time Markov Chains

3.1 Introduction

Several recent and classical biological applications of DTMC models are discussed in this chapter. The first application is a recent application of finite Markov chain theory to proliferating epithelial cells. Then a classical biological application of a DTMC is discussed, a random walk on a finite domain, often referred to as the gambler's ruin problem. Expressions are derived for the probability of absorption, the expected duration until absorption, and the entire probability distribution for absorption at the nth time step. Some of these results are extended to a random walk on a semi-infinite domain $\{0, 1, 2 \ldots\}$. Another classical application of DTMCs is to birth and death processes. A general discrete-time birth and death process is described, where the domain for the population is $\{0, 1, 2, \ldots\}$. The general birth and death process is applied to a logistic birth and death process, where the birth and death probabilities are nonlinear functions of the population size. In this model, it is assumed that there is a maximal population size, so that the process simplifies to a finite Markov chain, where a transition matrix can be defined. The theory developed from random walk models is useful to the analysis of birth and death processes. The probability of absorption or population extinction, the expected time until population extinction, and the distribution conditioned on nonextinction, known as the quasistationary distribution, are studied.

The last two sections are applications of DTMC theory to two different stochastic epidemic models. The first stochastic epidemic model is known as an SIS epidemic model, where the acronym SIS means Susceptible-Infectious-Susceptible due to the fact that susceptible individuals become infected and infectious but do not develop immunity so they immediately become susceptible again. The second stochastic epidemic model is a class of models known as chain binomial models. These types of models were first developed in the 1920s and 1930s by Reed, Frost, and Greenwood and are appropriately named Reed-Frost and Greenwood models.

3.2 Proliferating Epithelial Cells

Epithelial cells line the cavities and surfaces throughout the body and increase during cell division. Proliferating cells have similar shapes among many multicellular animals (e.g., epidermis on tadpoles, developing fruit fly wings). Epithelial cells are polygonal in shape with most cells being hexagonal (six-sided). In Gibson et al. (2006), an infinite Markov chain model $\{X_n\}_{n=0}^{\infty}$ is formulated such that X_n is the number of sides per cell created during the cell division process, $X_n \in \{4, 5, 6, \ldots\}$. Then this infinite Markov chain is approximated by a finite Markov chain that is positive recurrent. It is shown that the mean of the stationary distribution corresponding to the ergodic Markov chain is six and that the hexagonal shape has the highest probability among all of the polygonal shapes.

A k-sided cell is made up of k sides and k vertices which after cell division results in two daughter cells. The cell division does not result in half the number of sides per daughter cell because division occurs within a side, making two new vertices and three new sides per cell (Figure 3.1). From experimental observations, Gibson et al. (2006) formulated a DTMC model based on the following assumptions: (1) the minimal number of sides is four, (2) cells do not re-sort; vertices and sides are kept intact, (3) daughter cells retain a common side, (4) cells have approximately uniform cell cycle times, $n \to n + 1$, (5) dividing cells divide within a side, not at a vertex, and (6) mitosis randomly distributes tricellular junctions to both daughter cells. An infinite-dimensional transition matrix for the DTMC is defined in a two-step process.

In the first step, only cell division is considered. A single cell at time n has s_n sides. The random variable r_{n+1} is the number of sides distributed to one daughter upon division (next time step), leaving $s_n - r_{n+1}$ for the other daughter. Each daughter receives at least two sides from the parent, leaving $s_n - 4$ sides that are distributed among the daughters. Assuming the sides are distributed uniformly and randomly, the distribution of these remaining sides, $s_n - 4$, can be modeled as a binomial distribution $b(s_n - 4, 1/2)$. Each daughter also gains two new sides because of the new interface. Therefore, the probability of a transition from an i-sided cell to a j-sided cell can be computed as follows:

$$\text{Prob}\{r_{n+1} + 2 = j | s_n = i\} = p_{j,i} = \binom{i-4}{j-4} \frac{1}{2^{i-4}}.$$

A four-sided cell divides into two four-sided cells ($p_{4,4} = 1$). A five-sided cell divides into one four-sided cell and one five-sided cell ($p_{4,5} = 1/2$, $p_{5,5} = 1/2$). A six-sided cell divides into two five-sided cells or into a four-sided and a six-sided cell ($p_{4,6} = 1/4$, $p_{5,6} = 1/2$, $p_{6,6} = 1/4$) and so on. In the first step, counting the number of sides per cell after division gives the following transition matrix:

$$M = \begin{pmatrix} 1 & 1/2 & 1/4 & 1/8 & 1/16 & \cdots \\ 0 & 1/2 & 1/2 & 3/8 & 1/4 & \cdots \\ 0 & 0 & 1/4 & 3/8 & 3/8 & \cdots \\ 0 & 0 & 0 & 1/8 & 1/4 & \cdots \\ 0 & 0 & 0 & 0 & 1/16 & \cdots \\ \vdots & \vdots & \vdots & \vdots & \vdots & \vdots \end{pmatrix}.$$

In the second step, the change in number of sides for neighboring cells must be considered. Each neighbor gains one side after cell division because a new junction is created. One side is added after cell division, $p_{i+1,i} = 1$ and zero elsewhere. The transition matrix for the second step of the process is

$$S = \begin{pmatrix} 0 & 0 & 0 & 0 & \cdots \\ 1 & 0 & 0 & 0 & \cdots \\ 0 & 1 & 0 & 0 & \cdots \\ 0 & 0 & 1 & 0 & \cdots \\ \vdots & \vdots & \vdots & \vdots & \vdots \end{pmatrix}.$$

Then $p(n + 1) = SMp(n) = Pp(n)$. The product of S and M, $P = SM$, is the transition matrix for the DTMC. Matrix P has the same elements as M, except they are moved down one row. Thus, there are two communicating classes: $\{4\}$ is transient and $\{5, 6, \ldots\}$ is recurrent.

Matrix P is truncated to form a finite-dimensional $m \times m$ matrix $\tilde{P}$. Since the last column of the truncated matrix is missing the term 2^{1-m}, the truncated matrix is not a stochastic matrix. However, if $m > 15$, then $2^{1-m} < 1 \times 10^{-4}$. Approximating the elements of the truncated matrix $\tilde{P}$ to four decimal places yields a stochastic matrix of dimension $m \times m$, the transition matrix for a finite Markov chain model for proliferating epithelial cells. Theorem 2.6 and Corollary 2.4 can be applied. The finite Markov chain applied to the set $\{5, 6, \ldots, m\}$ is positive recurrent. Thus, there exists a unique stationary probability distribution. The approximate stationary probability distribution for cells up to 10 sides, $(5, 6, \ldots, 10)$ is

$$(0.2888, 0.4640, 0.2085, 0.0359, 0.0027, 0.0001)^{tr}.$$

The largest proportion of cells is hexagonal in shape. In addition, the expected value of π is $E(\pi) = 6$. See the MATLAB program in Exercise 1.

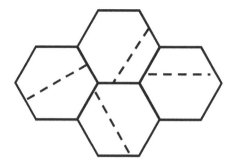

FIGURE 3.1: Cell division results in two new vertices and three new sides per cell.

3.3   Restricted Random Walk Models

A *restricted random walk* is a random walk model with at least one boundary, so that either the state space of the process is finite, $\{0, 1, 2 \ldots, N\}$, with two boundaries at 0 and N, or semi-infinite, $\{0, 1, 2, \ldots\}$, with one boundary at 0. In a random walk model, the states are positions and are denoted by the variable x, $x = 0, 1, 2, \ldots$ and time is denoted as n. It is assumed that p is the probability of moving to the right, x to $x + 1$, and q is the probability of moving to the left, x to $x - 1$.

Assumptions about movement at the boundaries, $x = 0$ or $x = N$, differ from movement at other positions. We shall discuss three types of boundary behavior: absorbing, reflecting, and elastic. An *absorbing boundary* at $x = 0$ assumes the one-step transition probability is

$$p_{00} = 1.$$

A *reflecting boundary* at $x = 0$ assumes the transition probabilities are

$$p_{00} = 1 - p \quad \text{and} \quad p_{10} = p, \quad 0 < p < 1.$$

An *elastic boundary* at $x = 0$ assumes the transition probabilities are

$$p_{21} = p, \quad p_{11} = sq, \quad p_{01} = (1 - s)q, \quad p + q = 1, \quad \text{and} \quad p_{00} = 1$$

for $0 < p, s < 1$. An elastic boundary is intermediate in relation to absorbing and reflecting boundaries. If $s = 0$ and $p = 0$, then $x = 1$ is an absorbing boundary, and if $s = 1$, then $x = 1$ is a reflecting boundary. When $0 < s < 1$, an object moving toward the boundary from position $x = 1$ will either reach $x = 0$ with probability $(1 - s)q$ or return to $x = 1$ with probability sq (elastic property).

In the next several sections, the restricted random walk with absorbing boundaries at $x = 0$ and $x = N$ is studied. This process is also known as the *gambler's ruin problem*, discussed briefly in Example 2.3.

3.4 Random Walk with Absorbing Boundaries

The random walk with absorbing boundaries is equivalent to the gambler's ruin problem on the set $\{0, 1, 2, \ldots, N\}$. The position x represents the gambler's capital, and each time step represents one game where the gambler may either increase his/her capital to $x+1$ or decrease it to $x-1$. After each game, there is either a gain or loss, never a tie. If the gambler's capital reaches zero, he/she is ruined, the opponent has won, and the games stop, whereas if the capital reaches N, he/she has won all of the capital (the opponent is ruined), and the games stop.

First, the transition matrix is defined. Let p be the probability of moving to the right (winning a game), q be the probability of moving to the left (losing a game), and $p + q = 1$. That is,

$$p_{ji} = \text{Prob}\{X_{n+1} = j | X_n = i\} = \begin{cases} p, & \text{if } j = i+1, \\ q, & \text{if } j = i-1, \\ 0, & \text{if } j \neq i+1, i-1, \end{cases}$$

for $i = 1, 2, \ldots, N-1$. The state space is $\{0, 1, 2, \ldots, N\}$. The boundaries 0 and N are absorbing,

$$p_{00} = 1 \quad \text{and} \quad p_{NN} = 1.$$

The transition matrix P is an $(N+1) \times (N+1)$ matrix of the following form:

$$P = \begin{pmatrix} 1 & q & 0 & \cdots & 0 & 0 & 0 \\ 0 & 0 & q & \cdots & 0 & 0 & 0 \\ 0 & p & 0 & \cdots & 0 & 0 & 0 \\ 0 & 0 & p & \cdots & 0 & 0 & 0 \\ \vdots & \vdots & \vdots & \cdots & \vdots & \vdots & \vdots \\ 0 & 0 & 0 & \cdots & 0 & q & 0 \\ 0 & 0 & 0 & \cdots & p & 0 & 0 \\ 0 & 0 & 0 & \cdots & 0 & p & 1 \end{pmatrix}.$$

There are three communication classes, $\{0\}$, $\{1, 2, \ldots, N-1\}$, and $\{N\}$. States 0 and N are absorbing (recurrent) and the other states are transient.

To understand the dynamics of the gambler's ruin problem, we investigate the probability that the gambler either loses or wins all of his/her money (probability of absorption) and the mean number of games until the gambler either wins all of the money or loses all of the money (expected duration of the games). Absorption occurs at either $x = 0$ (ruin) or at $x = N$ (jackpot). Beginning with a capital of k, the expected duration of the games (expected duration until absorption) is the sum of the following mean first passage times, $\mu_{0k} + \mu_{Nk}$.

Let

a_{kn} = probability of absorption at $x = 0$ on the nth game beginning with a capital of k, $\mathrm{Prob}\{X_n = 0 | X_0 = k\}$. The gambler has lost everything on the nth game.

b_{kn} = probability of absorption at $x = N$ on the nth game beginning with a capital of k, $\mathrm{Prob}\{X_n = N | X_0 = k\}$. The gambler has won everything on the nth game.

Assume that the beginning capital is restricted to $k = 1, \ldots, N-1$. Note that $a_{0n} = 1$, $a_{Nn} = 0$, $b_{0n} = 0$, and $b_{Nn} = 1$ because the games stop when the capital is either zero or N. As a mnemonic device, associate a with the left endpoint ($x = 0$) and b with the right endpoint ($x = N$). Note that $a_{kn} + b_{kn}$ is the probability of absorption at the nth game. Because absorption occurs at either $x = 0$ or $x = N$, $\{a_{kn} + b_{kn}\}_{n=0}^{\infty}$ represents the probability distribution associated with absorption,

$$\sum_{n=0}^{\infty} (a_{kn} + b_{kn}) = 1, \quad 1 \leq k \leq N-1.$$

Let A_k and B_k be the generating functions of the sequences $\{a_{kn}\}_{n=0}^{\infty}$ and $\{b_{kn}\}_{n=0}^{\infty}$

$$A_k(t) = \sum_{n=0}^{\infty} a_{kn} t^n \quad \text{and} \quad B_k(t) = \sum_{n=0}^{\infty} b_{kn} t^n, \quad |t| \leq 1.$$

The functions $A_k(t)$ and $B_k(t)$ by themselves are *not* probability generating functions. However, their sum $A_k(t) + B_k(t) = \sum_{n=0}^{\infty} (a_{kn} + b_{kn}) t^n$ is a probability generating function. Define

$$a_k = A_k(1) = \sum_{n=0}^{\infty} a_{kn}, \quad b_k = B_k(1) = \sum_{n=0}^{\infty} b_{kn},$$

and

$$\tau_k = A'(1) + B'(1) = \sum_{n=0}^{\infty} n(a_{kn} + b_{kn}).$$

Then a_k is the probability of absorption at $x = 0$ or the probability of ruin beginning with a capital of k, and b_k is the probability of absorption at $x = N$ or probability of winning all of the capital beginning with a capital of k. Finally, τ_k is the expected or mean duration of the games until absorption occurs either at $x = 0$ or at $x = N$, $\tau_k = \mu_{0k} + \mu_{Nk}$. In particular, if T_k denotes the random variable for the time until absorption, then $\tau_k = E(T_k)$. Note that

$$a_k + b_k = 1, \tag{3.1}$$

so that $b_k = 1 - a_k$. Expressions for a_k, τ_k, and a_{kn} will be derived in the next three sections.

3.4.1 Probability of Absorption

Two methods are discussed to compute the probability of absorption. The first method is analytical, applying methods from difference equations. The second method is computational, applying the fundamental matrix.

In the first method, an analytical expression for a_k is derived, the probability of absorption at $x = 0$ beginning with a capital of k. A difference equation relating a_{k-1}, a_k, and a_{k+1} is formulated. When boundary conditions are assigned at $k = 0$ and $k = N$, the difference equation can be solved for a_k.

An expression for the probability of ruin a_k on the interval $k \in [0, N]$ can be derived as follows. With a capital of k, the gambler may either win or lose the next game with probabilities p or q, respectively. If the gambler wins, the capital is $k + 1$ and the probability of ruin is a_{k+1}. If the gambler loses, the capital is $k - 1$ and the probability of ruin is a_{k-1}. This relationship is given by the following difference equation:

$$a_k = pa_{k+1} + qa_{k-1} \tag{3.2}$$

for $1 \leq k \leq N - 1$. Equation (3.2) is a second-order difference equation in a_k. The difference equation can be written as

$$pa_{k+1} - a_k + qa_{k-1} = 0, \quad k = 1, \ldots, N - 1. \tag{3.3}$$

This method of deriving equation (3.2) is referred to as a *first-step analysis* (Taylor and Karlin, 1998). In the derivation, we only consider what happens in the next step, then apply the Markov property. The equations are also referred to as the backward equations, rather than the forward equations because the equations are derived from the endpoint of the game rather than the starting point.

To solve the difference equation (3.3), the boundary conditions are

$$a_0 = 1 \quad \text{and} \quad a_N = 0.$$

If the capital is zero, then the probability of ruin equals one, and if the the capital is N, then the probability of ruin equals zero. The difference equation is linear and homogeneous, and the coefficients are constants. This type of difference equation can be solved (Elaydi, 1999). The method of solution is reviewed.

To solve the difference equation, let $a_k = \lambda^k \neq 0$, and substitute this value for a_k into the difference equation:

$$p\lambda^{k+1} - \lambda^k + q\lambda^{k-1} = 0.$$

Since $\lambda \neq 0$, this leads to the *characteristic equation*:

$$p\lambda^2 - \lambda + q = 0.$$

The roots of the characteristic equation are the *eigenvalues*

$$\lambda_{1,2} = \frac{1 \pm \sqrt{1 - 4pq}}{2p}.$$

The expression for the eigenvalues can be simplified by noticing that $(p+q)^2 = 1$. Expanding and rearranging terms,

$$1 = p^2 + 2pq + q^2$$
$$1 - 4pq = p^2 - 2pq + q^2 = (p - q)^2.$$

The radical in the expression for $\lambda_{1,2}$ simplifies to $\sqrt{1 - 4pq} = |p - q|$.

The solution to the difference equation (3.3) must be divided into two cases, depending on whether $p \neq q$ or $p = 1/2 = q$. In the first case, $p \neq q$, the eigenvalues are $\lambda_1 = 1$ and $\lambda_2 = q/p$. The general solution is

$$a_k = c_1 + c_2(q/p)^k.$$

The constants c_1 and c_2 are found by applying the boundary conditions: $a_0 = 1 = c_1 + c_2$ and $a_N = 0 = c_1 + c_2(q/p)^N$. Solving for c_1 and c_2 and substituting these values into the general solution yields the solution

$$a_k = \frac{(q/p)^N - (q/p)^k}{(q/p)^N - 1}, \quad p \neq q. \tag{3.4}$$

Since $a_k + b_k = 1$, the solution for b_k is

$$b_k = \frac{(q/p)^k - 1}{(q/p)^N - 1}, \quad p \neq q.$$

For the second case, $p = 1/2 = q$, note that $1 - 4pq = 0$, so that the characteristic equation has a root of multiplicity two, $\lambda_{1,2} = 1$. The general solution to the difference equation (3.3) is $a_k = c_1 + c_2 k$. Again applying the boundary conditions gives $a_0 = 1 = c_1$ and $a_N = 0 = c_1 + c_2 N$ so that the solution is

$$a_k = \frac{N - k}{N}, \quad p = 1/2 = q. \tag{3.5}$$

The solution to b_k is

$$b_k = \frac{k}{N}, \quad p = 1/2 = q.$$

Example 3.1. Suppose that the total capital is $N = 100$ and a gambler has $k = 50$ dollars. Table 3.1 gives values for the probability of losing all of the money, a_{50}, and winning all of the money, b_{50}, for different values of p and q. (Recall that $p = 1 - q$, where p is the probability of winning \$1 in each game.) Values for the expected duration, $\tau_{50} = A'(50) + B'(50)$, are also given in Table 3.1. Their derivation is discussed in Section 3.4.2. ∎

Table 3.1: Gambler's ruin problem with a beginning capital of $k = 50$ and a total capital of $N = 100$

Prob.	a_{50}	b_{50}	τ_{50}	$A'_{50}(1)$	$B'_{50}(1)$
$q = 0.50$	0.5	0.5	2500	1250	1250
$q = 0.51$	0.880825	0.119175	1904	1677	227
$q = 0.55$	0.999956	0.000044	500	499.93	0.07
$q = 0.60$	1.00000	0.00000	250	250	0

As the probability of losing a single game increases (q increases), the probability of ruin a_{50} also increases but a_{50} increases at a much faster rate than q. When $q = 0.6$, the probability of ruin starting with a capital of $k = 50$ is very close to one and the expected number of games until ruin equals 250.

The theory of difference equations was applied to find general solutions for a_k and b_k. Computational methods based on the fundamental matrix and the transition matrix can also be used to find solutions for a_k and b_k. The equations in (3.3) can be expressed as the following matrix equation, $\mathbf{a}^{tr} D = \mathbf{c}^{tr}$, where $\mathbf{a}^{tr} = (a_0, a_1, \ldots, a_N)$, $\mathbf{c}^{tr} = (1, 0, \ldots, 0)$, and

$$
D = \begin{pmatrix}
1 & | & q & 0 & 0 & \cdots & 0 & | & 0 \\
- & - & - & - & - & - & - & - & - \\
0 & | & -1 & q & 0 & \cdots & 0 & | & 0 \\
0 & | & p & -1 & q & \cdots & 0 & | & 0 \\
\vdots & | & \vdots & \vdots & \vdots & \ddots & \vdots & | & \vdots \\
0 & | & 0 & 0 & 0 & \cdots & -1 & | & 0 \\
- & - & - & - & - & - & - & - & - \\
0 & | & 0 & 0 & 0 & \cdots & p & | & 1
\end{pmatrix}
$$

$$
= \begin{pmatrix}
1 & | & A & | & 0 \\
- & - & - & - & - \\
\mathbf{0} & | & T - I & | & \mathbf{0} \\
- & - & - & - & - \\
0 & | & B & | & 1
\end{pmatrix}. \tag{3.6}
$$

Matrix D is an $(N + 1) \times (N + 1)$ matrix that is partitioned into block form according to the three communication classes, $\{0\}$, $\{1, 2, \ldots, N-1\}$, and $\{N\}$. The solution $\mathbf{a}$ is $\mathbf{a}^{tr} = \mathbf{c}^{tr} D^{-1}$, where

$$
D^{-1} = \left(\begin{array}{c|c|c}
1 & -A(T-I)^{-1} & 0 \\
\hline
0 & (T-I)^{-1} & 0 \\
\hline
0 & -B(T-I)^{-1} & 1
\end{array} \right).
$$

The probability of absorption into state zero from states $1, 2, \ldots, N-1$ equals AF, where $F = (I-T)^{-1}$ is the fundamental matrix, defined in (2.21). In addition, the probability of absorption into state N from states $1, 2, \ldots, N-1$ is BF.

3.4.2 Expected Time until Absorption

The duration of the games, beginning with a capital of k, lasts until all of the capital N is either gained or lost. The expected duration of the games can be calculated as in Section 2.7.1 with the fundamental matrix $F = (I-T)^{-1}$ or via difference equations. Applying the fundamental matrix, the expected duration is

$$
m_1 = \mathbf{1}^{tr}(I-T)^{-1} = \mathbf{1}^{tr} F.
$$

This method is useful computationally and can be used to calculate higher-order moments (MATLAB Program in the Appendix for Chapter 3 and Exercise 14).

An equivalent method using difference equations gives an analytical expression for the expected duration. Let $\tau_k = E(T_k)$ and apply a first-step analysis. Beginning with a capital of k, the gambler may either win or lose the next game with probabilities p or q, respectively. If the gambler wins, then the capital is $k+1$ and the duration of the games is $1 + \tau_{k+1}$ (counting the game just played), and if the gambler loses, then the capital is $k-1$ and the duration of the game is $1 + \tau_{k-1}$. The difference equation for τ_k has the following form:

$$
\tau_k = p(1 + \tau_{k+1}) + q(1 + \tau_{k-1}),
$$

for $k = 1, 2, \ldots, N-1$. Using the fact that $p + q = 1$, the difference equation can be expressed as follows:

$$
p\tau_{k+1} - \tau_k + q\tau_{k-1} = -1, \tag{3.7}
$$

a second-order linear, nonhomogeneous difference equation with constant coefficients. The boundary conditions are

$$
\tau_0 = 0 = \tau_N
$$

because if the capital is either zero or N, there can be no more games (absorption has occurred). The system (3.7) is equivalent to $\tau^{tr}(T-I) = -\mathbf{1}^{tr}$, where $\mathbf{1}$ is a column vector of ones. The difference equations in (3.7) can be solved in a manner similar to the difference equation for a_k. First, the general solution to the homogeneous difference equation is found, and then a particular solution to the nonhomogeneous is added to the homogeneous solution.

To solve for the homogeneous solution, let $\tau_k = \lambda^k \neq 0$ and substitute this value into the difference equation. The following characteristic equation is obtained:

$$p\lambda^2 - \lambda + q = 0.$$

The eigenvalues are $\lambda_1 = 1$ and $\lambda_2 = q/p$, if $p \neq q$ and $\lambda_1 = 1 = \lambda_2$, if $p = 1/2 = q$.

In the first case, $p \neq q$, the general solution to the homogeneous difference equation is $\tau_k = c_1 + c_2(q/p)^k$. To find a particular solution to the nonhomogeneous equation, let $\tau_k = ck$ for an arbitrary constant c and solve for c. Substituting $\tau_k = ck$ into the difference equation leads to $c = 1/(q-p)$. Thus, the general solution to the nonhomogeneous difference equation (3.7) is

$$\tau_k = c_1 + c_2(q/p)^k + \frac{k}{q-p}.$$

Applying the boundary conditions, the solution is

$$\tau_k = \frac{1}{q-p}\left[k - N\left(\frac{1 - (q/p)^k}{1 - (q/p)^N}\right)\right], \quad p \neq q. \tag{3.8}$$

A similar method is applied in the second case, where $p = 1/2 = q$. The general solution to the homogeneous difference equation (3.7) is $\tau_k = c_1 + c_2 k$ and a particular solution has the form ck^2. Substituting ck^2 into the difference yields $c = -1$, so that the general solution to the nonhomogeneous equation is $\tau_k = c_1 + c_2 k - k^2$. Applying the boundary conditions, the solution is

$$\tau_k = k(N - k), \quad p = 1/2 = q. \tag{3.9}$$

Example 3.2. Suppose $N = 100$ and $p = 1/2 = q$. Applying formula (3.9), $\tau_{50} = 2500$ (Table 3.1). But when $p = 0.45$ and $q = 0.55$, $\tau_{50} \approx 500$. If the capital is increased to $N = 1000$, then for $p = 1/2 = q$, $\tau_{500} = 250,000$. For this same capital, but $p = 0.45$ and $q = 0.55$, $\tau_{500} = 5000$. The duration of the game increases significantly as the capital increases. ∎

Figure 3.2 is a graph of the expected duration, τ_k, $k = 0, 1, 2, \ldots, N$, when $N = 100$ and $q = 0.55$. It can be seen from this graph that although $\tau_{50} \approx 500$, the maximum duration is approximately 800 games at an initial capital of $k = 85$ dollars.

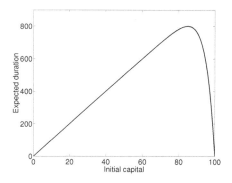

FIGURE 3.2: Expected duration of the games, τ_k for $k = 0, 1, 2, \ldots, 100$, when $q = 0.55$.

3.4.3 Probability Distribution for Absorption

In the previous section, a formula for the expected duration until absorption was derived. To calculate the entire probability distribution until absorption at either $x = 0$ or at $x = N$, expressions for a_{kn} and b_{kn} are required. The probability distribution until absorption is

$$\{a_{kn} + b_{kn}\}_{n=0}^{\infty}$$

with generating function

$$A_k(t) + B_k(t) = \sum_{n=0}^{\infty} (a_{kn} + b_{kn})t^n, \quad |t| \leq 1.$$

Closed form expressions can be derived for $A_k(t)$ and $B_k(t)$, but they do not give explicit expressions for a_{kn} and b_{kn}. However, assuming $A_k(t)$ and $B_k(t)$ have Maclaurin series expansions in t, these coefficients can be obtained by repeated differentiation; for example,

$$a_{kn} = \frac{1}{n!} \frac{d^n A_k(t)}{dt^n} \bigg|_{t=0}. \tag{3.10}$$

In practice, higher-order derivatives may be difficult to compute for large n.

A difference equation for A_k is derived using a first-step analysis, similar to the analysis used to derive expressions for a_k and τ_k in the previous sections. First, a difference equation is derived for a_{kn}. If absorption at $x = 0$ occurs in $n + 1$ steps from an initial capital of k, then in the next game if the gambler wins, the capital is $k + 1$ and absorption will occur with n more games. Otherwise, if the gambler loses the next game, the capital is $k-1$ and absorption will occur with n more games; that is,

$$a_{k,n+1} = pa_{k+1,n} + qa_{k-1,n}, \quad k \geq 1, \ n \geq 0.$$

The preceding equation is a partial difference equation, the difference equation version of a partial differential equation, since a_{kn} is a function of two variables, k and n. The boundary and initial conditions for this system of difference equations are

$$a_{0n} = 0 = a_{Nn}, \quad a_{00} = 1, \quad \text{and} \quad a_{k0} = 0$$

for $n = 1, 2, \ldots, N$ and $k = 1, 2, \ldots, N-1$. These conditions follow because if the beginning capital is zero, absorption has already occurred and no games are required, $a_{00} = 1$, and absorption cannot occur in $n > 0$ steps, $a_{0n} = 0$. In addition, if the gambler has all of the capital, his/her opponent has already lost and no games are required $a_{N0} = 0$ and absorption cannot occur in $n > 0$ steps, a_{Nn}. Finally, it takes at least k games to be ruined beginning with a capital of k, $a_{kn} = 0$ for $n < k$. These conditions give rise to simple expressions for the generating functions, $A_0(t)$ and $A_N(t)$:

$$A_0(t) = a_{00} + a_{01}t + a_{02}t^2 + \cdots = 1$$

and

$$A_N(t) = a_{N0} + a_{N1}t + a_{N2}t^2 + \cdots = 0.$$

To obtain a difference equation for A_k, the equation in a_{kn} is multiplied by t^{n+1} and summed from $n = 0$ to $n = \infty$. For $k \geq 1$, it follows that

$$\sum_{n=0}^{\infty} a_{k,n+1}t^{n+1} = \sum_{n=0}^{\infty} pa_{k+1,n}t^{n+1} + \sum_{n=0}^{\infty} qa_{k-1,n}t^{n+1}.$$

Because $a_{k0} = 0$ for $k \geq 1$, the preceding equation can be simplified:

$$\sum_{n=0}^{\infty} a_{kn}t^n = pt\sum_{n=0}^{\infty} a_{k+1,n}t^n + qt\sum_{n=0}^{\infty} a_{k-1,n}t^n$$
$$A_k(t) = ptA_{k+1}(t) + qtA_{k-1}(t).$$

For t fixed, $0 < t < 1$, the difference equation can be solved subject to the following boundary conditions:

$$A_0(t) = 1 \quad \text{and} \quad A_N(t) = 0.$$

Let $A_k(t) = \lambda^k \neq 0$ so that the characteristic equation is

$$pt\lambda^2 - \lambda + qt = 0.$$

The eigenvalues are

$$\lambda_{1,2} = \frac{1 \pm \sqrt{1 - 4pqt^2}}{2pt},$$

where $\lambda_1 > \lambda_2 > 0$. The two roots are real and distinct. The general solution is $A_k(t) = c_1\lambda_1^k + c_2\lambda_2^k$. The constants c_1 and c_2 are found by applying the boundary conditions, $c_1 + c_2 = 1$ and $c_1\lambda_1^N + c_2\lambda_2^N = 0$, so that

$$A_k(t) = \frac{\lambda_1^N\lambda_2^k - \lambda_2^N\lambda_1^k}{\lambda_1^N - \lambda_2^N}. \tag{3.11}$$

A closed form expression for $B_k(t)$ can be obtained in a similar manner. For example, $B_k(t)$ is a solution of the same difference equation as $A_k(t)$,

$$B_k(t) = ptB_{k+1}(t) + qtB_{k-1}(t)$$

for $k = 1, 2, \ldots, N - 1$, but the boundary conditions differ,

$$B_0(t) = 0 \quad \text{and} \quad B_N(t) = 1.$$

The solution $B_k(t)$ is

$$B_k(t) = \frac{\lambda_1^k - \lambda_2^k}{\lambda_1^N - \lambda_2^N}.$$

Thus, the p.g.f. for the duration of the games is given by $A_k(t) + B_k(t)$. Although the formula was derived for t restricted to $0 < t < 1$, if the p.g.f. is expressed as a Maclaurin series over this region,

$$A_k(t) + B_k(t) = \sum_{n=0}^{\infty}(a_{kn} + b_{kn})t^n,$$

then the series is absolutely convergent for $|t| < 1$ and all of its derivatives exist for $|t| < 1$ (Wade, 2000). Abel's convergence theorem (Chapter 2) can be applied to the series and the derivatives of the series to extend the domain to $t = 1$. For example, the derivative of the series is

$$A_k'(t) + B_k'(t) = \sum_{n=1}^{\infty} n(a_{kn} + b_{kn})t^{n-1},$$

which is finite for $|t| < 1$ and $n(a_{kn} + b_{kn}) \geq 0$. Therefore,

$$\lim_{t \to 1^-}\left[\sum_{n=1}^{\infty} n(a_{kn} + b_{kn})t^{n-1}\right] = L \leq \infty.$$

By Abel's convergence theorem, $\sum_{n=1}^{\infty} n(a_{kn} + b_{kn}) = L$. Therefore, the expected duration of the games, τ_k, can be derived from the generating function,

$$\tau_k = A_k'(1) + B_k'(1) = \lim_{t \to 1^-}[A_k'(t) + B_k'(t)].$$

Fortunately, the derivation in the previous section has already provided a solution for τ_k. Bailey (1990) derives a general formula for a_{kn} using a partial fraction expansion.

Example 3.3. Suppose $N = 10$ and $k = 5$. Then $A_5(t)$ can be computed from (3.11) so that

$$A_5(t) = \frac{q^5 t^5}{1 - 5pqt^2 + 5p^2 q^2 t^4}.$$

Through differentiation and application of (3.10), the values of a_{5n} for $n = 5, 6, \ldots,$ can be computed. A computer algebra system was used to compute a_{5n} for $n = 5, 7, 9, 11$,

$$a_{55} = q^5, \ a_{57} = 5q^6 p, \ a_{59} = 20q^7 p^2, \ a_{5,11} = 75q^8 p^3.$$

It can be seen that $a_{5n} = 0$ for $n < 5$ and for n even. ■

Example 3.4. Assume $N = 100$, $k = 50$, and $q = 0.55$. The expected duration of the games is $\tau_{50} \approx 500$ and the standard deviation is $\sigma_{50} \approx 222.4$. Approximations to these values were calculated from 1000 sample paths,

$$\text{mean} \approx 482.74, \quad \text{standard deviation} \approx 203.27.$$

The mean and variance of the distribution for the duration of the games, τ_{50} and σ_{50}^2, can be calculated directly from the p.g.f. $S_k(t) = A_k(t) + B_k(t)$ when $k = 50$,

$$\tau_{50} = S_{50}'(1) \quad \text{and} \quad \sigma_{50}^2 = S_{50}''(1) + S_{50}'(1) - [S_{50}'(1)]^2.$$

Three sample paths are graphed for this stochastic process in Figure 3.3. A Monte Carlo simulation that generated this figure is given in the Appendix for Chapter 3. ■

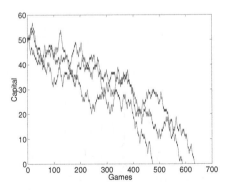

FIGURE 3.3: Three sample paths for the gambler's ruin problem when $N = 100$, $k = 50$, and $q = 0.55$.

3.5 Random Walk on a Semi-Infinite Domain

The random walk on a bounded domain can be extended to a semi-infinite domain $\{0, 1, 2 \ldots\}$. Think of the domain as unrestricted movement in one direction or as an unlimited amount of capital for a gambler, $N \to \infty$. The general solution to $A_k(t)$ has the same form as in the finite domain:

$$A_k(t) = c_1 \lambda_1^k + c_2 \lambda_2^k.$$

However, one of the boundary conditions changes,

$$A_0(t) = 1 \quad \text{and} \quad \lim_{k \to \infty} |A_k(t)| < \infty.$$

The boundedness condition must be satisfied because

$$|A_k(t)| \leq \sum_{n=0}^{\infty} a_{kn} = a_k \leq 1$$

for $|t| \leq 1$ and all k.

Recall that the two eigenvalues λ_1 and λ_2 satisfy

$$|\lambda_1| = \frac{1 + \sqrt{1 - 4pqt^2}}{2p|t|} \quad \text{and} \quad |\lambda_2| = \frac{1 - \sqrt{1 - 4pqt^2}}{2p|t|} \quad \text{for } |t| \leq 1.$$

In addition, the absolute values of the eigenvalues $|\lambda_i|$, $i = 1, 2$, are roots of the characteristic equation, $f(\lambda) = p|t|\lambda^2 - \lambda + q|t| = 0$. Because the graph of f is a parabola satisfying $f(0) = q|t| > 0$, $f(1) = |t| - 1 < 0$ for $0 < |t| < 1$, and $\lim_{\lambda \to \infty} f(\lambda) = \infty$, it follows that the two roots satisfy $0 < |\lambda_2| < 1 < |\lambda_1|$. Thus, the coefficient $c_1 = 0$, $c_2 = 1$, and

$$A_k(t) = \lambda_2^k = \left(\frac{1 - \sqrt{1 - 4pqt^2}}{2pt} \right)^k.$$

The probability of ruin or of absorption at $x = 0$, a_k, can be calculated directly from $A_k(t)$:

$$a_k = A_k(1) = \begin{cases} 1, & \text{if } p \leq q, \\ \left(\dfrac{q}{p} \right)^k, & \text{if } p > q. \end{cases} \tag{3.12}$$

Also, for $p \leq q$, the sequence $\{a_{kn}\}_{n=0}^{\infty}$ is a probability distribution and the expected duration until absorption at $x = 0$ can be computed. It is easy to see that for $p > q$ the expected duration is infinite. In the other cases, τ_k is computed from $A_k'(1)$:

$$\tau_k = \begin{cases} \dfrac{k}{q - p}, & \text{if } p < q, \\ \infty, & \text{if } p \geq q. \end{cases} \tag{3.13}$$

The probability of absorption in n time steps, a_{kn}, can be calculated by applying formula (3.10).

3.6 General Birth and Death Process

A general birth and death process is formulated as a DTMC. The Markov chain model is related to the gambler's ruin problem, but the probability of a birth (or winning) is not constant but depends on the size of the population and the probability of a death (or losing) also depends on the population size. To define a birth and death process, let X_n, denote the size of the population at time n, where the state space may be either finite or infinite, $\{0, 1, 2, \ldots, N\}$ or $\{0, 1, 2, \ldots\}$. The value N is the maximal population size in the finite case. The birth and death probabilities are b_i and d_i, respectively. In addition, $b_0 = 0 = d_0$, $b_i > 0$ and $d_i > 0$ for $i = 1, 2, \ldots$, except in the finite case, where $b_N = 0$. It is assumed that the time interval, $n \to n + 1$, is sufficiently small such that during this time interval at most one event occurs, either a birth or a death. Assume the transition probabilities are

$$p_{ji} = \text{Prob}\{X_{n+1} = j | X_n = i\}$$
$$= \begin{cases} b_i, & \text{if } j = i + 1 \\ d_i, & \text{if } j = i - 1 \\ 1 - (b_i + d_i), & \text{if } j = i \\ 0, & \text{if } j \neq i - 1, i, i + 1 \end{cases}$$

for $i = 1, 2, \ldots$, $p_{00} = 1$, and $p_{j0} = 0$ for $j \neq 0$. In the case of a finite state space, where N is the maximal population size, $p_{N+1,N} = b_N = 0$.

The transition matrix P for the finite Markov chain has the following form:

$$\begin{pmatrix} 1 & d_1 & 0 & \cdots & 0 & 0 \\ 0 & 1 - (b_1 + d_1) & d_2 & \cdots & 0 & 0 \\ 0 & b_1 & 1 - (b_2 + d_2) & \cdots & 0 & 0 \\ 0 & 0 & b_2 & \cdots & 0 & 0 \\ \vdots & \vdots & \vdots & \vdots & \vdots & \vdots \\ 0 & 0 & 0 & \cdots & 1 - (b_{N-1} + d_{N-1}) & d_N \\ 0 & 0 & 0 & \cdots & b_{N-1} & 1 - d_N \end{pmatrix}. \quad (3.14)$$

In general, for a finite or infinite state space, it is assumed that

$$\sup_{i \in \{1,2,\ldots\}} \{b_i + d_i\} \leq 1,$$

so that P is a stochastic matrix. During each time interval, n to $n + 1$, either the population size increases by one, decreases by one, or stays the same size. This is a reasonable assumption only if the time step is sufficiently small.

There are two communication classes, $\{0\}$ and $\{1, \ldots, N\}$ in the finite case. It is easy to see that zero is positive recurrent; all other states are transient. There exists a unique stationary probability distribution π, $P\pi = \pi$, where $\pi_0 = 1$ and $\pi_i = 0$ for $i = 1, 2, \ldots$. In the case of a finite Markov chain, it can be verified that eventually population extinction occurs from any initial state,

$$\lim_{n \to \infty} \text{Prob}\{X_n = 0\} = \lim_{n \to \infty} p_0(n) = 1.$$

In the notation of Section 3.4.1, $a_k = 1$ for $k = 0, 1, 2, \ldots, N$. The transition matrix in (3.14) can be partitioned according to the recurrent state and the transient states as follows:

$$P = \begin{pmatrix} 1 & | & A \\ -- & -- & -- \\ \mathbf{0} & | & T \end{pmatrix}.$$

3.6.1 Expected Time to Extinction

The techniques from the preceding sections are used to find an expression for the expected time until population extinction. Let τ_k denote the expected time until extinction for a population with initial size k. Then $\tau_0 = 0$ and the following relationship holds for τ_k, $k = 1, 2, \ldots$:

$$\tau_k = b_k(1 + \tau_{k+1}) + d_k(1 + \tau_{k-1}) + (1 - (b_k + d_k))(1 + \tau_k). \tag{3.15}$$

If the maximal population size is finite, then for $k = N$, $\tau_N = d_N(1 + \tau_{N-1}) + (1 - d_N)(1 + \tau_N)$. The difference equation can be simplified as follows:

$$d_k \tau_{k-1} - (b_k + d_k)\tau_k + b_k \tau_{k+1} = -1, \tag{3.16}$$

$k = 1, 2, \ldots$. If $k = N$, then $d_N \tau_{N-1} - d_N \tau_N = -1$. Because the coefficients for these difference equations are not constant, the same techniques cannot be employed as in the previous sections. However, when the maximal population size is finite, then the computational method can be employed to find τ. In particular,

$$\tau^{tr} = \mathbf{1}^{tr}(I - T)^{-1} = \mathbf{1}^{tr} F, \tag{3.17}$$

where F is the fundamental matrix. Nisbet and Gurney (1982) derived an analytical expression for τ_k. This expression is given in the next theorem.

THEOREM 3.1
Suppose $\{X_n\}_{n=0}^N$ is a general birth and death process with $X_0 = m \geq 1$ satisfying $b_0 = 0 = d_0$, $b_i > 0$ for $i = 1, 2, \ldots, N-1$, and $d_i > 0$ for $i = 1, 2, \ldots, N$. The expected time until population extinction is

$$\tau_m = \begin{cases} \dfrac{1}{d_1} + \displaystyle\sum_{i=2}^{N} \dfrac{b_1 \cdots b_{i-1}}{d_1 \cdots d_i}, & m = 1 \\[2ex] \tau_1 + \displaystyle\sum_{s=1}^{m-1} \left[\dfrac{d_1 \cdots d_s}{b_1 \cdots b_s} \displaystyle\sum_{i=s+1}^{N} \dfrac{b_1 \cdots b_{i-1}}{d_1 \cdots d_i} \right], & m = 2, \ldots, N. \end{cases} \tag{3.18}$$

Proof. For $k = 1, 2, \ldots, N - 1$, the equations (3.16) are solved recursively for $\tau_2, \ldots, \tau_N$ to obtain the formulas

$$\tau_m = \tau_1 + \sum_{k=1}^{m-1} \frac{d_1 \cdots d_k}{b_1 \cdots b_k} \left[\tau_1 - \frac{1}{d_1} - \sum_{i=2}^{k} \frac{b_1 \cdots b_{i-1}}{d_1 \cdots d_i} \right] \qquad (3.19)$$

for $m = 2, \ldots, N$. The second summation is zero when $k < 2$. Then applying the relation for $k = N$,

$$\tau_N = \frac{1}{d_N} + \tau_{N-1}$$

and equating the two values for τ_N, the following formula for τ_1 is obtained:

$$\tau_1 = \frac{1}{d_1} + \sum_{i=2}^{N} \frac{b_1 \cdots b_{i-1}}{d_1 \cdots d_i}.$$

Substituting τ_1 into (3.19), the formula for (3.18) follows. $\qquad \square$

Example 3.5. Suppose the maximal population size is $N = 20$ in a birth and death process, where $b_i \equiv bi$, for $i = 1, 2, \ldots, 19$, $d_i \equiv di$, for $i = 1, 2, \ldots, 20$, b and d are constants. This process is often referred to as a *simple* birth and death process. When $b > d$, there is population growth, and when $b < d$, there is population decline. Three cases are considered: (i) $b = 0.02 < 0.03 = d$, (ii) $b = 0.025 = d$, and (iii) $b = 0.03 > 0.02 = d$. The expected time until population extinction $\tau = (\tau_0, \ldots, \tau_{20})^{tr}$ is plotted in each of these three cases in Figure 3.4. The formula in Theorem 3.1 can be applied or τ can be calculated from (3.17). Note how much greater the expected duration is for $b > d$ than for $b < d$. Three sample paths of the simple birth and death process are also graphed in the case case $b = 0.025 = d$. Continuous-time birth and death processes are discussed in Chapter 6. $\qquad \blacksquare$

3.7 Logistic Growth Process

Assumptions are made on the general birth and death probabilities b_i and d_i so that the process has a logistic form. Recall that in the deterministic logistic model, if $y(t)$ is the population size at time t, then the rate of change of $y(t)$ is

$$\frac{dy}{dt} = \tilde{r} y \left(1 - \frac{y}{K} \right), \quad y(0) = y_0 > 0.$$

The right-hand side of the differential equation is a quadratic function of y and equals the birth rate minus the death rate. The parameter $\tilde{r}$ is the intrinsic growth rate and K is the carrying capacity. It is well known that the

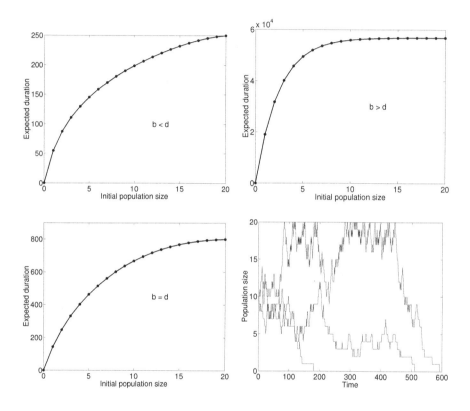

FIGURE 3.4: Expected time until population extinction τ when the maximal population size is $N = 20$ and $b = 0.02 < 0.03 = d$, $b = 0.025 = d$, or $b = 0.03 > 0.02 = d$. Three sample paths for the case $b = 0.025 = d$ with $X_0 = 10$.

unique solution $y(t)$ of this differential equation satisfies $\lim_{t\to\infty} y(t) = K$. The population size approaches the carrying capacity.

For a stochastic logistic growth process, assume

$$b_i - d_i = ri(1 - i/K) \tag{3.20}$$

for $i = 0, 1, 2, \ldots, N$, where $r = \tilde{r}\Delta t$, $N > K$, and the time interval Δt is chosen sufficiently small so that $\max_{i \in \{i=0,\ldots,N\}}\{b_i + d_i\} \leq 1$. (The time interval Δt is the interval n to $n+1$.) From equation (3.20), it follows that the birth probability will equal the death probability when the population size is zero ($i = 0$) or when the population size is at the carrying capacity ($i = K$). Due to the relationship (3.20), it is reasonable to assume that b_i and d_i are either linear or quadratic functions of i.

Two cases for the birth and death probabilities b_i and d_i are considered:

(a) $b_i = r\left(i - \dfrac{i^2}{2K}\right)$ and $d_i = r\dfrac{i^2}{2K}$, $i = 0, 1, 2, \ldots, 2K$

(b) $b_i = \begin{cases} ri, & i = 0, 1, 2, \ldots, N-1. \\ 0, & i \geq N \end{cases}$ and $d_i = r\dfrac{i^2}{K}$, $i = 0, 1, \ldots, N$.

In case (a), the maximal population size is $N = 2K$. Also, the birth probability increases when the population size is less than K but decreases when the population size is greater than K, whereas the death probability is an increasing function of the population size. In case (b), both the birth and death probabilities are increasing functions of the population size. Three sample paths of the DTMC logistic growth process are graphed against the solution of the deterministic logistic equation for case (b) in Figure 3.5.

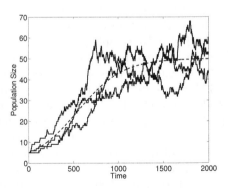

FIGURE 3.5: Three sample paths of the logistic growth process compared to the solution of the deterministic logistic equation in case (b), where $X_0 = 5$, $b_i = ri$, $d_i = ri^2/K$, $r = 0.004$, $\tilde{r} = 1$, $K = 50$, $\Delta t = 0.004$, and $N = 100$. Time is measured in increments of Δt, 2000 time increments.

Example 3.6. For the two cases (a) and (b), the expected time until population extinction is calculated. Let $r = 0.015$, $K = 10$, and $N = 20$. Graphs of the expected time to extinction in cases (a) and (b) are given in Figure 3.6. Note how much longer the population persists in case (a). ■

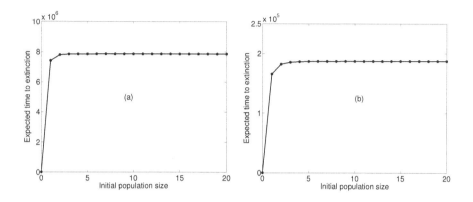

FIGURE 3.6: Expected time until population extinction when the birth and death rates satisfy (a) and (b) and the parameters are $r = 0.015$, $K = 10$, and $N = 20$.

There are some important differences in the deterministic versus the stochastic logistic equation. In the limit the stochastic population size does not approach the carrying capacity. The extinction state is an absorbing state rather than the carrying capacity. But as can be seen in Figure 3.5, for larger population sizes, $N = 100$ and $K = 50$, sample paths vary around K. Also, in the preceding example, if the maximal population size is doubled to $N = 40$ and $K = 20$ ($r = 0.0075$), the expected time to extinction increases to 7.7×10^{12} in case (a) and 3.6×10^9 in case (b). For large N, the stochastic model follows the deterministic model more closely. Prior to extinction (which may take a long time), the probability distribution is approximately stationary for a long period of time (quasistationary distribution). This can be seen in Figure 3.7, where the probability distribution $p(n) = (p_0(n), p_1(n), \ldots, p_{20}(n))^{tr}$ is graphed as a function of time $n = 0, 1, 2, \ldots, 1000$. An approximate stationary distribution has been reached by $n = 500$. A MATLAB program to generate the probability distribution is given in the Appendix for Chapter 3.

The unique stationary distribution corresponding to the simple birth and death process and the logistic growth process is $\pi = (1, 0, 0, \ldots, 0)^{tr}$. However, if the time to extinction is sufficiently long, the process approaches a quasistationary probability distribution, the distribution conditioned on

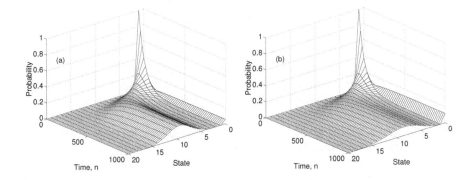

FIGURE 3.7: Probability distribution for the stochastic logistic model in cases (a) and (b) when $r = 0.015$, $K = 10$, $N = 20$, and $X_0 = 1$.

nonextinction. For large K and N, it will be seen that the mean of this quasistationary distribution is close to K (Figure 3.7).

3.8 Quasistationary Probability Distribution

When the expected duration until absorption is large, it is reasonable to examine the dynamics of the process prior to absorption. Let $\{X_n\}_{n=0}^{\infty}$ denote a general birth and death process with $p_i(n) = \text{Prob}\{X_n = i\}$, $i = 0, 1, 2, \ldots, N$. Define the conditional probability,

$$q_i(n) = \text{Prob}\{X_n = i | X_j \neq 0, j = 0, 1, 2, \ldots, n\}$$
$$= \frac{p_i(n)}{1 - p_0(n)}$$

for $i = 1, 2, \ldots, N$. The distribution $q(n) = (q_1(n), q_2(n), \ldots, q_N(n))^{tr}$ defines a probability distribution because

$$\sum_{i=1}^{N} q_i(n) = \frac{\sum_{i=1}^{N} p_i(n)}{1 - p_0(n)} = \frac{1 - p_0(n)}{1 - p_0(n)} = 1.$$

It is a conditional probability distribution. The probability $q_i(n)$ is conditioned on the population size not hitting zero by time n (i.e., conditional on nonextinction). Let this conditional DTMC be denoted as $\{Q_n\}_{n=0}^{\infty}$, where Q_n is the random variable for the population size at time n, conditional on nonextinction; $q_i(n) = \text{Prob}\{Q_n = i\}$. The stationary probability distribution for this process is denoted as q^*; q^* is referred to as the *quasistationary probability distribution* or *quasiequilibrium probability distribution*.

The forward equations for $q_i(n)$ can be derived based on those for $p_i(n)$. It will be seen that q^* cannot be calculated directly but must be calculated by an indirect method. An approximation to the process $\{Q_n\}_{n=0}^{\infty}$ yields an irreducible, positive recurrent, aperiodic Markov chain, $\{\tilde{Q}_n\}_{n=0}^{\infty}$, with associated probability distribution for $\tilde{Q}_n$ denoted $\tilde{q}(n)$. For this new process, a transition matrix, $\tilde{P}$, and the limiting positive stationary probability distribution $\tilde{q}^*$ can be defined. The stationary probability distribution $\tilde{q}^*$ is an approximation to the quasistationary probability distribution q^*.

The forward equations for $q_i(n+1)$ are derived from the identity $p(n+1) = Pp(n)$, where transition matrix P is defined in (3.14). From the definition of q_i,

$$q_i(n+1) = \frac{p_i(n+1)}{1 - p_0(n+1)}$$

$$= \left(\frac{p_i(n+1)}{1 - p_0(n)}\right)\left(\frac{1 - p_0(n)}{1 - p_0(n+1)}\right)$$

$$= \left(\frac{p_i(n+1)}{1 - p_0(n)}\right)\left(\frac{1 - p_0(n)}{1 - p_0(n) - d_1 p_1(n)}\right)$$

or

$$q_i(n+1)(1 - d_1 q_1(n)) = \left(\frac{p_i(n+1)}{1 - p_0(n)}\right).$$

Applying the identity $p_i(n+1) = b_{i-1}p_{i-1}(n) + (1 - b_i - d_i)p_i(n) + d_{i+1}p_{i+1}(n)$, yields the following equations:

$$q_i(n+1)[1 - d_1 q_1(n)] = b_{i-1}q_{i-1}(n) + (1 - b_i - d_i)q_i(n) + d_{i+1}q_{i+1}(n) \quad (3.21)$$

for $i = 1, 2, \ldots, N$, $b_0 = 0$, and $q_i(n) = 0$ for $i \notin \{1, 2, \ldots, N\}$. The difference equation for q_i is similar to the difference equation for p_i, except for an additional factor multiplying $q_i(n+1)$. An analytical solution to the stationary solution q^* cannot be found directly from these equations, since the coefficients depend on n. But q^* can be found via a numerical iterative method (Nåsell, 1999, 2001).

To approximate the quasistationary probability distribution q^* of the process $\{Q_n\}_{n=0}^{\infty}$ it is assumed that $d_1 = 0$. Equivalently, when the population size is reduced to one, the probability of dying is set equal to zero. If, over a long period of time, the probability is very small that the process is in state one, then this is a reasonable assumption. With this assumption, equation (3.21) simplifies to

$$\tilde{q}_i(n+1) = b_{i-1}\tilde{q}_{i-1}(n) + (1 - b_i - d_i)\tilde{q}_i(n) + d_{i+1}\tilde{q}_{i+1}(n),$$

$i = 2, \ldots, N - 1$, $\tilde{q}_1(n+1) = (1 - b_1)\tilde{q}_1(n) + d_2\tilde{q}_2(n)$, and $\tilde{q}_N(n+1) = b_{N-1}\tilde{q}_{N-1}(n) + (1 - d_N)\tilde{q}_N(n)$. The new transition matrix corresponding to this approximation is

$$
\tilde{P} = \begin{pmatrix}
1 - b_1 & d_2 & \cdots & 0 & 0 \\
b_1 & 1 - (b_2 + d_2) & \cdots & 0 & 0 \\
0 & b_2 & \cdots & 0 & 0 \\
\vdots & \vdots & \vdots & \vdots & \vdots \\
0 & 0 & \cdots & 1 - (b_{N-1} + d_{N-1}) & d_N \\
0 & 0 & \cdots & b_{N-1} & 1 - d_N
\end{pmatrix}.
$$

Matrix $\tilde{P}$ is a submatrix of matrix P, where the first column and first row are deleted and $d_1 = 0$. The DTMC $\{\tilde{Q}_n\}_{n=0}^{\infty}$, where $\tilde{q}(n+1) = \tilde{P}\tilde{q}(n)$, is ergodic (irreducible, positive recurrent, and aperiodic) and has a unique stationary probability distribution, $\tilde{q}^*$, $\tilde{P}\tilde{q}^* = \tilde{q}^*$, where $\tilde{q}^* = (\tilde{q}_1^*, \tilde{q}_2^*, \ldots, \tilde{q}_N^*)^{tr}$ is

$$
\tilde{q}_{i+1}^* = \frac{b_i \cdots b_1}{d_{i+1} \cdots d_2} \tilde{q}_1^* \quad \text{and} \quad \sum_{i=1}^{N} \tilde{q}_i^* = 1. \tag{3.22}
$$

Example 3.7. The approximate quasistationary probability distribution, $\tilde{q}^*$, is compared to the quasistationary probability distribution q^* when $r = 0.015$, $K = 10$, and $N = 20$ in cases (a) and (b) in Figure 3.8. The quasistationary distribution q^* is computed via an iterative method by applying equation (3.21). Both distributions have good agreement for $N = 20$, but when $N = 10$ and $K = 5$, then the two distributions differ, especially for values near zero [Figure 3.8 (c)].

The means and standard deviations for the quasistationary distribution q^*, graphed in Figures 3.8 (a) and (b), are

(a) $\mu_{q^*} = 9.435$ and $\sigma_{q^*} = 2.309$

(b) $\mu_{q^*} = 8.848$ and $\sigma_{q^*} = 3.171$.

∎

Some of the differences between the deterministic and stochastic models have been highlighted in the previous examples. The estimate for the expected time until population extinction, τ_k, and the probability of population extinction, $\lim_{n \to \infty} p_0(n) = 1$, are important in stochastic theory but they have no counterparts in deterministic theory. For large population sizes, K and N large, and initial conditions sufficiently large, the deterministic model is in much better agreement with the stochastic model than for K and N small. This can be seen in the stochastic logistic model graphed in Figure 3.9. The shape of the probability distribution over time is similar to that of the solution to the logistic differential equation. The mean of the stochastic process and the solution to the logistic differential equation show close agreement in Figure 3.9.

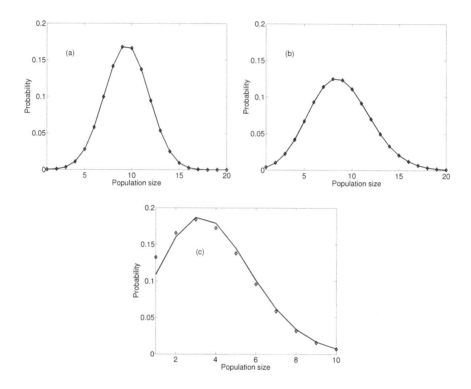

FIGURE 3.8: Quasistationary probability distribution, q^* (solid curve), and the approximate quasistationary probability distribution, $\tilde{q}^*$ (diamond marks), when $r = 0.015$, $K = 10$, and $N = 20$ in cases (a) and (b). In (c), $r = 0.015$, $K = 5$, $N = 10$, where $b_i = ri$ and $d_i = ri^2/K$.

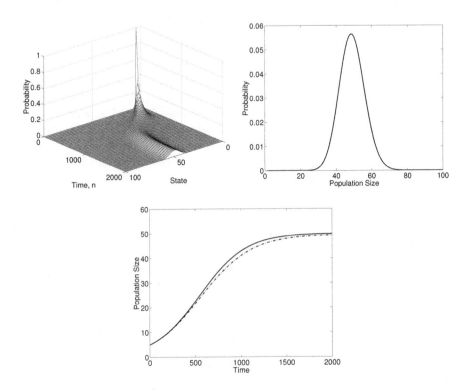

FIGURE 3.9: Stochastic logistic probability distribution for $n = 0, \ldots, 2000$ and the solution of the logistic differential equation are compared for parameter values and initial conditions: $r = 0.004$, $\tilde{r} = 1$, $\Delta t = 0.004$, $K = 50$, $N = 100$, $y(0) = 5 = X_0$ (top two figures). The mean of the stochastic logistic equation (dashed and dotted curve) and the solution of the deterministic model (solid curve) are also compared (bottom figure).

3.9 SIS Epidemic Model

In this section, a stochastic epidemic model is formulated. The model is referred to as an SIS epidemic model because susceptible individuals (S) become infected (I) but do not develop immunity after they recover. They can immediately become infected again, $S \to I \to S$. No latent period is included in the model; therefore, individuals that become infected are also infectious (i.e., they can transmit the infection to others). It is also assumed that there is no vertical transmission of the disease. That is, the disease is not passed from the mother to her offspring. In our simple model, having no vertical transmission means no individuals are born infected; newborns enter the susceptible class. The total population size remains constant for all time since the number of births equals the number of deaths, $S + I = N$. The compartmental diagram in Figure 3.10 illustrates the transitions between the two states, S and I.

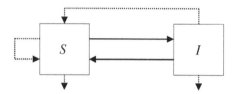

FIGURE 3.10: Compartmental diagram of the SIS epidemic model. A susceptible individual becomes infected with probability $\beta I/N$ and an infected individual recovers with probability γ (solid lines). Birth and death probabilities equal b (dotted lines).

Let the time interval from n to $n+1$ be sufficiently small so that at most one event occurs. Either a susceptible individual becomes infected, a susceptible individual gives birth (and a corresponding death of either a susceptible or infected individual), or an infected individual recovers. A susceptible individual becomes infected with probability $\beta I/N$. The constant β is the number of contacts made by one infected (and infectious) individual that results in infection during the time interval n to $n + 1$; only $\beta S/N$ of these contacts can result in a new infection, and the total number of new infections by the entire class of infected individuals is $\beta SI/N$. Susceptible and infected individuals die or are born with probability b and infected individuals recover with probability γ.

3.9.1 Deterministic Model

First, the dynamics of the deterministic SIS epidemic model are reviewed; then a DTMC SIS epidemic model is formulated and analyzed. Let S_n and I_n denote the number of susceptible individuals and infected and infectious individuals at time n. The dynamics of S_n and I_n, during the time interval Δt, can be modeled by a system of difference equations:

$$S_{n+1} = S_n - \beta S_n I_n / N + I_n(b + \gamma)$$
$$I_{n+1} = \beta S_n I_n / N + I_n(1 - b - \gamma),$$

where $n = 0, 1, 2, \ldots$, $S_0 > 0$, $I_0 > 0$, and $S_0 + I_0 = N$. For example, the number of new susceptible individuals at time $n + 1$ equals those individuals that did not become infected, $S_n[1 - \beta I_n/N]$, plus infected individuals that recovered, γI_n, plus newborns from the infected class, bI_n. The number of newborns from the susceptible class equals the number of susceptible individuals who die, bS_n, because the total population size is assumed to be constant. The parameters are positive and bounded above by one,

$$0 < \beta \leq 1, \quad 0 < b + \gamma \leq 1.$$

It can be seen that $S_n + I_n = N$. Therefore, it is sufficient to consider only the difference equation for I_n. Replacing S_n by $N - I_n$,

$$I_{n+1} = I_n \left(\beta \frac{N - I_n}{N} + 1 - b - \gamma \right)$$
$$= I_n \left(1 + \beta - b - \gamma - \beta \frac{I_n}{N} \right). \tag{3.23}$$

Because of the assumptions on the parameters and the initial conditions, it follows that $0 \leq I_n \leq N$ for all time. There exists two constant solutions $I_n = \bar{I}$ of (3.23). These solutions are known as *equilibrium solutions*, where $I_{n+1} = I_n = \bar{I}$; the solution does not change with time. The equilibria are

$$\bar{I} = 0 \quad \text{and} \quad \bar{I} = N \left(1 - \frac{b + \gamma}{\beta} \right). \tag{3.24}$$

It can be shown for model (3.23) that the dynamics depend on the following parameter $\mathcal{R}_0$, known as the *basic reproduction number*,

$$\mathcal{R}_0 = \frac{\beta}{b + \gamma}$$

(Allen, 1994). The parameter $\mathcal{R}_0$ has an epidemiological interpretation. If the entire population is susceptible and one infected and infectious individual is introduced into the population, then $\mathcal{R}_0$ represents the average number of successful contacts (β) during the period of infectivity ($1/[b + \gamma]$) that will result in a new infectious individual (Anderson and May, 1992). If $\mathcal{R}_0 > 1$, then one infected individual gives rise to more than one new infection, and if $\mathcal{R}_0 < 1$, then one infected individual gives rise to less than one new infection. Note that the second equilibrium in (3.24) is positive iff $\mathcal{R}_0 > 1$. It can be shown that if $\mathcal{R}_0 \le 1$, then $\lim_{n\to\infty} I_n = 0$ and if $\mathcal{R}_0 > 1$, then $\lim_{n\to\infty} I_n = N(1 - 1/\mathcal{R}_0)$, where this limit is the second equilibrium given in (3.24) (Allen, 1994; Allen and Burgin, 2000). The magnitude of $\mathcal{R}_0$ determines whether the epidemic persists in the population, that is, whether it becomes an endemic infection.

3.9.2 Stochastic Model

Let the discrete random variable I_n denote the number of infected and infectious individuals at time n. The state space of the random variable I_n is the set $\{0, 1, 2, \dots, N\}$. A transition matrix P is defined, an expression for the expected duration of the epidemic, τ_k, is derived, and an approximation to the probability of absorption, a_k (probability the epidemic dies out), is given for a large population size N.

Assume that Δt is sufficiently small such that during this time interval (Δt is the time interval $n \to n + 1$), there is at most one change in the random variable I_n. If $I_n = i$, then I_{n+1} may change to either $i + 1$ or $i - 1$ or remain at the state i. The one-step transition probabilities are

$$
\begin{aligned}
p_{i+1,i} &= \mathrm{Prob}\{I_{n+1} = i + 1 | I_n = i\} = \beta i(N - i)/N = \Pi_i \\
p_{i-1,i} &= \mathrm{Prob}\{I_{n+1} = i - 1 | I_n = i\} = (b + \gamma)i \\
p_{ii} &= \mathrm{Prob}\{I_{n+1} = i | I_n = i\} = 1 - \beta i(N - i)/N - (b + \gamma)i \\
&= 1 - \Pi_i - (b + \gamma)i,
\end{aligned}
$$

for $i = 1, 2, \dots, N - 1$ and $p_{ji} = 0$ if $j \ne i - 1, i, i + 1$. Also, $p_{00} = 1$, the zero state is an absorbing state. The transition matrix P has the following form:

$$
\begin{pmatrix}
1 & (b + \gamma) & 0 & \cdots & 0 \\
0 & 1 - \Pi_1 - (b + \gamma) & 2(b + \gamma) & \cdots & 0 \\
0 & \Pi_1 & 1 - \Pi_2 - 2(b + \gamma) & \cdots & 0 \\
\vdots & \vdots & \vdots & \cdots & \vdots \\
0 & 0 & 0 & \cdots & N(b + \gamma) \\
0 & 0 & 0 & \cdots & 1 - N(b + \gamma)
\end{pmatrix},
$$

where $\max_i \{\Pi_i + i(b + \gamma)\} \le 1$.

It can be seen from the transition matrix that there are two classes, $\{0\}$ and $\{1, 2, \dots, N\}$. The zero class is absorbing and the remaining states are all transient. Thus, $\lim_{n\to\infty} P^n p(0) = (1, 0, \dots, 0)^{tr}$. Eventually absorption occurs into the zero state, where there are no infected individuals.

This model is similar to a logistic growth process provided $\mathcal{R}_0 > 1$. Let $b_i = \Pi_i = \beta i(1 - i/N)$ and $d_i = (b + \gamma)i$. Then

$$b_i - d_i = i\left[\beta - (b + \gamma) - \beta i/N\right] = r\Delta t i\left[1 - i/K\right],$$

where $r\Delta t = \beta - (b + \gamma) > 0$ and $K = Nr/\beta = N(1 - 1/\mathcal{R}_0)$. The value of K represents the stable equilibrium of the deterministic model, equation (3.24). If $\mathcal{R}_0 \leq 1$, then $b_i - d_i \leq 0$ and only at $i = 0$ does the birth rate equal the death rate. The transition matrix can be used to calculate the probability distribution $p(n)$ and the submatrix corresponding to the transient states can be used to find the expected duration of an epidemic $\tau^{tr} = \mathbf{1}^{tr}F$, where F is the fundamental matrix.

Example 3.8. Suppose the population size $N = 100$, $\beta = 0.01$, $b = 0.0025 = \gamma$, and $\mathcal{R}_0 = 2$. Figure 3.11 shows the graphs of the probability distribution $p(n)$ when $I_0 = 1$ and the expected duration of the epidemic. ■

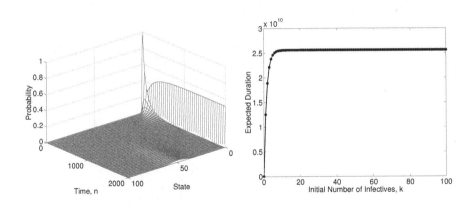

FIGURE 3.11: Probability distribution $p(n)$ for $n = 0, 1, \ldots, 2000$ when $I_0 = 1$ and the expected duration of the epidemic τ_k as a function of the initial number of infected individuals $k = I_0$. Parameters for the SIS epidemic model are $N = 100$, $\beta = 0.01$, $b = 0.0025 = \gamma$, and $\mathcal{R}_0 = 2$.

As illustrated in Figure 3.11, it may take a long time until the epidemic ends, especially if the population size N is large and if the initial number of infectives $I_0 = k$ is large. Instead of an epidemic, the disease becomes endemic. If N is sufficiently large and k is sufficiently small, the SIS model may behave similarly to a semi-infinite random walk model; either there is absorption with probability a_k or the size of the epidemic gets large and stays large for a long period of time (becomes endemic). The probability of absorption, a_k, can be used to estimate the probability that there is no epidemic [i.e., the value of $p_0(n) \approx$ constant as seen in Figure 3.11]. The probability of absorption at $x = 0$ for the semi-infinite random walk model is given by (3.12):

$$(q/p)^k, \text{ if } q < p \text{ and } 1, \text{ if } q \geq p,$$

where q is the probability of moving to the left ($x \to x-1$), p is the probability of moving to the right ($x \to x+1$), and k is the initial position. In the SIS model, the probability of moving to the left is $(b+\gamma)k$ and the probability of moving to the right is $\beta k(N-k)/N$. For N large, $q/p \approx (b+\gamma)/\beta = 1/\mathcal{R}_0$. Therefore, it follows from the semi-infinite random walk model that the probability that the disease does not become established, given initially k infected individuals, is $(q/p)^k \approx (1/\mathcal{R}_0)^k$. Summarizing,

$$p_0(n) \approx (1/\mathcal{R}_0)^k, \text{ if } \mathcal{R}_0 > 1 \text{ and } p_0(n) \approx 1, \text{ if } \mathcal{R}_0 \leq 1$$

(Allen and Burgin, 2000; Jacquez and Simon, 1993). In Figure 3.11, $I_0 = 1 = k$ and $\mathcal{R}_0 = 2$, so that the probability the disease does not persist is approximately $1/\mathcal{R}_0 = 1/2$. We can see in Figure 3.11 that $p_0(n)$ rises rapidly to $1/2$ and stays approximately constant. The increase in $p_0(n)$ after reaching $1/2$ is very slow; the average number of times steps until absorption is on the order of 10^{10} (Figure 3.11). If time is measured in seconds, then the mean time for the epidemic to end is about 800 years implying the quasistationary distribution is a reasonable approximation for the process.

Forward equations for the conditional distribution $q(n)$ can be derived in a manner similar to the stochastic logistic model. Let

$$q_i(n+1) = \frac{p_i(n+1)}{1 - p_0(n+1)}.$$

It follows that

$$q_i(n+1)(1 - (b+\gamma)q_1(n)) = \left(\frac{p_i(n+1)}{1 - p_0(n)}\right).$$

An approximation to the stationary distribution of $q(n)$ can be found by assuming that when there is one infected individual, that individual does not recover or give birth. The approximate quasistationary distribution $\tilde{q}^*(n)$ satisfies (3.22).

Consult the references for further information about stochastic SIS and other stochastic epidemic models (e.g., Allen, 2008; Allen and Burgin, 2000; Bailey, 1975; Bartlett, 1956; Daley and Gani, 1999; Gabriel et al., 1990; Jacquez and Simon, 1993; Nåsell, 1996, 1999, 2001). Models such as an Susceptible-Infected-Removed (SIR) epidemic model, where there are immune or removed individuals and the population size is constant, $S + I + R = N$, require a bivariate Markov process that includes two random variables, $\{(S_n, I_n)\}_{n=0}^{\infty}$. The SIR epidemic models studied in the next section are bivariate Markov processes. They are known as chain binomial epidemic models.

3.10   Chain Binomial Epidemic Models

Let S_n and I_n be discrete random variables for the number of susceptible and infected individuals at time n, respectively. The time interval n to $n+1$ is of length Δt and represents the latent period, the time period until individuals become infectious, $n = 0, 1, 2, \ldots$. The infectious period is contracted to a point. In other words, the number of infected individuals I_n represents *new infected individuals* who were latent during the time interval $n - 1$ to n. These new infected individuals are infectious. They will contact susceptible individuals at time n, who may then become infected at time $n + 1$. The infectious individuals I_n are removed or recovered in the next time interval $n + 1$. There are no births nor deaths; the number of susceptible individuals is nonincreasing over time. The newly infected individuals at time $n + 1$ and the susceptible individuals at time $n + 1$ represent all those individuals who were susceptible at time n:

$$S_{n+1} + I_{n+1} = S_n.$$

The epidemic ends when the number of infected individuals equals zero, $I_n = 0$, because in the next time interval no more individuals can become infected, $I_{n+1} = 0$. Thus, $S_{n+1} = S_n$. Consult Daley and Gani (1999) for further details about this model.

Two models based on the above assumptions are formulated. They are known as the Greenwood and Reed-Frost models, named after the individuals who developed the models. These models were developed in the 1931 and 1928, respectively (Abbey, 1952; Daley and Gani, 1999; Greenwood, 1931). Lowell Reed and Wade Hampton Frost, two medical researchers at John's Hopkins University, developed their model for the purpose of showing medical students the variability in the epidemic process. However, neither Reed nor Frost thought their model was worthy of publication; it was Abbey who published their results in 1952. Primarily, these two models have been applied to small epidemics, or to epidemics within a household, where an initial infected individual spreads the infection to other members of the household (e.g., Bailey, 1975; Daley and Gani, 1999; Gani and Mansouri, 1987). Both models are *bivariate* Markov chain models because they depend on two random variables, the number of susceptible individuals, S_n, and the number of infected individuals, I_n. (However, the Greenwood model will simplify to a univariate process.) The bivariate Markov process is denoted $\{(S, I)_n\}_{n=0}^{\infty}$. The state of the system at time $n + 1$ is determined only by the state of the system at the previous time n. The transition probability $p_{(s,i)_{n+1},(s,i)_n}$ specifies the one-step transition probability, $(s, i)_n \rightarrow (s, i)_{n+1}$. The lower case letters s and i or s_n and i_n represent values of the random variables, S_n and I_n, respectively, at time n.

Let α be the probability of a contact between a susceptible individual and an infected individual and β be the probability that the susceptible individual is infected after contact. Then the probability that a susceptible individual does not become infected is

$$1 - \alpha + \alpha(1 - \beta) = 1 - \alpha\beta = p.$$

The probability p is an important parameter in the Greenwood and Reed-Frost models. Please consult Daley and Gani (1999) and Bailey (1975) for additional information about the mathematical properties of these models and Ackerman, Elveback, and Fox (1984) for a discussion of numerical simulations based on the Reed-Frost model.

3.10.1   Greenwood Model

The Greenwood model assumes that the transition probability $p_{(s,i)_{n+1},(s,i)_n}$ is a binomial probability. The probability of a successful contact resulting in infection is $1-p$, and the probability of a contact not resulting in infection (not successful) is p. At time $n + 1$, if there are s_{n+1} susceptible individuals, s_{n+1} contacts were not successful and $i_{n+1} = s_n - s_{n+1}$ contacts were successful, so that

$$p_{(s,i)_{n+1},(s,i)_n} = \binom{s_n}{s_{n+1}} p^{s_{n+1}}(1 - p)^{s_n - s_{n+1}}. \qquad (3.25)$$

As shown in the preceding expression, the transition probability is independent of I_n. Because the transition probability can be expressed in terms of s_n and s_{n+1}, denote the transition probability as p_{s_{n+1},s_n}. To initiate an epidemic, $I_0 = i_0 > 0$. The state space for S_n and I_n is $\{0, 1, 2, \ldots, s_0\}$, where $S_0 = s_0 > 0$. The maximal number of infected individuals is s_0.

A particular *realization* or *sample path* of the process can be denoted as $\{s_0, s_1, \ldots, s_{t-1}, s_t\}$, where $i_t = 0$, or, alternately, $s_{t-1} = s_t$. The value t is the length of the sample path or the *duration* of the epidemic. Also, the *size* of the epidemic is the number of susceptible individuals who become infected during the epidemic or $s_0 - s_t$. It can be seen from the identity (3.25) that the random variable S_{n+1} has a binomial distribution, $b(S_n, p)$. It is for this reason that the Greenwood model is referred to as a *chain binomial model*. Using the facts that S_{n+1} has a binomial distribution and that $I_{n+1} = S_n - S_{n+1}$, the conditional expectation can be shown to satisfy

$$E(S_{n+1}|S_n = s_n) = ps_n$$

and

$$E(I_{n+1}|S_n = s_n) = s_n - ps_n = (1 - p)s_n.$$

[Recall that the mean of a binomial distribution $b(s_n, p)$ is $\mu = ps_n$.]

A transition matrix for the Greenwood model can be expressed in terms of the initial condition s_0. It is a matrix of size $(s_0 + 1) \times (s_0 + 1)$. The transition matrix is given by

$$
P = \begin{pmatrix}
1 & (1-p) & (1-p)^2 & \cdots & (1-p)^{s_0} \\
0 & p & 2p(1-p) & \cdots & \binom{s_0}{1} p(1-p)^{s_0-1} \\
0 & 0 & p^2 & \cdots & \binom{s_0}{2} p^2(1-p)^{s_0-2} \\
\vdots & \vdots & \vdots & \cdots & \vdots \\
0 & 0 & 0 & \cdots & p^{s_0}
\end{pmatrix}.
$$

The transition matrix does not tell the whole story. Once there has been a transition of the type $s_n \to s_n$, p_{s_n,s_n}, the epidemic ends because $s_n = s_{n+1}$ and $i_n = i_{n+1}$.

Example 3.9. Suppose that, initially, there are three susceptible individuals and one infective. The epidemic with sample path $\{s_0, s_1\} = \{3, 3\}$ has probability $p_{33} = p^3$. The duration of the epidemic in this case is $T = 1$ and the size of the epidemic is $W = s_0 - s_1 = 0$. The epidemic with sample path $\{s_0, s_1, s_2\} = \{3, 1, 1\}$ has probability $p_{13}p_{11} = [3p(1-p)^2]p = 3(1-p)^2p^2$ with duration $T = 2$ and size $W = s_0 - s_2 = 2$. The other sample paths and their corresponding probabilities are given in Table 3.2 (e.g., Daley and Gani, 1999). ∎

Table 3.2: Sample paths, duration, and size for the Greenwood and Reed-Frost models when $s_0 = 3$ and $i_0 = 1$

Sample Path $\{s_0, s_1, \ldots, s_{t-1}, s_t\}$	Duration T	Size W	Greenwood	Reed-Frost
3 3	1	0	p^3	p^3
3 2 2	2	1	$3(1-p)p^4$	$3(1-p)p^4$
3 2 1 1	3	2	$6(1-p)^2p^4$	$6(1-p)^2p^4$
3 1 1	2	2	$3(1-p)^2p^2$	$3(1-p)^2p^3$
3 2 1 0 0	4	3	$6(1-p)^3p^3$	$6(1-p)^3p^3$
3 2 0 0	3	3	$3(1-p)^3p^2$	$3(1-p)^3p^2$
3 1 0 0	3	3	$3(1-p)^3p$	$3(1-p)^3p(1+p)$
3 0 0	2	3	$(1-p)^3$	$(1-p)^3$

Figure 3.12 illustrates four sample paths for the Greenwood model when $s_0 = 6$ and $i_0 = 1$.

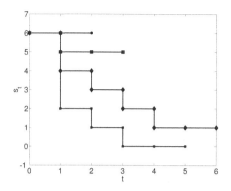

FIGURE 3.12: Four sample paths for the Greenwood chain binomial model when $s_0 = 6$ and $i_0 = 1$, $\{6, 6\}$, $\{6, 5, 5\}$, $\{6, 4, 3, 2, 1, 1\}$, and $\{6, 2, 1, 0, 0\}$.

3.10.2 Reed-Frost Model

In the Reed-Frost model, a susceptible individual at time n will still be susceptible at time $n + 1$ if there is no contact with an infected individual. If the number of infected individuals at time n is i_n, it is assumed that the probability that there is no successful contact of a susceptible individual with any of the i_n infected individuals is p^{i_n}. The Reed-Frost model has the form of the Greenwood model except that p is replaced by p^{i_n}. The transition probabilities in the Reed-Frost model are binomial probabilities:

$$p_{(s,i)_{n+1},(s,i)_n} = \binom{s_n}{s_{n+1}} (p^{i_n})^{s_{n+1}} (1 - p^{i_n})^{s_n - s_{n+1}}.$$

The one-step transition probability depends on i_n, s_n, and s_{n+1}, and, therefore, it cannot be expressed just in terms of the values of s_n and s_{n+1} as was done in the Greenwood model. Recall that $s_n + i_n = s_{n-1}$ or $i_n = s_{n-1} - s_n$. Although the transition probability depends on i_n, s_n and s_{n+1}, for simplicity of notation, denote the transition probability for the Reed-Frost model as p_{s_{n+1}, s_n}. It follows from the form of the transition probability that the random variable S_{n+1} has a binomial distribution, $b(S_n, p^{I_n})$. Hence, the Reed-Frost model is also referred to as a *chain binomial model*. Using the facts that S_{n+1} has a binomial distribution and $I_{n+1} = S_n - S_{n+1}$, it can be shown that the conditional expectation

$$E(S_{n+1}|(S,I)_n = (s_n, i_n)) = s_n p^{i_n}$$

and

$$E(I_{n+1}|(S,I)_n = (s_n, i_n)) = s_n - s_n p^{i_n} = s_n(1 - p^{i_n}).$$

Example 3.10. Suppose that, initially, there are three susceptible individuals and one infected individual, $s_0 = 3$, $i_0 = 1$. In the Reed-Frost epidemic model, the sample path $\{3,3\}$ has probability $p_{33} = p^3$, the sample path $\{3,2,2\}$ has probability $p_{23}p_{22} = [3p^2(1-p)]p^2 = 3(1-p)p^4$, and the sample path $\{3,1,1\}$ has probability $p_{13}p_{11} = [3p(1-p)^2]p^2 = 3(1-p)^2p^3$. This last sample path has a probability that is different from the Greenwood model. In general, if there is more than one infected individual in a time interval, the Greenwood and Reed-Frost models will differ. ∎

3.10.3   Duration and Size

Let T denote the duration of an epidemic and let W denote the size of an epidemic or the total number of susceptible individuals who become infected. For example, for a sample path $\{s_0, s_1, \ldots, s_{t-1}, s_t\}$, $T = t$ and $W = s_0 - s_t$. For a given number of initial susceptible and infected individuals, $s_0 > 0$ and $i_0 > 0$, the maximum value of T is $s_0 + 1$, $T \in \{1, 2, \ldots, s_0 + 1\}$ and the maximum value of W is s_0, $W \in \{0, 1, \ldots, s_0\}$. The epidemic may end in one time step if no one gets infected, $S_1 = s_0$ and $I_1 = 0$ ($T = 1$ and $W = 0$), or it may end after $s_0 + 1$ time steps when one individual gets infected each time step ($T = s_0 + 1$ and $W = s_0$). The variables T and W are random variables whose probability distributions can be computed from the probabilities of the sample paths (Table 3.2).

Example 3.11. The probability distribution corresponding to the duration an epidemic, T, in the Greenwood model for $s_0 = 3$ and $i_0 = 1$ can be computed from Table 3.2:

$$\text{Prob}\{T = 1\} = p^3$$
$$\text{Prob}\{T = 2\} = (1-p)^3 + 3p^2(1-p)^2 + 3p^4(1-p)$$
$$\text{Prob}\{T = 3\} = 3p(1+p)(1-p)^3 + 6p^4(1-p)^2$$
$$\text{Prob}\{T = 4\} = 6p^3(1-p)^3.$$

∎

Another method can be applied to find the probability distributions for T and W in the Greenwood model. This method is described by Daley and Gani (1999) and is briefly presented here. First, partition the transition matrix P of the Greenwood model into two matrices, $P = U + D$, where U is a strictly upper triangular matrix with zeros along the diagonal and D is a diagonal matrix–that is, $D = \text{diag}(1, p, p^2, \ldots, p^{s_0})$ and

$$U = \begin{pmatrix} 0 & (1-p) & (1-p)^2 & \cdots & (1-p)^{s_0} \\ 0 & 0 & 2p(1-p) & \cdots & \binom{s_0}{1}p(1-p)^{s_0-1} \\ 0 & 0 & 0 & \cdots & \binom{s_0}{2}p^2(1-p)^{s_0-2} \\ \vdots & \vdots & \vdots & \cdots & \vdots \\ 0 & 0 & 0 & \cdots & 0 \end{pmatrix}.$$

Note that the matrix U represents those transitions that do not return to the same state in one time step with probability p_{ij} for $i \neq j$, whereas D represents those transitions that do return to the same state in one time step with probability p_{ii}. When $s_t = s_{t-1}$ or $p_{s_t,s_t}^{(n)} > 0$, there is a positive probability that the epidemic ends at time n. The elements of the matrix U^{n-1} represent the probability of transition between states i and j in $n-1$ time steps, $p_{ij}^{(n-1)}$, where $j \to i$, $j \neq i$. Let $p(n) = (p_0(n), p_1(n), \ldots, p_{s_0}(n))^{tr}$, denote the probability distribution for the state of susceptible individuals at time n, then $U^{n-1}p(0)$ represents the probability distribution $p(n-1)$ given that the epidemic has not ended at time $n-1$. Multiplying by D, $DU^{n-1}p(0)$ gives the probability distribution vector that the epidemic has ended exactly at time n. The sum of the elements of the probability distribution vector $DU^{n-1}p(0)$ is the probability that the epidemic has ended at time n, Prob$\{T = n\}$. Let $E = (1,1,1,\ldots,1)$ be a row vector of ones. Then

$$\text{Prob}\{T = n\} = EDU^{n-1}p(0).$$

Since the epidemic could end at states $0, 1, 2, \ldots, s_0$, the p.g.f. for the random variable T, the duration of the epidemic, is given by

$$\sum_{n=1}^{s_0+1} EDU^{n-1}p(0)t^n = \sum_{n=1}^{s_0+1} \text{Prob}\{T = n\}t^n.$$

The computer algebra system Maple^{TM} can be used to calculate the probability distribution T with the matrix formulas above (Appendix for Chapter 3).

In a similar manner, the probability generating function for the random variable W, the size of the epidemic, can be derived (Daley and Gani, 1999). For the Greenwood model, let $U(t)$ be defined as follows:

$$U(t) = \begin{pmatrix} 0 & (1-p)t & [(1-p)t]^2 & \cdots & & [(1-p)t]^{s_0} \\ 0 & 0 & 2p(1-p)t & \cdots & \binom{s_0}{1} & p[(1-p)t]^{s_0-1} \\ 0 & 0 & 0 & \cdots & \binom{s_0}{2} & p^2[(1-p)t]^{s_0-2} \\ \vdots & \vdots & \vdots & \cdots & & \vdots \\ 0 & 0 & 0 & \cdots & & 0 \end{pmatrix}.$$

The matrix U defined previously is then $U \equiv U(1)$. The elements $p_{ij}(t)$ of $U(t)$ are $p_{ij}(t) = p_{ij}t^{j-i}$. The elements of $U(t)$, $p_{ij}(t)$, $i \neq j$, represent generating elements for the probability the epidemic has not ended in one-step transitions. Since the number of susceptible individuals has gone from j to i, the size of the epidemic is $j - i$. In addition, the elements in $U^2(t)$ are $p_{ij}^{(2)}(t) = p_{ij}^{(2)}t^{j-i}$. If, in two time steps, the number of susceptible individuals has gone from j to i, then the size of the epidemic is $j - i$. Thus, $EDU^{n-1}(t)p(0)$ represents the generating function for the size of the epidemic when it has ended in n time steps. Since the epidemic can end in $1, 2, \ldots, s_0 + 1$ time steps, the *probability generating function* for W is

$$\sum_{n=1}^{s_0+1} EDU(t)^{n-1} p(0) = \sum_{k=0}^{s_0} \text{Prob}\{W = k\} t^k.$$

The coefficient of t^k in the expansion of the left-hand side is $\text{Prob}\{W = k\}$ (Daley and Gani, 1999).

Example 3.12. It can be shown using the probability generating function for W (Appendix for Chapter 3) or directly from Table 3.2 that if $s_0 = 3$ and $i_0 = 1$, then

$$\text{Prob}\{W = 0\} = p^3$$
$$\text{Prob}\{W = 1\} = 3p^4(1 - p)$$
$$\text{Prob}\{W = 2\} = 3p^2(1 + 2p^2)(1 - p)^2$$
$$\text{Prob}\{W = 3\} = (1 - p)^3(1 + 3p + 3p^2 + 6p^3)$$

The probability distributions for W and T are graphed in Figure 3.13 in the case $p = 0.5$, $s_0 = 3$, and $i_0 = 1$, $E(W) \approx 2.156$ and $E(T) \approx 2.438$. ■

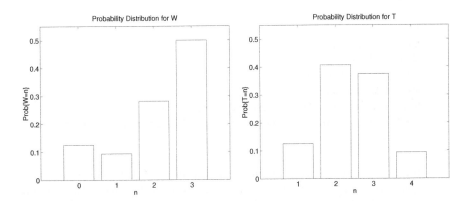

FIGURE 3.13: Probability distributions for the size of the epidemic, W, and the duration of the epidemic, T, when $p = 0.5$, $s_0 = 3$, and $i_0 = 1$.

3.11 Exercises for Chapter 3

1. The infinite Markov chain model for the number of sides of proliferating epithelial cells is approximated by a finite Markov chain model,
   assuming there are a maximum number of sides. Draw a digraph corresponding to the states $\{4, 5, 6, \ldots\}$ and show that state 4 is transient.
   After one cell division, there are no four-sided cells. Use the following
   MATLAB program to find the stationary distribution, assuming a cell
   has a maximum number of sides. Then show that the mean number
   of sides of a proliferating epithelial cell is six, a hexagon (mean of the
   stationary distribution).

```
%Proliferating epithelial cells
%Gibson et al. Nature 2006
%d is matrix dimension of truncated P
%states {4,5,...,d+3}
d=16;
C=abs(pascal(d,1));
E=zeros(d); S=zeros(d);
for i=1:d
    E(i,i)=1/2^(i-1);
end
for i=2:d
    S(i,i-1)=1;
end
M=C'*E;
P=S*M;
[V,D]=eig(P);
G=diag(D);
ival=find(G==max(G));
W=abs(V(:,ival));
W=W/norm(W,1)  %Stationary Distribution
Me=Dot(W,[4:1:d+3]') %Mean of the Stationary Distribution
```

2. Suppose there are N different virus strains. A mutation occurs each
   generation and changes one virus strain into another one with probability $a > 0$. This process can be modeled as a finite Markov chain with
   N states: $p_{ii} = 1 - a$ and $p_{ji} = a/(N-1)$ for $i \neq j$ (Norris, 1997).
   Initially, suppose there is a single virus strain of type i.

   (a) Find the probability that in the nth generation the virus strain is
       type i, that is, the probability $p_{ii}^{(n)}$.

(b) Note that matrix P is doubly stochastic. Find an expression for the stationary distribution. In particular, note that matrix P^n is a symmetric matrix with the property $p_{ii}^{(n)} = p_{11}^{(n)}$ and $p_{ij}^{(n)} = p_{12}^{(n)}$ if $i \neq j$. Write a first-order difference equation for $p_{11}^{(n)}$ and solve this equation with the initial condition $p_{11}^{(0)} = 1$.

3. A simple DTMC model for molecular evolution was described in Chapter 2, Example 2.1, the Jukes-Cantor model for molecular evolution. The transition matrix corresponding to substitution of one the four DNA bases $(A, G, C, \text{or } T)$ per generation is

$$P = \begin{pmatrix} 1 - 3a & a & a & a \\ a & 1 - 3a & a & a \\ a & a & 1 - 3a & a \\ a & a & a & 1 - 3a \end{pmatrix}$$

(Allman and Rhodes, 2004). The probability that one base is substituted for another (a mutation) is $p_{ij} = \text{Prob}\{X_{n+1} = i | X_n = j\} = a$, $0 < a < 1/3$ for $i, j = A, G, C, T$ and $i \neq j$.

(a) Use the method of Exercise 2 to show that

$$p_{ii}^{(n)} = \frac{1}{4} + \frac{3}{4}(1 - 4a)^n.$$

(b) Find the stationary probability distribution.

4. The Jukes-Cantor model for molecular evolution in Exercise 3 assumes all base substitutions are equally likely. Let $p(0)$ be the proportion of the bases $A, G, C,$ and T in a segment of DNA. After n generations, the proportion of the bases in this segment is $p(n)$. In part (a) of Exercise 3, it is shown that the probability that there has been no substitutions in base i in n generations is

$$p_{ii}^{(n)} = \frac{1}{4} + \frac{3}{4}(1 - 4a)^n.$$

Thus, the probability that a base substitution has occurred is

$$q = 1 - p_{ii}^{(n)} = 3/4[1 - (1 - 4a)^n]$$
$$= 3/4[1 - (1 - 4a/3)^n],$$

where $\alpha = 3a$ is the mutation rate (probability of a base substitution per site per generation) (Allman and Rhodes, 2004).

(a) The expected number of base substitutions per site during n generations is called the Jukes-Cantor distance which is defined as

$d_{JC} = n\alpha$. The Jukes-Cantor distance is used to compare ancestral and descendant DNA sequences and to construct phylogenetic trees. Use the equation for q to solve for n, then use the approximation $\ln(1 - 4\alpha/3) \approx -4\alpha/3$, assuming α is small, to obtain the following approximation for the Jukes-Cantor distance:

$$d_{JC} = -\frac{3}{4} \ln\left(1 - \frac{4}{3}q\right).$$

This latter measure does not require the values for the mutation rate α nor the number of generations n.

(b) Suppose two DNA segments consisting of 100 bases are from an ancestor and its descendant. Suppose the ancestral and descendant sequences differ in a total of 5 bases. Compute d_{JC}.

(c) Given the value of d_{JC} in part (b), substitute the formula for q into the right side of d_{JC}, so that it is written in terms of n and α. Use this expression to estimate the mutation rate α if $n = 100$.

5. Another DTMC model for molecular evolution of the four DNA bases A, G, C, and T makes different assumptions regarding base substitutions. The mutation process assumes that $p_{A,G} = p_{12} = a$, $0 < a < 1/3$ and $p_{A,C} = p_{13} = b$, $0 \leq b < 1/3$ (Allman and Rhodes, 2004). The remaining transitions are given in matrix P:

$$P = \begin{pmatrix} p_{11} & a & b & b \\ a & p_{22} & b & b \\ b & b & p_{33} & a \\ b & b & a & p_{44} \end{pmatrix}.$$

This model for molecular evolution is known as the Kimura 2-parameter model.

(a) What are the values of p_{ii}, the probabilities of no mutations.

(b) Let $b > 0$. Compute the limit: $\lim_{n\to\infty} P^n p(0)$.

(c) When $b = 0$, the Basic Limit Theorem for aperiodic Markov chains does not apply. But, in this case, there are two communicating classes that are aperiodic and irreducible for which the theorem does apply. If $b = 0$ and $p(0) = (p_A(0), p_G(0), 0, 0)^T$ find $\lim_{n\to\infty} P^n p(0)$. If $b = 0$ and $p(0) = (0, 0, p_C(0), p_T(0))^T$ find $\lim_{n\to\infty} P^n p(0)$. Are the two limits equal? Explain the differences in the mutation process if $b > 0$ versus $b = 0$? $a > b$ versus $a < b$?

6. Modify the gambler's ruin problem so that the probability of winning is p, losing is q, and a tie is r, $p + q + r = 1$.

(a) Show that the probability of ruin (or absorption at $x = 0$), beginning with a capital of k, is

$$a_k = \frac{(q/p)^N - (q/p)^k}{(q/p)^N - 1}, \quad p \neq q.$$

(b) Show that the expected duration of the games is

$$\tau_k = \frac{1}{(q - p)}\left[k - N\frac{(q/p)^k - 1}{(q/p)^N - 1}\right], \quad p \neq q. \qquad (3.26)$$

Note that this expression is the same as the one for $r = 0$, but for $r > 0$ the values of p and q are smaller, so the expected duration will have a larger value.

(c) Find the probability of ruin a_k and expected duration τ_k when $p = q$.

7. For the semi-infinite random walk model in one dimension, derive the formulas for the probability of absorption, a_k, equation (3.12), and the expected duration until absorption, τ_k, equation (3.13), directly from the random walk model with absorbing boundaries. Let the right-hand endpoint of the domain approach infinity, $N \to \infty$ in (3.4), (3.5), (3.8), and (3.9).

8. Compare the dynamics of the gambler's ruin problem on a bounded domain $\{0, 1, 2, \ldots, N\}$ with a semi-infinite domain $\{0, 1, 2, \ldots\}$. In particular, consider $A_k(t)$ for $N = 10$ and $k = 5$ (Example 3.3). Show that $a_{5,n}$ for $n \leq 13$ has the same value for both domains. In general, verify that $a_{k,n}$ has the same value for both domains provided $n \leq 2(N-1)-k$. Explain why.

9. Consider a restricted random walk in one dimension with an absorbing barrier at $x = 0$ and an elastic barrier at $x = N$. Assume p is the probability of moving to the right and q is the probability of moving to the left, $p + q = 1$ and $p \neq q$. Find the probability of absorption at $x = 0$.

10. Verify the following statements for the gambler's ruin problem.

(a) If $n < k$, then $a_{kn} = 0$, and if $n = k$, then $a_{kk} = q^k$.

(b) The probability $a_{k,k+2i+1} = 0$ for $i = 0, 1, 2, \ldots$.

(c) The probability $a_{k,k+2} = kq^{k+1}p$.

(d) What are the values of b_{kn}, if $n < N - k$? What are the values of $b_{k,N-k}$ and $b_{k,N-k+2i+1}$ for $i = 0, 1, 2, \ldots$?

11. Consider the random walk model on $\{0, 1, 2, \ldots, N\}$. Suppose state N is absorbing, $p_{N,N} = 1$, and state zero is reflecting, $p_{00} = 1 - p$ and $p_{10} = p$. Let b_k be the probability of absorption at $x = N$ beginning at state $x = k$.

(a) Show that $b_N = 1$ and $b_0 = b_1$.

(b) Derive a difference equation for the probability of absorption, b_k; then show that $b_k = 1$ (i.e., the probability of absorption is one).

12. In a simple birth and death process the transition matrix is

$$
P = \begin{pmatrix}
1 & d & 0 & \cdots & 0 \\
0 & 1 - b - d & 2d & \cdots & 0 \\
0 & b & 1 - 2(b + d) & \cdots & 0 \\
0 & 0 & 2b & \cdots & 0 \\
\vdots & \vdots & \vdots & \vdots & \vdots \\
0 & 0 & 0 & \cdots & dN \\
0 & 0 & 0 & \cdots & 1 - dN
\end{pmatrix}.
$$

(See Example 3.5.)

(a) Show that the unique stationary distribution for this process is $\pi = (1, 0, 0, \ldots, 0)^{tr}$.

(b) Assume $N = 20$ and $b = 0.025 = d$. Suppose the initial probability distribution for X_0 is $X_0 = 5$ [i.e., $p_5(0) = 1$]. Find $p(1), p(2), \ldots, p(100)$. Then graph the probability distribution over time, $n = 0, 5, 10, \ldots, 2000$. Modify the MATLAB logistic growth program in the Appendix for Chapter 3.

(c) Graph $p_0(n)$ for $n = 0, 1, 2, \ldots, 2000$. What is $\lim_{n \to \infty} p_0(n)$?

13. The formula for the expected duration of the game, given by equation (3.26), can be checked numerically by performing some numerical simulations. Let $N = 100$, $k = 50$, $q = 0.5$, $r = 0.2$, and $p = 0.3$. Write a computer program for this gambler's ruin problem and simulate a sufficient number of sample paths (total sample paths$\geq$ 1000) to find the time until the game ends. Then compute the mean duration for all of the sample paths. Compare the computer-generated mean duration with formula (3.26).

14. In Exercise 24 of Chapter 2, formulas for the higher-order moments until absorption are derived, $E(T_k^r)$, $r = 2, 3, \ldots$. Let the second-order moment (with initial value k) be denoted as

$$
\tau_k^2 = E(T_k^2).
$$

Then the vector of second moments, $m_2 = (\tau_1^2, \tau_2^2, \ldots, \tau_{N-1}^2)$, can be expressed in terms of the vector of first moments, $m_1 = (\tau_1, \tau_2, \ldots, \tau_{N-1})$, as follows:

$$m_2 = m_1(2F - I),$$

where $F = (I - T)^{-1}$ is the fundamental matrix and I is the identity matrix.

(a) Let $N = 100$, $q = 0.55$, and $p = 0.45$. Write the form of the transient matrix T. Use the preceding formulas to calculate the first and second moments for the duration of the game (use a computer algebra system or write a computer program). See the MATLAB program in the Appendix.

(b) Then use part (a) to calculate the variance for the duration of the game. Plot the mean and standard deviation for the duration of the game for initial capital $k = 0, 1, \ldots, N$. Compare with Figure 3.2.

15. For the gambler's ruin problem, calculate the mean τ_k and standard deviation σ_k of the the duration of the games. Use the p.g.f. $S_k(t) = A_k(t) + B_k(t)$ when $N = 100$, $k = 50$ and for values of $p = 2/5$, $9/20$, and $1/2$.

16. In a simple birth and death process, the birth and death rates are $b_i = bi$ for $i = 1, \ldots, N - 1$, $d_i = di$ for $i = 1, \ldots, N$ and zero elsewhere (Example 3.5 and Exercise 12). The parameters b and d are positive and satisfy $(b + d)N \leq 1$. The mean of the population size at time n, denoted as $\mu(n)$, is

$$\mu(n) = \sum_{i=0}^{N} i p_i(n).$$

(a) Use the transition matrix P to compute $p_i(n + 1)$, $p(n + 1) = Pp(n)$. Then show that $\mu(n)$ is a solution of the following first-order difference equation:

$$\mu(n + 1) = (1 + b - d)\mu(n) - bN p_N(n).$$

(b) Use the fact that $0 \leq \mu(n + 1) \leq (1 + b - d)\mu(n)$ to show

$$\mu(n) \leq (1 + b - d)^n \mu(0).$$

If $b < d$, find $\lim_{n \to \infty} \mu(n)$.

17. For the logistic growth process,

(a) Show that the approximate quasistationary probability distribution, $\tilde{q}^*$, satisfies the relation given in equation (3.22).

(b) Use this relation to compute $\tilde{q}^*$ for $N = 50$ and $r = 0.01$ when $b_i = r(i - i^2/N)$ and $d_i = ri^2/N$ $(K = 25)$ for $i = 1, 2, \ldots, 50$. Graph $\tilde{q}_i^*$ for $i = 1, 2, \ldots, 50$.

18. For the deterministic SIS epidemic model (3.23), verify the following.

(a) If $\mathcal{R}_0 \leq 1$, then $\lim_{n \to \infty} I_n = 0$. First, verify that $I_{n+1} < I_n$.

(b) If $\mathcal{R}_0 > 1$, then $\lim_{n \to \infty} I_n = N(1 - 1/\mathcal{R}_0)$.

19. Consider the stochastic SIS epidemic model with the following parameters.

(a) Let $N = 20$, $\beta = 0.01$, $b = 0.0025 = \gamma$, and $\mathcal{R}_0 = 2$. Calculate the expected duration of an epidemic $\tau_k = E(T_k)$; then sketch a graph of τ_k for $k = 0, 1, \ldots, 20$. A numerical method may be used to solve the linear system.

(b) Let $N = 20$, $\beta = 0.005$, $b = 0.0025 = \gamma$, and $\mathcal{R}_0 = 1$. Calculate the expected duration of an epidemic $\tau_k = E(T_k)$; then sketch a graph of τ_k.

20. Calculate an approximate quasistationary distribution $\tilde{q}^*$ for the stochastic SIS epidemic model when $N = 20$, $\beta = 0.01$, and $b = 0.0025 = \gamma$ (i.e., the solution to $\tilde{P}\tilde{q}^* = \tilde{q}^*$, where $\sum \tilde{q}_i^* = 1$). Sketch $\tilde{q}_i^*$ for $i = 1, 2, \ldots, 20$.

21. In the SIS epidemic model, an estimate was obtained for the probability that the epidemic ends quickly. It was found that the probability of extinction, $p_0(n)$, reaches a plateau or constant value that is less than one prior to ultimate extinction, $p_0(n) \approx \text{constant} = \hat{p}_0 < 1$. In particular,

$$p_0(n) \approx \hat{p}_0 \approx (1/\mathcal{R}_0)^k$$

for $\mathcal{R}_0 > 1$ and $I_0 = k$. This latter estimate was obtained from the formula for the probability of absorption at $x = 0$ in the semi-infinite random walk model, $a_k = (q/p)^k$. This estimate can also be obtained from the product $\Pi_{i=1}^k (d_i/b_i)$. We shall use this latter formula to estimate the probability of population extinction $\hat{p}_0$ in the stochastic logistic model.

(a) Consider the stochastic logistic model with $b_i = ri(1 - i/(2K))$ and $d_i = ri^2/(2K)$, for $i = 0, 1, 2, \ldots, 2K$. Suppose $r = 0.015$ and $K = 20$. Use the formula $\Pi_{i=1}^k (d_i/b_i)$ to estimate the probability of population extinction, $\hat{p}_0$, when $I_0 = 1$ and $I_0 = 2$. Compare these estimates with the values obtained from the probability distribution $p_0(n)$ in Figure 3.14. For $1000 < n \leq 2000$, $p_0(n)$ is approximately constant, $p_0(n) \approx \hat{p}_0$. For $I_0 = 1$, $\hat{p}_0 \approx 0.0256$, and for $I_0 = 2$, $\hat{p}_0 \approx 0.00135$.

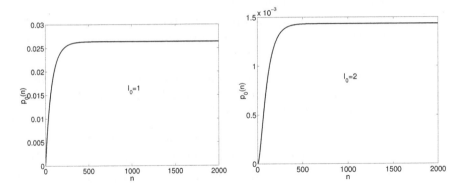

FIGURE 3.14: Probability of population extinction for the stochastic logistic model when $b_i = ri(1 - i/(2K))$ and $d_i = ri^2/(2K)$, for $i = 0, 1, 2, \ldots, 2K$, $r = 0.015$ and $K = 20$.

(b) Use the formula $\Pi_{i=1}^{k}(d_i/b_i)$ to estimate $\hat{p}_0$ when $b_i = ri$ and $d_i = ri^2/K$ for $i = 0, 1, 2, \ldots, N$, $N = 2K$, $b_N = 0$, and $r = 0.005$ and compare the estimates with those in Table 3.3. The values in Table 3.3 are the values of $p_0(n)$ for $2000 < n \le 6000$.

Table 3.3: Estimates of the probability of rapid population extinction for the stochastic logistic model when $N = 2K$, $b_N = 0$, $r = 0.005$, $b_i = ri$, and $d_i = ri^2/K$ for $i = 0, 1, 2, \ldots, N$

K	I_0	$\hat{p}_0$
20	1	0.05
20	2	0.005
30	1	0.033
30	2	0.0022

22. Consider a chain binomial epidemic model with initially one infective and two susceptible individuals, $s_0 = 2$ and $i_0 = 1$.

(a) Calculate the probabilities for the different types of chains in the Reed-Frost and Greenwood chain binomial models and show that both models have the same probabilities.

(b) Find the probability distribution for the duration time, T; that is, Prob$\{T = n\}$, $n = 1, 2, 3$. Graph this distribution for $p = 0.2, 0.5$

and 0.8. Then compute $E(T)$.

(c) Find the probability distribution for the size of the epidemic, W: Prob$\{W = n\}$, $n = 0, 1, 2$. Graph this distribution for $p = 0.2$, 0.5 and 0.8. Then compute $E(W)$.

23. Consider the Reed-Frost chain binomial epidemic model with initially one infective and three susceptible individuals, $s_0 = 3$ and $i_0 = 1$. Use Table 3.2 for the different types of chains.

(a) Find the probability distribution for the duration time, T.

(b) Find the probability distribution for the size of the epidemic, W.

(c) Sketch the probability distributions in (a) and (b) when $p = 0.2$ and $p = 0.8$.

24. Suppose the size of a population remains constant from generation to generation; the size equals N. The dynamics of a particular gene in this population is modeled. Suppose the gene has two alleles, A and a. Therefore, individual genotypes are either AA, Aa, or aa. Let the random variable X_n denote the number of A alleles in the population in the nth generation, $n = 0, 1, 2, \ldots$. Then $X_n \in \{0, 1, 2, \ldots, 2N\}$. Assume random mating of individuals so that the genes in generation $n+1$ are found by sampling with replacement from the genes in generation n (Ewens, 1979). Then the one-step transition probability has a binomial probability distribution with the probability of success $X_n/(2N)$, i.e., if $X_n = i$, then the one-step transition probability is the binomial p.d.f $b(2N, i/2N)$,

$$ p_{ji} = \binom{2N}{j} \left(\frac{i}{2N}\right)^j \left(1 - \frac{i}{2N}\right)^{2N-j}, $$

$i, j = 0, 1, 2, \ldots, 2N$ (Ewens, 1979; Schinazi, 1999). This model is known as the Wright-Fisher model.

(a) Show that states 0 and $2N$ are absorbing and states $\{1, 2, \ldots, 2N-1\}$ are transient.

(b) Given $X_n = k$, show that the mean of X_{n+1} satisfies $\mu_{X_{n+1}} = E(X_{n+1}|X_n = k) = k$. A discrete-time Markov process with the property $E(X_{n+1}|X_n = k) = k$ is called a *martingale*.

(c) Show that the gambler's ruin problem with $1 \leq k \leq N-1$, $p_{k+1,k} = p$ and $p_{k-1,k} = q$ is a martingale iff $p = q$, i.e., $E(X_{n+1}|X_n = k) = k$ iff $p = q$. In game theory, a martingale is a "fair game". On the average, there is no gain nor loss with each game that is played.

(d) It was shown that the probability of winning in the gambler's ruin problem in the case of a fair game, $p = q$, is k/N. Show that the probability of fixation of allele A is $k/2N$, i.e., Prob$\{X_n = 2N|X_0 = k\} = k/2N$.

25. The Moran model, like the Wright-Fisher model, has been applied to population genetics. In the Moran model, every generation, one individual from the population is randomly chosen to give birth (to the same genotype) and another individual is randomly chosen to die (which may be the same one who gives birth). Let X_n denote the number of A alleles in the population in generation n, where $X_n \in \{0, 1, 2, \ldots, 2N\}$. If $X_n = i$, the one-step transition probabilities are

$$p_{i-1,i} = \frac{i(2N - i)}{4N^2}, \quad p_{ii} = \frac{i^2 + (2N - i)^2}{4N^2}, \quad p_{i+1,i} = \frac{i(2N - i)}{4N^2},$$

for $i = 0, 1, \ldots, 2N$ (Ewens, 1979).

(a) Show that states 0 and $2N$ are absorbing and states $\{1, 2, \ldots, 2N - 1\}$ are transient.

(b) Given $X_n = k$, show that the mean of X_{n+1} is $E(X_{n+1}|X_n = k) = k$. Thus, the Moran model is also a martingale.

(c) Show that the probability of fixation of allele A is $k/2N$, i.e., $\text{Prob}\{X_n = 2N|X_0 = k\} = k/2N$.

3.12 References for Chapter 3

Abbey, H. 1952. An examination of the Reed-Frost theory of epidemics. *Hum. Biology.* 24: 201–233.

Ackerman, E., L. R. Elveback, and J. P. Fox. 1984. *Simulation of Infectious Disease Epidemics.* Charles C. Thomas, Springfield, Ill.

Allen, L. J. S. 2008. An introduction to stochastic epidemic models. In: *Mathematical Epidemiology, Lecture Notes in Mathematics.* Brauer, F., P. van den Driessche, and J. Wu (eds.), Vol. 1945, pp. 81–130, Springer-Verlag, New York.

Allen, L. J. S. 1994. Some discrete-time SI, SIR and SIS epidemic models. *Math. Biosci.* 124: 83–105.

Allen, L. J. S. and A. Burgin. 2000. Comparison of deterministic and stochastic SIS and SIR models in discrete time. *Math. Biosci.* 163: 1–33.

Allman, E. S. and J. A. Rhodes. 2004. *Mathematical Models in Biology: An Introduction.* Cambridge Univ. Press, Cambridge, U. K.

Anderson, R. M. and R. M. May. 1992. *Infectious Diseases of Humans: Dynamics and Control.* Oxford Univ. Press, Oxford.

Bailey, N. T. J. 1975. *The Mathematical Theory of Infectious Diseases and Its Applications.* Charles Griffin, London.

Bailey, N. T. J. 1990. *The Elements of Stochastic Processes with Applications to the Natural Sciences.* John Wiley & Sons, New York.

Bartlett, M. S. 1956. Deterministic and stochastic models for recurrent epidemics. *Proc. Third Berkeley Symp. Math. Stat. and Prob.* 4: 81–109.

Daley, D. J. and J. Gani. 1999. *Epidemic Modelling: An Introduction.* Cambridge Studies in Mathematical Biology: 15. Cambridge Univ. Press, Cambridge, U. K.

Elaydi, S. N. 1999. *An Introduction to Difference Equations.* 2nd ed. Springer-Verlag, New York.

Elaydi, S. N. 2000. *Discrete Chaos.* Chapman & Hall/CRC, Boca Raton, Fla.

Ewens, W. J. 1979. *Mathematical Population Genetics.* Springer-Verlag, Berlin, Heidelberg, New York.

Gabriel, J. -P., C. Lefèvre, and P. Picard (eds.) 1990. *Stochastic Processes in Epidemic Theory.* Lecture Notes in Biomathematics, Springer-Verlag, New York.

Gani, J. and H. Mansouri. 1987. Fitting chain binomial models to the common cold. *Math. Scientist* 12: 31–37.

Gibson, M. C., A. B. Patel, R. Nagpal, and N. Perrimon. 2006. The emergence of geometric order in proliferating metazoan epithelia. *Nature* 442: 1038–1041.

Greenwood, M. 1931. On the statistical measure of infectiousness. *J. Hyg.* 31: 336–351.

Jacquez, J. A. and C. P. Simon. 1993. The stochastic SI model with recruitment and deaths I. Comparison with the closed SIS model. *Math. Biosci.* 117: 77–125.

Kimura, M. 1980. A simple method for estimating evolutionary rates of base substitutions through comparative studies of nucleotide sequences. *J. Molec. Evol.* 16: 111–120.

Nåsell, I. 1996. The quasi-stationary distribution of the closed endemic SIS model. *Adv. Appl. Prob.* 28: 895–932.

Nåsell, I. 1999. On the quasi-stationary distribution of the stochastic logistic epidemic. *Math. Biosci.* 156: 21–40.

Nåsell, I. 2001. Extinction and quasi-stationarity in the Verhulst logistic model. *J. Theor. Biol.* 211: 11–27.

Nisbet, R. M. and W. S. C. Gurney. 1982. *Modelling Fluctuating Populations.* John Wiley & Sons, Chichester and New York.

Norris, J. R. 1997. *Markov Chains.* Cambridge Series in Statistical and Probabilistic Mathematics. Cambridge Univ. Press, Cambridge, U. K.

Schinazi, R. B. 1999. *Classical and Spatial Stochastic Processes.* Birkhäuser, Boston.

Taylor, H. M. and S. Karlin. 1998. *An Introduction to Stochastic Modeling.* 3rd ed. Academic Press, New York.

Wade, W. R. 2000. *An Introduction to Analysis.* 2nd ed. Prentice Hall, Upper Saddle River, N. J.

3.13 Appendix for Chapter 3

3.13.1 MATLAB® Programs

The MATLAB programs calculate the expected duration discussed in Section 3.4.2 (Figure 3.2), several sample paths for the gambler's ruin problem, and the probability distribution for logistic growth as in Figure 3.9.

```
% Expected duration for the gambler's ruin problem
clear all
N=100; q=0.55; p=1-q;
L=sparse(2:N-1,1:N-2,p*ones(1,N-2),N-1,N-1); % Subdiagonal.
U=sparse(1:N-2,2:N-1,q*ones(1,N-2),N-1,N-1); % Superdiagonal.
ImT=eye(N-1)-L-U;% I-T
d=ones(1,N-1);
tau=d/ImT; % Expected Duration
plot([1:N-1],tau,'k-','LineWidth',2);
xlabel('Initial capital'); ylabel('Expected duration');
max(tau) % Maximum value of the expected duration.
```

Notes: The command "sparse" is used for greater efficiency.

```
% Monte Carlo simulation for  the gambler's ruin problem.
clear all
sim=1000; q=0.55;
for j=1:sim
    clear r
```

```
    r(1)=50;
    i=1;
    while r(i)>0 & r(i)<100
        y=rand;
        if y<=q
            r(i+1)=r(i)-1;
        else
            r(i+1)=r(i)+1;
        end
    i=i+1;
    end
    t(j)=i; % Time until absorption.
    if j<=3 % Plots three sample paths.
        l1=stairs([0:1:i-1],r)
        set(l1,'LineWidth',2);
        hold on
    end
end
meandur=mean(t) % Mean duration
stdevdur=std(t) % Standard deviation
xlabel('Games'); ylabel('Capital')
hold off

% Probability distribution for  logistic growth.
clear all
time=2000;
K=50; N=2*K; r=0.004;
en=25; % Plot every en time interval.
T=zeros(N+1,N+1); % T is the transition matrix
p=zeros(time+1,N+1);
p(1,6)=1;
v=linspace(0,N,N+1);
b1=r*v.*(1-v/(2*K));
d1=r*v.^ 2/(2*K);
b2=r*v;
d2=r*v.*v/K;
b2(N+1)=0;
for i=2:N
    T(i,i)=1-b1(i)-d1(i);
    T(i,i+1)=d1(i+1);
    T(i+1,i)=b1(i);
end
T(1,1)=1; T(1,2)=d1(2); T(N+1,N+1)=1-d1(N+1);
for t=1:time
    y=T*p(t,:)';
```

```
      p(t+1,:)=y';
end
pm(1,:)=p(1,:);
for t=1:time/en;
      pm(t+1,:)=p(en*t,:);
end
mesh([0:1:N],[0:en:time],pm); % Three dimensional plot.
xlabel('State'); ylabel('Time, n'); zlabel('Probability');
view(140,30)
```

3.13.2 Maple™ Program

The following Maple program can be used to find p.g.f.s for the duration time and size of the epidemic in the Greenwood chain binomial epidemic model.

```
with(linalg):
P:=t->matrix(4,4,[1,(1-p)*t,((1-p)*t)^2,((1-p)*t)^3,0,p,
2*p*(1-p)*t,3*p*((1-p)*t)^2,0,0,p^2,3*p^2*(1-p)*t,0,0,0,p^3]):
Di:=diag(1,p,p^2,p^3):
U:=evalm(P(1)-Di):
E:=vector([1,1,1,1]):
p0:=vector([0,0,0,1]):
T:=t*dotprod(evalm(E&*Di),p0):
for k from 2 to 4 do
 T:=T+factor(dotprod(evalm(E&*Di&*U^(k-1)),p0))*t^k;
od:
T:=sort(T,t):
T1:=coeff(T,t,1);T2:=factor(coeff(T,t,2));
T3:=factor(coeff(T,t,3));T4:=factor(coeff(T,t,4));
```

$$T1 := p^3$$
$$T2 := (1-p)(3p^4 - 3p^3 + 4p^2 - 2p + 1)$$
$$T3 := 3p(2p^3 - p^2 + 1)(1-p)^2$$
$$T4 := 6p^3(1-p)^3$$

```
Ut:=evalm(P(t)-Di):
W:=dotprod(evalm(E&*Di),p0):
for k from 2 to 4 do
 W:=W+factor(dotprod(evalm(E&*Di&*Ut^(k-1)),p0));
od:
W:=sort(W,t):
W0:=coeff(W,t,0);W1:=factor(coeff(W,t,1));
W2:=factor(coeff(W,t,2));W3:=factor(coeff(W,t,3));
```

$$W0 := p^3$$
$$W1 := 3p^4(1 - p)$$
$$W2 := 3p^2(2p^2 + 1)(1 - p)^2$$
$$W3 := (6p^3 + 3p^2 + 3p + 1)(1 - p)^3$$

Chapter 4

Discrete-Time Branching Processes

4.1 Introduction

The study of branching processes has a long history. The subject of branching processes began in 1845 with Irénée-Jules Bienaymé, a probabilist and statistician, and was advanced in the 1870s with the work of Reverend Henry William Watson, a clergyman and mathematician, and Francis Galton, a biometrician (Mode, 1971). These individuals were interested in studying the survival of family names. Galton in 1873 submitted a problem to the *Educational Times* (Mode, 1971) stating the following: Suppose adult males (N in number) in a population each have different surnames. Suppose in each generation, a_0 percent of the adult males have no male children who survive to adulthood; a_1 have one such child; a_2 have two, and so on up to a_5, who have five. Then Galton posed two questions (Mode, 1971):

(1) Find what proportion of the surnames become extinct after r generations.

(2) Find how many instances there are of the same surname being held by m persons.

Galton did not receive satisfactory solutions to his problems and sought help from Watson. Watson rephrased the problems in terms of probability generating functions. Even Galton and Watson did not completely solve the problems, and it wasn't until the 1930s that complete solutions were found. Fisher, Haldane, Erlang, and Steffenson contributed to the solution of these problems (Mode, 1971). Thus, appropriately, these discrete-time processes are known as Galton-Watson branching processes. Schinazi (1999) also notes Bienaymé's contributions and refers to the theory as Bienaymé-Galton-Watson branching processes. Branching processes have been applied to electron multipliers, neutron chain reactions, population growth, cancer growth, and the survival of mutant genes.

In the next section, some notation and preliminary results are given. The main result regarding population extinction is stated in Section 4.4. Branching processes with environmental variation are discussed in Section 4.6. The results are extended to multivariate branching processes, known as multitype branching processes. A well-known application of multitype branching process is described, a discrete age-structured population which is related to the Leslie matrix model. Excellent references to branching processes with applications to biology include the books by Haccou et al. (2005), Harris (1963), Jagers (1975), Kimmel and Axelrod (2002), and Mode (1971). A brief summary of some of the methods and applications of branching processes in ecology can be found in Allen (In press).

4.2   Definitions and Notation

Discrete-time branching processes are DTMCs; the time variable and state space are discrete and the state of the system at time $n + 1$ depends only on the state of the system at time n. Frequently, branching processes are studied separately from Markov chains. One reason for this separate study is the wide variety of applications in branching processes. Another reason is that different techniques from transition matrices are used to analyze their behavior. Multiple births for each individual at each time step make setting up a transition matrix impractical. Techniques that employ probability generating functions are useful in the study of branching processes.

The following assumptions are made in studying Galton-Watson branching processes. Let X_0 denote the total size of the population at the zeroth generation and X_n the size of the population at the nth generation. The process $\{X_n\}_{n=0}^{\infty}$ has state space $\{0, 1, 2, \ldots\}$.

DEFINITION 4.1   *Three basic assumptions define a* Galton-Watson branching process.

(i) *Each individual in generation n gives birth to Y_n offspring in the next generation, where Y_n is a random variable that takes values in $\{0, 1, 2 \ldots\}$ whose offspring distribution is $\{p_k\}_{k=0}^{\infty}$,*

$$\text{Prob}\{Y_n = k\} = p_k, \quad k = 0, 1, 2, \ldots.$$

(ii) *Each individual gives birth independently from all other individuals.*

(iii) *The same offspring distribution applies to all n generations, $Y_n = Y$.*

Figure 4.1 indicates why the process is referred to as a branching process; one sample path is graphed in the case $X_0 = 1$.

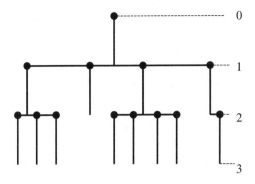

FIGURE 4.1: Sample path of a branching process $\{X_n\}_{n=0}^{\infty}$. In the first generation, four individuals are born, $X_1 = 4$. The four individuals in generation one give birth to three, zero, four, and one individuals, respectively, making a total of eight individuals in generation two, $X_2 = 8$.

If, in any generation n, the population size is zero, $X_n = 0$, then the process stops, $X_{n+k} = 0$ for $k = 1, 2, \ldots$. The zero state is absorbing (i.e., the one-step transition probability $p_{00} = 1$) and therefore, positive recurrent. It is verified later that all of the other states are transient.

4.3 Probability Generating Function of X_n

Some preliminary results are needed before the p.g.f. of the random variable X_n can be defined. Let $\{Y_i\}_{i=1}^{m}$ be a collection of independent and identically distributed (i.i.d.) discrete random variables. Suppose the p.g.f. for Y_i is $f(t)$, $i = 1, 2, \ldots, m$. Then the p.g.f. for $Z = \sum_{i=1}^{m} Y_i$, where m is a fixed number, is

$$
\begin{aligned}
P_Z(t) &= E\left(t^{\sum_{i=1}^{m} Y_i}\right) \\
&= E\left(t^{Y_1} t^{Y_2} \cdots t^{Y_m}\right) \\
&= E\left(t^{Y_1}\right) E\left(t^{Y_2}\right) \cdots E\left(t^{Y_m}\right) \\
&= [f(t)]^m
\end{aligned}
\tag{4.1}
$$

This fact is used to define the p.g.f for X_n which is the sum of a random number (not fixed number) of i.i.d. discrete random variables.

Suppose $X_0 = 1$. Let the p.g.f. of the offspring distribution Y be $f(t) = \sum_{k=0}^{\infty} p_k t^k$. Denote the p.g.f. of X_0 as h_0 and the p.g.f. of X_n as h_n. It follows that $h_0(t) = t$. In the next generation, each individual gives birth to

k individuals with probability p_k. Thus, the p.g.f. of $X_1 = Y$ is

$$h_1(t) = f(t). \tag{4.2}$$

It will be shown that the p.g.f. of $X_n = Y_1 + \cdots + Y_{X_{n-1}}$ is the n-fold composition of f,

$$h_n(t) = f^{n-1}(f(t)) = f(f(\cdots(f(t))\cdots)) = f^n(t), \tag{4.3}$$

where $\{Y_i\}_{i=1}^{\infty}$ is a collection of i.i.d. discrete random variables with p.g.f. $f(t)$.

THEOREM 4.1

Suppose $X_0 = 1$ and $\{Y_i\}_{i=1}^{\infty}$ is a collection of i.i.d. discrete random variables with p.g.f. $f(t) = \sum_{k=0}^{\infty} p_k t^k$. In addition, suppose $X_1 = Y_1$, $X_2 = Y_1 + \cdots + Y_{X_1}$ and, in general, $X_n = Y_1 + \cdots + Y_{X_{n-1}}$. Then the p.g.f. of X_1 is $h_1(t) = f(t)$ and the p.g.f. of X_n is $h_n(t) = f^{n-1}(f(t)) = f^n(t)$.

Proof. It has been shown that $h_0(t) = t$ and $h_1(t) = f(t)$, equation (4.2). Assume that $h_{n-1}(t) = f^{n-1}(t)$. Let $h_{n-1}(t) = \sum_{k=0}^{\infty} r_k t^k$, where $r_k = \text{Prob}\{X_{n-1} = k\}$. The p.g.f. of $X_n = Y_1 + \cdots + Y_{X_{n-1}}$ is

$$
\begin{aligned}
h_n(t) &= E\left(t^{\sum_{i=1}^{X_{n-1}} Y_i}\right) \\
&= \sum_{j=0}^{\infty} t^j \text{Prob}\left\{\sum_{i=1}^{X_{n-1}} Y_i = j\right\} \\
&= \sum_{j=0}^{\infty} t^j \sum_{m=0}^{\infty} \text{Prob}\left\{\sum_{i=1}^{X_{n-1}} Y_i = j | X_{n-1} = m\right\} \text{Prob}\{X_{n-1} = m\} \\
&= \sum_{j=0}^{\infty} t^j \sum_{m=0}^{\infty} r_m \text{Prob}\left\{\sum_{i=1}^{X_{n-1}} Y_i = j | X_{n-1} = m\right\} \\
&= \sum_{m=0}^{\infty} r_m \sum_{j=0}^{\infty} \text{Prob}\left\{\sum_{i=1}^{m} Y_i = j\right\} t^j \tag{4.4} \\
&= \sum_{m=0}^{\infty} r_m [f(t)]^m = h_{n-1}(f(t)) = f^{n-1}(f(t)), \tag{4.5}
\end{aligned}
$$

where in lines (4.4) and (4.5) the summations have been interchanged (assuming absolute convergence of the summations) and the sum is over a fixed number m of random variables $\{Y_i\}_{i=1}^{m}$ which means the p.g.f. is $[f(t)]^m$ [identity (4.1)]. Thus, h_n is just the n-fold composition of f as given in (4.3). $\square$

The derivation of the generating function h_n is based on the fact that $X_0 = 1$. If $X_0 = N$, where N is a positive integer, then $h_0(t) = t^N$ and the process begins with N independent branches (Figure 4.2).

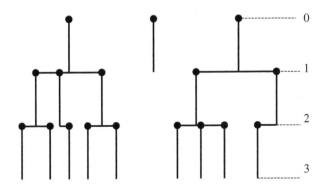

FIGURE 4.2: Sample path of a branching process $\{X_n\}_{n=0}^{\infty}$, where $X_0 = 3$, $X_1 = 5$, and $X_2 = 9$.

When $X_0 = N$, the process may be considered as N independent branching processes, $X_n = \sum_{i=1}^{N} X_{i,n}$, where $\{X_{in}\}_{i=1}^{N}$ are i.i.d. for each n. That is, each of the random variables $X_{i,n}$ has the same p.g.f. $f^n(t)$. Because N is a fixed number, it follows from (4.1) and Theorem 4.1 that the p.g.f. of X_n is

$$h_n(t) = [f^n(t)]^N, \quad \text{if} \quad X_0 = N.$$

Example 4.1. Suppose a branching process $\{X_n\}_{n=0}^{\infty}$ with $X_0 = 1$ has an offspring distribution with $p_0 = 1/4$, $p_1 = 3/4$, and $p_k = 0$, $k = 2, 3, \ldots$. There are either no births or just a single birth. Then the p.g.f. for X_1 is

$$f(t) = \frac{1}{4} + \frac{3}{4}t.$$

The p.g.f for X_2 is

$$f^2(t) = f(f(t)) = \frac{1}{4}\left(1 + \frac{3}{4}\right) + \left(\frac{3}{4}\right)^2 t.$$

In general, the p.g.f. for X_n is

$$f^n(t) = \frac{1}{4}\left(1 + \frac{3}{4} + \cdots + \left(\frac{3}{4}\right)^{n-1}\right) + \left(\frac{3}{4}\right)^n t$$

$$= 1 - \left(\frac{3}{4}\right)^n + \left(\frac{3}{4}\right)^n t.$$

If $p_k(n)$ is the probability that the population size is k in generation n, then

$$p_0(n) = 1 - \left(\frac{3}{4}\right)^n \quad \text{and} \quad p_1(n) = \left(\frac{3}{4}\right)^n.$$

If $X_0 = N$, then the p.g.f. for X_n is

$$[f^n(t)]^N = \binom{N}{0} (p_0(n))^N + \binom{N}{1} (p_0(n))^{N-1} p_1(n) t$$
$$+ \cdots + \binom{N}{N} (p_1(n))^N t^N. \tag{4.6}$$

■

Do not confuse the notation p_k and $p_k(n)$. The notation p_k refers to the probability an individual gives birth to k individuals, and $p_k(n)$ refers to the probability that the total population size is k at generation n. This latter notation is consistent with previous chapters. Note that when $X_0 = 1$, $p_k = p_k(1)$ for $k = 0, 1, 2, \ldots$.

Galton's question (1) can be addressed for this example. After r generations, the probability that all surnames have gone extinct is $[p_0(r)]^N$, the probability that $N - 1$ surnames have gone extinct is $N[p_0(r)]^{N-1} p_1(r)$, and so on; so that the probability that no surnames have gone extinct is $[p_1(r)]^N$. It can then be shown that the expected proportion of surnames that have gone extinct by generation r is $p_0(r)$. In Example 4.1, $p_0(r) = 1 - (3/4)^r$. Notice, for this example,

$$\lim_{r \to \infty} p_0(r) = 1.$$

Galton's question (2) is not addressed in general, but note that in a single branching process, $X_0 = 1$, where adult males are followed, the probability that there are exactly m males with the same surnames in generation r is $p_m(r)$. In Example 4.1, the probability that there are two or more surnames the same in any generation is zero.

In the next section, the probability of population extinction as $n \to \infty$ is studied for the branching process $\{X_n\}_{n=0}^{\infty}$; that is, $\lim_{n \to \infty} \text{Prob}\{X_n = 0\} = \lim_{n \to \infty} p_0(n)$.

4.4 Probability of Population Extinction

Denote the p.g.f. of X_n by

$$h_n(t) = \sum_{k=0}^{\infty} p_k(n) t^k,$$

where the probability the total population size is k in generation n is given by $p_k(n)$. Denote the probability distribution of the process at time zero by $p(0) = (p_0(0), p_1(0), \ldots)^{tr}$ and the probability distribution of the process at time n as $p(n) = (p_0(n), p_1(n), \ldots)^{tr}$. The probability of total population extinction in the nth generation is $p_0(n) = h_n(0)$. If $X_0 = 1$, then $h_n(0) = f^n(0)$, and if $X_0 = N$, then $h_n(0) = [f^n(0)]^N$.

The following assumptions are made regarding the offspring distribution $\{p_k\}_{k=0}^{\infty}$:

$$0 < p_0 \quad \text{and} \quad 0 < p_0 + p_1 < 1. \tag{4.7}$$

Assumptions (4.7) imply that there are positive probabilities of no offspring and of more than one offspring. If $p_0 = 0$ or $p_1 = 1$, then in every generation there is at least one birth and there is no chance of extinction, $p_0(n) = 0$. The probability of ultimate extinction is zero in these cases. In the case that the p.g.f. is linear, $f(t) = p_0 + p_1 t$, $p_0 > 0$, it can be seen from Example 4.1 that the p.g.f. $f^n(t) = 1 - (p_1)^n + p_1^n t^n$, where $p_0(n) = 1 - (p_1)^n$. Thus, $\lim_{n \to \infty} p_0(n) = 1$. The assumptions (4.7) exclude these few cases for which the asymptotic results have already been verified.

The p.g.f. for the offspring distribution is

$$f(t) = \sum_{k=0}^{\infty} p_k t^k. \tag{4.8}$$

Denote the mean number of births as m, where

$$m = f'(1) = \lim_{t \to 1^-} f'(t) = \sum_{k=1}^{\infty} k p_k. \tag{4.9}$$

Assume the p.g.f. has the following five properties:

(1) $f(0) = p_0 > 0$, $f(1) = 1$, and $p_0 + p_1 < 1$.

(2) $f(t)$ is continuous for $t \in [0, 1]$.

(3) $f(t)$ is infinitely differentiable for $t \in [0, 1)$.

(4) $f'(t) = \sum_{k=1}^{\infty} k p_k t^{k-1} > 0$ for $t \in (0, 1]$, where $f'(1)$ is defined by (4.9).

(5) $f''(t) = \sum_{k=2}^{\infty} k(k-1) p_k t^{k-2} > 0$ for $t \in (0, 1)$.

The five properties imply that the function f is continuous, strictly increasing, and its first derivative is strictly increasing (concave upward) on $[0, 1]$. The graph of the p.g.f. $y = f(t)$ may intersect $y = t$ in either one or two points on the interval $[0, 1]$. These points are fixed points of the p.g.f. f. (See Figure 4.3.) Properties (1)–(5) are used to prove two lemmas and the main result concerning ultimate extinction of the branching process $\{X_n\}_{n=0}^{\infty}$.

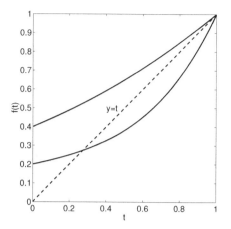

FIGURE 4.3: The p.g.f. $y = f(t)$ intersects $y = t$ in either one or two points on $[0, 1]$.

LEMMA 4.1

Assume the p.g.f. f satisfies properties (1)–(5). Then $m \leq 1$ if and only if $f'(t) < 1$ for $t \in [0, 1)$.

Proof. Since f' is strictly increasing on $[0, 1]$ [property (5)] and $m = f'(1) \leq 1$, it follows that $f'(t) < 1$ for $t \in [0, 1)$. The converse is straightforward. $\square$

LEMMA 4.2

Assume the p.g.f. f satisfies properties (1)–(5). If $m \leq 1$, then f has a unique fixed point at $t = 1$ on the interval $[0, 1]$.

Proof. To show that the fixed point is unique, note that Lemma 4.1 implies $f'(t) < 1$ for $t \in [0, 1)$. Integration from t to 1 yields $1 - f(t) < 1 - t$ or $t < f(t)$ for $t \in [0, 1)$. Thus, the only fixed point of f on $[0, 1]$ is $t = 1$. $\square$

THEOREM 4.2

Assume the p.g.f. f of the branching process $\{X_n\}_{n=0}^{\infty}$ satisfies properties (1)–(5). In addition, assume $X_0 = 1$. If $m \leq 1$, then

$$\lim_{n \to \infty} \text{Prob}\{X_n = 0\} = \lim_{n \to \infty} p_0(n) = 1$$

and if $m > 1$, then there exists a unique q, $0 < q < 1$ such that $f(q) = q$ and

$$\lim_{n \to \infty} \text{Prob}\{X_n = 0\} = \lim_{n \to \infty} p_0(n) = q.$$

If $m \leq 1$, then Theorem 4.2 states that the probability of ultimate extinction is one. If $m > 1$, then there is a positive probability $1 - q$ that the branching process does not become extinct (e.g., a family name does not become extinct, a mutant gene becomes established, a population does not die out). For a proof with the assumption $p_0 + p_1 < 1$ instead of (4.7) (Schinazi, 1999).

The branching process can be divided into three cases: $m > 1$, $m = 1$, and $m < 1$. The case $m > 1$ is referred to as *supercritical*, $m = 1$ is referred to as *critical*, and $m < 1$ is referred to as *subcritical*.

Proof. First, it is shown that $\{p_0(n)\}_{n=1}^{\infty}$ is a monotone increasing sequence:

$$p_0 = p_0(1) < p_0(2) < p_0(3) < \cdots < p_0(n) < \cdots \leq 1.$$

By property (4), f is strictly increasing on [0,1], so that $0 < f(0) = p_0 < 1$ implies $f(0) < f(p_0) < f(1) = 1$. But $p_0 = p_0(1)$ and $f(p_0) = f(f(0)) = p_0(2)$ so that $p_0(1) < p_0(2) < 1$. Assume $p_0(n-1) < p_0(n) < 1$. Then again, since f is strictly increasing and $f(p_0(k)) = p_0(k+1)$, it follows $f(p_0(n-1)) < f(p_0(n)) < f(1)$ or $p_0(n) < p_0(n+1) < 1$. The monotonicity of this sequence can also be deduced logically since $p_0(n)$ is the probability of extinction by the time n that includes the event of extinction at times $1, 2, \ldots, n$.

The sequence $\{p_0(n)\}_{n=1}^{\infty}$ is monotone increasing and bounded above by one. Therefore, it has a limit. Let

$$q = \lim_{n \to \infty} p_0(n).$$

Thus, by the continuity of f,

$$q = \lim_{n \to \infty} f(p_0(n-1)) = f(q).$$

The limit q is a fixed point of f, $f(q) = q$, where $q \leq 1$.

Suppose $m \leq 1$. By Lemma 4.2, the only fixed point of f on $[0, 1]$ is one, so that $q = 1$. The graph of f lies above $y = t$. See Figure 4.3.

Suppose $m > 1$. It is shown that f has only two fixed points on $[0, 1]$, q and 1, where $0 < q < 1$. Since f' is strictly increasing and continuous on $[0, 1]$, there exists $0 < r < 1$ such that if $r < s < 1$, then $1 < f'(s) < f'(1) = m$. Integration of $f'(t)$ from s to 1 yields $1 - f(s) > 1 - s$ or $s > f(s)$ for $r < s < 1$ [the graph of $y = f(t)$ lies below $y = t$ for $t \in (r, 1)$]. See Figure 4.3.

Let $s \in (r, 1)$. Consider the function $g(t) = f(t) - t$. Then $g(0) = p_0 > 0$ and $g(s) = f(s) - s < 0$. By the intermediate value theorem, there exists a $q \in (0, s)$ such that $g(q) = 0$ or $f(q) = q$. Now, it is shown that q is the unique fixed point on (0,1). There can be no fixed point on the interval $(r, 1)$ since $f(t) < t$ for $t \in (r, 1)$. Suppose there exists another fixed point $u \in (0, 1)$. Either $u \in (0, q)$ or $u \in (q, 1)$. In either case, $g(q) = 0$, $g(u) = 0$ and $g(1) = 0$. By Rolle's theorem, there exist numbers u_1 and u_2 such that $0 < u < u_1 < q < u_2 < 1$ if $u \in (0, q)$ or $q < u_1 < u < u_2 < 1$ if $u \in (q, 1)$ such that $g'(u_1) = 0 = g'(u_2)$. Then $f'(u_1) = 1 = f'(u_2)$. This is a contradiction because f' is strictly increasing on $(0,1)$. Thus, f has only two fixed points on $[0, 1]$–namely, q and 1.

Next, it is shown that $\lim_{n\to\infty} p_0(n) = q < 1$. Suppose

$$\lim_{n\to\infty} p_0(n) = 1.$$

Then, for sufficiently large n, $p_0(n) > r$. But on the interval $(r, 1)$, the graph of $f(t)$ lies below the line $y = t$ so that

$$p_0(n) > f(p_0(n)) = p_0(n + 1).$$

This contradicts the fact that $\{p_0(n)\}_{n=0}^\infty$ is an increasing sequence. Hence, $\lim_{n\to\infty} p_0(n) = q < 1$. $\qquad\square$

Although the special case where the p.g.f. is linear, $f(t) = p_0 + p_1 t$, $p_0 > 0$, was verified separately, the results of Theorem 4.2 can be applied. In this case, $m = p_1 \leq 1$ so that $\lim_{n\to\infty} \text{Prob}\{X_n = 0\} = 1$.

Example 4.2. Suppose the offspring probabilities are

$$p_0 = 1/5, \quad p_1 = 1/2, \quad p_2 = 3/10,$$

and $p_k = 0$ for $k = 3, 4, \ldots$. Then $m = 1/2 + 2(3/10) = 11/10 > 1$, so that the probability of ultimate extinction is the fixed point of $f(t) = 1/5 + t/2 + 3t^2/10$ on $(0,1)$. The solutions to $f(t) = t$ are $t = 1$ and $t = 2/3$:

$$f(t) - t = \frac{1}{5} - \frac{1}{2}t + \frac{3}{10}t^2 = \frac{1}{10}(3t - 2)(t - 1) = 0.$$

The probability of ultimate extinction is $2/3$. $\qquad\blacksquare$

Example 4.3. (Schinazi, 1999) Alfred Lotka in 1931 assumed a zero-modified geometric distribution to fit the offspring of the 1920s American male population. It was found that the number of sons a male has in his lifetime closely fit the following probabilities:

$$p_0 = 1/2 \quad \text{and} \quad p_k = \left(\frac{3}{5}\right)^{k-1}\frac{1}{5} \quad \text{for} \quad k = 1, 2, \ldots.$$

The p.g.f. of the offspring distribution is

$$f(t) = \frac{1}{2} + \frac{1}{5}\sum_{k=1}^\infty \left(\frac{3}{5}\right)^{k-1} t^k = \frac{1}{2} + \frac{1}{5}\left(\frac{t}{1 - 3t/5}\right)$$

with mean number of offspring,

$$m = f'(1) = \frac{1/5}{(1 - 3/5)^2} = \frac{5}{4} > 1.$$

The fixed points of f are found by solving

$$\frac{1}{2} + \frac{t}{5 - 3t} = t \quad \text{or} \quad 6t^2 - 11t + 5 = 0.$$

This latter equation factors into $(6t - 5)(t - 1) = 0$, so that $q = 5/6$. A male has a probability of 5/6 that his line of descent becomes extinct and a probability of 1/6 that his descendants will continue forever. ∎

The zero-modified geometric distribution is one of the few offspring distributions for which an explicit formula can be computed for the p.g.f. of the corresponding branching process $\{X_n\}_{n=0}^{\infty}$. The p.g.f. for the offspring distribution is

$$f(t) = p_0 + \frac{bt}{1 - pt},$$

where $p_j = bp^{j-1}$, $j = 1, 2, \ldots$. The mean number of offspring is

$$m = f'(1) = \frac{b}{(1 - p)^2}.$$

There are two fixed points of f, q and 1, when $m \neq 1$ and only one fixed point when $m = 1$. However, the fixed point $q > 1$ when $m < 1$ and $0 < q < 1$ when $m > 1$. It can be shown that the n-fold composition of f is given by

$$f^n(t) = \begin{cases} \dfrac{(m^n q - 1)t + q(1 - m^n)}{(m^n - 1)t + q - m^n}, & \text{if } m \neq 1 \\[2mm] \dfrac{[(n+1)p - 1]t - np}{npt - (n-1)p - 1}, & \text{if } m = 1 \end{cases}$$

(Exercise 8). Thus, in Lotka's model, Example 4.3, where $p_0 = 1/2$, $b = 1/5$, and $p = 3/5$, the probability of extinction in the nth generation if $X_0 = 1$ is

$$f^n(0) = \frac{q(m^n - 1)}{m^n - q} = \frac{(5/6)((5/4)^n - 1)}{(5/4)^n - 5/6}.$$

See Figure 4.4.

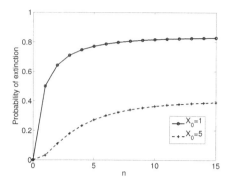

FIGURE 4.4: Probability of extinction in Lotka's model in generation n, if $X_0 = 1$ ($q \approx 0.83$) and $X_0 = 5$ ($q^5 \approx 0.40$).

Theorem 4.2 can be extended to the case $X_0 = N > 1$. Recall that the p.g.f. of X_n in this case is $[f^n(t)]^N$. Thus, the probability of extinction at the nth generation is $[f^n(0)]^N = [p_0(n)]^N$. The result is stated in the following corollary.

COROLLARY 4.1
Assume the p.g.f. f of the branching process $\{X_n\}_{n=0}^{\infty}$ satisfies properties (1)–(5). In addition, assume $X_0 = N$. If $m \leq 1$, then

$$\lim_{n \to \infty} \text{Prob}\{X_n = 0\} = \lim_{n \to \infty} [p_0(n)]^N = 1.$$

If $m > 1$, then

$$\lim_{n \to \infty} \text{Prob}\{X_n = 0\} = \lim_{n \to \infty} [p_0(n)]^N = q^N < 1.$$

The corollary also holds in the case of a linear p.g.f., $f(t) = p_0 + p_1 t$, $p_0 > 0$, because as shown in Example 4.1, $[f^n(0)]^N = [1 - p_1^n]^N$. The mean $m = p_1 \leq 1$ and $0 < p_1$, $\lim_{n \to \infty} \text{Prob}\{X_n = 0\} = \lim_{n \to \infty} [p_0(n)]^N = 1$. In the two special cases, $p_0 = 0$ or $p_0 = 1$, it is impossible for ultimate extinction to occur, $\lim_{n \to \infty} \text{Prob}\{X_n = 0\} = 0$.

The zero state of the branching process is positive recurrent. Next, it is shown that the remaining states are transient. In addition, it is shown that either the total population size X_n approaches zero or infinity (Harris, 1963).

THEOREM 4.3
Assume the p.g.f. f of the branching process $\{X_n\}_{n=0}^{\infty}$ satisfies properties (1)–(5). In addition, assume $X_0 = 1$. Then the states $1, 2, \ldots$, are transient.

In addition, if the mean $m > 1$, then $\lim_{n \to \infty} \text{Prob}\{X_n = 0\} = q$, where $0 < q < 1$ is the unique fixed point of the p.g.f., $f(q) = q$, and $\text{Prob}\{\lim_{n \to \infty} X_n = \infty\} = 1 - q$.

Proof. Consider the first return to state k, where $k \in \{1, 2, \ldots\}$. The process begins in state $k = 1$, $X_0 = 1$. Let m be the first time such that $X_m = k$ for $k \neq 1$. If there exists no such time m, then state k is automatically transient. On the other hand, if there exists such a time m, then extend the definition of first return to state k for $k \neq 1$ as follows. Define the first return to state k, $k \neq 1$, at the nth generation as

$$f_{kk}^{(n)} = \text{Prob}\{X_{m+n} = k, X_{m+j} \neq k, j = 1, 2, \ldots, n - 1 | X_m = k\},$$

where $f_{kk}^{(0)} = 0$. Then define

$$f_{kk} = \sum_{n=0}^{\infty} f_{kk}^{(n)}.$$

Thus, the first return probability is defined for all states $k = 1, 2, \ldots$. Recall that a state k is transient iff $f_{kk} < 1$.

Let p_{0k} be the probability that beginning in state k, the process is in state 0 in the next generation; that is,

$$p_{0k} = \text{Prob}\{X_{m+1} = 0 | X_m = k\}.$$

Since zero is an absorbing state, $X_n = 0$ for $n \geq m + 1$; the process cannot leave the zero state. Therefore, there is a positive probability of at least p_{0k} that the process never returns to state k. Hence,

$$f_{kk} \leq 1 - p_{0k} < 1.$$

Therefore, every state k is transient, where $k \in \{1, 2, \ldots\}$.

For a transient state k, the n step transition probability $p_{kj}^{(n)} \to 0$ as $n \to \infty$, for any state j; that is,

$$\lim_{n \to \infty} \text{Prob}\{X_n = k\} = 0$$

(Lemma 2.2). Because X_n cannot approach any finite state k as $n \to \infty$, either X_n approaches 0 or X_n approaches infinity. From Theorem 4.2, it follows that $\lim_{n \to \infty} \text{Prob}\{X_n = 0\} = q$. If $m \leq 1$, then $q = 1$ and if $m > 1$, then $0 < q < 1$. $\square$

If $m > 1$, then, as $n \to \infty$, the population size approaches zero with probability q and approaches infinity with probability $1 - q$. Theorem 4.3 also holds in the case $p_0 = 0$. But in this case, $q = 0$; the process approaches infinity with probability one.

Example 4.4. (Bailey, 1990) Suppose the population size is very large. A new mutant gene appears in N individuals of the population; the remaining individuals in the population do not carry the mutant gene. All individuals reproduce according to a branching process. Suppose the mean number of individuals produced by those with a mutant gene is m. If $m \leq 1$, the mutant gene will not persist in the population. Instead, suppose a normal individual has a mean reproductive potential of one, but the mean for individuals with a mutant gene is slightly larger than 1,

$$m = 1 + \epsilon, \quad \epsilon > 0.$$

There is still a probability q that the subpopulation with the mutant gene will become extinct. The value of q can be approximated from the mean and variance of the offspring distribution without knowing the p.g.f. f. Let $q = e^\theta$. Then $e^\theta = f(e^\theta) = M(\theta)$. Also,

$$\theta = \ln M(\theta) = K(\theta)$$
$$= 0 + m\theta + \sigma^2 \frac{\theta^2}{2!} + \cdots$$

because $K(0) = 0$, $K'(0) = m$, and $K''(0) = \sigma^2$. Now, $q = e^\theta \approx 1$ and θ is small and negative. Thus, the preceding Maclaurin series can be truncated:

$$\theta \approx m\theta + \sigma^2 \frac{\theta^2}{2}.$$

Solving for θ, $\theta \approx -\frac{2}{\sigma^2}\epsilon$, leads to the following approximation for q: $q \approx e^{-\frac{2}{\sigma^2}\epsilon}$. For an initial size of N mutants, the chance of extinction is q^N. In the case of a Poisson distribution, $m = \sigma^2 = 1 + \epsilon$ with $\epsilon = 0.01$,

$$q^N \approx e^{-\frac{2N}{1.01}(0.01)} = e^{-N0.01980\ldots} \approx (0.980)^N.$$

The probability that the mutant gene becomes established in the population is $1 - q^N$. See Table 4.1. ∎

Table 4.1: Approximations to the probability that a mutant gene becomes extinct, q^N, or becomes established, $1 - q^N$, with initally N mutant genes. Offspring distribution is Poisson with $m = 1.01 = \sigma^2$

N	q^N	$1 - q^N$
1	0.980	0.020
100	0.138	0.862
200	0.019	0.981
300	0.003	0.997

4.5   Mean and Variance of X_n

Generating functions can be used to find the mean and variance of X_n, the random variable for the total population size in generation n. Recall from Chapter 1 the properties of the p.g.f., $f(t)$, moment generating function (m.g.f.), $M(t) = f(e^t)$, and cumulant generating function (c.g.f.), $K(t) = \ln M(t)$. These functions satisfy

$$f(1) = 1, \quad f'(1) = m = E(X), \quad f''(1) = E(X(X-1)).$$

$$M(0) = 1, \quad M'(0) = m, \quad M''(0) = E(X^2).$$

$$K(0) = 0, \quad K'(0) = m, \quad K''(0) = \sigma^2 = E[(X-m)^2].$$

Denote the mean and variance of X_n as m_n and σ_n^2, respectively, and the three generating functions associated with X_n as $f^n(t)$, $M_n(t)$, and $K_n(t)$, respectively. In the first generation, the random variable X_1 has mean $m_1 = m$ and variance $\sigma_1^2 = \sigma^2$. That is,

$$m_1 = m = \sum_{k=1}^{\infty} kp_k \quad \text{and} \quad \sigma_1^2 = \sigma^2 = \sum_{k=1}^{\infty} k^2 p_k - m^2.$$

The three generating functions of X_1 are $f^1 = f$, $M_1 = M$, and $K_1 = K$, respectively. The following theorem gives the mean and variance of the branching process when $X_0 = 1$.

THEOREM 4.4
Let $\{X_n\}_{n=0}^{\infty}$ be a branching process. Assume $X_0 = 1$. The mean of the random variable X_n is

$$m_n = E(X_n) = m^n$$

and the variance is

$$\sigma_n^2 = E[(X_n - m_n)^2] = \begin{cases} \dfrac{m^{n-1}(m^n - 1)}{m-1}\sigma^2, & m \neq 1 \\ n\sigma^2, & m = 1. \end{cases}$$

Before the proof is given, recall some properties of the three generating functions.

$$M_n(t) = f^n(e^t) = f^{n-1}(f(e^t)) = f^{n-1}(M(t))$$
$$= f^{n-1}(e^{\ln M(t)}) = M_{n-1}(\ln M(t)).$$

Thus, $M_n(t) = M_{n-1}(K(t))$. Taking natural logarithms of this latter identity, leads to

$$K_n(t) = K_{n-1}(K(t)).$$

The first and second derivatives of the preceding identity yield two relationships that are used to verify Theorem 4.4:

$$K'_n(t) = K'_{n-1}(K(t))K'(t) \tag{4.10}$$
$$K''_n(t) = K''_{n-1}(K(t))[K'(t)]^2 + K'_{n-1}(K(t))K''(t). \tag{4.11}$$

Proof. (Proof of Theorem 4.4) The proof follows Bailey (1990). First, identity (4.10) is evaluated at $t = 0$,

$$K'_n(0) = K'_{n-1}(K(0))K'(0),$$
$$m_n = m_{n-1}m$$

because $K_n(0) = 0$, $K'_n(0) = m_n$, and $m_1 = m$. The equation $m_n - mm_{n-1} = 0$ is a first-order, homogeneous, constant coefficient, difference equation in m_n. The solution is

$$m_n = m^n.$$

Second, identity (4.11) is evaluated at $t = 0$,

$$K''_n(0) = K''_{n-1}(K(0))[K'(0)]^2 + K'_{n-1}(K(0))K''(0)$$
$$\sigma_n^2 = \sigma_{n-1}^2 m^2 + m_{n-1}\sigma^2.$$

Substituting $m_{n-1} = m^{n-1}$, then $\sigma_n^2 - m^2\sigma_{n-1}^2 = m^{n-1}\sigma^2$ is a first-order, nonhomogeneous, constant coefficient, difference equation in σ_n^2. The general solution to this difference equation is a sum of the general solution to the homogeneous equation and a particular solution. The general solution to the homogeneous equation is cm^{2n}. Assume the particular solution has the form $\sigma_n^2 = km^{n-1}$, $m \neq 1$. Substituting this value into the difference equation yields

$$m^{n-1}[k - km - \sigma^2] = 0$$

which leads to $k = \sigma^2/(1 - m)$ provided $m \neq 1$. The general solution to the nonhomogeneous difference equation is

$$\sigma_n^2 = cm^{2n} + \frac{\sigma^2 m^{n-1}}{1 - m}, \quad m \neq 1.$$

The constant c is found by setting $\sigma_1^2 = \sigma^2$. Then $c = \sigma^2/[m(m - 1)]$. The solution is

$$\sigma_n^2 = \frac{m^{n-1}(m^n - 1)}{m - 1}\sigma^2, \quad m \neq 1.$$

In the case $m = 1$, the particular solution has the form kn. Substitution of this solution into the difference equation yields

$$kn - k(n - 1) = \sigma^2$$

or $k = \sigma^2$. The general solution to the nonhomogeneous difference equation is $\sigma_n^2 = c + n\sigma^2$. Application of $\sigma_1^2 = \sigma^2$ yields $c = 0$. Thus, the solution to the difference equation is

$$\sigma_n^2 = n\sigma^2, \quad m = 1.$$

The proof is complete. □

In the trivial case $m = 0$, $p_0 = 1$ and $\sigma^2 = 0$. The population becomes extinct in one generation. In general, in the subcritical case, $m < 1$, the mean decays geometrically. In the critical case, $m = 1$, the mean is constant, and in the supercritical case, the mean increases geometrically.

Alternately, the mean of X_n can be derived as follows. The conditional expectation in the case $X_0 = 1$ can be expressed in terms of the mean (Karlin and Taylor, 1975):

$$E(X_{n+1}|X_n) = E\left(\sum_{i=1}^{X_n} Y_i|X_n\right) = E(X_n Y_i|X_n) = X_n E(Y_i) = mX_n,$$

which follows because all of the random variables Y_i are i.i.d. with mean m. Thus,

$$E(X_{n+1}) = E[E(X_{n+1}|X_n)] = mE(X_n).$$

In general, it follows from the Markov property and by induction that

$$E(X_{n+r}|X_n) = m^r X_n. \tag{4.12}$$

Thus, $E(X_n) = m^n E(X_0)$.

Example 4.5. Consider the p.g.f. in Example 4.2, $\mathcal{P}(t) = f(t) = 1/5 + t/2 + 3t^2/10$. The mean $m = f'(1) = 1.1$ and variance $\sigma^2 = f''(1) + f'(1) - [f'(1)]^2 = 0.6 + 1.1 - (1.1)^2 = 0.49$. Therefore, the mean and variance for X_n are

$$m_n = (1.1)^n \quad \text{and} \quad \sigma_n^2 = \left[(1.1)^{2n-1} - (1.1)^{n-1}\right] 4.9.$$

∎

4.6   Environmental Variation

The offspring distribution may change from generation to generation due to environmental variations. Then assumption (iii) in the Galton-Watson branching process no longer holds. Suppose the other two assumptions still hold for the branching process. That is,

(i) Each individual in the population in generation k gives birth to Y_k offspring of the same type in the next generation, where Y_k is a discrete random variable.

(ii) Each individual in the population gives birth independently of all other individuals.

If the random variables for the offspring distribution, $\{Y_k\}_{k=1}^{\infty}$, can be chosen according to a particular distribution, then the branching process may in some cases be treated like a Galton-Watson branching process. For example, suppose the environment varies periodically, a good year followed by a bad year, so that the offspring random variables follow sequentially as Y_1, Y_2, Y_1, Y_2, and so on. The population dynamics can be studied via a Galton-Watson branching process by grouping the two years into one. Alternately, the environment may vary randomly between good and bad years so that a good year occurs with probability $1/2$ and a bad one with probability $1/2$.

In general, if m_n is the expected number of offspring for generation n, conditioning on the previous generation, an expression for the expectation can be obtained,

$$E(X_n) = E[E(X_n|X_{n-1})] = m_{n-1}E(X_{n-1}).$$

Applying this same identity to $E(X_{n-1})$ and continuing this process,

$$E(X_n) = m_{n-1}m_{n-2}\cdots m_0 E(X_0) = \prod_{i=0}^{n-1} m_i E(X_0).$$

Suppose the environment changes periodically with period T so that the expected number of offspring in successive generations is m_i, $i = 1, 2, \ldots, T$. Then the expected size of the population in generation nT is

$$E(X_{nT}) = (m_T \cdots m_1)^n E(X_0)$$

and the expected population growth rate in each generation is $(m_T \cdots m_1)^{1/T}$, the geometric mean of $\{m_i\}_{i=1}^{T}$. Because the geometric mean is less than the arithmetic mean,

$$(m_T \cdots m_1)^{1/T} \leq \frac{1}{T} \sum_{i=1}^{T} m_i,$$

the expected population growth rate is less than the average of the growth rates. If $(m_T \cdots m_1)^{1/T} < 1$, then the expected population size will approach zero, extinction.

Suppose the environment varies randomly, so that the random variable for the offspring distribution in generation k is Y_k. Let $m_k = E(Y_k)$ be the expected number of offspring in generation k. Suppose $\{m_k\}_{k=0}^{\infty}$ are i.i.d. random variables with expectation and variance $E(m_k) = \mu$ and $Var(m_k) = \sigma^2$.

The expectation of the growth rates, μ, is not the same as the expected population growth rate. Instead the expected population growth rate is determined by the expectation of the collection $\{\ln m_k\}_{k=0}^{\infty}$. The expected growth rate of the population after n generations is

$$
\begin{aligned}
(m_{n-1}m_{n-2}\cdots m_0)^{1/n} &= e^{(1/n)\ln[m_{n-1}m_{n-2}\cdots m_0]} \\
&= e^{(1/n)[\ln m_{n-1}+\ln m_{n-2}+\cdots \ln m_0]}.
\end{aligned}
$$

By the Law of Large Numbers, a sequence $\{\ln m_i\}_{i=0}^{\infty}$ of i.i.d. random variables with finite expectation $\ln \mu_r$ $(\mu_r \neq \mu)$ satisfies

$$
\lim_{n \to \infty} \frac{1}{n} \sum_{i=0}^{n-1} \ln m_i = \ln \mu_r = E(\ln m_k).
$$

That is,

$$
\lim_{n \to \infty} (m_{n-1}m_{n-2}\cdots m_0)^{1/n} = e^{E(\ln m_k)}.
$$

In a random environment, the branching process is called *subcritical* if $E(\ln m_k) < 0$, *critical* if $E(\ln m_k) = 0$, and *supercritical* if $E(\ln m_k) > 0$. The expected population growth rate in a random environment (stochastic growth rate) is generally less than the expectation of the growth rates,

$$
\mu_r = e^{\ln \mu_r} = e^{E(\ln m_k)} \leq E(e^{\ln m_k}) = E(m_k) = \mu,
$$

a consequence of Jensen's inequality for the convex function e^x. Thus, a population subject to a random environment may not survive (subcritical), even though the expected number of offspring for any generation is greater than one (supercritical), $\mu_r < 1 < \mu$.

Of course, the expected population growth rate is affected by the expectation and the variance of the growth rates, $E(m_k) = \mu$ and $Var(m_k) = \sigma^2$. Expanding $\ln(m_k)$ about μ in a Taylor series leads to

$$
\ln(m_k) = \ln(\mu) + \frac{1}{\mu}(m_k - \mu) - \frac{1}{2\mu^2}(m_k - \mu)^2 + \cdots.
$$

Next, taking the expectation of both sides, yields the following approximation:

$$
E(\ln(m_k)) \approx \ln(\mu) - \frac{Var(m_k)}{2\mu^2} = \ln \mu - \frac{\sigma^2}{2\mu^2} \tag{4.13}
$$

which shows that, in a random environment, the variance negatively impacts the expected population growth rate.

Example 4.6. Suppose in a good year the offspring probability distribution for Y_1 is $\{p_i\}_{i=0}^{3}$, $p_i = 1/4$, $i = 0, 1, 2, 3$ and in a bad year Y_2 is $\{p_0, p_1\}$ with $p_0 = 0.4$ and $p_1 = 0.6$. The mean number of offspring in a good year is $m_1 = 1.5$ and in a bad year $m_2 = 0.6$. In the case that the two years follow sequentially, $Y_1, Y_2, Y_1, Y_2, \ldots$, the expected population growth rate is $\sqrt{m_1 m_2} = 0.949 < 1$. Alternately, suppose a good year occurs with probability $1/2$ and a bad year with probability $1/2$, then $m_k = E(Y_k)$. The mean growth rates for each year $\{m_k\}_{k=0}^{\infty}$ are i.i.d. with mean and variance

$$E(m_k) = \mu = 1.05 \text{ and } Var(m_k) = \sigma^2 = 0.2025.$$

The population growth process is subcritical in both cases since

$$E(\ln(m_k)) = \ln(0.949) = -0.527 < 0.$$

Applying the approximation (4.13) yields

$$\ln \mu - \frac{\sigma^2}{2\mu^2} = -0.0430.$$

For additional applications, please consult some of the references (Benaïm and Schreiber, 2009; Caswell, 2001; Chesson, 1982; Cushing et al., 2003; Ellner, 1984; Tuljapurkar, 1990),

4.7 Multitype Branching Processes

In a multitype Galton-Watson process, it is assumed that each individual behaves independently of any other individual. In the "single-type" Galton-Watson process, each individual gives birth and is replaced by its progeny. Every individual is of the same "type"–that is, gives birth to new individuals, each with the same probability distribution from generation to generation. In a multitype branching process, each individual may give birth to different types of individuals in the population. There is an offspring distribution corresponding to each of these different types of individuals. For example, a population may be divided according to age, size, or developmental stage, and in each generation, individuals age, grow, or develop into another age, size, or stage. Another example from genetics is classification of genes as wild or mutant; mutations change a wild type into a mutant type.

The notation differs slightly from the previous section. Denote a Galton-Watson multitype branching process as $\{X(n)\}_{n=0}^{\infty}$, where $X(n)$ is a vector of random variables,

$$X(n) = (X_1(n), X_2(n), \ldots, X_k(n))^{tr},$$

with k different types of individuals. Here, the subscript i in $X_i(n)$ denotes the ith component of the vector random variable $X(n)$ and n is the time step.

In addition, each random variable $X_i(n)$ has k associated random variables, $\{Y_{1i}, Y_{2i}, \ldots, Y_{ki}\}$ where Y_{ji} is a random variable for the offspring of type j from a parent of type i. For example if $X_i(0) = x$, for each $l = 1, 2, \ldots, x$, there is associated a set $\{Y_{1i}^l, Y_{2i}^l, \ldots, Y_{ki}^l\}$ of random variables, where each random variable Y_{ji}^l has the same offspring distribution as Y_{ji} for $j = 1, \ldots, k$ and they are independent for $l = 1, 2, \ldots, x$. For simplicity, denote this set as $\{Y_{i1}, Y_{2i}, \ldots, Y_{ki}\}$.

Suppose $X_i(0) = 1$. Let $p_i(s_1, s_2, \ldots, s_k)$ denote the offspring probability of an individual of type i. That is, an individual of type i gives birth to s_1 individuals of type 1, s_2 individuals of type 2, $\ldots$, and s_k individuals of type k,

$$p_i(s_1, s_2, \ldots, s_k) = \text{Prob}\{Y_{1i} = s_1, Y_{2i} = s_2, \ldots, Y_{ki} = s_k\},$$

where $s_j = 0, 1, 2, \ldots$ and $i, j = 1, 2, \ldots, k$. The p.g.f. associated with the offspring distribution of $X_i(0) = 1$ is $f_i : [0, 1]^k \to [0, 1]$, defined as follows:

$$f_i(t_1, t_2, \ldots, t_k) = \sum_{s_k=0}^{\infty} \cdots \sum_{s_2=0}^{\infty} \sum_{s_1=0}^{\infty} p_i(s_1, s_2, \ldots, s_k) t_1^{s_1} t_2^{s_2} \cdots t_k^{s_k},$$

where $i = 1, 2, \ldots, k$.

Let δ_i denote a k-vector with the ith component being one and the remaining components zero, $\delta_i = (\delta_{1i}, \delta_{2i}, \ldots, \delta_{ki})^{tr}$, where δ_{ij} is the Kronecker delta symbol. Then $X(0) = \delta_i$ means there is initially one individual of type i in the population. The p.g.f. for $X_i(0)$ given $X(0) = \delta_i$ is $f_i^0(t_1, t_2, \ldots, t_k) = t_i$. Denote the p.g.f. for $X_i(n)$ as $f_i^n(t_1, t_2, \ldots, t_k)$. For example, if $n = 1$, $f_i^1(t_1, t_2, \ldots, t_k) = f_i(t_1, t_2, \ldots, t_k)$. Let

$$F \equiv F(t_1, \ldots, t_k) = (f_1(t_1, \ldots, t_k), \ldots, f_k(t_1, \ldots, t_k))$$

denote the vector of p.g.f.s, $F : [0, 1]^k \to [0, 1]^k$. The function F has a fixed point at $(1, 1, \ldots, 1)$, since for each i, $f_i(1, 1, \ldots, 1) = 1$. Ultimate extinction of the population depends on whether F has another fixed point in $[0, 1]^k$. The following theorem on extinction (Theorem 4.5) is an extension of Theorem 4.2, and as in Theorem 4.2, the probability of extinction depends on the value of the mean. The analogue of the mean for a multitype branching process is defined next.

The mean number of births of a j-type of an individual by an i-type individual is defined. Let m_{ji} denote the expected number of "births" of a type j individual by a type i individual; that is,

$$m_{ji} = E(X_j(1)|X(0) = \delta_i) \text{ for } i, j = 1, 2, \ldots, k.$$

The means m_{ji} can be defined in terms of the p.g.f.s,

$$m_{ji} = \left. \frac{\partial f_i(t_1, \ldots, t_k)}{\partial t_j} \right|_{t_1=1, \ldots, t_k=1}.$$

Define the $k \times k$ expectation matrix as

$$M = \begin{pmatrix} m_{11} & m_{12} & \cdots & m_{1k} \\ m_{21} & m_{22} & \cdots & m_{2k} \\ \vdots & \vdots & \cdots & \vdots \\ m_{k1} & m_{k2} & \cdots & m_{kk} \end{pmatrix}.$$

If matrix M is regular (i.e., some power of M is strictly positive, $M^p > 0$, for some $p > 0$), then M has a simple eigenvalue of maximum modulus (Perron Theorem). Denote this eigenvalue as λ. The main theorem regarding ultimate extinction in a multitype branching process assumes that M is a nonnegative regular matrix. Extinction depends on the magnitude of λ. We state the theorem but do not include a proof. A proof in the two-dimensional case when M is positive can be found in Karlin and Taylor (1975), and for the more general case, see Harris (1963) or Mode (1971).

THEOREM 4.5
Assume the p.g.f. F, where

$$F(t_1, \ldots, t_k) = (f_1(t_1, \ldots, t_k), \ldots, f_k(t_1, \ldots, t_k))$$

is a nonlinear function of the variables $t_1, \ldots, t_k$ and the expectation matrix M is regular with dominant eigenvalue λ. If $\lambda \leq 1$, then

$$\lim_{n \to \infty} \mathrm{Prob}\{X(n) = \mathbf{0} | X(0) = \delta_i\} = 1,$$

$i = 1, 2, \ldots, k$. *If $\lambda > 1$, then there exists unique vector $q = (q_1, q_2, \ldots, q_k)$, $0 < q_i < 1$, $i = 1, 2, \ldots, k$ such that*

$$\lim_{n \to \infty} \mathrm{Prob}\{X(n) = \mathbf{0} | X(0) = \delta_i\} = q_i,$$

where q is the fixed point of the generating functions f_i, $f_i(q_1, q_2, \ldots, q_n) = q_i$, $i = 1, 2 \ldots, k$.

Theorem 4.5 excludes the case that each of the functions f_i are linear,

$$f_i(t_1, \ldots, t_k)) \neq p_i(0, 0, \ldots, 0) + p_i(1, 0, \ldots, 0)t_1 + \cdots + p_i(0, 0, \ldots, 1)t_k$$

for all i. In the case of a single branching process, the p.g.f. is linear when $p_0 + p_1 = 1$. This case was studied separately. In Theorems 4.2 and 4.3 and Corollary 4.1, the case of a linear p.g.f. was also excluded and it was assumed that $0 < p_0 + p_1 < 1$. In the particular case $f_i(0, 0, \ldots, 0) = 0$ for all i, meaning that every individual has at least one offspring, then the fixed point of F is at the origin, $q_i = 0$ for all i (Exercise 15).

Corollary 4.1 can be extended to the multitype branching case. If $X(0) = (r_1, r_2, \ldots, r_k)^{tr}$, $r_i \geq 0$, $i = 1, 2, \ldots, k$, then

$$\lim_{n \to \infty} \text{Prob}\{X(n) = \mathbf{0} | X(0) = (r_1, r_2, \ldots, r_k)^{tr}\} = q_1^{r_1} q_2^{r_2} \cdots q_k^{r_k} = \prod_{i=1}^{k} q_i^{r_i}.$$

For the multitype branching process, the zero state is an absorbing state. It can be shown under the hypotheses of Theorem 4.5 that all other states are transient (Harris, 1963).

Example 4.7. This example shows why it is important for M to be regular. Consider a two-dimensional multitype branching process, where $f_1(t_1, t_2) = t_1 t_2$ and $f_2(t_1, t_2) = 1$. The expectation matrix is

$$M = \begin{pmatrix} 1 & 0 \\ 1 & 0 \end{pmatrix}.$$

Matrix M is not regular. If $X(0) = \delta_1 = (1,0)^{tr}$, then $X(n) = (1,1)^{tr}$, for $n \geq 1$, and if $X(0) = \delta_2 = (0,1)^{tr}$, then $X(n) = (0,0)^{tr}$, for $n \geq 1$. According to the generating functions, an individual of type 1 gives birth to an individual of type 1 and type 2 with probability 1, but an individual of type 2 gives birth to zero individuals of type 1 or type 2. The chain is reducible. Only states $(0,0)^{tr}$ and $(1,1)^{tr}$ can be reached when $X(0) = \delta_i$, $i = 1, 2$. State $(1,1)^{tr}$ is not transient. ∎

Example 4.8. Consider a two-dimensional multitype branching process. Suppose the p.g.f.s are

$$f_1(t_1, t_2) = \frac{1}{4}(1 + t_1 + t_2^2 + t_1^2 t_2) \quad \text{and} \quad f_2(t_1, t_2) = \frac{1}{4}(1 + t_1 + t_2^2 + t_1 t_2^2).$$

For example, an individual of type 1 gives birth to a single individual of the same type or two individuals of type 2 or two individuals of type 1 and one individual of type 2, each with probability $1/4$. The expectation matrix is

$$M = \begin{pmatrix} 3/4 & 1/2 \\ 3/4 & 1 \end{pmatrix}.$$

The expectation matrix is regular with dominant eigenvalue $\lambda = 3/2$ and the p.g.f.s satisfy the hypotheses of the theorem. Since $\lambda > 1$, there exists $q_1, q_2 \in [0, 1)$ such that $f_1(q_1, q_2) = q_1$ and $f_2(q_1, q_2) = q_2$. The fixed point is $q_1 = \sqrt{2} - 1 \approx 0.4142$ and $q_2 = \sqrt{2} - 1 \approx 0.4142$. If $X(0) = (r_1, r_2)^{tr}$, then $\lim_{n \to \infty} \text{Prob}\{X(n) = \mathbf{0}\} \approx (0.4142)^{r_1 + r_2}$. ∎

The identity for the conditional expectation (4.12) can be extended to multitype branching processes (Karlin and Taylor, 1975; Harris, 1963). The conditional expectation is

$$E(X(n+1)|X(n)) = MX(n); \tag{4.14}$$

that is, the expectation of $X(n+1)$ given the value of $X(n)$ is the expectation matrix times $X(n)$. In general,

$$E(X(n + r)|X(n)) = M^r X(n).$$

Thus,

$$E(X_n) = E[E(X_n|X_0)] = M^n E(X_0).$$

See Exercises 11, 12, and 13.

Example 4.9. Consider the multitype branching process in Example 4.8. If $X(0) = (1, 1)^{tr}$, then

$$E(X(1)|X(0)) = \begin{pmatrix} 3/4 & 1/2 \\ 3/4 & 1 \end{pmatrix} \begin{pmatrix} 1 \\ 1 \end{pmatrix} = \begin{pmatrix} 5/4 \\ 7/4 \end{pmatrix}$$

and

$$E(X(2)|X(0)) = \begin{pmatrix} 3/4 & 1/2 \\ 3/4 & 1 \end{pmatrix} \begin{pmatrix} 5/4 \\ 7/4 \end{pmatrix} = \begin{pmatrix} 29/16 \\ 43/16 \end{pmatrix}.$$

∎

4.7.1 An Example: Age-Structured Model

Suppose there are k age classes, $i = 1, 2 \ldots, k$. The first age class, type 1, represents newborns. An individual of age i gives birth to individuals of type 1, then survives, with a given probability, to the next age class becoming an individual of type $i + 1$. Age class k is the oldest age class, and individuals in this class do not survive past age k. Assume an individual of type i at time n either survives to become a type $i+1$ individual at time $n+1$ with probability $p_{i+1,i} > 0$ or dies with probability $1 - p_{i+1,i}$, $i = 1, 2, \ldots, k - 1$. Probability $p_{k+1,k} = 0$ because age k is the oldest age class. In addition, a type i individual gives birth to r individuals of type 1 at time $n + 1$ with probability $b_{i,r}$. The offspring distribution for an individual of type i is $\{b_{i,r}\}_{r=0}^{\infty}$, where

$$b_{i,r} \geq 0, \quad \text{and} \quad \sum_{r=0}^{\infty} b_{i,r} = 1, \quad i = 1, 2, \ldots, k.$$

Denote the mean by

$$b_i = \sum_{r=1}^{\infty} r b_{i,r}.$$

The expectation matrix M can be computed from these probability distributions:

$$M = \begin{pmatrix} b_1 & b_2 & \cdots & b_{k-1} & b_k \\ p_{21} & 0 & \cdots & 0 & 0 \\ 0 & p_{32} & \cdots & 0 & 0 \\ \vdots & \vdots & \ddots & \vdots & \vdots \\ 0 & 0 & \cdots & p_{k,k-1} & 0 \end{pmatrix}. \tag{4.15}$$

The form of matrix M is known as a *Leslie matrix* or a *projection matrix* (Caswell, 2001; Cushing, 1998; Leslie, 1945). The name honors the contributions of Patrick Holt Leslie to demography (Caswell, 2001). In the deterministic Leslie matrix model, the value of each of the age classes at time $n + 1$, $X(n+1)$, is found after multiplication by M; that is, $X(n+1) = MX(n)$. In particular, the first age group $x_1(n + 1)$ consists of offspring from all of the other age groups; that is,

$$x_1(n + 1) = b_1 x_1(n) + b_2 x_2(n) + \cdots + b_k x_k(n) = \sum_{i=1}^{k} b_i x_i(n).$$

The $i + 1$st age group, $i = 1, 2, \ldots, k - 1$, $x_{i+1}(n + 1)$, consists of individuals from age group i who survived and became age $i + 1$:

$$x_{i+1}(n + 1) = p_{i+1,i} x_i(n).$$

In the stochastic model, the conditional expectation satisfies a similar identity, equation (4.14).

The expectation matrix M can be determined directly from the p.g.f.s (Exercise 14). The p.g.f.s are

$$f_i(t_1, t_2, \ldots, t_k) = [p_{i+1,i} t_{i+1} + (1 - p_{i+1,i})] \sum_{r=0}^{\infty} b_{i,r} t_1^r, \quad i = 1, \ldots, k. \quad (4.16)$$

Note that $f_i(1, 1, \ldots, 1) = 1$.

Example 4.10. Consider an age-structured branching process with two ages, whose generating functions are

$$f_1(t_1, t_2) = [(1/2)t_2 + 1/2][1/2 + (1/6)t_1 + (1/6)t_1^2 + (1/6)t_1^3]$$
$$f_2(t_1, t_2) = 1/4 + (1/4)t_1 + (1/4)t_1^2 + (1/4)t_1^3.$$

The offspring probabilities are

$$b_{1,r} = \begin{cases} 1/2, & r = 0 \\ 1/6, & r = 1, 2, 3 \\ 0, & r \neq 0, 1, 2, 3 \end{cases}, \qquad b_{2,r} = \begin{cases} 1/4, & r = 0, 1, 2, 3 \\ 0, & r \neq 0, 1, 2, 3 \end{cases}.$$

The mean numbers of births for each age are

$$b_1 = 1 = \sum_{r=1}^{\infty} r b_{1,r} \quad \text{and} \quad b_2 = 3/2 = \sum_{r=1}^{\infty} r b_{2,r}.$$

The expectation matrix,

$$M = \begin{pmatrix} 1 & 3/2 \\ 1/2 & 0 \end{pmatrix},$$

has a dominant eigenvalue equal to $\lambda = 3/2$. There exists a unique fixed point (q_1, q_2) such that $f_1(q_1, q_2) = q_1$ and $f_2(q_1, q_2) = q_2$,

$$(q_1, q_2) \approx (0.446, 0.433).$$

For example, if there are initially two individuals in stage 1 and three individuals in stage 2, then the probability of ultimate extinction of the total population is approximately $(0.446)^2 (0.433)^3 \approx 0.016$. ∎

A multitype branching process simplifies to a single type branching process when the number of age classes is reduced to one, $k = 1$. In this case, there is one p.g.f. given by

$$f_1(t) = \sum_{r=0}^{\infty} b_{1,r} t^r,$$

where $b_{1,r} = p_r$ is the probability of r births. If the mean number of births $m = f_1'(1) > 1$, then there exists a fixed point $q \in [0, 1)$ of f_1 such that $\lim_{n \to \infty} \text{Prob}\{X_n = 0 | X_0 = 1\} = q$.

4.7.2 Environmental Variation

In the case of a random environment, the expectation matrix M may change from generation to generation in a deterministic fashion or randomly according to some distribution. As shown in the case of a single type Galton-Watson branching process, if the environment varies in a periodic fashion, a good year followed by a bad year, where M_1 is the expectation matrix in a good year and M_2 in a bad year, then

$$E(X(2n)) = (M_2 M_1)^n E(X(0)).$$

The expected population growth rate is $\sqrt{\lambda_r}$, where λ_r is the dominant eigenvalue of $M_2 M_1$. In general, $\lambda_r \neq \lambda_1 \lambda_2$, where λ_1 and λ_2 are the dominant eigenvalues of M_1 and M_2, respectively. Suppose the environment varies randomly so that the expectation matrices $\{M_i\}_{i=1}^{\infty}$ are drawn from a set of i.i.d. regular matrices, where $E(M_k)$ has a dominant eigenvalue of λ. Then there exists a stochastic growth rate, the expected population growth rate λ_r, $\lambda_r \neq \lambda$. The expected total population size,

$$E(N(n)) = \sum_{i=1}^{k} E(X_i(n)) = \|E(X(n))\|_1 = \|M_{n-1} \cdots M_0 E(X(0))\|_1,$$

grows according to this stochastic growth rate,

$$\lim_{n \to \infty} \frac{1}{n} \ln[E(N(n))] = \lim_{n \to \infty} \frac{1}{n} \ln \|M_{n-1} \cdots M_0 E(X(0))\|_1 = \ln \lambda_r.$$

The stochastic growth rate, λ_r, of the population in a random environment is generally less than the expected growth rate, λ. If $\lambda_r < 1$, then the expected population size approaches zero, extinction.

Examples on environmental variation in discrete-time processes can be found in Caswell (2001), Cushing et al. (2003), Haccou et al. (2005), Mangel (2006), and Tuljapurkar (1990). Cushing et al. (2003) formulate discrete-time stochastic models for flour beetles which have three developmental stages, larvae, pupae, and adult, denoted as L, P, and A, respectively. Demographic or environmental variation are added to the population variables at the next time step after the terms are appropriately scaled. The scaling allows the models to be easily fit to time series data. Two examples of models of the form

$$X(n + 1) = M_{n+1}X(n),$$

where matrix M is a random matrix and a periodically varying matrix are discussed in Exercises 20 and 21, respectively. We end this chapter with an example of a size-dependent branching process.

The single-type and multitype branching processes discussed thus far exhibit either exponential growth or decline. Eventually, either the total population size approaches zero or infinity. This is due to the fact that the offspring distribution is constant over time. If the offspring distribution depends on the population size, then the branching process is size-dependent. Suppose population growth is based on the *Ricker model*:

$$x_{n+1} = x_n \exp(r - \gamma x_n), \quad 0 < r, \ 0 < \gamma$$

(May, 1976; Ricker, 1954). The Ricker model has been used frequently in biological applications; it has interesting behavior for values of the parameter r. For example, if $0 < r < 2$, then x_n converges to a stable fixed point r/γ:

$$\lim_{n \to \infty} x_n = \frac{r}{\gamma}.$$

But if $2 < r < 2.526$, then solutions converge to a stable two-cycle; solutions x_n oscillate between two values. For increasing values of r, solutions exhibit what is known as period-doubling behavior (May, 1976; Elaydi, 2000). See Block and Allen (2000) for a size-dependent structured branching process and Allen et al. (2005) for a size-dependent branching process with an Allee effect.

Example 4.11. A size-dependent Ricker branching process formulated by Högnäs (1997) is described. Let X_n be the random variable for the total population size. Let $\{p_k\}_{k=0}^\infty$ denote the size-independent offspring distribution, where the probability an individual produces k offspring is p_k, $\sum_{k=0}^\infty p_k = 1$. In addition, assume the mean is

$$m = \sum_{k=1}^\infty k p_k = e^r > 1.$$

Each individual produces offspring independently of any other individual. For a population of size $x \in \{0, 1, 2, \ldots\}$, let the size-dependent offspring distribution be defined as follows: The probability that an individual produces k

offspring is $p_k \exp(-\gamma x)$ for $k = 1, 2, \ldots$ and the probability that an individual produces no offspring is $1 - \exp(-\gamma x)(1 - p_0)$. Let $\{Y_j\}_{j=1}^{\infty}$ be a set of i.i.d. random variables with the size-dependent offspring distribution. Then if $X_n = x$,

$$X_{n+1} = \begin{cases} \sum_{j=1}^{x} Y_j, & x = 1, 2, \ldots, \\ 0, & x = 0. \end{cases}$$

It can be shown that the following conditional expectation of the branching process has the same behavior as the deterministic model:

$$E(X_{n+1}|X_n = x) = x \exp(r - \gamma x).$$

Numerical simulations comparing the deterministic and stochastic Ricker models are graphed in Figure 4.5 for two cases, $r = 1.2528$ and $r = 2.1972$. ∎

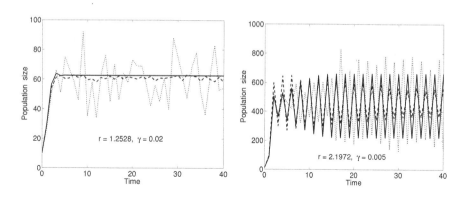

FIGURE 4.5: Solution of the deterministic model (solid curve), one sample path of the size-dependent stochastic Ricker model (dotted curve), and the the mean of 100 sample paths (dashed curve). In the figure on the left, the size-independent offspring distribution is $p_k = 1/6$ for $k = 1, 2, \ldots, 6$ with $r = 1.2528 < 2$ and $\gamma = 0.02$; the deterministic solution and stochastic mean converge to $r/\gamma = 62.64$. In the figure on the right, the size-independent offspring distribution is $p_8 = 0.2$, $p_9 = 0.6$, and $p_{10} = 0.2$ with $r = 2.1972 > 2$ and $\gamma = 0.005$; the deterministic solution and stochastic mean oscillate between two values.

4.8 Exercises for Chapter 4

1. Consider Galton's problem in the context of Example 4.1, where $X_0 = N$ adult males each have different surnames. In each generation, a proportion p_0 of the adult males have no male children who reach adult life and p_1 have one such child, $p_0 + p_1 = 1$. Show that the expected proportion of surnames that has gone extinct in generation r is $p_0(r)$. Notice that the proportion that has gone extinct is either $1, (N-1)/N, (N-2)/N, \ldots, 1/N$, or 0. The probabilities for each of these proportions are given by the coefficients in the expansion of $[f^r(t)]^N$, equation (4.6). See Example 4.1.

2. Suppose a branching process with $X_0 = 1$ has a zero modified geometric offspring distribution, where $p_k = ab^{k-1}$, $k = 1, 2, \ldots$ and

$$p_0 = 1 - \sum_{k=1}^{\infty} p_k, \quad 0 < b < a + b < 1.$$

(a) Show that the p.g.f is

$$f(t) = \frac{1 - (a+b)}{1-b} + \frac{at}{1-bt}.$$

(b) Assume $a > (b-1)^2$. Compute $\lim_{n \to \infty} \text{Prob}\{X_n = 0\}$.

3. The p.g.f. of a branching process with $X_0 = 1$ is $f(t) = at^2 + bt + c$, where $a, b,$ and c are positive and $f(1) = 1$ (Karlin and Taylor, 1975). Assume $f'(1) > 1$. Then show that

$$\lim_{n \to \infty} \text{Prob}\{X_n = 0\} = \frac{c}{a}.$$

4. Suppose the p.g.f. of a branching process is $f(t) = p_0 + p_1 t$, $p_0 > 0$, $p_1 > 0$, and $p_0 + p_1 = 1$.

(a) Show that $f^n(t) = 1 - p_1^n + p_1^n t$. See Example 4.1.

(b) Suppose $X_0 = N$ so that the p.g.f. of X_n is $[f^n(t)]^N$. Let T be the random variable for the first time to extinction; that is, the smallest n such that $X_n = 0$ (i.e., the *first* passage time into the state 0). Then $\text{Prob}\{T = 1\} = (1-p_1)^N$. Use the fact that $\text{Prob}\{T \leq n\} = [f^n(0)]^N$ to show that $\text{Prob}\{T = n\} = (1 - p_1^n)^N - (1 - p_1^{n-1})^N$.

5. Suppose a Galton-Watson branching process with $X_0 = 1$ has a Poisson offspring distribution,

$$p_k = \frac{e^{-\lambda} \lambda^k}{k!}, \quad k = 0, 1, 2, \ldots.$$

(a) Find the p.g.f. f. Then compute the mean and variance of X_1, $m = E(X_1)$, and $\sigma^2 = \text{Var}(X_1)$.

(b) Find the mean and variance of X_n, $m_n = E(X_n)$, and $\sigma_n^2 = \text{Var}(X_n)$.

(c) For $\lambda = 1.5$ and $\lambda = 2$, compute $\lim_{n \to \infty} \text{Prob}\{X_n = 0\}$.

6. Suppose a Galton-Watson branching process $\{X_n\}_{n=0}^{\infty}$ has an offspring p.g.f.

$$f(t) = \frac{0.1}{1 - 0.9t} = p_0 + p_1 t + p_2 t^2 + \cdots = \sum_{k=0}^{\infty} p_k t^k.$$

(a) Find the two probabilities p_0 and p_1 (no children, one child).

(b) Find the mean m and variance σ^2 of the offspring distribution.

(c) If $X_0 = 5$, find $\lim_{n \to \infty} \text{Prob}\{X_n = 0\}$.

7. Suppose $Z_k = \sum_{n=0}^{k} X_n$ for $k = 0, 1, 2, \ldots$ and $Z = \sum_{n=0}^{\infty} X_n$, where $\{X_n\}$ is a branching process with $X_0 = 1$. Suppose the mean m of the offspring distribution satisfies $0 < m < 1$. Show that $E(Z_k) = \sum_{n=0}^{k} m^n$ and $E(Z) = (1 - m)^{-1}$ (Taylor and Karlin, 1998).

8. If the offspring distribution is a zero modified geometric distribution, then the p.g.f. has the form

$$f(t) = p_0 + \frac{bt}{1 - pt},$$

where $p_j = bp^{j-1}$, $j = 1, 2, \ldots$ and $m = f'(1)$. The p.g.f. in the nth generation can be computed in this case (Athreya and Ney, 1972).

(a) Show for any two points u and v that

$$\frac{f(t) - f(u)}{f(t) - f(v)} = \frac{t - u}{t - v} \frac{1 - pv}{1 - pu}.$$

(b) If $m \neq 1$, then f has two distinct fixed points, q and 1. Let $u = q$ and $v = 1$ in the identity in part (a). Then verify the following limit

$$\lim_{t \to 1} \left[\frac{f(t) - q}{t - q} \right] \left[\frac{t - 1}{f(t) - 1} \right] = \frac{1}{m}.$$

Hence, the constant

$$\frac{1 - p}{1 - pq} = \frac{1}{m}.$$

(c) Let $u = q$ and $v = 1$ in the identities in parts (a) and (b). Then

$$\frac{f(t) - q}{f(t) - 1} = \frac{1}{m}\frac{t - q}{t - 1}.$$

Replace t by $f(t)$ and obtain an equation with $f^2(t)$ and $f(t)$. Continue this process to obtain the following identity:

$$\frac{f^n(t) - q}{f^n(t) - 1} = \frac{1}{m^n}\frac{t - q}{t - 1}.$$

Solve for $f^n(t)$ to obtain the p.g.f. of X_n:

$$f^n(t) = \frac{(m^n q - 1)t + q(1 - m^n)}{(m^n - 1)t + q - m^n}, \quad \text{if } m \neq 1.$$

(d) If $m = 1$, use the equation for the p.g.f. $f(t)$ to obtain a formula for the p.g.f. of X_n:

$$f^n(t) = \frac{[(n+1)p - 1]t - np}{npt - (n-1)p - 1}, \quad \text{if } m = 1.$$

9. Suppose a branching process with $X_0 = 1$ has an offspring distribution with mean $m > 0$. Let $Z_n = X_n/m^n$. Show that $E(Z_{n+1}|Z_n = k) = k$ (Taylor and Karlin, 1998).

10. Let I_n be the number of new cases of a disease at time n. Assume $I_0 = k \geq 1$ and $\{I_n\}_{n=0}^{\infty}$ is a Galton-Watson branching process, where the p.g.f. associated with this process is $f(t) = \exp(-\mathcal{R}_0(1 - t))$, a p.g.f. of a Poisson distribution (Antia et al., 2003). The parameter $\mathcal{R}_0$ is generally referred to as the basic reproduction number in disease dynamics, the number of secondary infections caused by one infectious individual in an entirely susceptible population. If $\mathcal{R}_0 = 1.5$ and $k = 5$, find the probability the disease becomes endemic. If $\mathcal{R}_0$ is sufficiently large or if $\mathcal{R}_0 > 1$ and k is sufficiently large, show that the probability the disease becomes endemic is approximately one. Why is this latter result epidemiologically reasonable?

11. Suppose a multitype branching process $X(n) = (X_1(n), X_2(n))^{tr}$ has p.g.f.s

$$f_1(t_1, t_2) = \frac{1}{4}(1 + t_2 + 2t_1^2) \quad \text{and} \quad f_2(t_1, t_2) = \frac{2}{3}\left(\frac{1}{2} + t_1^2\right).$$

(a) Calculate the expectation matrix M and show that M is a regular matrix.

(b) Compute the expectation $E\left(X(n)|X(0) = (3,4)^{tr}\right)$ for $n = 1, 2, 3$.

(c) Compute the following limit: $\lim_{n \to \infty} \text{Prob}\{X(n) = (0,0)^{tr} | X(0) = (3,4)^{tr}\}$.

12. Suppose a multitype branching process $X(n) = (X_1(n), X_2(n), X_3(n))^{tr}$ has p.g.f.s

$$f_1(t_1, t_2, t_3) = \frac{1}{3} + \frac{1}{3}t_2(t_1 + t_3), \quad f_2(t_1, t_2, t_3) = \frac{1}{2} + \frac{1}{2}t_1 t_2 t_3,$$

and

$$f_3(t_1, t_2, t_3) = \frac{1}{4}(1 + t_1^2 + t_2^2 + t_3^2).$$

(a) Find the expectation matrix M and show that M is a regular matrix.

(b) Compute $\lim_{n \to \infty} \text{Prob}\{X(n) = 0 | X(0) = (r_1, r_2, r_3)^{tr}\}$.

(c) If $X(0) = (1, 1, 1)^{tr}$, find the expectation $E(X(n)|X(0))$ for $n = 1, 2, 3$.

13. Suppose a multitype branching process $X(n) = (X_1(n), X_2(n), X_3(n))^{tr}$ has p.g.f.s

$$f_1(t_1, t_2, t_3) = \frac{1}{2}t_1 t_2 t_3 + \frac{1}{2}t_2^2, \quad f_2(t_1, t_2, t_3) = 1, \quad \text{and} \quad f_3(t_1, t_2, t_3) = t_3.$$

(a) Find the expectation matrix M. Is M regular?

(b) If $X(0) = \delta_1 = (1, 0, 0)^{tr}$, find the expectation $E(X(n)|X(0))$ for $n = 1, 2, 3$.

(c) If $X(0) = \delta_2 = (0, 1, 0)^{tr}$, find the probability distribution for $X(n)$, $n \geq 1$.

(d) If $X(0) = \delta_3 = (0, 0, 1)^{tr}$, find the probability distribution for $X(n)$, $n \geq 1$.

14. Use the p.g.f.s given by (4.16) to verify that the expectation matrix M of the age-structured example is given by (4.15).

15. Suppose a multitype branching process for an age-structured population model has an offspring distribution $b_{i,0} = 0$ and $b_{i,r} = 0$ for $r > 5$ and $i = 1, 2, \ldots, k$. Show that a fixed point of the generating function F is the origin [i.e., $f_i(0, 0, \ldots, 0) = 0$–the probability of population extinction is zero].

16. Suppose

$$M_1 = \begin{pmatrix} b_1 & 0 & 0 & b_4 \\ p_{21} & 0 & 0 & 0 \\ 0 & p_{32} & 0 & 0 \\ 0 & 0 & p_{43} & 0 \end{pmatrix} \quad \text{and} \quad M_2 = \begin{pmatrix} 0 & b_2 & 0 & b_4 \\ p_{21} & 0 & 0 & 0 \\ 0 & p_{32} & 0 & 0 \\ 0 & 0 & p_{43} & 0 \end{pmatrix}$$

are two expectation matrices for an age-structured branching process of Leslie type, equation (4.15). The elements $p_{i+1,i} > 0$ and $b_i > 0$. For an age-structured expectation matrix of Leslie type, criteria of Sykes (1969) can be applied to show M is regular. Assume $p_{i+1,i} > 0$ for $i = 1, \ldots, k - 1$ and $b_k \geq 0$. Then M is regular if and only if the greatest common divisor of the set of indices i, where $b_i > 0$, is 1, g.c.d.$\{i | b_i > 0\} = 1$ (Sykes, 1969). Apply Sykes criteria to show that M_1 is regular and M_2 is not regular.

17. Suppose the offspring distribution for an age-structured branching process is

$$b_{1,r} = \begin{cases} 1/2, & r = 0 \\ 1/2, & r = 2 \\ 0, & r \neq 0, 2, \end{cases} \qquad b_{2,r} = \begin{cases} 1/6, & r = 0, 2 \\ 2/3, & r = 1 \\ 0, & r \neq 0, 1, 2 \end{cases}.$$

In addition, suppose the probability of surviving from the first to the second age class is $p_{21} = 3/4$.

(a) Calculate the mean birth rates, b_1 and b_2.

(b) Calculate the expectation matrix M. Show that M is regular. Then find the dominant eigenvalue of M.

(c) Write the form of the two p.g.f.s, $f_1(t_1, t_2)$ and $f_2(t_1, t_2)$. Then find the probability of population extinction given $X(0) = (1, 2)^{tr}$; that is, $\lim_{n \to \infty} \text{Prob}\{X(n) = \mathbf{0} | X(0) = (1, 2)^{tr}\}$.

18. A simple example of a multitype branching process related to cellular dynamics is discussed by Jagers (1975). Cell division results in two identical daughter cells containing the same number of chromosomes as the original cell ($2n$ for a diploid cell). Sometimes a mistake occurs and only one cell is produced having twice the number of chromosomes ($4n$ chromosomes), referred to as *endomitosis*. When this abnormal cell divides again, it will produce two daughter cells with twice the number of chromosomes ($4n$ chromosomes). Endomitosis can occur again for a cell having $4n$ chromosomes to produce a cell with $8n$ chromosomes and, in general, endomitosis occurring in a cell with $2^i n$ chromosomes produces a cell with $2^{i+1} n$ chromosomes. Cells with more than two copies of the genes are known as *polyploid cells*. The incidence of higher ploidies than four is small. Therefore, it is reasonable to consider a cellular model with only two types: diploid cells and polyploid cells (Jagers, 1975). Let p be the probability of endomitosis, $0 < p < 1/2$. The p.g.f.s for the two types are

$$f_1(t_1, t_2) = (1 - p)t_1^2 + pt_2 \quad \text{and} \quad f_2(t_1, t_2) = pt_2 + (1 - p)t_2^2.$$

(a) Calculate the expectation matrix M and the dominant eigenvalue of M. Is M regular?

(b) Compute all of the fixed points of (f_1, f_2) on the interval $[0, 1] \times [0, 1]$.

19. (a) Consider the size-dependent Ricker branching process $\{X_n\}_{n=0}^{\infty}$ discussed in Example 4.11. Show that the conditional expectation $E(X_{n+1}|X_n = x)$ is

$$E(X_{n+1}|X_n = x) = x \exp(r - \gamma x).$$

(b) Formulate a size-dependent branching process similar to the one described in Example 4.11 but one based on the following discrete-time population model:

$$x_{n+1} = \frac{r x_n}{1 + \gamma x_n}, \quad 0 < r, \ 0 < \gamma.$$

This model is known as the Beverton-Holt model (Caswell, 2001). Assume the size-independent offspring distribution $\{p_k\}_{k=0}^{\infty}$ satisfies $\sum_{k=1}^{\infty} k p_k = r > 1$. Then show that the model satisfies the following conditional expectation:

$$E(X_{n+1}|X_n = x) = \frac{r x}{1 + \gamma x}.$$

20. Suppose a population has two stages and the birth rates for each stage are constant, $b_1 > 0$ and $b_2 > 0$, but the survival probability from stage 1 to 2 is an environmentally determined, time-dependent random variable p_n, where $0 < \underline{p} \le p_n \le \overline{p} \le 1$, for $n = 1, 2, \ldots$. The stochastic model is

$$X(n + 1) = \begin{pmatrix} b_1 & b_2 \\ p_{n+1} & 0 \end{pmatrix} X(n), \quad b_i > 0, \quad i = 1, 2$$

(Tuljapurkar, 1990). The ratio of stage 2 to stage 1 over time is denoted as $U_n = X_2(n)/X_1(n)$, where $X(n) = (X_1(n), X_2(n))^{tr}$.

(a) Show that
$$U_{n+1} = \frac{p_{n+1}}{b_1 + b_2 U_n}.$$

(b) Use the fact that $\underline{p}/(b_1 + b_2 U_n) \le U_{n+1} \le \overline{p}/(b_1 + b_2 U_n)$ to show that there exist constants $0 < c_1 < c_2$ such that $c_1 \le U_n \le c_2$ for all time n (i.e., it is possible to obtain bounds on the ratio of the age structure).

21. A structured population model for the common dandelion, *Taraxacum officinale*, based on four seasons of growth and reproduction was formulated by Vavrek et al. (1997). The seasonal model is a matrix model, where
$$X(n + 1) = SuSpWAX(n).$$

The vector $X = (x_1, x_2, x_3, x_4)^{tr}$ is not random, but represents the average number of plants in each of the four size classes defined according to total leaf area: size class 1: 0.01-8.50 cm^2; size class 2: 8.51-23.00 cm^2; size class 3: 23.01-51.00 cm^2; size class 4: > 51 cm^2. Matrices Su, Sp, W, and A represent a combination of births, survival, and growth during summer, spring, winter, and autumn, respectively. The four matrices are

$$Su = \begin{pmatrix} 0.1719 & 0.2155 & 0.0536 & 0.1429 \\ 0.0912 & 0.250 & 0.1607 & 0.0953 \\ 0.0363 & 0.2457 & 0.3036 & 0.2381 \\ 0.0058 & 0.0905 & 0.3214 & 0.4286 \end{pmatrix}$$

$$Sp = \begin{pmatrix} 0.5975 & 0.5908 & 1.0845 & 2.2515 \\ 0.1055 & 0.3901 & 0.3171 & 0.3770 \\ .0053 & 0.1277 & 0.1098 & 0.1967 \\ 0 & 0 & 0.0244 & 0.0328 \end{pmatrix},$$

$$W = \begin{pmatrix} 1.0226 & 0 & 0 & 0 \\ 0.1647 & 0 & 0 & 0 \\ 0.1049 & 0 & 0 & 0 \\ 0.1580 & 1.0 & 0 & 0 \end{pmatrix},$$

and

$$A = \begin{pmatrix} 2.1191 & 0.7500 & 1.0054 & 1.4436 \\ 0 & 0 & 0 & 0.0086 \\ 0 & 0 & 0 & 0 \\ 0 & 0 & 0 & 0 \end{pmatrix}.$$

For example, transitions in the autumn from size classes 1, 2, 3, and 4 to size class 1 are 2.1191, 0.75, 1.0054, and 1.4436, and a transition in the summer from size class 2 to 3 is 0.2457.

(a) Calculate the dominant eigenvalue of each of the four matrices, λ_A, λ_W, λ_{Sp} and λ_{Su} and the dominant eigenvalue λ_P of $P = SuSpWA$. Calculate the eigenvector V corresponding to λ_P. The eigenvalue λ_P is the average annual population growth rate.

(b) Show that $\lambda_{Su}\lambda_{Sp}\lambda_W\lambda_A \neq \lambda_P$. Does the population grow or decline on the average?

(c) Let $X(0) = (80, 60, 40, 20)^{tr}$. Graph the population size $\|X(n)\|_1$ for $n = 0, 1, 2, 3, 4, 5, 6$ and calculate the ratio $x_1(n+1)/x_1(n)$.

(d) Show that, in general, $\lim_{n \to \infty} x_i(n+1)/x_i(n) = \lambda_P$, $i = 1, 2, 3, 4$, and $\lim_{n \to \infty} X(n)/\|X(n)\|_1 = V/\|V\|_1$. (See the Perron Theorem in the Appendix for Chapter 2.)

4.9 References for Chapter 4

Allen, L. J. S. Branching Processes. In: *Sourcebook in Theoretical Ecology.* Hastings, A. and L. Gross (eds.), University of California Press, Berkeley, CA. In press.

Allen, L. J. S., J. F. Fagan, G. Hognas, and H. Fagerholm. 2005. Population extinction in discrete-time stochastic population models with an Allee effect. *J. Difference Eqns and Appl.* 11: 273–293.

Antia, R., R. R. Regoes, J. C. Koella, and C. T. Bergstrom. 2003. The role of evolution in the emergence of infectious diseases. *Nature* 426: 658–661.

Athreya, K. B. and P. E. Ney. 1972. *Branching Processes.* Springer-Verlag, Berlin.

Bailey, N. T. J. 1990. *The Elements of Stochastic Processes with Applications to the Natural Sciences.* John Wiley & Sons, New York.

Benaïm, M. and S. J. Schreiber. 2009. Persistence of structured populations in random environments. *Theor. Pop. Biol.* 76: 19–34.

Block, G. L. and L. J. S. Allen. 2000. Population extinction and quasistationary behavior in stochastic density-dependent structured models. *Bull. Math. Biol.* 62: 199–228.

Caswell, H. 2001. *Matrix Population Models: Construction, Analysis and Interpretation.* 2nd ed. Sinauer Assoc. Inc., Sunderland, MA.

Chesson, P. L. 1982. The stabilizing effect of a random environment. *J. Math. Biol.* 15: 1–36.

Cushing, J. M. 1998. *An Introduction to Structured Population Dynamics.* CBMS-NSF Regional Conf. Series in Applied Mathematics # 71. SIAM, Philadelphia.

Cushing, J. M., R. F. Costantino, B. Dennis, R. Desharnais, and S. M. Henson. 2003. *Chaos in Ecology Experimental Nonlinear Dynamics.* Theoretical Ecology Series. Academic Press, San Diego.

Elaydi, S. N. 2000. *Discrete Chaos.* Chapman & Hall/CRC, Boca Raton.

Ellner, S. P. 1984. Asymptotic behavior of some stochastic difference equation population models. *J. Math. Biol.* 19: 169–200.

Haccou, P., P. Jagers, and V. A Vatutin. 2005. *Branching Processes Variation, Growth, and Extinction of Populations.* Cambridge Studies in Adaptive Dynamics. Cambridge Univ. Press, Cambridge, U. K.

Harris, T. E. 1963. *The Theory of Branching Processes.* Springer-Verlag, Berlin.

Högnäs, G. 1997. On the quasi-stationary distribution of a stochastic Ricker model. *Stoch. Proc. Appl.* 70: 243–263.

Jagers, P. 1975. *Branching Processes with Biological Applications.* John Wiley & Sons, London.

Karlin, S. and H. Taylor. 1975. *A First Course in Stochastic Processes.* 2nd ed. Academic Press, New York.

Kimmel, M. and D. Axelrod. 2002. *Branching Processes in Biology.* Springer-Verlag, New York.

Leslie, P. H. 1945. On the use of matrices in certain population mathematics. *Biometrika* 21: 1–18.

May, R. M. 1976. Simple mathematical models with very complicated dynamics. *Nature.* 261: 459–467.

Mangel, M. 2006. *The Theoretical Biologist's Toolbox.* Cambridge Univ. Press, Cambridge, U. K.

Mode, C. J. 1971. *Multitype Branching Processes Theory and Applications.* Elsevier, New York.

Ricker, W. E. 1954. Stock and recruitment. *J. Fish. Res. Bd. Can.* 11: 559–623.

Schinazi, R. B. 1999. *Classical and Spatial Stochastic Processes.* Birkhäuser, Boston.

Sykes, Z. M. 1969. On discrete stable population theory. *Biometrics* 25: 285–293.

Taylor, H. M. and S. Karlin. 1998. *An Introduction to Stochastic Modeling.* 3rd ed. Academic Press, New York.

Tuljapurkar, S. 1990. *Population Dynamics in Variable Environments.* Springer-Verlag, Berlin and New York.

Vavrek, M. C., J. B. McGraw, and H. S. Yang. 1997. Within-population variation in demography of *Taraxacum officinale*: season- and size-dependent survival, growth and reproduction. *J. Ecology* 85: 277–287.

Chapter 5

Continuous-Time Markov Chains

5.1 Introduction

In Chapters 2, 3, and 4, the stochastic processes were discrete in time. In this chapter, continuous-time Markov chains are introduced, where time is continuous, $t \in [0, \infty)$, but the random variables are discrete. Notation and basic definitions are introduced and a corresponding discrete-time process is defined, known as the embedded Markov chain. As a preliminary example, one of the simplest continuous-time Markov chains, the Poisson process is defined. The Poisson process serves as an illustration for the techniques that will be used in this chapter and lays the foundation for the construction of more general birth and death and population processes discussed in later chapters.

Methods are presented for studying continuous-time Markov chains, including the forward and backward Kolmogorov differential equations and generating function techniques. Generating function techniques are procedures to derive partial differential equations satisfied by the generating functions. In simple cases, where the birth and death processes are linear in the state variable, the partial differential equations are also linear and first-order. In these simple cases, an explicit solution for the generating function can be obtained via the method of characteristics. The time between events or jumps in a continuous-time Markov chain is shown to be exponential. Based on this exponential distribution, a simple numerical method is described for generating stochastic realizations of the process.

5.2 Definitions and Notation

Let $\{X(t) : t \in [0, \infty)\}$ be a collection of discrete random variables with values in a finite or infinite set, $\{0, 1, 2, \ldots, N\}$ or $\{0, 1, 2, \ldots\}$. The index set is continuous, $t \in [0, \infty)$.

DEFINITION 5.1 *The stochastic process $\{X(t) : t \in [0, \infty)\}$, is called a* continuous-time Markov chain (CTMC) *if it satisfies the following condition: For any sequence of real numbers satisfying* $0 \le t_0 < t_1 < \cdots < t_n < t_{n+1}$,

$$\text{Prob}\{X(t_{n+1}) = i_{n+1} | X(t_0) = i_0, X(t_1) = i_1, \ldots, X(t_n) = i_n\}$$
$$= \text{Prob}\{X(t_{n+1}) = i_{n+1} | X(t_n) = i_n\}.$$

The latter condition in Definition 5.1 is known as the *Markov property*. The transition to state i_{n+1} at time t_{n+1} depends only on the value of the state at the most recent time t_n and does not depend on the history of the process. Each random variable $X(t)$ has an associated probability distribution $\{p_i(t)\}_{i=0}^{\infty}$, where

$$p_i(t) = \text{Prob}\{X(t) = i\}.$$

Denote the vector of probabilities as $p(t) = (p_0(t), p_1(t), \ldots)^{tr}$. A relation between the random variable $X(s)$ and the random variable $X(t)$, $s < t$, is defined by the transition probabilities. Define the transition probabilities as

$$p_{ji}(t, s) = \text{Prob}\{X(t) = j | X(s) = i\}, \quad s < t$$

for $i, j = 0, 1, 2, \ldots$. If the transition probabilities do not depend explicitly on s or t but depend only on the length of the time interval, $t - s$, they are called *stationary* or *homogeneous* transition probabilities; otherwise the transition probabilities are referred to as nonstationary or nonhomogeneous. Unless otherwise stated, it shall be assumed that the transition probabilities are stationary; that is,

$$p_{ji}(t - s) = \text{Prob}\{X(t) = j | X(s) = i\} = \text{Prob}\{X(t - s) = j | X(0) = i\}$$

for $s < t$. Denote the matrix of transition probabilities or the *transition matrix* as

$$P(t) = (p_{ji}(t)).$$

In most cases, the transition probabilities have the following properties: $p_{ji}(t) \ge 0$ and

$$\sum_{j=0}^{\infty} p_{ji}(t) = 1, \quad t \ge 0. \tag{5.1}$$

The probability that there is a transition from state i to some other state in time $[0, t]$ equals one. Hence, matrix $P(t)$ is a stochastic matrix for all $t \ge 0$. In addition, the transition probabilities are solutions of the *Chapman-Kolmogorov* equations

$$\sum_{k=0}^{\infty} p_{jk}(s) p_{ki}(t) = p_{ji}(t + s).$$

In matrix form,

$$P(s)P(t) = P(s + t)$$

for all $s, t \in [0, \infty)$. Notice that these definitions are the continuous analogues of the definitions given for DTMCs. There are some exceptional cases known as explosive processes when condition (5.1) does not hold for all times. These exceptional cases can occur only when the state space is infinite. An explosive process is defined, but first the concept of a jump time is defined.

The distinction between a DTMC and a CTMC is that in discrete-time chains, there is a "jump" to a new state at times $1, 2, \ldots$, but in continuous-time chains the "jump" to a new state may occur at any time $t \geq 0$. In a continuous-time chain, with beginning state $X(0)$, the process stays in state $X(0)$ for a random amount of time W_1 until it jumps to a new state, $X(W_1)$. Then it stays in state $X(W_1)$ for a random amount of time until it jumps to a new state at time W_2, $X(W_2)$, and so on. In general, W_i is the random variable for the time of the ith jump. Define $W_0 = 0$. The collection of random variables $\{W_i\}_{i=0}^{\infty}$ is referred to as the *jump times* or *waiting times* of the process (Norris, 1997; Taylor and Karlin, 1998). In addition, the random variables $T_i = W_{i+1} - W_i$ are referred to as the *interevent times* or *holding times* or *sojourn times* (Norris, 1997; Taylor and Karlin, 1998). The waiting times W_i and interevent times T_i are illustrated in Figure 5.1.

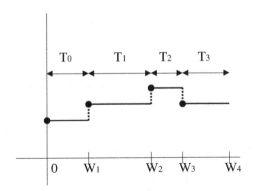

FIGURE 5.1: Sample path of a CTMC, illustrating waiting times and interevent times.

In an explosive process, the value of the state approaches infinity at a finite time, $\lim_{t \to T^-} X(t) = \infty$ for $T < \infty$. Then $p_{ji}(T) = 0$ for all $i, j = 0, 1, 2, \ldots$, which means $\sum_{j=0}^{\infty} p_{ji}(T) = 0$. Hence, condition (5.1) does not hold. See Norris (1997) or Karlin and Taylor (1975) for further discussion on explosive processes. Most of the well-known birth and death processes are nonexplosive. In particular, all finite CTMCs are nonexplosive. Explosive birth processes are discussed in Chapter 6.

One sample path of an explosive process is graphed in Figure 5.2. The values of the waiting times are approaching a positive constant, $W = \sup\{W_i\}$, and the values of the states are approaching infinity, $\lim_{i \to \infty} X(W_i) = \infty$; the process is explosive. Notice that the sample path in Figure 5.2 is a piecewise constant function that is continuous from the right. However, for ease in sketching sample paths, they are drawn as connected rectilinear curves (as in Figure 5.3).

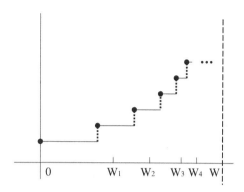

FIGURE 5.2: Sample path of an explosive CTMC.

As before, our notation differs from the standard notation in that the transition probability from state $i \to j$ is defined as p_{ji}, rather than the more commonly used notation p_{ij} (e.g., Bailey, 1990; Karlin and Taylor, 1975, 1981; Norris, 1997; Schinazi, 1999; Stewart, 1994; Taylor and Karlin, 1998). This notation is consistent with the notation for DTMCs in Chapters 2, 3, and 4. In our notation, the element in the ith row and jth column of $P(t)$ is p_{ij}, which represents the transition $j \to i$.

5.3 The Poisson Process

The *Poisson process* is a CTMC $\{X(t) : t \in [0, \infty)\}$ with state space the nonnegative integers, $\{0, 1, 2, \ldots\}$. The following two properties define the Poisson process:

(1) For $t = 0$, $X(0) = 0$.

(2) For Δt sufficiently small, the transition probabilities are

$$p_{i+1,i}(\Delta t) = \text{Prob}\{X(t + \Delta t) = i + 1 | X(t) = i\} = \lambda \Delta t + o(\Delta t)$$
$$p_{ii}(\Delta t) = \text{Prob}\{X(t + \Delta t) = i | X(t) = i\} = 1 - \lambda \Delta t + o(\Delta t)$$
$$p_{ji}(\Delta t) = \text{Prob}\{X(t + \Delta t) = j | X(t) = i\} = o(\Delta t), \quad j \geq i + 2$$
$$p_{ji}(\Delta t) = 0, \quad j < i,$$

where the notation $o(\Delta t)$ ("little oh Δt") is the Landau order symbol.

In general, the function $f(\Delta t) = o(\Delta t)$ or $f(\Delta t)$ is $o(\Delta t)$ means as $\Delta t \to 0$ f has the following property:

$$\lim_{\Delta t \to 0} \frac{f(\Delta t)}{\Delta t} = 0.$$

Therefore, in part (2) of the definition of the Poisson process, the functions $p_{i+1,i}(\Delta t) - \lambda \Delta t$, $p_{ii}(\Delta t) - 1 + \lambda \, \Delta t$, and $p_{ji}(\Delta t)$ are $o(\Delta t)$ as $\Delta t \to 0$. In particular,

$$\lim_{\Delta t \to 0} \frac{p_{i+1,i}(\Delta t) - \lambda \Delta t}{\Delta t} = 0 = \lim_{\Delta t \to 0} \frac{p_{ii}(\Delta t) - 1 + \lambda \Delta t}{\Delta t}$$

and

$$\lim_{\Delta t \to 0} \frac{p_{ji}(\Delta t)}{\Delta t} = 0, \quad j \geq i + 2.$$

Frequently, continuous-time Markov processes are defined by conditions such as those given in (2). From these "infinitesimal" transition probabilities, other properties of the process can be derived.

In a small time interval Δt, the Poisson process can either stay in the same state or move to the next larger state, $i \to i + 1$; it cannot move to a smaller state. The probability that the Poisson process moves up two or more states is a very small probability and approaches zero when $\Delta t \to 0$. Note that the transition probabilities $p_{ji}(\Delta t)$ are independent of i and j and only depend on the length of the interval Δt. If the intervals $[s, s + \Delta t]$ and $[t, t + \Delta t]$ are nonoverlapping, $s + \Delta t \leq t$, then property (2) and the Markov property imply that the following random variables from the Poisson process, $X(t + \Delta t) - X(t)$ and $X(s + \Delta t) - X(s)$, are independent and have the same probability distributions (i.e., the Poisson process has stationary and independent increments). Equivalently, the Poisson process $\{X(t) : t \in [0, \infty)\}$ is defined as a process with stationary independent increments and for each fixed t, $X(t)$ has a Poisson distribution with parameter λt (Norris, 1997). This latter property will be verified by applying properties (1) and (2).

The assumptions (1) and (2) are used to derive a system of differential equations for $p_i(t)$ for $i = 0, 1, 2, \ldots$. The solutions are then shown to represent a Poisson distribution with parameter λt. Because in the Poisson process

$X(0) = 0$, it follows that $p_{i0}(t) = p_i(t)$. Let $p_{00}(t + \Delta t) = p_0(t + \Delta t)$. Then

$$
\begin{aligned}
p_0(t + \Delta t) &= \text{Prob}\{X(t + \Delta t) = 0\} \\
&= \text{Prob}\{X(t) = 0, X(t + \Delta t) - X(t) = 0\} \\
&= \text{Prob}\{X(t) = 0\}\text{Prob}\{X(t + \Delta t) - X(t) = 0\} \\
&= \text{Prob}\{X(t) = 0\}\text{Prob}\{X(\Delta t) = 0\},
\end{aligned}
$$

where $X(t) - X(0) = X(t)$ and $X(t + \Delta t) - X(t)$ are independent. Therefore,

$$
p_0(t + \Delta t) = p_0(t)\left[1 - \lambda\Delta t + o(\Delta t)\right].
$$

Subtracting $p_0(t)$ from both sides of the preceding equation and dividing by Δt yields

$$
\frac{p_0(t + \Delta t) - p_0(t)}{\Delta t} = -\lambda p_0(t) + p_0(t)\frac{o(\Delta t)}{\Delta t}.
$$

Then taking the limit as $\Delta t \to 0$, leads to the following differential equation:

$$
\frac{dp_0(t)}{dt} = -\lambda p_0(t). \tag{5.2}
$$

Given that $p_0(0) = 1 = \text{Prob}\{X(0) = 0\}$, the solution of this linear first-order differential equation is

$$
p_0(t) = e^{-\lambda t}.
$$

The differential equations for $i \geq 1$ are derived in a similar manner. Let $p_{i0}(t + \Delta t) = p_i(t + \Delta t)$. Then

$$
\begin{aligned}
p_i(t + \Delta t) &= \text{Prob}\{X(t + \Delta t) = i\} \\
&= \text{Prob}\{X(t) = i, \Delta X(t) = 0\} \\
&\quad + \text{Prob}\{X(t) = i - 1, \Delta X(t) = 1\} \\
&\quad + \sum_{k=2}^{k \leq i} \text{Prob}\{X(t) = i - k, \Delta X(t) = k\},
\end{aligned}
$$

where $\Delta X(t) = X(t + \Delta t) - X(t)$. The latter summation is $o(\Delta t)$ since

$$
\sum_{k=2}^{k \leq i} \text{Prob}\{X(t) = i - k, \Delta X(t) = k\} = \sum_{k=2}^{k \leq i} p_{i-k}(t)o(\Delta t) = o(\Delta t).
$$

Applying the definition of the transition probabilities and the independence of the increments,

$$
p_i(t + \Delta t) = p_i(t)[1 - \lambda\Delta t + o(\Delta t)] + p_{i-1}(t)[\lambda\Delta t + o(\Delta t)] + o(\Delta t). \tag{5.3}
$$

Note that the equations given in (5.3) can be derived directly from the infinitesimal transition probabilities. If the process is in state i at time $t + \Delta t$, then at the previous time t it was either in state i or $i - 1$ [the probability it was in some other state is $o(\Delta t)$]. If the process is in state i at time t, the process stays in state i with probability $1 - \lambda \Delta t + o(\Delta t)$, and if the process is in state $i - 1$ at time t, it moves to state i with probability $\lambda \Delta t + o(\Delta t)$.

Subtracting $p_i(t)$ from both sides of (5.3) and dividing by Δt leads to

$$\frac{p_i(t + \Delta t) - p_i(t)}{\Delta t} = -\lambda p_i(t) + \lambda p_{i-1}(t) + \frac{o(\Delta t)}{\Delta t},$$

where all terms with a factor of $o(\Delta t)$ are put in the $o(\Delta t)$ expression. Taking the limit as $\Delta t \to 0$, then

$$\frac{dp_i(t)}{dt} = -\lambda p_i(t) + \lambda p_{i-1}(t), \quad i \geq 1. \tag{5.4}$$

The preceding equations represent a system of differential-difference equations, difference equations in the variable i and differential equations in the variable t.

The system of differential-difference equations (5.4) can be solved sequentially. Replace $p_0(t)$ by $e^{-\lambda t}$ and apply the initial conditions $p_i(0) = 0$, $i \geq 1$. The differential equation for $p_1(t)$ is a linear first-order differential equation,

$$\frac{dp_1(t)}{dt} + \lambda p_1(t) = \lambda e^{-\lambda t}, \quad p_1(0) = 0.$$

Multiplying by the factor $e^{\lambda t}$ (known as an *integrating factor*) yields

$$\frac{d\left[e^{\lambda t} p_1(t)\right]}{dt} = \lambda. \tag{5.5}$$

Integrating both sides of (5.5) from 0 to t and applying the initial condition yields the solution

$$p_1(t) = \lambda t e^{-\lambda t}.$$

Next, the differential equation for $p_2(t)$ is

$$\frac{dp_2(t)}{dt} + \lambda p_2(t) = \lambda^2 t e^{-\lambda t}, \quad p_2(0) = 0.$$

Applying the same technique to $p_2(t)$ as for $p_1(t)$, yields the solution

$$p_2(t) = (\lambda t)^2 \frac{e^{-\lambda t}}{2!}.$$

By induction, it can be shown that

$$p_i(t) = (\lambda t)^i \frac{e^{-\lambda t}}{i!}, \quad i = 0, 1, 2, \ldots.$$

The probability distribution, $\{p_i(t)\}_{i=0}^{\infty}$, represents a Poisson distribution with parameter λt. The mean and variance of this Poisson distribution are

$$m(t) = \lambda t = \sigma^2(t).$$

For more general CTMCs, it can be difficult to solve a system of differential-difference equations in a sequential manner and obtain a general formula for $p_i(t)$. A pattern may not emerge as in the case of the Poisson process. Therefore, other techniques for obtaining information about the probabilities $p_i(t)$ will be applied such as solving a partial differential equation for the probability or moment generating function.

The probability $p_0(t) = e^{-\lambda t}$ in the Poisson process can be thought of as a waiting-time probability (i.e., the probability that the first event $0 \to 1$ occurs at a time greater than t). Let W_1 be the random variable for the time until the process reaches state 1, the holding time until the first jump. Then

$$\text{Prob}\{W_1 > t\} = e^{-\lambda t} \quad \text{or} \quad \text{Prob}\{W_1 \le t\} = 1 - e^{-\lambda t}.$$

This latter expression is the cumulative distribution function for an exponential random variable with parameter λ. Thus, W_1 is an exponential random variable with parameter λ, with c.d.f. $F(t) = 1 - e^{-\lambda t}$ and p.d.f. $f(t) = F'(t) = \lambda e^{-\lambda t}$. In general, it can be shown that it takes an exponential amount of time to move from state i to state $i + 1$ (i.e., the random variable for the time between jumps i and $i + 1$, $W_{i+1} - W_i$, has an exponential distribution with parameter λ). In fact, sometimes in the definition of the Poisson process it is stated that the interevent times, $T_i = W_{i+1} - W_i$, are independent exponential random variables with parameter λ (Norris, 1997; Schinazi, 1999). Interevent times for more general CTMCs are discussed in Section 5.10, where it is shown that they are also exponential random variables.

Figure 5.3 is a sample path or realization of a Poisson process when $\lambda = 1$. A general method will be given for generating sample paths of birth and death processes based on this exponential distribution for the interevent time.

5.4 Generator Matrix Q

The transition probabilities p_{ji} are used to derive transition rates q_{ji}. The transition rates form a matrix known as the infinitesimal generator matrix Q. Matrix Q defines a relationship between the rates of change of the transition probabilities.

Assume the transition probabilities $p_{ji}(t)$ are continuous and differentiable for $t \ge 0$ and at $t = 0$ they equal

$$p_{ji}(0) = 0, \quad j \ne i, \quad \text{and} \quad p_{ii}(0) = 1.$$

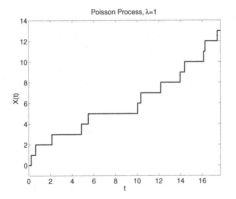

FIGURE 5.3: Sample path of a Poisson process with $\lambda = 1$.

Define

$$q_{ji} = \lim_{\Delta t \to 0^+} \frac{p_{ji}(\Delta t) - p_{ji}(0)}{\Delta t} = \lim_{\Delta t \to 0^+} \frac{p_{ji}(\Delta t)}{\Delta t}, \quad i \neq j. \tag{5.6}$$

Notice that $q_{ji} \geq 0$ since $p_{ji}(\Delta t) \geq 0$. In addition, define

$$q_{ii} = \lim_{\Delta t \to 0^+} \frac{p_{ii}(\Delta t) - p_{ii}(0)}{\Delta t} = \lim_{\Delta t \to 0^+} \frac{p_{ii}(\Delta t) - 1}{\Delta t}. \tag{5.7}$$

From equation (5.1), $\sum_{j=0}^{\infty} p_{ji}(\Delta t) = 1$, it follows that

$$1 - p_{ii}(\Delta t) = \sum_{j=0, j \neq i}^{\infty} p_{ji}(\Delta t) = \sum_{j=0, j \neq i}^{\infty} [q_{ji}\Delta t + o(\Delta t)].$$

Thus,

$$q_{ii} = \lim_{\Delta t \to 0^+} \frac{-\sum_{j=0, j \neq i}^{\infty} [q_{ji}\Delta t + o(\Delta t)]}{\Delta t} \tag{5.8}$$

$$= -\sum_{j=0, j \neq i}^{\infty} q_{ji},$$

where it is assumed that $\sum_{j \neq i} o(\Delta t) = o(\Delta t)$. This is certainly true if the summation contains only a finite number of terms (finite state space). If the summation contains an infinite number of terms (state space is infinite), Karlin and Taylor (1981) show that the limit (5.8) does exist on the extended interval, $[-\infty, 0]$, $0 \leq -q_{ii} \leq \infty$. In either case $q_{ii} \leq 0$. If q_{ii} is finite, then $\sum_{j=0}^{\infty} q_{ji} = 0$ and it follows from (5.6) and (5.7) that

$$p_{ji}(\Delta t) = \delta_{ji} + q_{ji}\Delta t + o(\Delta t), \tag{5.9}$$

where δ_{ji} is Kronecker's delta symbol.

The individual limits (5.6) and (5.7) can be expressed more simply as the entries in a matrix limit. Let $P(\Delta t) = (p_{ji}(\Delta t))$ be the infinitesimal transition matrix and let $\mathbb{I}$ be the matrix of the same dimension as $P(\Delta t)$ but with ones along the diagonal and zeros elsewhere (identity matrix in the finite case). Then matrix Q equals

$$Q = \lim_{\Delta t \to 0+} \frac{P(\Delta t) - \mathbb{I}}{\Delta t}. \tag{5.10}$$

DEFINITION 5.2 *The matrix of transition rates* $Q = (q_{ji})$, *defined in (5.6), (5.7), and (5.10), is known as the* infinitesimal generator matrix,

$$Q = \begin{pmatrix} q_{00} & q_{01} & q_{02} & \cdots \\ q_{10} & q_{11} & q_{12} & \cdots \\ q_{20} & q_{21} & q_{22} & \cdots \\ \vdots & \vdots & \vdots & \end{pmatrix}$$

$$= \begin{pmatrix} -\sum_{i=1}^{\infty} q_{i0} & q_{01} & q_{02} & \cdots \\ q_{10} & -\sum_{i=0, i\neq 1}^{\infty} q_{i1} & q_{12} & \cdots \\ q_{20} & q_{21} & -\sum_{i=0, i\neq 2}^{\infty} q_{i2} & \cdots \\ \vdots & \vdots & \vdots & \end{pmatrix}.$$

Sometimes the terms *transition rate matrix* or *infinitesimal matrix* or simply *generator matrix* are used when referring to matrix Q. Matrix Q has the property that each column sum is zero and the ith diagonal element is the negative of the sum of the off-diagonal elements in that column.

The next example shows that the system of differential-difference equations for the Poisson process (5.4) can be expressed in terms of the generator matrix Q.

Example 5.1. The generator matrix for the Poisson process (5.4) can be easily calculated. The infinitesimal transition matrix

$$P(\Delta t) = \begin{pmatrix} 1 - \lambda \Delta t & 0 & 0 & \cdots \\ \lambda \Delta t & 1 - \lambda \Delta t & 0 & \cdots \\ 0 & \lambda \Delta t & 1 - \lambda \Delta t & \cdots \\ 0 & 0 & \lambda \Delta t & \cdots \\ \vdots & \vdots & \vdots & \end{pmatrix} + \mathbf{o}(\Delta t),$$

where $\mathbf{o}(\Delta t)$ is a lower triangular matrix with all entries equal to $o(\Delta t)$.

Applying the limit (5.10) leads to the generator matrix

$$Q = \begin{pmatrix} -\lambda & 0 & 0 & \cdots \\ \lambda & -\lambda & 0 & \cdots \\ 0 & \lambda & -\lambda & \cdots \\ 0 & 0 & \lambda & \cdots \\ \vdots & \vdots & \vdots & \end{pmatrix}.$$

The system of differential-difference equations for the Poisson process, equations (5.2) and (5.4), can be expressed in terms of the generator matrix Q:

$$\boxed{\frac{dp(t)}{dt} = Qp(t).}$$

This relationship will be shown to hold for more general processes as well. ∎

The generator matrix Q will be shown to be important for several reasons. In particular, Q is used to define the forward and backward Kolmogorov equations, which express the transition matrix $P(t)$ in terms of a differential equation, $dP/dt = QP$ and $dP/dt = PQ$. In addition, the generator matrix Q is used to define a transition matrix for the embedded Markov chain. These two concepts are discussed in the next two sections.

5.5 Embedded Markov Chain and Classification of States

Every sample path or realization of a CTMC remains in a particular state (stays constant) for a random amount of time before making a jump to a new state. Recall that the waiting times are denoted as W_i, $i = 0, 1, 2, \ldots$, and the interevent times as $T_i = W_{i+1} - W_i$, $i = 0, 1, 2, \ldots$. For example, in the Poisson process, the states $0, 1, 2, \ldots$, are visited sequentially with an exponential amount of time between jumps.

DEFINITION 5.3 *Let Y_n denote the random variable for the state of a CTMC $\{X(t) : t \in [0, \infty)\}$ at the nth jump,*

$$Y_n = X(W_n), \quad n = 0, 1, 2, \ldots.$$

The set of random variables $\{Y_n\}_{n=0}^{\infty}$ is known as the embedded Markov chain *or the* jump chain *associated with the CTMC $\{X(t) : t \in [0, \infty)\}$.*

The embedded Markov chain is a DTMC, useful for classifying states in the corresponding CTMC. Define a transition matrix $T = (t_{ji})$ for the embedded

Markov chain, $\{Y_n\}_{n=0}^{\infty}$, where $t_{ji} = \text{Prob}\{Y_{n+1} = j | Y_n = i\}$. As an illustration, the transition matrix is defined for the embedded Markov chain of the Poisson process.

Example 5.2. Consider the Poisson process, where $X(0) = X(W_0) = 0$ and $X(W_n) = n$ for $n = 1, 2, \ldots$. The embedded Markov chain $\{Y_n\}_{n=0}^{\infty}$ has the property that $Y_n = n$, $n = 0, 1, 2, \ldots$. The transition from state n to $n+1$ occurs with probability 1. It is easily follows that the transition matrix for the embedded Markov chain is

$$
T = \begin{pmatrix} 0 & 0 & 0 & \cdots \\ 1 & 0 & 0 & \cdots \\ 0 & 1 & 0 & \cdots \\ 0 & 0 & 1 & \cdots \\ \vdots & \vdots & \vdots & \end{pmatrix}.
\tag{5.11}
$$

∎

In general, the transition matrix $T = (t_{ji})$ can be defined using the generator matrix Q. First, note that the transition probability t_{ii} is zero, because an assumption inherent in the definition of the embedded Markov chain is that the state must change, unless state i is absorbing. A state i in the continuous-time chain is *absorbing* if $q_{ii} = 0$ (i.e., the rate of change is zero since the state does not change). Thus,

$$
t_{ii} = \begin{cases} 0, \text{ if } q_{ii} \neq 0 \\ 1, \text{ if } q_{ii} = 0. \end{cases}
\tag{5.12}
$$

As motivation for the definition of the transition probability t_{ji}, recall that $q_{ji} = \lim_{\Delta t \to 0^+} p_{ji}(\Delta t)/\Delta t$ and $-q_{ii} = \lim_{\Delta t \to 0^+} (1 - p_{ii}(\Delta t))/\Delta t$. Thus,

$$
-\frac{q_{ji}}{q_{ii}} = \lim_{\Delta t \to 0^+} \frac{p_{ji}(\Delta t)}{1 - p_{ii}(\Delta t)}.
$$

This latter probability is the probability of a transfer from state i to j, given the process does not remain in state i. Define $t_{ji} = -q_{ji}/q_{ii}$, $q_{ii} \neq 0$. The transition probability t_{ji} for $j \neq i$ is

$$
t_{ji} = \begin{cases} \dfrac{q_{ji}}{\displaystyle\sum_{k=0, k\neq i}^{\infty} q_{ki}} = -\dfrac{q_{ji}}{q_{ii}}, \text{ if } q_{ii} \neq 0 \\ 0, \qquad\qquad\qquad \text{ if } q_{ii} = 0. \end{cases}
\tag{5.13}
$$

DEFINITION 5.4 *The matrix $T = (t_{ji})$, where the elements t_{ji} are defined in (5.12) and (5.13), is the* transition matrix of the embedded Markov chain $\{Y_n\}_{n=0}^{\infty}$. *In particular, for $q_{ii} \neq 0$, $i = 0, 1, 2, \ldots$,*

$$
T = \begin{pmatrix}
0 & -\dfrac{q_{01}}{q_{11}} & -\dfrac{q_{02}}{q_{22}} & \cdots \\
-\dfrac{q_{10}}{q_{00}} & 0 & -\dfrac{q_{12}}{q_{22}} & \cdots \\
-\dfrac{q_{20}}{q_{00}} & -\dfrac{q_{21}}{q_{11}} & 0 & \cdots \\
\vdots & \vdots & \vdots &
\end{pmatrix}.
$$

If any $q_{ii} = 0$, the (i, i) element of T is one and the remaining elements in that column are zero.

Matrix T is a stochastic matrix; the column sums equal one. The transition probabilities are homogeneous (i.e., independent of n). In addition, $T^n = (t_{ji}^{(n)})$, where $t_{ji}^{(n)} = \text{Prob}\{Y_n = j | Y_0 = i\}$. Using the generator matrix Q defined in Example 5.1 for the Poisson process, it can be seen that the transition matrix of the embedded Markov chain has the form given by (5.11).

Example 5.3. Suppose a continuous-time, finite Markov chain has a generator matrix given by

$$
Q = \begin{pmatrix}
-1 & 0 & 0 & 1 \\
1 & -1 & 0 & 0 \\
0 & 1 & -1 & 0 \\
0 & 0 & 1 & -1
\end{pmatrix}.
\tag{5.14}
$$

The transition matrix of the corresponding embedded Markov chain is

$$
T = \begin{pmatrix}
0 & 0 & 0 & 1 \\
1 & 0 & 0 & 0 \\
0 & 1 & 0 & 0 \\
0 & 0 & 1 & 0
\end{pmatrix}.
\tag{5.15}
$$

From the embedded Markov chain, it is easy to see that the states communicate in the following manner: $1 \to 2 \to 3 \to 4 \to 1$. ∎

The classification schemes for states in CTMCs are the same as for DTMCs. The transition probabilities $P(t) = (p_{ji}(t))$ and the transition matrix for the embedded Markov chain $T = (t_{ji})$ are used to define these classification schemes. Definitions for communication class and irreducible in CTMCs can be defined in a manner similar to those for DTMCs.

DEFINITION 5.5 *State j can be reached from state i, $i \to j$, if $p_{ji}(t) > 0$ for some $t \geq 0$. State i communicates with state j, $i \leftrightarrow j$, if $i \to j$ and $j \to i$. The set of states that communicate is called a communication class.*

If every state can be reached from every other state, the Markov chain is irreducible; *otherwise, it is said to be* reducible. *A set of states C is* closed *if it is impossible to reach any state outside of C from a state inside C,* $p_{ji}(t) = 0$ *for* $t \geq 0$ *if* $i \in C$ *and* $j \notin C$.

In the case that $p_{ji}(\Delta t)$ equals $\delta_{ji} + q_{ji}\Delta t + o(\Delta t)$, then $p_{ji}(\Delta t) > 0$ iff $q_{ji} > 0$ iff $t_{ij} > 0$ for $j \neq i$. Therefore, $i \leftrightarrow j$ in the CTMC iff $i \leftrightarrow j$ in the embedded Markov chain. The generator matrix Q in the CTMC is irreducible iff the transition matrix T in the embedded Markov chain is irreducible.

Definitions for recurrent and transient states in CTMCs can be defined in a manner similar to DTMCs. Let T_{ii} be the first time the chain is in state i after leaving state i. The random variable T_{ii} is known as the *first return time*. The first return can occur for $t > 0$; T_{ii} is a continuous random variable.

DEFINITION 5.6   *State i is* recurrent (transient) *in a CTMC* $\{X(t) : t \in [0, \infty)\}$, *if the first return time is finite (infinite),*

$$\text{Prob}\{T_{ii} < \infty | X(0) = i\} = 1 \ (< 1). \tag{5.16}$$

These definitions are similar to the definitions of recurrence and transience in DTMCs. Recall that state i is said to be *recurrent (transient)* in a DTMC $\{Y_n\}_{n=0}^{\infty}$, with $Y_0 = i$, if

$$\sum_{n=0}^{\infty} f_{ii}^{(n)} = 1 \ (< 1),$$

where $f_{ii}^{(n)}$ is the probability that the first return to state i is at step n. The following theorem relates recurrent and transient states in CTMCs to recurrent and transient states in the corresponding embedded Markov chains. For a proof of this result, please consult Norris (1997) or Schinazi (1999).

THEOREM 5.1
State i in a CTMC $\{X(t) : t \in [0, \infty)\}$ *is recurrent (transient) iff state i in the corresponding embedded Markov chain* $\{Y_n\}_{n=0}^{\infty}$ *is recurrent (transient).*

Recurrence or transience in a CTMC can be determined from the properties of the transition matrix T of the embedded Markov chain. For example, a state i in a CTMC $\{X(t) : t \in [0, \infty)\}$ is *recurrent (transient)* iff

$$\sum_{n=0}^{\infty} t_{ii}^{(n)} = \infty \ (< \infty),$$

where $t_{ii}^{(n)}$ is the (i, i) element in the transition matrix of T^n of the embedded Markov chain $\{Y_n\}_{n=0}^{\infty}$. Other properties that determine recurrence and transience in DTMCs can be applied to CTMCs. For example, in a finite

Markov chain, all states cannot be transient and if the finite Markov chain is irreducible, it is recurrent.

Note that the transition matrix of the embedded Markov chain for the Poisson process (Example 5.2) satisfies $\lim_{n\to\infty} T^n = \mathbf{0}$. For sufficiently large n and all i, $t_{ii}^{(n)} = 0$, which implies $\sum_{n=0}^{\infty} t_{ii}^{(n)} < \infty$. Therefore, every state is transient in the Poisson process. This is an obvious result since each state $X(W_i) = i$ can only advance to state $i+1$, $X(W_{i+1}) = i+1$; a return to state i is impossible.

Unfortunately, the concepts of null recurrence and positive recurrence for a continuous-time chain cannot be defined in terms of the embedded Markov chain. Positive recurrence depends on the waiting times $\{W_i\}_{i=0}^{\infty}$ so that the embedded Markov chain alone is not sufficient to define positive recurrence. See Schinazi (1999) for an example of an embedded Markov chain that is null recurrent but the corresponding CTMC is positive recurrent.

Recall the definitions of positive recurrence and null recurrence in DTMCs. State i is *positive recurrent (null recurrent)* in the DTMC $\{Y_n\}_{n=0}^{\infty}$ if the mean recurrence time is finite (infinite), $\sum_{n=1}^{\infty} nf_{ii}^{(n)} < \infty$ $(= \infty)$. Positive and null recurrence in a CTMC depend on the expected value of the random variable T_{ii}.

DEFINITION 5.7 *State i is* positive recurrent (null recurrent) *in the CTMC* $\{X(t) : t \in [0, \infty)\}$ *if the* mean recurrence time *is finite (infinite); that is,*

$$\mu_{ii} = E(T_{ii}|X(0) = i) < \infty \quad (= \infty).$$

This definition is not very useful to show positive or null recurrence. Instead, the next theorem gives a method that can be used to determine μ_{ii}, and hence, positive or null recurrence. There are a number of limit theorems for CTMCs that give results similar to those for DTMCs. Recall the basic limit theorem for aperiodic DTMCs: A DTMC $\{Y_n\}_{n=0}^{\infty}$ that is recurrent, irreducible, and aperiodic with transition matrix $T = (t_{ji})$ satisfies

$$\lim_{n\to\infty} t_{ji}^{(n)} = \frac{1}{\mu_{jj}},$$

where μ_{jj} is the mean recurrence time for the DTMC. There is no concept of aperiodic and periodic in CTMCs because the interevent time is random. Therefore, the basic limit theorem for CTMCs is simpler. See Norris (1997) for a proof of the following result.

THEOREM 5.2 Basic Limit Theorem for CTMCs
If the CTMC $\{X(t) : t \in [0, \infty)\}$ *is nonexplosive and irreducible, then for all i and j,*

$$\lim_{t\to\infty} p_{ij}(t) = -\frac{1}{q_{ii}\mu_{ii}}, \tag{5.17}$$

where μ_{ii} is the mean recurrence time, $0 < \mu_{ii} \leq \infty$. In particular, a finite, irreducible CTMC is nonexplosive and the limit (5.17) exists and is positive.

Matrix Q is irreducible iff matrix T is irreducible . The result (5.17) differs slightly from DTMCs since an additional term $-q_{ii}$ is needed to define the limit. The additional term is needed because the units of μ_{ii} are time and the units of q_{ii} are 1/time. The value of $q_{ii} < 0$ in the limit (5.17) because the Markov chain is irreducible which means state i cannot be absorbing (if $q_{ii} = 0$, then $q_{ji} = 0$ which means state i is absorbing). However, μ_{ii} can be infinite if the Markov chain is either null recurrent or transient. If the Markov chain is null recurrent and irreducible, then $\lim_{t \to \infty} p_{ij}(t) = 0$. On the other hand, if the Markov chain is positive recurrent and irreducible, then the limit in (5.17) is positive. For finite Markov chains, all that is needed to show the existence of a positive limit (5.17) is to show that the generator matrix Q is irreducible. The following result is reminiscent of the results for a finite DTMC.

COROLLARY 5.1
A finite, irreducible CTMC is positive recurrent.

Example 5.4. Consider Example 5.3. Matrix T given in equation (5.15) and matrix Q given in equation (5.14) are irreducible. All states are positive recurrent. Notice that the embedded Markov chain is periodic with period 4. However, the CTMC is not periodic because periodicity is not defined for a CTMC. ∎

5.6 Kolmogorov Differential Equations

The forward and backward Kolmogorov differential equations are expressions for the rate of change of the transition probabilities. The transition probability $p_{ji}(t + \Delta t)$ can be expanded by applying the Chapman-Kolmogorov equations,

$$p_{ji}(t + \Delta t) = \sum_{k=0}^{\infty} p_{jk}(\Delta t) p_{ki}(t).$$

Given that the generator matrix Q exists, the identity (5.9) can be applied,

$$p_{ji}(t + \Delta t) = \sum_{k=0}^{\infty} p_{ki}(t) \left[\delta_{jk} + q_{jk} \Delta t + o(\Delta t) \right].$$

Subtract $p_{ji}(t)$, divide by Δt, and apply the identity $\sum_{k=0}^{\infty} p_{ki}(t) = 1$,

$$\frac{p_{ji}(t + \Delta t) - p_{ji}(t)}{\Delta t} = \sum_{k=0}^{\infty} p_{ki}(t) \left[q_{jk} + \frac{o(\Delta t)}{\Delta t} \right] = \sum_{k=0}^{\infty} p_{ki}(t) q_{jk} + \frac{o(\Delta t)}{\Delta t}.$$

Let $\Delta t \to 0$. Then

$$\frac{dp_{ji}(t)}{dt} = \sum_{k=0}^{\infty} q_{jk} p_{ki}(t), \quad i, j = 0, 1, \ldots. \tag{5.18}$$

DEFINITION 5.8   *The system of equations (5.18) is known as the forward Kolmogorov differential equations. Expressed in matrix form, they are*

$$\frac{dP(t)}{dt} = QP(t), \tag{5.19}$$

where $P(t) = (p_{ji}(t))$ is the matrix of transition probabilities and $Q = (q_{ji})$ is the generator matrix.

In physics and chemistry, the system (5.19) is often referred to as the *master equation* (or chemical master equation). One of the first references to the "master equation" was in a paper by Nordsieck, Lamb, and Uhlenbeck in 1940. A differential equation was derived for probability of the energy distribution of the particles which emerge after an electron with a fixed initial energy falls on a layer of matter. The name "master" was used because from this equation "all other equations can be derived" (Nordsieck et al., 1940).

In the case that the initial distribution of the process is $X(0) = k$ ($p_i(0) = \delta_{ik}$), then the transition probability $p_{ik}(t)$ is the same as the state probability $p_i(t) = \text{Prob}\{X(t) = i | X(0) = k\}$. Therefore, in this case, the state probabilities are solutions of the forward Kolmogorov differential equations,

$$\frac{dp(t)}{dt} = Qp(t), \tag{5.20}$$

where $p(t) = (p_0(t), p_1(t), \ldots)^{tr}$.

The system of differential equations (5.20) can be approximated by a system of difference equations, the forward equations corresponding to the DTMC: $p(n + 1) = Pp(n)$. This approximation shows the relationship between the Kolmogorov differential equations and DTMC models. In particular, if the derivative $dp(t)/dt$ is approximated by the finite difference scheme, $[p(t+\Delta t) - p(t)]/\Delta t$, then the differential equation (5.20) can be expressed as

$$p(t + \Delta t) \approx [Q\Delta t + I] p(t),$$

where I is an infinite dimensional identity matrix, $I = \text{diag}(1, 1, \ldots)$. Suppose time is measured in units of Δt and $1 + q_{ii}\Delta t > 0$. Then it can be shown that the matrix $P = Q\Delta t + I$ is a stochastic matrix and

$$p(n + 1) \approx Pp(n),$$

where the unit length of time n to $n+1$ is Δt (Exercise 9).

The backward Kolmogorov differential equations can be derived in a manner similar to the forward Kolmogorov equations. Apply the Chapman-Kolmogorov equations and make the following substitutions:

$$p_{ji}(t + \Delta t) = \sum_{k=0}^{\infty} p_{ki}(\Delta t) p_{jk}(t) = \sum_{k=0}^{\infty} [\delta_{ki} + q_{ki}\, \Delta t + o(\Delta t)]\, p_{jk}(t).$$

Simplifications similar to the derivation of the forward equations and the assumption $\sum_{k=0}^{\infty} p_{jk}(t) < \infty$ yield the following system of differential equations:

$$\frac{dp_{ji}(t)}{dt} = \sum_{k=0}^{\infty} p_{jk}(t) q_{ki}, \quad i, j = 0, 1, \ldots. \tag{5.21}$$

DEFINITION 5.9   *The system of equations (5.21) is known as the* backward Kolmogorov differential equations. *Expressed in matrix form, they are*

$$\frac{dP(t)}{dt} = P(t)Q,$$

where $P(t) = (p_{ji}(t))$ is the matrix of transition probabilities and $Q = (q_{ji})$ is the generator matrix.

The backward Kolmogorov differential equations are useful in first passage time problems such as the distribution for the time to reach a particular state or the probability of reaching a particular state for the first time. Both of these problems depend on the initial state of the process. Such types of problems were investigated in Chapter 3, e.g., mean first passage time or expected duration. In these problems, backward equations were also applied.

The Kolmogorov differential equations depend on the existence of the generator matrix Q. For finite-dimensional systems or finite Markov chains, Q always exists. The solution $P(t)$ can be found via the forward or backward equations. In birth and death chains and other applications, the transition matrix $P(t)$ is defined in such a way that the forward and backward Kolmogorov differential equations can be derived.

5.7   Stationary Probability Distribution

The Kolmogorov differential equations can be used to define a stationary probability distribution. A constant solution to (5.20) is a stationary probability distribution. A formal definition is given next.

DEFINITION 5.10 *Let $\{X(t) : t \in [0, \infty)\}$ be a CTMC with generator matrix Q. Suppose $\pi = (\pi_0, \pi_1, \ldots)^{tr}$ is nonnegative, i.e., $\pi_i \geq 0$ for $i = 0, 1, 2, \ldots,$*

$$Q\pi = 0, \quad \text{and} \quad \sum_{i=0}^{\infty} \pi_i = 1.$$

Then π is called a stationary probability distribution *of the CTMC.*

A stationary probability distribution π can be defined in terms of the transition matrix $P(t)$ as well. A constant solution π is called a *stationary probability distribution* if

$$P(t)\pi = \pi, \quad \text{for} \;\; t \geq 0, \;\; \sum_{i=0}^{\infty} \pi_i = 1, \;\; \text{and} \;\; \pi_i \geq 0$$

for $i = 0, 1, 2 \ldots$. This latter definition can be applied if the transition matrix $P(t)$ is known and the process is nonexplosive.

The two definitions involving $P(t)$ and Q for the stationary probability distribution are equivalent if the transition matrix $P(t)$ is a solution of the forward and backward Kolmogorov differential equations [i.e., $dP(t)/dt = QP(t)$ and $dP(t)/dt = P(t)Q$]. This is always the case for finite Markov chains. The equivalence of these two definitions can be seen as follows. If $Q\pi = 0$, then for a finite Markov chain

$$\left[\frac{dP(t)}{dt}\right]\pi = \frac{d[P(t)\pi]}{dt} = P(t)Q\pi = 0,$$

which implies $P(t)\pi = \text{constant}$ for all t. But $P(0) = I$ implies $P(t)\pi = \pi$. On the other hand, if $P(t)\pi = \pi$ in a finite Markov chain, then

$$0 = \frac{d[P(t)\pi]}{dt} = \left[\frac{dP(t)}{dt}\right]\pi = QP(t)\pi = Q\pi.$$

An explicit solution $P(t)$ cannot be found for many continuous-time Markov processes. Definition 5.10 is the one that will be applied most often to find the stationary probability distribution.

It can be shown that the limiting distribution in the Basic Limit Theorem is a stationary probability distribution if the CTMC is nonexplosive, positive recurrent, and irreducible (Norris, 1997). That is, Theorem 2.3, the Basic Limit Theorem for aperiodic DTMC can be extended to CTMC. The theorem is stated below. A proof can be found in Norris (1997).

THEOREM 5.3

Let $\{X(t) : t \in [0, \infty)\}$ be a nonexplosive, positive recurrent, and irreducible CTMC with transition matrix $P(t) = (p_{ij}(t))$ and generator matrix $Q = (q_{ij})$.

Then there exists a unique positive stationary probability distribution π, $Q\pi = 0$, such that

$$\lim_{n \to \infty} p_{ij}(t) = \pi_i, \quad i, j = 1, 2, \ldots.$$

It follows that the mean recurrence time can be computed from the stationary distribution:

$$\boxed{\pi_i = -\frac{1}{q_{ii}\mu_{ii}} > 0.} \tag{5.22}$$

Example 5.5. The CTMC in Example 5.3 is irreducible. There exists a unique positive stationary probability distribution, $\pi = 1/4(1,1,1,1)^{tr}$. It will be shown for a finite Markov chain that irreducibility is sufficient for uniqueness of the stationary distribution and for applying identity (5.22). The mean recurrence times are $\mu_{ii} = 4$, $i = 1, 2, 3, 4$. ∎

5.8   Finite Markov Chains

For a finite, irreducible CTMC, the assumptions of nonexplosiveness and positive recurrence are not required to obtain the limiting distribution. The following result is a consequence of Theorem 5.2 (Basic Limit Theorem), Corollary 5.1, and Theorem 5.3.

COROLLARY 5.2

Let $\{X(t) : t \in [0, \infty)\}$ be a finite, irreducible CTMC with transition matrix $P(t) = (p_{ij}(t))$ and generator matrix $Q = (q_{ij})$. Then there exists a unique stationary probability distribution π, $Q\pi = 0$, such that

$$\lim_{n \to \infty} p_{ij}(t) = \pi_i = -\frac{1}{q_{ii}\mu_{ii}}, \quad i, j = 1, 2, \ldots, N.$$

Example 5.6. Suppose the generator matrix of a CTMC with two states is

$$Q = \begin{pmatrix} -a & b \\ a & -b \end{pmatrix},$$

where $a > 0$ and $b > 0$. Matrix Q is irreducible, so the CTMC is irreducible. The unique stationary distribution is $\pi = (b/(a+b), a/(a+b))^{tr}$. Hence, the mean recurrence times are

$$\mu_{ii} = -\frac{1}{q_{ii}\pi_i} = \frac{a+b}{ab}, \quad i = 1, 2.$$

∎

For some finite Markov chains it is possible to find an explicit solution $P(t)$ to the forward and backward Kolmogorov differential equations. In the case that $X(0) = k$ or $p_i(0) = \delta_{ik}$, then the probability distribution $p(t)$ is the solution of the forward Kolmogorov equation so that $p(t) = P(t)p(0)$.

Assume the state space of a finite Markov chain is $\{0, 1, 2, \ldots, N\}$. The forward and backward Kolmogorov differential equations are $dP/dt = QP$ and $dP/dt = PQ$, respectively, where $P(0) = I$. The systems are linear, and the unique solution of each of them is given by

$$P(t) = e^{Qt}P(0) = e^{Qt},$$

where e^{Qt} is the matrix exponential,

$$e^{Qt} = I + Qt + Q^2\frac{t^2}{2!} + Q^3\frac{t^3}{3!} + \cdots = \sum_{k=0}^{\infty} Q^k\frac{t^k}{k!}.$$

This result can be verified easily. Let $P(t) = e^{Qt}$. Then differentiation yields

$$\frac{dP(t)}{dt} = \sum_{k=1}^{\infty} Q^k\frac{t^{k-1}}{(k-1)!} = Q\sum_{k=0}^{\infty} Q^k\frac{t^k}{k!} = Qe^{Qt} = QP(t).$$

However, it is also true that

$$\frac{dP(t)}{dt} = \left(\sum_{k=0}^{\infty} Q^k\frac{t^k}{k!}\right)Q = e^{Qt}Q = P(t)Q.$$

Then e^{Qt} is the solution of the forward and backward Kolmogorov differential equations, $dP/dt = QP$ and $dP/dt = PQ$, respectively. Uniqueness follows from the theory of differential equations (Brauer and Nohel, 1969).

Example 5.7. Suppose the generator matrix of a CTMC is the same as in the previous example,

$$Q = \begin{pmatrix} -a & b \\ a & -b \end{pmatrix},$$

$a, b > 0$. Note that $Q^2 = -(a+b)Q$ and, in general, $Q^n = [-(a+b)]^{n-1}Q$. Then $P(t) = e^{Qt} = I + \sum_{n=1}^{\infty}\frac{(Qt)^n}{n!}$. Applying the identity for Q^n, it follows that

$$P(t) = I - \frac{Q}{a+b}\sum_{n=1}^{\infty}\frac{[-(a+b)t]^n}{n!} = I - \frac{Q}{a+b}[e^{-(a+b)t} - 1]. \qquad (5.23)$$

The limit

$$\lim_{t\to\infty} P(t) = \begin{pmatrix} \dfrac{b}{a+b} & \dfrac{b}{a+b} \\ \dfrac{a}{a+b} & \dfrac{a}{a+b} \end{pmatrix},$$

where the columns of the limiting matrix are the stationary probability distribution π. ∎

Numerous methods exist for obtaining an expression for the matrix exponential (see e.g., Leonard, 1996; Moler and Van Loan, 1978; Waltman, 1986). Analytical methods similar to the methods discussed in Chapter 2, Section 2.13.3 are discussed in the Appendix for Chapter 5. In addition, computational methods or a computer algebra system can be used to obtain an expression for e^{Qt}.

Example 5.8. In Example 5.3, the forward Kolmogorov differential equations $dP/dt = QP$ have a generator matrix Q and transition matrix T of the embedded Markov chain equal to

$$Q = \begin{pmatrix} -1 & 0 & 0 & 1 \\ 1 & -1 & 0 & 0 \\ 0 & 1 & -1 & 0 \\ 0 & 0 & 1 & -1 \end{pmatrix} \quad \text{and} \quad T = \begin{pmatrix} 0 & 0 & 0 & 1 \\ 1 & 0 & 0 & 0 \\ 0 & 1 & 0 & 0 \\ 0 & 0 & 1 & 0 \end{pmatrix},$$

respectively. The matrix exponential e^{Qt} can be obtained by any of the preceding methods or via a computational method. There are built-in commands for many software programs that will compute matrix exponentials. For example, the computer algebra system Maple^{TM} was used to compute

$$e^{Qt} = \frac{1}{4}E + \frac{1}{2}e^{-t} \begin{pmatrix} \cos(t) & -\sin(t) & -\cos(t) & \sin(t) \\ \sin(t) & \cos(t) & -\sin(t) & -\cos(t) \\ -\cos(t) & \sin(t) & \cos(t) & -\sin(t) \\ -\sin(t) & -\cos(t) & \sin(t) & \cos(t) \end{pmatrix}$$

$$+ \frac{1}{4}e^{-2t} \begin{pmatrix} 1 & -1 & 1 & -1 \\ -1 & 1 & -1 & 1 \\ 1 & -1 & 1 & -1 \\ -1 & 1 & -1 & 1 \end{pmatrix},$$

where E is a 4×4 matrix of ones. Thus,

$$\lim_{t \to \infty} P(t) = \frac{1}{4}E.$$

The columns of the limit are the stationary probability distribution $\pi = \frac{1}{4}(1,1,1,1)^{tr}$. As shown in Example 5.5, the mean recurrence times are $\mu_{ii} = 4$, $i = 1,2,3,4$. ∎

The next example is a finite CTMC with all states transient, except for a single absorbing state.

Example 5.9. Suppose the generator matrix Q of a CTMC and transition matrix T of the embedded Markov chain are

$$Q = \begin{pmatrix} 0 & 1 & 0 & 0 \\ 0 & -1 & 2 & 0 \\ 0 & 0 & -2 & 3 \\ 0 & 0 & 0 & -3 \end{pmatrix} \quad \text{and} \quad T = \begin{pmatrix} 1 & 1 & 0 & 0 \\ 0 & 0 & 1 & 0 \\ 0 & 0 & 0 & 1 \\ 0 & 0 & 0 & 0 \end{pmatrix}.$$

Matrices T and Q are reducible because the first state is absorbing. The eigenvalues of Q are $0, -1, -2$, and -3. A closed form expression can be obtained for e^{Qt}. Matrix

$$e^{Qt} = \begin{pmatrix} 1 & 1-e^{-t} & 1-2e^{-t}+e^{-2t} & 1-3e^{-t}+3e^{-2t}-e^{-3t} \\ 0 & e^{-t} & 2e^{-t}-2e^{-2t} & 3e^{-t}-6e^{-2t}+3e^{-3t} \\ 0 & 0 & e^{-2t} & 3e^{-2t}-3e^{-3t} \\ 0 & 0 & 0 & e^{-3t} \end{pmatrix}.$$

Although Corollary 5.2 does not apply because Q is reducible, the limiting distribution is the unique stationary probability distribution. The columns of e^{Qt} approach $(1,0,0,0)^{tr}$; all states are absorbed into state one (Figure 5.4). This example is a special case of a simple death process that will be discussed more fully in Chapter 6. ∎

FIGURE 5.4: Directed graph of the embedded Markov chain $\{Y_n\}_{n=0}^{\infty}$.

5.9 Generating Function Technique

In this section, another method is proposed for obtaining information about the probability distribution associated with the CTMC. In this method, a partial differential equation is derived so that the solution of the equation is a generating function. Depending on the equation, the solution is either a probability generating function (p.g.f.), a moment generating function (m.g.f.) or a cumulant generating function (c.g.f.). Denote the p.g.f. of a CTMC $\{X(t) : t \in [0, \infty)\}$ as

$$\mathcal{P}(z,t) = \sum_{i=0}^{\infty} p_i(t) z^i,$$

the m.g.f. as

$$M(\theta,t) = \sum_{i=0}^{\infty} p_i(t) e^{\theta i},$$

and the c.g.f. as $K(\theta,t) = \ln M(\theta,t)$. Notice that the generating functions depend on two continuous variables, z and t or θ and t, where $t \geq 0$, and the domain of z or θ consists of the values where the summation converges [e.g., $|z| < 1$ for $P(z,t)$].

The mean $m(t)$ of the process at time t is

$$m(t) = \left.\frac{\partial \mathcal{P}(z,t)}{\partial z}\right|_{z=1} = \sum_{i=0}^{\infty} i p_i(t)$$

Recall from Chapter 1 that

$$m(t) = \left.\frac{\partial M(\theta,t)}{\partial \theta}\right|_{\theta=0} = \left.\frac{\partial K(\theta,t)}{\partial \theta}\right|_{\theta=0}.$$

In addition, the variance $\sigma^2(t)$ is

$$\sigma^2(t) = \left.\frac{\partial^2 \mathcal{P}(z,t)}{\partial z^2}\right|_{z=1} + \left.\frac{\partial \mathcal{P}(z,t)}{\partial z}\right|_{z=1} - \left(\left.\frac{\partial \mathcal{P}(z,t)}{\partial z}\right|_{z=1}\right)^2$$

or, in terms of $M(\theta,t)$ and $K(\theta,t)$,

$$\sigma^2(t) = \left.\frac{\partial^2 M(\theta,t)}{\partial \theta^2}\right|_{\theta=0} - \left(\left.\frac{\partial M(\theta,t)}{\partial \theta}\right|_{\theta=0}\right)^2 = \left.\frac{\partial^2 K(\theta,t)}{\partial \theta^2}\right|_{\theta=0}.$$

A partial differential equation is derived from the forward Kolmogorov equations. The p.g.f. is a solution of this differential equation. First, the technique is discussed; then an example is given using the Poisson process. When the initial distribution is a fixed value, then the forward Kolmogorov differential equations can be expressed in terms of the state probabilities, $dp/dt = Qp$. That is,

$$\frac{dp_i(t)}{dt} = \sum_{k=0}^{\infty} q_{ik} p_k(t), \quad i = 0, 1, 2, \ldots.$$

Multiplying each of these equations by z^i and summing over i,

$$\sum_{i=0}^{\infty} \frac{dp_i(t)}{dt} z^i = \sum_{i=0}^{\infty} \sum_{k=0}^{\infty} q_{ik} p_k(t) z^i.$$

Interchanging the summation and differentiation and the order of the summation (possible for values of t and z, where the summation converges absolutely), yields

$$\frac{\partial \mathcal{P}(z,t)}{\partial t} = \sum_{k=0}^{\infty} \left[\sum_{i=0}^{\infty} q_{ik} p_k(t) z^i\right].$$

If for each i, q_{ik} is zero, except for finitely many k, then the right-hand side is the sum of a finite number of terms that are either $\mathcal{P}$ or the first-, second-, or higher-order derivatives of $\mathcal{P}$ with respect to z.

A partial differential equation for the m.g.f. can be derived in a similar manner. Instead of multiplying by z^i, the forward Kolmogorov differential equation is multiplied by $e^{i\theta}$. Alternately, a differential equation for the m.g.f.

can be derived directly from the differential equation for the p.g.f. by making a change of variable. Recall that $M(\theta, t) = \mathcal{P}(e^\theta, t)$, so that $z = e^\theta$. In addition, a differential equation for the c.g.f. can be obtained from the one for the m.g.f. by letting $K(\theta, t) = \ln M(\theta, t)$. The generating function technique will be used frequently with birth and death chains and other applications. If the differential equations are first-order, they can be solved by the method of characteristics. The last section of this chapter is devoted to a brief review of the method of characteristics. The generating function technique is illustrated in the next example for the Poisson process.

Example 5.10. The forward Kolmogorov differential equations for the Poisson process given in equations (5.2) and (5.4) are

$$\frac{dp_i\,(t)}{dt} = -\lambda p_i(t) + \lambda p_{i-1}(t), \quad i \geq 1,$$

$$\frac{dp_0\,(t)}{dt} = -\lambda p_0(t).$$

Multiplying by z^i, then summing over i, leads to the equation

$$\sum_{i=0}^{\infty} \frac{dp_i\,(t)}{dt} z^i = -\lambda \sum_{i=0}^{\infty} p_i(t) z^i + \lambda \sum_{i=1}^{\infty} p_{i-1}(t) z^i.$$

Interchanging differentiation and summation yields the differential equation,

$$\frac{\partial \mathcal{P}(z,t)}{\partial t} = -\lambda \mathcal{P}(z,t) + z\lambda \mathcal{P}(z,t) = \lambda(z-1)\mathcal{P}(z,t).$$

Because there is no differentiation with respect to z, the variable z can be treated as a constant. The solution of this differential equation is an exponential function of t,

$$\mathcal{P}(z,t) = \mathcal{P}(z,0)e^{\lambda(z-1)t}.$$

Recall that $X(0) = 0$. Thus, $p_0(0) = \text{Prob}\{X(0) = 0\} = 1$ and $p_i(0) = \text{Prob}\{X(0) = i\} = 0$ for $i = 1, 2, \ldots$. Hence, $\mathcal{P}(z,0) = 1$ and the p.g.f. is

$$\mathcal{P}(z,t) = e^{\lambda t(z-1)}.$$

Replacing z by e^θ yields the m.g.f. $M(\theta, t) = e^{\lambda t(e^\theta - 1)}$, and taking logarithms yields the c.g.f, $K(\theta, t) = \lambda t(e^\theta - 1)$. As expected, the p.g.f., m.g.f., and c.g.f. are the generating functions corresponding to a Poisson distribution with parameter λt. ■

5.10 Interevent Time and Stochastic Realizations

To calculate sample paths of a CTMC $\{X(t) : t \in [0, \infty)\}$, it is necessary to know the distribution for the time between successive events or the *interevent* time. Recall that the random variable for the interevent time is continuous and nonnegative, $T_i = W_{i+1} - W_i \geq 0$, where W_i is the time of the ith jump (Figure 5.5). In applications, the event may be a birth, death, immigration, or any other event that changes the value of the state variable. It will be shown that the interevent time T_i is an exponential random variable.

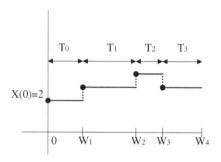

FIGURE 5.5: Sample path $X(t)$ of a CTMC, illustrating the jump times $\{W_i\}_{i=0}^{\infty}$ and the interevent times $\{T_i\}_{i=0}^{\infty}$, $X(0) = 2$, $X(W_1) = 3$, $X(W_2) = 4$, and $X(W_3) = 3$.

Assume the value of the state at the ith jump is n, $X(W_i) = n$. Let $\alpha(n)\Delta t + o(\Delta t)$ be the probability that the process moves to a state different from n in the time period Δt. That is,

$$\sum_{j \neq n} p_{jn}(\Delta t) = \alpha(n)\Delta t + o(\Delta t).$$

If $p_{jn}(\Delta t) = \alpha_j(n)\Delta t + o(\Delta t)$, then the functions $\alpha_j(n)$ are known as *propensity functions* in biochemical reactions, $\alpha(n) = \sum_{j, j \neq n} \alpha_j(n)$ (Gillespie, 1976; 1977; 2001). The probability of no change in state is $1 - \alpha(n)\Delta t + o(\Delta t)$; that is,

$$p_{nn}(\Delta t) = 1 - \alpha(n)\Delta t + o(\Delta t).$$

Let $G_i(t)$ be the probability that the process remains in state n for a time of length t, that is, for a time of length $[W_i, W_i + t]$. Then $G_i(t)$ can be expressed in terms of the interevent time T_i,

$$G_i(t) = \text{Prob}\{t + W_i < W_{i+1}\} = \text{Prob}\{T_i > t\}.$$

If state n is not an absorbing state, so that there is a positive probability of moving to another state. Then, at $t = 0$, $G_i(0) = \text{Prob}\{T_i > 0\} = 1$. For Δt sufficiently small,

$$G_i(t + \Delta t) = G_i(t)p_{nn}(\Delta t) = G_i(t)(1 - \alpha(n)\Delta t + o(\Delta t)). \qquad (5.24)$$

Subtracting $G_i(t)$ from both sides of the preceding equation and dividing by Δt, then taking the limit as $\Delta t \to 0$, it follows that

$$\frac{dG_i(t)}{dt} = -\alpha(n)G_i(t).$$

The differential equation is first-order and homogeneous with initial condition $G_i(0) = 1$. The solution is

$$G_i(t) = \text{Prob}\{T_i > t\} = e^{-\alpha(n)t}.$$

Thus, the probability that $T_i \leq t$ is

$$\text{Prob}\{T_i \leq t\} = 1 - G_i(t) = 1 - e^{-\alpha(n)t} = F_i(t), \quad t \geq 0.$$

The function $F_i(t)$ is the cumulative distribution function for the interevent time T_i, which corresponds to an exponential random variable with parameter $\alpha(n)$. The p.d.f. for T_i is $F_i'(t) = f_i(t) = \alpha(n)e^{-\alpha(n)t}$. Recall that the mean and variance for an exponential random variable with parameter λ are $E(T_i) = 1/\alpha(n)$ and $\text{Var}(T_i) = 1/[\alpha(n)]^2$. These results are summarized in the next theorem.

THEOREM 5.4 Interevent time
Let $\{X(t) : t \in [0, \infty)\}$ be a CTMC with transition matrix $P(t) = (p_{ij}(t))$ such that

$$\sum_{j \neq n} p_{jn}(\Delta t) = \alpha(n)\Delta t + o(\Delta t)$$

and

$$p_{nn}(\Delta t) = 1 - \alpha(n)\Delta t + o(\Delta t)$$

for Δt sufficiently small. Then the interevent time, $T_i = W_{i+1} - W_i$, given $X(W_i) = n$, is an exponential random variable with parameter $\alpha(n)$. The c.d.f. for T_i is $F_i(t) = 1 - e^{-\alpha(n)t}$ so that the mean and variance of T_i are

$$E(T_i) = \frac{1}{\alpha(n)} \quad \text{and} \quad \text{Var}(T_i) = \frac{1}{[\alpha(n)]^2}.$$

For example, in a birth process with birth probability $b_n \Delta t + o(\Delta t)$, given $X(W_i) = n$, there will be a mean waiting time of $E(T_i) = 1/b_n$, until another birth, $X(W_{i+1}) = n+1$. Suppose there is more than one event, such as a birth or a death and $X(W_i) = n$. If the death probability is $d_n \Delta t + o(\Delta t)$, there will be a mean waiting time of $E(T_i) = 1/(b_n + d_n)$. When an event occurs, it will be a birth with probability $b_n/(b_n + d_n)$ and a death with probability $d_n/(b_n + d_n)$. In this latter case, there are two propensity functions, b_n and d_n, where $b_n + d_n = \alpha(n)$. For more rigorous mathematical justification of the preceding arguments, please consult Karlin and Taylor (1981) or Norris (1997).

An assumption used in the preceding derivation for $G_i(t)$ is an inherent property of the exponential distribution. It was assumed in (5.24) that $\text{Prob}\{T_i \geq t + \Delta t\} = \text{Prob}\{T_i \geq t\}\text{Prob}\{T_i \geq \Delta t\}$. This property can be written as

$$\text{Prob}\{T_i \geq t + \Delta t | T_i \geq t\} = \text{Prob}\{T_i \geq \Delta t\}.$$

It is a property of the exponential distribution referred to as a *memoryless property*. It is due to this memoryless property that Markov processes have an interevent time that is exponential.

The set of interevent times $\{T_i\}_{i=0}^{\infty}$ have an important probabilistic relationship. The interevent times $\{T_i\}_{i=0}^{\infty}$ are independent if conditioned on the successive states visited by the Markov chain. In particular, given $X(W_i) = Y_i$, then the interevent time T_i is independent of T_{i-1}, $i = 1, 2, \ldots$ (Schinazi, 1999).

Example 5.11. A simple birth process is defined. Let $\Delta X(t) = X(t + \Delta t) - X(t)$. For Δt sufficiently small, the transition probabilities are

$$p_{i+j,i}(\Delta t) = \text{Prob}\{\Delta X(t) = j | X(t) = i\}$$
$$= \begin{cases} bi\Delta t + o(\Delta t), & j = 1 \\ 1 - bi\Delta t + o(\Delta t), & j = 0 \\ o(\Delta t), & j \geq 2 \\ 0, & j < 0. \end{cases}$$

Denote the expected time to reach state k, $k \geq 2$, as τ_k. In general, if $X(W_i) = k$, then the interevent time T_i has p.d.f. bke^{-bkt} and the expected time to reach state $k + 1$ from state k is $E(T_i | X(W_i) = k) = 1/(bk)$. If, for example, $X(0) = 1$, then $X(W_1) = 2$, and, in general, $X(W_k) = k + 1$. The expected time to reach state k beginning from state 1 can be easily computed. First, $\tau_2 = E(T_0) = 1/b$. Then

$$\tau_3 = E(T_0) + E(T_1) = \frac{1}{b} + \frac{1}{2b} = \frac{1}{b}\left[1 + \frac{1}{2}\right],$$

and, in general,

$$\tau_k = \sum_{i=0}^{k-2} E(T_i) = \frac{1}{b}\sum_{i=1}^{k-1}\frac{1}{i}.$$

∎

Next, it is shown that the random variable T_i can be expressed in terms of the distribution function $F_i(t)$ and a uniform random variable U. This relationship is very useful for computational purposes.

THEOREM 5.5

Let U be a uniform random variable defined on $[0, 1]$ and T be a continuous random variable defined on $[0, \infty)$. Then $T = F^{-1}(U)$, where F is the cumulative distribution of the random variable T.

Proof. Since $\text{Prob}\{T \leq t\} = F(t)$, we want to show that $\text{Prob}\{F^{-1}(U) \leq t\} = F(t)$. First note that $F : [0, \infty) \to [0, 1)$ is strictly increasing, so that F^{-1} exists. In addition, for $t \in [0, \infty)$,

$$\text{Prob}\{F^{-1}(U) \leq t\} = \text{Prob}\{F(F^{-1}(U)) \leq F(t)\}$$
$$= \text{Prob}\{U \leq F(t)\}.$$

Because U is a uniform random variable, $\text{Prob}\{U \leq y\} = y$ for $y \in [0, 1]$. Thus, $\text{Prob}\{U \leq F(t)\} = F(t)$. $\qquad\square$

In the Poisson process, the only change in state is a birth that occurs with probability $\lambda \Delta t + o(\Delta t)$ in a small interval of time Δt. Because $\alpha(n) = \lambda$, the distribution function for the interevent time is $F_i(t) = \text{Prob}\{T_i \leq t\} = 1 - \exp(-\lambda t)$. But because λ is independent of the state of the process, the interevent time is the same for every jump i, $T_i \equiv T$. The interevent time T, expressed in terms of the uniform random variable U, is $T = F^{-1}(U)$. The function $F^{-1}(U)$ is found by solving $F(T) = 1 - \exp(-\lambda T) = U$ for U:

$$T = F^{-1}(U) = -\frac{\ln(1 - U)}{\lambda}.$$

However, because U is a uniform random variable on $[0, 1]$, so is $1 - U$. It follows that the interevent time can be expressed in terms of a uniform random variable U as follows:

$$T = -\frac{\ln(U)}{\lambda}. \tag{5.25}$$

For more general processes, the formula given in (5.25) for the interevent time depends on the state of the process. In particular, given $X(W_i) = n$, the interevent time T_i is

$$\boxed{T_i = -\frac{\ln(U)}{\alpha(n)},} \tag{5.26}$$

where U is a uniform random variable on $[0,1]$.

The formula given in (5.26) for the interevent time is applied to three simple birth and death processes, known as simple birth, simple death, and simple birth and death processes. In each of these processes, probabilities of births and deaths are linear functions of the population size. These processes will be considered in more detail in Chapter 6. In all cases, $X(t)$ is the random variable for the total population size at time t.

Example 5.12 (Simple birth process). Consider the simple birth process defined in Example 5.11. Given $X(W_i) = n$, the probability of a change in the population size is $bn\Delta t + o(\Delta t)$. Thus, $\alpha(n) = bn$. The interevent time T_i is

$$T_i = -\frac{\ln(U)}{bn},$$

where U is a uniform random variable. The next event is a birth $n \to n+1$, $X(W_{i+1}) = n+1$. The *deterministic* analogue of this simple birth process is the exponential growth model $dn/dt = bn$, $n(0) = N$, whose solution is $n(t) = Ne^{bt}$. ∎

The MATLAB® program in the Appendix for Chapter 5 generates three sample paths or realizations of a simple birth process when $b = 1$ and $X(0) = 1$. Three sample paths are graphed in Figure 5.6.

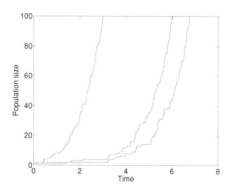

FIGURE 5.6: Three sample paths of the simple birth process with $X(0) = 1$ and $b = 1$.

Example 5.13 (Simple death process). In a simple death process, the only event is a death, that is, state $i \to i - 1$. For Δt sufficiently small, the transition probabilities are

$$
\begin{aligned}
p_{i+j,i}(\Delta t) &= \text{Prob}\{\Delta X(t) = j | X(t) = i\} \\
&= \begin{cases}
di\Delta t + o(\Delta t), & j = -1 \\
1 - di\Delta t + o(\Delta t), & j = 0 \\
o(\Delta t), & j \le -2 \\
0, & j > 0.
\end{cases}
\end{aligned} \tag{5.27}
$$

Given $X(W_i) = n$, then $\alpha(n) = dn$. Therefore, the interevent time T_i is

$$
T_i = -\frac{\ln(U)}{dn}.
$$

The next event is a death $n \to n - 1$, $X(W_{i+1}) = n - 1$. The *deterministic* analogue of this simple death process is the differential equation $dn/dt = -dn$, $n(0) = N$, with solution

$$
n(t) = Ne^{-dt}.
$$

Figure 5.7 is a graph of three sample paths of this process when $X(0) = 100$ and $d = 0.5$. ■

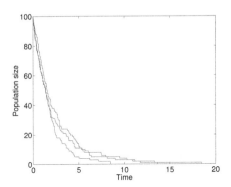

FIGURE 5.7: Three sample paths of the simple death process with $X(0) = 100$ and $d = 0.5$.

Example 5.14 (Simple birth and death process). In the simple birth and death process, an event can be a birth or a death, $i \to i+1$ or $i \to i-1$. For Δt sufficiently small, the transition probabilities are

$$
p_{i+j,i}(\Delta t) = \text{Prob}\{\Delta X(t) = j | X(t) = i\}
$$
$$
= \begin{cases}
di\Delta t + o(\Delta t), & j = -1 \\
bi\Delta t + o(\Delta t), & j = 1 \\
1 - (b+d)i\Delta t + o(\Delta t), & j = 0 \\
o(\Delta t), & j \ne -1,\, 0,\, 1.
\end{cases}
$$

Given $X(W_i) = n$, $\alpha(n) = (b+d)n$. Therefore, the interevent time T_i is

$$
T_i = -\frac{\ln(U)}{(b+d)n}.
$$

The next event is either a birth or a death; a birth occurs with probability $b/(b+d)$ and a death with probability $d/(b+d)$. The *deterministic* analogue of this simple birth and death process is the differential equation $dn/dt = (b-d)n$, $n(0) = N$, with solution

$$
n(t) = Ne^{(b-d)t}.
$$

Three sample paths are graphed in Figure 5.8 when $X(0) = 5$, $b = 1$, and $d = 0.5$. One sample path hits zero before $t = 3$. A MATLAB program for the simple birth and death process is given in the Appendix for Chapter 5. ■

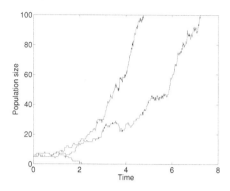

FIGURE 5.8: Three sample paths of the simple birth and death process with $X(0) = 5$, $b = 1$, and $d = 0.5$.

The stochastic simulation algorithm used to compute the sample paths for these simple birth and death processes applies what is referred to as the direct method in biochemical applications. The algorithm is also referred to as Gillespie's algorithm, named for the contributions of Daniel T. Gillespie (1976, 1977). In the direct method, two uniform random variables are needed per iteration, the first to simulate the time to the next event and the second to choose the event. The direct method works well when population sizes are small, on the order of 100, but when population sizes are large, on the order of 10^4 or larger, alternate methods are required because the time between events becomes so small that each sample path takes an extremely long time to compute. A more efficient but approximate numerical method is known as tau-leaping (Gillespie, 2001; Gillespie and Petzold, 2006). In this latter method, a fixed time τ is chosen in which more than one event occurs during this "leap". The value of τ must be chosen sufficiently small so that no propensity function ($\alpha_j(n)$) changes by a significant amount (Gillespie and Petzold, 2006). The Poisson distribution with parameter $\alpha_j(n)\tau$ is used to determine the number of times an event occurs. For more details on tau-leaping and other numerical methods for simulating sample paths please consult work by Gillespie, Petzold, and colleagues, e.g., Gillespie (2001), Gillespie and Petzold (2006) and Li et al. (2008).

5.11 Review of Method of Characteristics

For the simple birth, simple death, and simple birth and death processes, the generating functions will be first-order linear partial differential equations.

To solve these partial differential equations, the method of characteristics can be applied. For more information on the method of characteristics, a textbook on partial differential equations can be consulted (Farlow, 1982; John, 1975; Schovanec and Gilliam, 2000). The method of characteristics is illustrated with an example.

Example 5.15. Let $P(z,t)$ be a solution of the partial differential equation

$$\frac{\partial P}{\partial t} + (z+1)\frac{\partial P}{\partial z} = 1, \quad P(z,0) = \phi(z). \tag{5.28}$$

The domains for t and z are $t \in [0,\infty)$ and $z \in (-\infty,\infty)$.

In the method of characteristics, it is assumed that the partial differential equation can be expressed as a system of ordinary differential equations along characteristic curves, curves expressed in terms of auxiliary variables s and τ. Assume that $P(z,t) \equiv P(z(s,\tau),t(s,\tau)) \equiv P(s,\tau)$. Along the characteristic curves, the variable $s = $ constant, so that $P(z(s,\tau),t(s,\tau)) = P(z(\tau),t(\tau))$. The characteristic curves are found by solving the following ordinary differential equations:

$$\frac{dt}{d\tau} = 1, \quad \frac{dz}{d\tau} = z+1, \quad \text{and} \quad \frac{dP}{d\tau} = 1,$$

with initial conditions

$$t(s,0) = 0, \quad z(s,0) = s, \quad \text{and} \quad P(s,0) = \phi(s).$$

The reason this method works is that along characteristic curves, solutions of the ordinary differential equations are also solutions of the partial differential equation,

$$\frac{dP}{d\tau} = \frac{\partial P}{\partial z}\frac{dz}{d\tau} + \frac{\partial P}{\partial t}\frac{dt}{d\tau} = (z+1)\frac{\partial P}{\partial z} + \frac{\partial P}{\partial t}.$$

The system of ordinary differential equations is solved and P is expressed in terms of s and τ. Then the variables s and τ are expressed in terms of the original variables z and t.

The solution of the system of ordinary differential equations along the characteristic curves is

$$t(s,\tau) = \tau, \quad z+1 = (s+1)e^\tau, \quad \text{and} \quad P(s,\tau) = \tau + \phi(s).$$

The variables τ and s can be expressed in terms of z and t as follows: $\tau = t$ and $s = (z+1)e^{-t} - 1$. Substituting these values for τ and s into P gives the solution in terms of z and t,

$$P(z,t) = t + \phi((z+1)e^{-t} - 1).$$

For example, if $\phi(z) = z^3$, then

$$P(z,t) = t + \left[(z+1)e^{-t} - 1\right]^3.$$

The solution can be verified by checking that it solves the partial differential equation and the initial condition given in (5.28). ∎

5.12 Exercises for Chapter 5

1. The transition matrix of a simple birth process with states $\{1, 2, \ldots\}$ has the following form:

$$P(\Delta t) = \begin{pmatrix} 1 - b\Delta t & 0 & 0 & 0 & \cdots \\ b\Delta t & 1 - 2b\Delta t & 0 & 0 & \cdots \\ 0 & 2b\Delta t & 1 - 3b\Delta t & 0 & \cdots \\ 0 & 0 & 3b\Delta t & 1 - 4b\Delta t & \cdots \\ \vdots & \vdots & \vdots & \vdots & \cdots \end{pmatrix} + o(\Delta t).$$

Apply Definition 5.2 to compute the generator matrix Q for this CTMC.

2. Suppose the generator matrix of a finite CTMC is

$$Q = \begin{pmatrix} -a - b & c & e \\ a & -c - d & f \\ b & d & -e - f \end{pmatrix},$$

where the constants a, b, c, d, e, and f are positive. Compute the transition matrix T of the embedded Markov chain. Are the states recurrent or transient?

3. Suppose the generator matrix Q of a finite CTMC is

$$Q = \begin{pmatrix} -a & d & 0 & 0 \\ a & -b - d - a & 2d & 0 \\ 0 & b + a & -2b - 2d - a & 3d \\ 0 & 0 & 2b + a & -3d \end{pmatrix}.$$

(a) Suppose $b = d = a > 0$. Compute the corresponding transition matrix T for the embedded Markov chain. Are the states recurrent or transient?

(b) Suppose $a = 0$ and $b = d > 0$. Compute the corresponding transition matrix T for the embedded Markov chain. Are the states recurrent or transient?

4. Suppose the generator matrix Q of a finite CTMC is

$$Q = \begin{pmatrix} -1 & 2 \\ 1 & -2 \end{pmatrix}.$$

(a) Show that $Q^n = (-3)^{n-1}Q$.

(b) Use (a) to compute $P(t) = e^{Qt} = \sum_{n=0}^{\infty}(Qt)^n/n!$.

(c) Show that Q is irreducible. Find the following limit: $\lim_{t \to \infty} P(t)$. Then apply (5.17) to compute μ_{ii}, for $i = 1, 2$.

(d) Show that there exists a unique probability stationary distribution π, $Q\pi = 0$.

(e) Verify that the limit $\pi = \lim_{t \to \infty} P(t)p(0)$ equals the unique stationary probability distribution.

5. Consider the generator matrix Q given in Example 5.8,

$$Q = \begin{pmatrix} 0 & 1 & 0 & 0 \\ 0 & -1 & 2 & 0 \\ 0 & 0 & -2 & 3 \\ 0 & 0 & 0 & -3 \end{pmatrix}.$$

Elements of the matrix exponential, $P(t) = e^{Qt}$, can be computed using the method discussed in the Appendix. The eigenvalues of Q are $0, -1, -2,$ and -3, $Q = H\mathrm{diag}(0, -1, -2, -3)H^{-1}$ so that

$$e^{Qt} = H\mathrm{diag}(1, e^{-t}, e^{-2t}, e^{-3t})H^{-1}.$$

Compute $p_{44}(t)$ from equation (5.33):

$$p_{44}(t) = a_1 + a_2 e^{-t} + a_3 e^{-2t} + a_4 e^{-3t}.$$

Apply the initial conditions, $p_{44}(0) = 1$, $p'_{44}(0) = q_{44}$, $p''_{44}(0) = q_{44}^{(2)}$, and $p'''_{44}(0) = q_{44}^{(3)}$, to obtain four linear equations for the coefficients a_i, $i = 1, 2, 3, 4$. Solve for the a_i.

6. Suppose the generator matrix of a finite CTMC is

$$Q = \begin{pmatrix} -2 & 1 & 2 \\ 1 & -1 & 1 \\ 1 & 0 & -3 \end{pmatrix}. \tag{5.29}$$

Elements of the matrix exponential, $P(t) = e^{Qt}$, can be computed using the method discussed in the Appendix.

(a) Compute the three eigenvalues λ_i, $i = 1, 2, 3$ of matrix Q. Then express

$$p_{11}(t) = a_1 e^{\lambda_1 t} + a_2 e^{\lambda_2 t} + a_3 e^{\lambda_3 t}.$$

(b) Use the values of $p_{11}(0) = 1$, $p'_{11}(0) = q_{11}$, and $p''_{11}(0) = q_{11}^{(2)}$ to find a linear system for the coefficients a_i, $i = 1, 2, 3$. Solve for the coefficients to find $p_{11}(t)$.

7. Suppose the generator matrix of a finite CTMC is given by (5.29).

(a) Compute the corresponding transition matrix T and show that T is irreducible.

(b) Apply any method to find the matrix exponential, $P(t) = e^{Qt}$.

(c) Find the following limit: $\lim_{t \to \infty} P(t)$.

(d) Show that the limit of $P(t)$ is the stationary distribution of Q (Corollary 5.2).

(e) Compute the mean recurrence times μ_{ii}, $i = 1, 2, 3$ (Corollary 5.2).

8. Suppose the generator matrix of a finite CTMC is

$$Q = \begin{pmatrix} -1 & 4 & 2 \\ 0 & -4 & 1 \\ 1 & 0 & -3 \end{pmatrix}.$$

(a) Compute the corresponding transition matrix T of the embedded Markov chain. Is the chain irreducible or reducible?

(b) Find the following limit: $\lim_{t \to \infty} P(t)p(0)$ (Corollary 5.2).

9. When the initial distribution of the process is a fixed value, then the probability distribution $p(t)$ is a solution of the forward Kolmogorov differential equations,

$$\frac{dp(t)}{dt} = Qp(t).$$

(a) Show that the differential equation can be approximated by

$$p(n + 1) = Pp(n),$$

where $n = t$, the unit length of time n to $n + 1$ is Δt, and $P = Q\Delta t + I$.

(b) Assume the elements q_{ii} are finite and $1 + q_{ii}\Delta t \geq 0$ for $i = 0, 1, 2, \ldots$. Show that P is a stochastic matrix.

10. Suppose $\{X(t) : t \in [0, \infty)\}$ is a birth process with values in $\{1, 2, \ldots\}$. Let $\Delta X(t) = X(t + \Delta t) - X(t)$. Assume

$$p_{i+j,i}(\Delta t) = \text{Prob}\{\Delta X(t) = j | X(t) = i\}$$
$$= \begin{cases} b_i \Delta t + o(\Delta t), & j = 1 \\ 1 - b_i \Delta t + o(\Delta t), & j = 0 \\ o(\Delta t), & j \geq 2 \\ 0, & j < 0, \end{cases}$$

for Δt sufficiently small and $i = 1, 2, \ldots$. Suppose $X(0) = 1$, $b_0 = 0$, $b_1 = 1$, $b_2 = 3$, and $b_3 = 6$. Write the differential equations satisfied by the probabilities $p_1(t)$, $p_2(t)$, and $p_3(t)$. Then solve for $p_1(t)$, $p_2(t)$, and $p_3(t)$.

11. Suppose in a simple death process $X(0) = 100$ and $d > 0$.

    (a) Compute the expected time to reach a population size of zero.

    (b) For $d = 0.5$, use part (a) to compute the expected time to reach a population size of zero.

12. Suppose in a simple birth process $X(0) = 1$ and $b > 0$.

    (a) Compute the expected time to reach a population size of 100.

    (b) For $b = 0.5$, use part (a) to find the expected time to reach a population size of 100.

13. Modify the MATLAB program for a simple birth process (in the Appendix) and graph three sample paths in the case $b = 0.5$ and $X(0) = 1$.

14. Write a MATLAB program for a simple death process and graph three sample paths in the case $d = 0.25$ and $X(0) = 100$.

15. Modify the MATLAB program for a simple birth and death process (in the Appendix) and graph three sample paths in the case $b = 2$, $d = 1$, and $X(0) = 1$.

16. For the simple birth and death process in Example 5.14 write the form of the transition matrix $P(\Delta t)$ for the states $\{0, 1, 2, \ldots\}$. Then use Definition 5.2 to write the generator matrix Q for this process.

17. Consider the simple death process described by the transition probabilities in equation (5.27).

    (a) Derive the differential equations for the probabilities,

$$p_i(t) = \text{Prob}\{X_t = i\},$$

    in the same manner as for the Poisson process, to show that

$$\frac{dp_i}{dt} = d(i + 1)p_{i+1}(t) - dip_i(t), \quad i < N.$$

    What is the differential equation satisfied by $p_N(t)$?

    (b) Suppose there are initially N individuals, $p_N(0) = 1$. Use the generating function technique to show that the probability generating function $\mathcal{P}(z, t) = \sum_{i=0}^{N} p_i(t)z^i$ satisfies

$$\frac{\partial \mathcal{P}}{\partial t} = d(1 - z)\frac{\partial \mathcal{P}}{\partial z}, \quad \mathcal{P}(z, 0) = z^N. \tag{5.30}$$

18. Assume the p.g.f. of a process satisfies equation (5.30).

(a) Apply the method of characteristics to show that the solution of (5.30) is

$$P(z,t) = \left[1 - e^{-dt} + ze^{-dt}\right]^{N}.$$

(b) Note that $P(z,t)$ is a p.g.f. of a binomial distribution, $b(n,p)$, where $n = N$, $p = e^{-dt}$, and $q = 1 - p$. Find the mean $m(t)$ and the variance $\sigma^2(t)$ of the simple death process.

19. Consider the simple death process described by the transition probabilities in equation (5.27).

   (a) Use the differential equation satisfied by the p.g.f., equation (5.30), and make a change of variable $z = e^\theta$ to find a differential equation satisfied by the m.g.f., $M(\theta,t) = P(e^\theta,t)$.

   (b) Use the differential equation satisfied by the m.g.f. in part (a) to find a differential equation satisfied by the c.g.f., $K(\theta,t) = \ln M(\theta,t)$.

20. Suppose the m.g.f. $M(\theta,t)$ of a continuous-time Markov process is a solution of the following first-order partial differential equation:

$$\frac{\partial M}{\partial t} + \frac{e^{-\theta} - 1}{e^{-\theta}} \frac{\partial M}{\partial \theta} = 0,$$

with corresponding initial condition

$$M(\theta,0) = e^{5\theta}.$$

Apply the method of characteristics to show that the solution $M(\theta,t)$ is

$$M(\theta,t) = [1 + e^t(e^{-\theta} - 1)]^{-5}.$$

21. Modify the Monte Carlo simulation for the Probability of Population Extinction for the simple birth and death process (in the Appendix). The MATLAB code counts the number of times out of 10,000 that the population either hits 25 or hits zero. If the population size reaches 25, it is assumed that the population will not become extinct.

   (a) Let $X(0) = 1$, $d = 1$, and $b = 2$. Use the MATLAB program to demonstrate that the probability of hitting zero (probability of population extinction) is $1/2$.

   (b) Simulate four different cases:
   (i) $X(0) = 1$, $d = 0.5$, and $b = 1$.
   (ii) $X(0) = 2$, $d = 1$, and $b = 2$.
   (iii) $X(0) = 3$, $d = 1$, and $b = 2$.
   (iv) $X(0) = 2$, $d = 2$, and $b = 1$.
   In each case, use the MATLAB program to estimate the probability of population extinction.

(c) Conjecture a value for the probability of population extinction if $X(0) = a$ and $d/b = p < 1$. Conjecture a value for the probability of population extinction if $d/b = p > 1$. The probability of population extinction in a simple birth and death process will be estimated in Section 6.4.3.

5.13   References for Chapter 5

Bailey, N. T. J. 1990. *The Elements of Stochastic Processes with Applications to the Natural Sciences.* John Wiley & Sons, New York.

Brauer, F. and J. A. Nohel. 1969. *The Qualitative Theory of Ordinary Differential Equations An Introduction.* Dover Pub., New York.

Farlow, S. J. 1982. *Partial Differential Equations for Scientists & Engineers.* John Wiley & Sons, New York.

Gillespie, D. T. 1976. A general method for numerically simulating the stochastic time evolution of coupled chemical reactions. *J. Computational Physics* 22: 403-434.

Gillespie, D. T. 1977. Exact stochastic simulation of coupled chemical reactions. *J. Chemical Physics* 81: 2340–2361.

Gillespie, D. T. 2001. Approximate accelerated stochastic simulation of chemically reacting systems. *J. Chemical Physics* 115: 1716–1733.

Gillespie, D. T. and L. Petzold. 2006. Numerical simulation for biochemical kinetics. In: *System Modeling in Cellular Biology From Concepts to Nuts and Bolts.* Szallasi, Z., J. Stelling, and V. Periwal (eds.), pp. 331–353, MIT Press, Cambridge, MA.

John, F. 1975. *Partial Differential Equations.* 2nd ed. Applied Mathematical Sciences, Vol. 1. Springer-Verlag, New York.

Karlin, S. and H. Taylor. 1975. *A First Course in Stochastic Processes.* 2nd ed. Academic Press, New York.

Karlin, S. and H. Taylor. 1981. *A Second Course in Stochastic Processes.* Academic Press, New York.

Leonard, I. E. 1996. The matrix exponential. *SIAM Review* 39: 507–512.

Li, H., Y. Cao, L. R. Petzold, and D. T. Gillespie. 2008. Algorithms and software for stochastic simulation of biochemical reacting systems. *Biotechnology Progress* 24: 56-61.

Moler, C. and C. Van Loan. 1978. Nineteen dubious ways to compute the exponential of a matrix. *SIAM Review* 20: 801–836.

Nordsieck, A., W. E. Lamb Jr., G. E. Uhlenbeck. 1940. On the theory of cosmic-ray showers I The furry model and the fluctuation problem. *Physica* VII: 344-360.

Norris, J. R. 1997. *Markov Chains*. Cambridge Series in Statistical and Probabilistic Mathematics, Cambridge Univ. Press, Cambridge, U. K.

Schinazi, R. B. 1999. *Classical and Spatial Stochastic Processes*. Birkhäuser, Boston.

Schovanec, L. and D. Gilliam. 2000. Classroom Notes for Ode/Pde Class. `http://texas.math.ttu.edu/~gilliam/ttu/ode_pde_pdf/odepde.html` (accessed June 3, 2010).

Stewart, W. J. 1994. *Introduction to the Numerical Solution of Markov Chains*. Princeton Univ. Press, Princeton, N. J.

Taylor, H. M. and S. Karlin. 1998. *An Introduction to Stochastic Modeling*. 3rd ed. Academic Press, New York.

Waltman, P. 1986. *A Second Course in Elementary Differential Equations*. Academic Press, New York.

5.14   Appendix for Chapter 5

5.14.1   Calculation of the Matrix Exponential

Suppose matrix Q is an $n \times n$ diagonalizable matrix with eigenvalues, λ_i, $i = 1, 2, \ldots, n$. Then an expression for Q^k can be obtained by the method presented in Chapter 2, Section 2.13.3,

$$Q^k = H \Lambda^k H^{-1},$$

where $\Lambda = \text{diag}(\lambda_1, \lambda_2, \ldots, \lambda_n)$ and the columns of H are the right eigenvectors of Q. The expression for e^{Qt} simplifies to

$$e^{Qt} = H \sum_{k=0}^{\infty} \Lambda^k \frac{t^k}{k!} H^{-1} = H \text{diag}(e^{\lambda_1 t}, e^{\lambda_2 t}, \ldots, e^{\lambda_n t}) H^{-1}. \tag{5.31}$$

Differentiation of $P(t) = e^{Qt}$ can be used to generate information about the derivatives of P evaluated at $t = 0$. Notice that $P'(0) = Q$, $P''(0) = Q^2$ and, in general, $d^k P(t)/dt^k|_{t=0} = Q^k$ or

$$\frac{d^k p_{ji}(t)}{dt^k}\bigg|_{t=0} = q_{ji}^{(k)}, \tag{5.32}$$

where $q_{ji}^{(k)}$ is the element in the jth row and ith column of Q^k. The identity (5.31) shows that the elements of $P(t)$ are

$$p_{ji}(t) = a_1 e^{\lambda_1 t} + a_2 e^{\lambda_2 t} + \cdots + a_n e^{\lambda_n t}. \tag{5.33}$$

Using the initial conditions (5.32), the coefficients a_k, $k = 1, 2, \ldots, n$ can be determined. Alternately, first computing H, then H^{-1},

$$P(t) = H \mathrm{diag}(e^{\lambda_1 t}, e^{\lambda_2 t}, \ldots, e^{\lambda_n t}) H^{-1}.$$

Another method for computing e^{Qt} is discussed by Leonard (1996). Suppose Q is an $n \times n$ matrix with characteristic polynomial,

$$det(\lambda I - Q) = \lambda^n + a_{n-1}\lambda^{n-1} + \cdots + a_0 = 0.$$

This polynomial equation is also a characteristic polynomial of an nth-order scalar differential equation of the form

$$x^{(n)}(t) + a_{n-1}x^{(n-1)}(t) + \cdots a_0 x(t) = 0.$$

To find a formula for e^{Qt} it is necessary to find n linearly independent solutions to this nth order scalar differential equation, $x_1(t), x_2(t), \ldots, x_n(t)$, with initial conditions

$$\left.\begin{array}{c} x_1(0) = 1 \\ x_1'(0) = 0 \\ \vdots \\ x_1^{(n-1)}(0) = 0 \end{array}\right\}, \quad \left.\begin{array}{c} x_2(0) = 0 \\ x_2'(0) = 1 \\ \vdots \\ x_2^{(n-1)}(0) = 0 \end{array}\right\}, \quad \cdots, \quad \left.\begin{array}{c} x_n(0) = 0 \\ x_n'(0) = 0 \\ \vdots \\ x_n^{(n-1)}(0) = 1 \end{array}\right\}.$$

Then

$$e^{Qt} = x_1(t)I + x_2(t)Q + \cdots + x_n(t)Q^{n-1}, \quad -\infty < t < \infty. \tag{5.34}$$

Example 5.16. The matrix exponential of Q, defined in Example 5.7,

$$Q = \begin{pmatrix} -a & b \\ a & -b \end{pmatrix},$$

$a, b > 0$, is computed using the identity (5.34). The characteristic polynomial of Q is $\lambda^2 + (a+b)\lambda = 0$. Therefore, the eigenvalues of Q are $\lambda_{1,2} = 0, -(a+b)$. The general solution of this second-order differential equation $x''(t) + (a +$

b)$x'(t) = 0$ is $x(t) = c_1 + c_2 e^{-(a+b)t}$. Applying the initial conditions to find the constants c_1 and c_2, the solutions $x_1(t)$ and $x_2(t)$ are $x_1(t) = 1$ and $x_2(t) = (1 - e^{-(a+b)t})/(a+b)$, respectively. Applying the identity (5.34) gives the solution

$$e^{Qt} = x_1(t)I + x_2(t)Q = \frac{1}{a+b}\begin{pmatrix} b + ae^{-(a+b)t} & b - be^{-(a+b)t} \\ a - ae^{-(a+b)t} & a + be^{-(a+b)t} \end{pmatrix}.$$

This latter formula agrees with the solution given in (5.23). ∎

5.14.2 MATLAB® Programs

The following MATLAB programs generate three stochastic realizations for the simple birth process and for the simple birth and death process and estimate the probability of population extinction for a simple birth and death process.

```
% MatLab program: simple birth process
clear
b=1; %Parameter values
x0=1; % initial population size
xe=100; % ending population size
tot=xe-x0+1;
x=[0:.2:10];
y=x0*exp(x); % Deterministic solution
plot(x,y,'k--','Linewidth',2);
hold on
n=linspace(x0,xe,tot); % Defines the population vector.
for j=1:3; % Three sample paths.
    t(1)=0;
    for i=1:tot-1
        t(i+1)=t(i)-log(rand)/(b*n(i));
    end
    stairs(t,n,'r-','Linewidth',2); % Draws stairstep graph.
end
hold off
xlabel('Time'); ylabel('Population Size');


%MatLab program: simple birth and death process
clear
x0=5; % Initial population size
b=1; d=0.5; % parameter values
x=[0:.1:8];
y=x0*exp((b-d).*x);% Deterministic solution
plot(x,y,'k--','Linewidth',2);
```

```
hold on
for j=1:3 % Three sample paths
   clear n t
   t(1,j)=0;
   tt=1;
   n(tt)=x0;
   while n(tt)>0 & tt<400
      y1=rand; y2=rand;
      t(tt+1,j)=-log(y1)/(b*n(tt)+d*n(tt))+t(tt,j);
      tt=tt+1;
      if y2< b/(b+d)
         n(tt)=n(tt-1)+1;
      else
         n(tt)=n(tt-1)-1;
      end
   end
   s=stairs(t(:,j),n,'r-','Linewidth',2);
end
hold off
xlabel('Time'); ylabel('Population size');


% Monte Carlo simulation: Probability of Population Extinction
clear
x0=1;   d=1; b=2;
sim=10000; count=0;
for j=1:sim
   clear n
   n=x0;
   while n>0 & n<25;
      r=rand;
      if r< b/(b+d)
         n=n+1;
      else
         n=n-1;
      end
   end
   if n==0
       count=count+1;
   end
end
propext=count/sim % Approximate probability of extinction
```

Chapter 6

Continuous-Time Birth and Death Chains

6.1 Introduction

A variety of continuous-time birth and death processes are studied in this chapter. First, a general birth and death process is formulated. Then conditions are stated for existence of a unique positive stationary probability distribution of this general birth and death process. It is shown that if the process is nonexplosive, then the general birth and death process converges to this stationary probability distribution. Some simple but classical birth and death processes are presented in Section 6.4: birth, death, birth and death, and birth and death with immigration processes. Explicit formulas are derived for the moment generating functions. In addition, for the simple birth and simple death processes, explicit formulas are derived for their probability distributions. Queueing processes are discussed in Section 6.5, an important application of birth and death processes, where births and deaths are arrivals and departures in the queue.

For many birth and death processes in biology, a positive stationary probability distribution may not exist because the zero state (extinction) is absorbing. For such types of processes, the probability of extinction and the expected time until population extinction are investigated. A classical example of a birth and death process with an absorbing state at zero is logistic growth. This density-dependent birth and death process is formulated in Section 6.8. A quasistationary probability distribution is defined, where the process is conditioned on nonextinction. The last two sections of this chapter cover two types of processes that have not been considered previously, an explosive birth process and a nonhomogeneous birth and death process.

241

6.2 General Birth and Death Process

Some notation is introduced that will be used throughout this chapter. Let $\Delta X(t)$ denote the change in state of the stochastic process from t to $t + \Delta t$. That is,

$$\Delta X(t) = X(t + \Delta t) - X(t).$$

When birth and death processes were introduced at the end of Chapter 5, the birth and death rates were denoted as b_i and d_i, respectively, given $X(t) = i$. However, in this chapter, notation that has become almost standard in birth and death processes will be applied. If $X(t) = i$, then the birth and death rates will be denoted as

$$\lambda_i = \text{birth rate} \quad \text{and} \quad \mu_i = \text{death rate}$$

(see, e.g., Bailey, 1990; Karlin and Taylor, 1975; Norris, 1997; Schinazi, 1999; Taylor and Karlin, 1998).

The continuous-time birth and death Markov chain $\{X(t) : t \in [0, \infty)\}$ may have either a finite or infinite state space $\{0, 1, 2, \ldots, N\}$ or $\{0, 1, \ldots\}$. Assume the infinitesimal transition probabilities for this process are

$$p_{i+j,i}(\Delta t) = \text{Prob}\{\Delta X(t) = j | X(t) = i\}$$
$$= \begin{cases} \lambda_i \Delta t + o(\Delta t), & j = 1 \\ \mu_i \Delta t + o(\Delta t), & j = -1 \\ 1 - (\lambda_i + \mu_i)\Delta t + o(\Delta t), & j = 0 \\ o(\Delta t), & j \neq -1, 0, 1 \end{cases} \quad (6.1)$$

for Δt sufficiently small, where $\lambda_i \geq 0$, $\mu_i \geq 0$ for $i = 0, 1, 2, \ldots$ and $\mu_0 = 0$. It is often the case that $\lambda_0 = 0$, except when there is immigration. The initial conditions are $p_{ji}(0) = \delta_{ji}$, where $P(0) = (p_{ji}(0))$ is the identity matrix $P(0) = I$. In a small interval of time Δt, at most one change in state can occur, either a birth, $i \to i + 1$ or a death, $i \to i - 1$.

The forward Kolmogorov differential equations for $p_{ji}(t)$ can be derived directly from the assumptions in (6.1). For Δt sufficiently small, consider the transition probability $p_{ji}(t + \Delta t)$:

$$p_{ji}(t + \Delta t) = p_{j-1,i}(t)[\lambda_{j-1}\Delta t + o(\Delta t)] + p_{j+1,i}(t)[\mu_{j+1}\Delta t + o(\Delta t)]$$
$$+ p_{ji}(t)[1 - (\lambda_j + \mu_j)\Delta t + o(\Delta t)] + \sum_{k \neq -1, 0, 1}^{\infty} p_{j+k,i}(t)o(\Delta t)$$
$$= p_{j-1,i}(t)\lambda_{j-1}\Delta t + p_{j+1,i}(t)\mu_{j+1}\Delta t$$
$$+ p_{ji}(t)[1 - (\lambda_j + \mu_j)\Delta t] + o(\Delta t),$$

which holds for all i and j in the state space with the exception of $j = 0$ and $j = N$ (if the population size is finite). If $j = 0$, then

$$p_{0i}(t + \Delta t) = p_{1i}(t)\mu_1 \Delta t + p_{0i}(t)[1 - \lambda_0 \Delta t] + o(\Delta t).$$

If $j = N$ is the maximum population size, then

$$p_{Ni}(t + \Delta t) = p_{N-1,i}(t)\lambda_{N-1}\Delta t + p_{Ni}(t)[1 - \mu_N \Delta t] + o(\Delta t),$$

where $\lambda_N = 0$ and $p_{kN}(t) = 0$ for $k > N$. Subtracting $p_{ji}(t)$, $p_{0i}(t)$, and $p_{Ni}(t)$ from the preceding three equations, respectively, dividing by Δt and taking the limit as $\Delta t \to 0$, yields the forward Kolmogorov differential equations for the general birth 'and death process,

$$\frac{dp_{ji}(t)}{dt} = \lambda_{j-1}p_{j-1,i}(t) - (\lambda_j + \mu_j)p_{ji}(t) + \mu_{j+1}p_{j+1,i}(t)$$

$$\frac{dp_{0i}(t)}{dt} = -\lambda_0 p_{0i}(t) + \mu_1 p_{1i}(t)$$

for $i \geq 0$ and $j \geq 1$. For a finite state space, the differential equation for $p_{Ni}(t)$ is

$$\frac{dp_{Ni}(t)}{dt} = \lambda_{N-1}p_{N-1,i}(t) - \mu_N p_{Ni}(t).$$

Written in matrix form the forward Kolmogorov differential equations are $dP/dt = QP$, where Q is the generator matrix. For the infinite state space,

$$Q = \begin{pmatrix} -\lambda_0 & \mu_1 & 0 & 0 & \cdots \\ \lambda_0 & -\lambda_1 - \mu_1 & \mu_2 & 0 & \cdots \\ 0 & \lambda_1 & -\lambda_2 - \mu_2 & \mu_3 & \cdots \\ 0 & 0 & \lambda_2 & -\lambda_3 - \mu_3 & \cdots \\ \vdots & \vdots & \vdots & \vdots & \end{pmatrix}, \tag{6.2}$$

and for the finite state space,

$$Q = \begin{pmatrix} -\lambda_0 & \mu_1 & 0 & \cdots & 0 \\ \lambda_0 & -\lambda_1 - \mu_1 & \mu_2 & \cdots & 0 \\ 0 & \lambda_1 & -\lambda_2 - \mu_2 & \cdots & 0 \\ \vdots & \vdots & \vdots & \cdots & \vdots \\ 0 & 0 & 0 & \cdots & \mu_N \\ 0 & 0 & 0 & \cdots & -\mu_N \end{pmatrix}. \tag{6.3}$$

If $X(0) = i$ is a fixed value, the state probabilities $p(t) = (p_0(t), p_1(t), \ldots)^{tr}$, $p_i(t) = \text{Prob}\{X(t) = i\}$, are solutions of the forward Kolmogorov differential equations, $dp/dt = Qp$. In addition, these differential equations can be derived from a limiting argument by applying the infinitesimal transition probabilities in (6.1):

$$p_i(t + \Delta t) = p_{i-1}(t)\lambda_{i-1}\Delta t + p_{i+1}(t)\mu_{i+1}\Delta t$$
$$+ p_i(t)[1 - (\lambda_i + \mu_i)\Delta t] + o(\Delta t).$$

Subtracting $p_i(t)$, dividing by Δt and letting $\Delta t \to 0$ leads to

$$\frac{dp_i}{dt} = \lambda_{i-1}p_{i-1} - (\lambda_i + \mu_i)p_i(t) + \mu_{i+1}p_{i+1}.$$

The transition matrix $T = (t_{ji})$ for the embedded Markov chain $\{Y_n\}_{n=0}^{\infty}$ can be defined from the generator matrices (6.2) and (6.3). For the generator matrix (6.2), the transition matrix of the embedded Markov chain is

$$T = \begin{pmatrix} 0 & \mu_1/(\lambda_1 + \mu_1) & 0 & 0 & \cdots \\ 1 & 0 & \mu_2/(\lambda_2 + \mu_2) & 0 & \cdots \\ 0 & \lambda_1/(\lambda_1 + \mu_1) & 0 & \mu_3/(\lambda_3 + \mu_3) & \cdots \\ 0 & 0 & \lambda_2/(\lambda_2 + \mu_2) & 0 & \cdots \\ \vdots & \vdots & \vdots & \vdots & \end{pmatrix}$$

and for (6.3), it is

$$T = \begin{pmatrix} 0 & \mu_1/(\lambda_1 + \mu_1) & 0 & \cdots & 0 \\ 1 & 0 & \mu_2/(\lambda_2 + \mu_2) & \cdots & 0 \\ 0 & \lambda_1/(\lambda_1 + \mu_1) & 0 & \cdots & 0 \\ \vdots & \vdots & \vdots & \cdots & \vdots \\ 0 & 0 & 0 & \cdots & 1 \\ 0 & 0 & 0 & \cdots & 0 \end{pmatrix},$$

provided $\lambda_i + \mu_i > 0$ for $i = 0, 1, 2, \ldots$. If $\lambda_0 = 0$, then state 0 is absorbing.

The embedded Markov chain can be thought of as a generalized random walk model with a reflecting boundary at zero if $\lambda_0 > 0$ (and at N in the finite case). The probability of moving right (or a birth) is $t_{i+1,i} = \lambda_i/(\lambda_i + \mu_i)$ and the probability of moving left (or a death) is $t_{i-1,i} = \mu_i/(\lambda_i + \mu_i)$. See the directed graph in Figure 6.1. It can be verified from the transition matrix T or from the directed graph that the chain is irreducible iff $\lambda_i > 0$ and $\mu_{i+1} > 0$ for $i = 0, 1, 2, \ldots$.

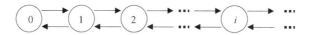

FIGURE 6.1: Directed graph for the embedded Markov chain of the general birth and death process when $\lambda_0 > 0$ and $\lambda_i + \mu_i > 0$ for $i = 1, 2, \ldots$.

6.3 Stationary Probability Distribution

A formula is derived for the stationary probability distribution of a general birth and death chain. Recall that a stationary probability distribution $\pi = (\pi_0, \pi_1, \pi_2, \ldots)^{tr}$ of a CTMC with generator matrix Q satisfies

$$Q\pi = 0, \quad \sum_{i=0}^{\infty} \pi_i = 1, \quad \text{and} \quad \pi_i \geq 0$$

for $i = 0, 1, 2, \ldots$.

THEOREM 6.1

Suppose the CTMC $\{X(t) : t \in [0, \infty)\}$ is a general birth and death chain with infinitesimal transition probabilities given by (6.1). If the state space is infinite, $\{0, 1, 2, \ldots\}$, a unique positive stationary probability distribution π exists iff

$$\mu_i > 0 \quad \text{and} \quad \lambda_{i-1} > 0 \quad \text{for} \quad i = 1, 2, \ldots, \tag{6.4}$$

and

$$\sum_{i=1}^{\infty} \frac{\lambda_0 \lambda_1 \cdots \lambda_{i-1}}{\mu_1 \mu_2 \cdots \mu_i} < \infty. \tag{6.5}$$

The stationary probability distribution equals

$$\pi_i = \frac{\lambda_0 \lambda_1 \cdots \lambda_{i-1}}{\mu_1 \mu_2 \cdots \mu_i} \pi_0, \quad i = 1, 2, \ldots \tag{6.6}$$

and

$$\pi_0 = \frac{1}{1 + \sum\limits_{i=1}^{\infty} \dfrac{\lambda_0 \lambda_1 \cdots \lambda_{i-1}}{\mu_1 \mu_2 \cdots \mu_i}}. \tag{6.7}$$

If the state space is finite, $\{0, 1, 2, \ldots, N\}$, then a unique positive stationary probability distribution π exists iff

$$\mu_i > 0 \quad \text{and} \quad \lambda_{i-1} > 0 \quad \text{for} \quad i = 1, 2, \ldots, N.$$

The stationary probability distribution is given by (6.6) and (6.7), where the index and the summation extend from $i = 1, \ldots, N$.

Proof. The explicit equations for the stationary distribution $Q\pi = 0$ are

$$0 = \lambda_{i-1}\pi_{i-1} - (\lambda_i + \mu_i)\pi_i + \mu_{i+1}\pi_{i+1}, \quad i = 1, 2, \ldots,$$
$$0 = -\lambda_0 \pi_0 + \mu_1 \pi_1.$$

These equations can be solved recursively. First,

$$\pi_1 = \frac{\lambda_0}{\mu_1}\pi_0.$$

Then

$$\mu_2\pi_2 = (\lambda_1 + \mu_1)\pi_1 - \lambda_0\pi_0$$
$$= \left[\frac{(\lambda_1 + \mu_1)\lambda_0}{\mu_1} - \lambda_0\right]\pi_0$$
$$\pi_2 = \frac{\lambda_0\lambda_1}{\mu_1\mu_2}\pi_0.$$

Applying the induction hypothesis, assume π_j has been defined for $j = 1, 2, \ldots, i$,

$$\pi_i = \frac{\lambda_0\lambda_1\cdots\lambda_{i-1}}{\mu_1\mu_2\cdots\mu_i}\pi_0.$$

Then

$$\mu_{i+1}\pi_{i+1} = (\lambda_i + \mu_i)\pi_i - \lambda_{i-1}\pi_{i-1}$$
$$= \left[\frac{\lambda_0\lambda_2\cdots\lambda_{i-1}(\lambda_i + \mu_i)}{\mu_1\mu_2\cdots\mu_i} - \frac{\lambda_0\lambda_1\cdots\lambda_{i-1}}{\mu_1\mu_2\cdots\mu_{i-1}}\right]\pi_0$$
$$= \frac{\lambda_0\lambda_1\cdots\lambda_{i-1}}{\mu_1\mu_2\cdots\mu_{i-1}}\left[\frac{\lambda_i + \mu_i}{\mu_i} - 1\right]\pi_0$$
$$\pi_{i+1} = \frac{\lambda_0\lambda_1\cdots\lambda_i}{\mu_1\mu_2\cdots\mu_{i+1}}\pi_0.$$

For the infinite case, applying the additional constraint, $\sum_{i=0}^{\infty}\pi_i = 1$ or $\pi_0(1 + \sum_{i=1}^{\infty}\pi_i/\pi_0) = 1$, it follows that

$$\pi_0 = \frac{1}{1 + \sum\limits_{i=1}^{\infty}\dfrac{\lambda_0\lambda_1\cdots\lambda_{i-1}}{\mu_1\mu_2\cdots\mu_i}}.$$

A unique positive stationary distribution exists iff the conditions in (6.4) and (6.5) are satisfied. For the finite case, the assumption (6.4) is not required. □

Example 6.1. Suppose a continuous-time birth and death Markov chain has birth and death rates, $\lambda_i = b$ and $\mu_i = id$ for $i = 0, 1, 2, \ldots$. Applying Theorem 6.1,

$$\frac{\lambda_0\lambda_1\cdots\lambda_{i-1}}{\mu_1\mu_2\cdots\mu_i} = \frac{b^i}{d^i i!} = \frac{(b/d)^i}{i!}.$$

Since

$$1 + \sum_{i=1}^{\infty}\frac{(b/d)^i}{i!} = e^{b/d}$$

the unique stationary probability distribution is a Poisson distribution with parameter b/d,

$$\pi_i = \frac{(b/d)^i}{i!} e^{-b/d}$$

for $= 0, 1, 2, \ldots$. ∎

Example 6.2. Suppose a continuous-time birth and death Markov chain has birth and death rates $\mu_i = q > 0$, $i = 1, 2, \ldots$, and $\lambda_i = p > 0$, $i = 0, 1, 2, \ldots$, where $p + q = 1$. The embedded Markov chain is a semi-infinite random walk model with reflecting boundary conditions at zero. The transition matrix for the embedded Markov chain has a directed graph, which is illustrated in Figure 6.1. The chain has a unique stationary probability distribution iff

$$\sum_{j=1}^{\infty} \left(\frac{p}{q}\right)^j < \infty$$

iff $p < q$. The stationary probability distribution is a geometric probability distribution,

$$\pi_0 = \frac{q - p}{q} \quad \text{and} \quad \pi_i = \left(\frac{p}{q}\right)^i \pi_0$$

for $i = 0, 1, 2, \ldots$. ∎

6.4 Simple Birth and Death Processes

Four classic continuous-time birth and death processes are described, simple birth, simple death, simple birth and death, and simple birth and death with immigration. The dynamics of these processes are analyzed by applying techniques from Chapter 5. Bailey (1990) refers to these simple birth and death processes as pure birth and death processes.

6.4.1 Simple Birth

In the simple birth process, the only event is a birth. Let $X(t)$ represent the population size at time t and $X(0) = N$, so that $p_i(0) = \delta_{iN}$. Because there are only births, the population size can only increase in size. For Δt sufficiently small, the transition probabilities are

$$p_{i+j,i}(\Delta t) = \text{Prob}\{\Delta X(t) = j | X(t) = i\}$$
$$= \begin{cases} \lambda i \, \Delta t + o(\Delta t), & j = 1 \\ 1 - \lambda i \, \Delta t + o(\Delta t), & j = 0 \\ o(\Delta t), & j \geq 2 \\ 0, & j < 0. \end{cases}$$

The probabilities $p_i(t) = \text{Prob}\{X(t) = i\}$ are solutions of the forward Kolmogorov differential equations, $dp/dt = Qp$, where

$$\frac{dp_i(t)}{dt} = \lambda(i-1)p_{i-1}(t) - \lambda i p_i(t), \quad i = N, N+1, \ldots,$$

$$\frac{dp_i(t)}{dt} = 0, \quad i = 0, 1, \ldots, N-1,$$

with initial conditions $p_i(0) = \delta_{iN}$. These differential equations can be solved in a sequential manner as was done for the Poisson process in Chapter 5. For example, note that $p_i(t) = 0$ for $i < N$. Then $dp_N/dt = -\lambda N p_N$ so that the solution is $p_N(t) = e^{-\lambda Nt}$. The generating function technique will be used to compute the remaining solutions $p_i(t)$ for $i > N$. This technique has wider applicability and is often easier to apply than solving the system of differential equations sequentially.

Observe that the state space for this process is $\{N, N+1, \ldots\}$; all of the states are transient and there is no stationary probability distribution (Exercise 4). It will be shown that the generating functions for the simple birth process correspond to a negative binomial distribution.

To derive the partial differential equation for the p.g.f., multiply the differential equations by z^i and sum over i,

$$\frac{\partial P(z,t)}{\partial t} = \lambda \sum_{i=N+1}^{\infty} p_{i-1}(i-1)z^i - \lambda \sum_{i=N}^{\infty} p_i i z^i$$

$$= \lambda z^2 \sum_{i=N}^{\infty} i p_i z^{i-1} - \lambda z \sum_{i=N}^{\infty} i p_i z^{i-1}$$

$$= \lambda z(z-1)\frac{\partial P}{\partial z}.$$

The initial condition is $P(z,0) = z^N$.

The partial differential equation for the m.g.f. can be derived by a change of variable. Let $z = e^{\theta}$. Then $P(e^{\theta}, t) = M(\theta, t)$ and

$$\frac{\partial P}{\partial z} = \frac{\partial M}{\partial \theta}\frac{d\theta}{dz} = \frac{1}{z}\frac{\partial M}{\partial \theta},$$

so that the m.g.f is a solution of the following partial differential equation:

$$\frac{\partial M}{\partial t} = \lambda(e^{\theta} - 1)\frac{\partial M}{\partial \theta}$$

with initial condition $M(\theta, 0) = e^{N\theta}$.

The method of characteristics is applied to find the solution $M(\theta, t)$. Rewriting the differential equation for $M(\theta, t)$,

$$\frac{\partial M}{\partial t} + \lambda(1 - e^{\theta})\frac{\partial M}{\partial \theta} = 0.$$

Along characteristic curves, s and τ, $t(s,\tau)$, $\theta(s,\tau)$, and $M(s,\tau)$ are solutions of

$$\frac{dt}{d\tau} = 1, \quad \frac{d\theta}{d\tau} = \lambda(1 - e^{\theta}) \quad \text{and} \quad \frac{dM}{d\tau} = 0,$$

with initial conditions

$$t(s,0) = 0, \quad \theta(s,0) = s, \quad \text{and} \quad M(s,0) = e^{Ns}.$$

Separating variables and simplifying leads to

$$\frac{d\theta}{1 - e^{\theta}} = \lambda d\tau \quad \text{or} \quad \frac{e^{-\theta} d\theta}{e^{-\theta} - 1} = \lambda d\tau.$$

Integrating yields the relations

$$t = \tau, \quad \ln(e^{-\theta} - 1) = -\lambda\tau + c, \quad \text{and} \quad M(s,\tau) = e^{Ns}.$$

Applying the initial condition $\theta(s,0) = s$, the second expression can be written as $e^{-\theta} - 1 = (e^{-s} - 1)e^{-\lambda\tau}$. Finally, the solution M must be expressed in terms of the original variables θ and t. Using the preceding formulas, e^{-s} can be expressed in terms of θ and t, $e^{-s} = 1 - e^{\lambda t}(1 - e^{-\theta})$. Since $e^{Ns} = [e^{-s}]^{-N}$, the m.g.f. for the simple birth process is

$$M(\theta,t) = [1 - e^{\lambda t}(1 - e^{-\theta})]^{-N}.$$

The p.g.f can be found directly from the m.g.f. by making the change of variable, $\theta = \ln z$,

$$\begin{aligned}
\mathcal{P}(z,t) &= [1 - e^{\lambda t}(1 - z^{-1})]^{-N} \\
&= z^N e^{-\lambda Nt}[ze^{-\lambda t} - (z-1)]^{-N} \\
&= \frac{z^N e^{-\lambda Nt}}{[1 - z(1 - e^{-\lambda t})]^N}.
\end{aligned}$$

Letting $p = e^{-\lambda t}$ and $q = 1 - e^{-\lambda t}$, the p.g.f. is

$$\mathcal{P}(z,t) = \frac{(pz)^N}{(1 - zq)^N}. \tag{6.8}$$

The p.g.f. and m.g.f. for the simple birth process correspond to a negative binomial distribution. The probabilities $p_i(t)$ in (6.8) can be written as

$$p_{i+N}(t) = \binom{N+i-1}{i} p^N q^i, \quad i = 0, 1, \ldots.$$

Let $i + N = n$ and replace p and q by $e^{-\lambda t}$ and $1 - e^{-\lambda t}$, respectively,

$$p_n(t) = \binom{n-1}{n-N} e^{-\lambda Nt}(1 - e^{-\lambda t})^{n-N}, \quad n = N, N+1, \ldots.$$

The mean and variance for the simple birth process are

$$m(t) = N/p = Ne^{\lambda t} \quad \text{and} \quad \sigma^2(t) = Nq/p^2 = Ne^{2\lambda t}(1 - e^{-\lambda t}).$$

The moments can be calculated directly from one of the generating functions. It is interesting to note that the mean of the simple birth process corresponds to exponential growth with $X(0) = N$. The variance also increases exponentially with time. See Figure 6.2.

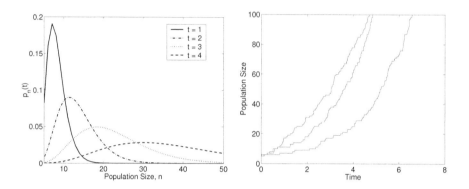

FIGURE 6.2: Probability distributions of $X(t)$ for the simple birth process for $t = 1, 2, 3, 4$ and three sample paths when $\lambda = 0.5$ and $X(0) = 5$. The mean and variance are $m(t) = 5e^{0.5t}$, $\sigma^2(t) = 5(e^t - e^{0.5t})$.

6.4.2 Simple Death

In the simple death process, the only event is a death. Let $X(0) = N$. The infinitesimal transition probabilities are

$$p_{i+j,i}(\Delta t) = \text{Prob}\{\Delta X(t) = j | X(t) = i\}$$
$$= \begin{cases} \mu i \, \Delta t + o(\Delta t), & j = -1 \\ 1 - \mu i \, \Delta t + o(\Delta t), & j = 0 \\ o(\Delta t), & j \leq -2 \\ 0, & j > 0. \end{cases}$$

Since the process begins in state N, the state space is $\{0, 1, 2, \ldots, N\}$.

The forward Kolmogorov equations are

$$\frac{dp_i(t)}{dt} = \mu(i+1)p_{i+1}(t) - \mu i p_i(t)$$
$$\frac{dp_N(t)}{dt} = -\mu N p_N(t),$$

$i = 0, 1, 2, \ldots, N - 1$ with initial conditions $p_i(0) = \delta_{iN}$. It can be easily seen that zero is an absorbing state and the unique stationary probability distribution is $\pi = (1, 0, 0, \ldots, 0)$.

The generating function technique and the method of characteristics are used to find the probability distribution. Multiplying the forward Kolmogorov equations by z^i and summing over i, the partial differential equation for the p.g.f. is

$$\frac{\partial \mathcal{P}}{\partial t} = \mu(1 - z)\frac{\partial \mathcal{P}}{\partial z}, \quad \mathcal{P}(z, 0) = z^N.$$

Substituting $z = e^\theta$, the partial differential for the m.g.f. is

$$\frac{\partial M}{\partial t} = \mu(e^{-\theta} - 1)\frac{\partial M}{\partial \theta}, \quad M(\theta, 0) = e^{N\theta}.$$

Applying the method of characteristics, the solutions are

$$\mathcal{P}(z, t) = [1 - e^{-\mu t}(1 - z)]^N = (1 - e^{-\mu t} + e^{-\mu t}z)^N$$

and

$$M(\theta, t) = [1 - e^{-\mu t}(1 - e^\theta)]^N.$$

Let $p = e^{-\mu t}$ and $q = 1 - e^{-\mu t}$. Then the p.g.f. has the form $\mathcal{P}(z, t) = (q + pz)^N$, corresponding to a binomial distribution, $b(N, p)$. The probabilities

$$p_i(t) = \binom{N}{i} p^i q^{N-i} = \binom{N}{i} e^{-i\mu t}(1 - e^{-\mu t})^{N-i}$$

for $i = 0, 1, \ldots, N$. The mean and variance of a binomial distribution $b(N, p)$ are $m = Np$ and $\sigma^2 = Npq$. Expressed in terms of the parameters for the death process,

$$m(t) = Ne^{-\mu t} \quad \text{and} \quad \sigma^2(t) = Ne^{-\mu t}(1 - e^{-\mu t}).$$

The mean corresponds to exponential decay. Also, the variance decreases exponentially with time. Compare the sample paths of the simple birth and simple death processes in Figures 6.2 and 6.3.

6.4.3 Simple Birth and Death

In the simple birth and death process, an event can be a birth or a death. Let $X(0) = N$. The infinitesimal transition probabilities are

$$p_{i+j,i}(\Delta t) = \text{Prob}\{\Delta X(t) = j | X(t) = i\}$$
$$= \begin{cases} \mu i \, \Delta t + o(\Delta t), & j = -1 \\ \lambda i \, \Delta t + o(\Delta t), & j = 1 \\ 1 - (\lambda + \mu)i \, \Delta t + o(\Delta t), & j = 0 \\ o(\Delta t), & j \neq -1, 0, 1. \end{cases}$$

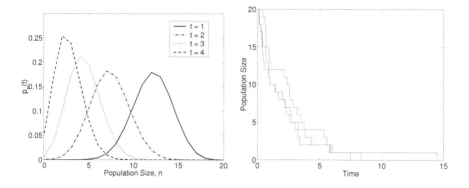

FIGURE 6.3: Probability distributions of $X(t)$ for the simple death process for $t = 1, 2, 3, 4$ and three sample paths when $\mu = 0.5$ and $X(0) = 20$. The mean and variance are $m(t) = 20e^{-0.5t}$ and $\sigma^2(t) = 20(e^{-0.5t} - e^{-t})$.

The forward Kolmogorov differential equations are

$$\frac{dp_i(t)}{dt} = \lambda(i-1)p_{i-1}(t) + \mu(i+1)p_{i+1}(t) - (\lambda + \mu)ip_i(t)$$

$$\frac{dp_0(t)}{dt} = \mu p_1(t)$$

for $i = 1, 2, \ldots$ with initial conditions $p_i(0) = \delta_{iN}$. As in the simple death process, $\lambda_0 = 0$, so that zero is an absorbing state and $\pi = (1, 0, 0, \ldots)^{tr}$ is the unique stationary probability distribution.

Applying the generating function technique and the method of characteristics (Appendix for Chapter 6), the m.g.f. is

$$M(\theta, t) = \begin{cases} \left(\dfrac{e^{t(\mu-\lambda)}(\lambda e^\theta - \mu) - \mu(e^\theta - 1)}{e^{t(\mu-\lambda)}(\lambda e^\theta - \mu) - \lambda(e^\theta - 1)} \right)^N, & \text{if } \lambda \neq \mu \\[4mm] \left(\dfrac{1 - (\lambda t - 1)(e^\theta - 1)}{1 - \lambda t(e^\theta - 1)} \right)^N, & \text{if } \lambda = \mu. \end{cases}$$

Making the change of variable $\theta = \ln z$, the p.g.f. is

$$P(z, t) = \begin{cases} \left(\dfrac{e^{t(\mu-\lambda)}(\lambda z - \mu) - \mu(z - 1)}{e^{t(\mu-\lambda)}(\lambda z - \mu) - \lambda(z - 1)} \right)^N, & \text{if } \lambda \neq \mu \\[4mm] \left(\dfrac{1 - (\lambda t - 1)(z - 1)}{1 - \lambda t(z - 1)} \right)^N, & \text{if } \lambda = \mu. \end{cases}$$

Obtaining a formula for the probabilities $p_i(t)$ is not as straightforward as it was for the simple birth and simple death processes because the generating functions cannot be associated with a well-known probability distribution. However, recall that

$$P(z,t) = \sum_{i=0}^{\infty} p_i(t)z^i \quad \text{and} \quad p_i(t) = \frac{1}{i!}\frac{\partial^i P}{\partial z^i}\bigg|_{z=0}.$$

A computer algebra system may be helpful in finding the terms in the series expansion. The first term in the expansion of $\mathcal{P}(z,t)$ is $p_0(t) = \mathcal{P}(0,t)$:

$$p_0(t) = \begin{cases} \left(\dfrac{\mu - \mu e^{(\mu-\lambda)t}}{\lambda - \mu e^{(\mu-\lambda)t}}\right)^N, & \text{if } \lambda \neq \mu \\[4mm] \left(\dfrac{\lambda t}{1 + \lambda t}\right)^N, & \text{if } \lambda = \mu. \end{cases}$$

The probability of extinction, $p_0(t)$, has a simple expression when $t \to \infty$. Taking the limit,

$$p_0(\infty) = \lim_{t\to\infty} p_0(t) = \begin{cases} 1, & \text{if } \lambda \leq \mu \\[2mm] \left(\dfrac{\mu}{\lambda}\right)^N, & \text{if } \lambda > \mu, \end{cases} \tag{6.9}$$

This latter result is reminiscent of a semi-infinite random walk with an absorbing barrier at $x = 0$, that is, the gambler's ruin problem, where the probability of losing a game is μ and the probability of winning a game is λ. When the probability of losing (death) is greater than or equal to the probability of winning (birth), then, in the long run ($t \to \infty$), the probability of losing all of the initial capital N approaches one. However, if the probability of winning is greater than the probability of losing, then, in the long run, the probability of losing all of the initial capital is $(\mu/\lambda)^N$.

The mean and variance of the simple birth and death process can be derived from the generating functions. For $\lambda \neq \mu$,

$$m(t) = Ne^{(\lambda-\mu)t} \quad \text{and} \quad \sigma^2(t) = N\frac{(\lambda+\mu)}{(\lambda-\mu)}e^{(\lambda-\mu)t}(e^{(\lambda-\mu)t} - 1).$$

The mean corresponds to exponential growth when $\lambda > \mu$ and exponential decay when $\lambda < \mu$. For the case $\lambda = \mu$, the mean and variance are

$$m(t) = N \quad \text{and} \quad \sigma^2(t) = 2N\lambda t.$$

Three sample paths for the simple birth and death process are graphed in Figure 6.4 for parameter values $\lambda = 1 = \mu$ and initial population size $X(0) = 50$. The mean, variance, and p.g.f. for the simple birth, simple death, and simple birth and death processes are summarized in Table 6.1.

Table 6.1: Mean, variance, and p.g.f. for the simple birth, simple death, and simple birth and death processes, where $X(0) = N$ and $\rho = e^{(\lambda-\mu)t}$, $\lambda \neq \mu$

	Simple Birth	Simple Death	Simple Birth and Death
$m(t)$	$Ne^{\lambda t}$	$Ne^{-\mu t}$	$Ne^{(\lambda-\mu)t}$
$\sigma^2(t)$	$Ne^{2\lambda t}(1 - e^{-\lambda t})$	$Ne^{-\mu t}(1 - e^{-\mu t})$	$N\dfrac{\lambda + \mu}{\lambda - \mu}\rho(\rho - 1)$
$\mathcal{P}(z,t)$	$\dfrac{(pz)^N}{(1 - z(1 - p))^N}$ Negative binomial $p = e^{-\lambda t}$	$(1 - p + pz)^N$ Binomial $b(N,p)$ $p = e^{-\mu t}$	$\left(\dfrac{\rho^{-1}(\lambda z - \mu) - \mu(z - 1)}{\rho^{-1}(\lambda z - \mu) - \lambda(z - 1)}\right)^N$

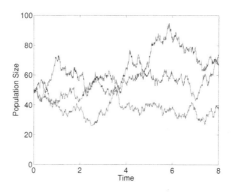

FIGURE 6.4: Three sample paths for the simple birth and death process when $\lambda = 1 = \mu$ and $X(0) = 50$.

6.4.4 Simple Birth and Death with Immigration

Suppose immigration is included in the simple birth and death process at a constant rate ν. Let $X(0) = N$. The infinitesimal transition probabilities for this process are

$$p_{i+j,i}(\Delta t) = \text{Prob}\{\Delta X(t) = j | X(t) = i\}$$

$$= \begin{cases} \mu i\, \Delta t + o(\Delta t), & j = -1 \\ (\nu + \lambda i)\Delta t + o(\Delta t), & j = 1 \\ 1 - [\nu + (\lambda + \mu)i]\,\Delta t + o(\Delta t), & j = 0 \\ o(\Delta t), & j \neq -1,\ 0,\ 1. \end{cases}$$

Because of the immigration term, the value of $\lambda_0 = \nu > 0$.

The forward Kolmogorov equations are

$$\frac{dp_i}{dt} = [\lambda(i-1) + \nu]p_{i-1} + \mu(i+1)p_{i+1} - (\lambda i + \mu i + \nu)p_i$$

$$\frac{dp_0}{dt} = -\nu p_0 + \mu p_1$$

for $i = 1, 2, \ldots$ with initial conditions $p_i(0) = \delta_{iN}$. Applying the generating function technique, it can be shown that the m.g.f. $M(\theta, t)$ is a solution of

$$\frac{\partial M}{\partial t} = \left[\lambda(e^\theta - 1) + \mu(e^{-\theta} - 1)\right]\frac{\partial M}{\partial \theta} + \nu(e^\theta - 1)M$$

with initial condition $M(\theta, 0) = e^{N\theta}$. Bailey (1990) shows the solution is

$$M(\theta, t) = \frac{(\lambda - \mu)^{\nu/\lambda}\left[\mu(e^{(\lambda-\mu)t} - 1) - e^\theta(\mu e^{(\lambda-\mu)t} - \lambda)\right]^N}{\left[(\lambda e^{(\lambda-\mu)t} - \mu) - \lambda(e^{(\lambda-\mu)t} - 1)e^\theta\right]^{N+\nu/\lambda}}.$$

The moments of the probability distribution $X(t)$ can be found by differentiating with respect to θ and evaluating at $\theta = 0$. The mean is

$$m(t) = \begin{cases} \dfrac{e^{(\lambda-\mu)t}(N\mu - N\lambda - \nu) + \nu}{\mu - \lambda}, & \text{if } \lambda \neq \mu \\ \nu t + N, & \text{if } \lambda = \mu. \end{cases} \tag{6.10}$$

The mean increases exponentially in time when $\lambda > \mu$ and linearly when $\lambda = \mu$. However, in the case $\lambda < \mu$, the mean approaches a constant:

$$m(\infty) = \frac{\nu}{\mu - \lambda}, \quad \lambda < \mu.$$

Thus, for the case $\lambda < \mu$, the process is nonexplosive and irreducible. In addition, Theorem 6.1 can be applied to show that the process has a unique positive stationary distribution. Norris (1997) shows that the conditions of being nonexplosive, irreducible, and having a positive stationary distribution imply the process is positive recurrent. Then, Theorem 5.3 implies the limiting distribution exists and equals the stationary distribution. The mean, variance, and p.g.f. are summarized for the simple birth and death process with immigration in Table 6.2.

Table 6.2: Mean, variance, and p.g.f. for the simple birth and death with immigration process, where $X(0) = N$ and $\rho = e^{(\lambda-\mu)t}$, $\lambda \neq \mu$

	Simple Birth and Death with Immigration
$m(t)$	$\dfrac{\rho[N(\lambda - \mu) + \nu] - \nu}{\lambda - \mu}$
$\sigma^2(t)$	$N\dfrac{(\lambda^2 - \mu^2)\rho[\rho - 1]}{(\lambda - \mu)^2} + \nu\dfrac{\mu + \rho(\lambda\rho - \mu - \lambda)}{(\lambda - \mu)^2}$
$P(z,t)$	$\dfrac{(\lambda - \mu)^{\nu/\lambda}\left[\mu(\rho - 1) - z(\mu\rho - \lambda)\right]^N}{[\lambda\rho - \mu - \lambda(\rho - 1)z]^{N+\nu/\lambda}}$

Example 6.3. Consider a simple birth and death process with immigration. Let $\lambda = 0.5 = \nu$, and $\mu = 1$ so that $\lambda_i = 0.5(i + 1)$ and $\mu_i = \mu i = i$. To find the stationary probability distribution π, apply Theorem 6.1. Then

$$\pi_{i+1} = \frac{\lambda_i}{\mu_{i+1}}\pi_i, \quad \pi_i = \frac{\lambda_0\lambda_1 \cdots \lambda_{i-1}}{\mu_1\mu_2 \ldots \mu_i}\pi_0, \quad \text{and} \quad \sum_{i=0}^{\infty} \pi_i = 1.$$

Thus,

$$\pi_{i+1} = \frac{0.5(i + 1)}{i + 1}\pi_i = 0.5\pi_i \quad \text{and} \quad \pi_i = (0.5)^i\pi_0.$$

Also,

$$\sum_{i=0}^{\infty} \pi_i = \pi_0 \sum_{i=0}^{\infty} \left(\frac{1}{2}\right)^i = 2\pi_0.$$

This implies $\pi_0 = 1/2$. The stationary distribution is a geometric distribution,

$$\pi_i = \left(\frac{1}{2}\right)^{i+1}, \quad i = 0, 1, 2, \ldots.$$

The m.g.f. of this geometric distribution (Appendix for Chapter 1) is $M(\theta) = (2 - e^\theta)^{-1}$. The mean and variance of the stationary distribution are $m = 1$ and $\sigma^2 = 2$. The stationary probability distribution for this example is graphed in Figure 6.5. ∎

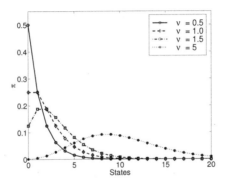

FIGURE 6.5: Stationary probability distributions for the simple birth, death and immigration process with $\lambda = 0.5$, $\mu = 1$, and $\nu = 0.5$, 1.0, 1.5, or 5.

A general formula for the m.g.f. of the stationary distribution π can be found by taking the limit of $M(\theta, t)$ as $t \to \infty$,

$$\lim_{t \to \infty} M(\theta, t) = \left(\frac{1 - \lambda/\mu}{1 - (\lambda/\mu)e^{\theta}} \right)^{\nu/\lambda} .$$

If $\nu/\lambda = n$ is an integer, then the m.g.f. corresponds to a negative binomial distribution with parameter $p = 1 - \lambda/\mu$ (Appendix for Chapter 1). In the case $n = 1$, the negative binomial distribution is the same as the geometric distribution (Example 6.3). The mean, variance, and probability distribution of a negative binomial distribution are

$$m = \frac{nq}{p} = \frac{\nu}{\mu - \lambda}, \quad \text{and} \quad \sigma^2 = \frac{nq}{p^2} = \frac{\nu\mu}{(\mu - \lambda)^2},$$

and

$$\pi_i = \binom{i + n - 1}{n - 1} p^n (1 - p)^i, \quad i = 0, 1, 2, \dots .$$

Figure 6.5 shows the graphs of the stationary probability distributions for four different sets of parameter values; $\lambda = 0.5$, $\mu = 1$ and the immigration rates are $\nu = 0.5, 1, 1.5$, and 5, respectively. The mean values for the stationary probability distributions corresponding to each parameter set are $m = 1, 2, 3$, and 10, respectively. A set of four sample paths for each set of parameter values are graphed in Figure 6.6. After time $t = 25$ the sample paths appear to vary around their respective mean values.

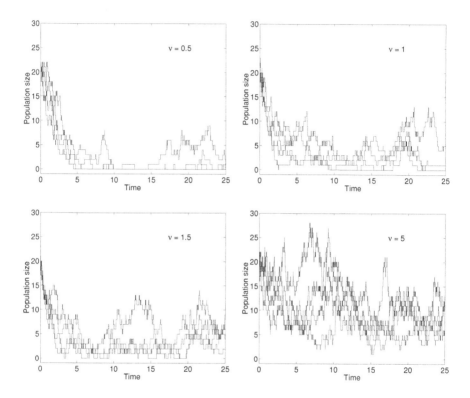

FIGURE 6.6: Four sample paths corresponding to the birth, death, and immigration process when $X(0) = 20$, $\lambda = 0.5$, $\mu = 1$, and $\nu = 0.5$, 1.0, 1.5, 5. The mean values of the respective stationary distributions are $m = 1$, 2, 3, and 10.

6.5 Queueing Process

An important application related to birth and death processes is queueing processes, processes that involve waiting in queues or lines. The arrival and departure processes are similar to birth and death processes. The formal study of queueing processes began during the early part of the twentieth century with the work of the Danish engineer A. K. Erlang. The field of queueing theory has expanded tremendously because of the diversity of applications, which include scheduling of patients, traffic regulation, telephone routing, aircraft landing, and restaurant service. Here, we give a very brief introduction to some simple queueing processes. For a more thorough but elementary introduction to queueing systems, please consult Bharucha-Reid (1997), Hsu (1997), or Taylor and Karlin (1998). A more in-depth treatment of queueing processes and networks can be found in Kleinrock (1975) and Chao et al. (1999).

A queueing process involves three components: (1) arrival process, (2) queue discipline, and (3) service mechanism (Figure 6.7). The arrival process involves the arrival of customers for service and specifies the sequence of arrival times for the customers. The queue discipline is a rule specifying how customers form a queue and how they behave while waiting (e.g., first-come, first-served basis). The service mechanism involves how customers are serviced and specifies the sequence of service times. The notation $A/B/s/K$ is used to denote the type of queue. The variables A = arrival process, B = service time distribution, s = number of servers, and K = capacity of the system. If the queue has unlimited capacity, then it is denoted as $A/B/s$. We shall consider a Poisson arrival process so that the interarrival time is exponential (Markov process) and the service-time distribution is exponential (Markov process). In this case, the queue is denoted as $M/M/s$ or $M/M/s/K$.

FIGURE 6.7: A simple queueing system.

Consider a queueing system of type $M/M/1$. Assume the arrival process is Poisson with parameter λ (mean arrival or birth rate). The service time is exponentially distributed with parameter μ (mean departure or death rate). After being serviced, individuals leave the system. Let $X(t)$ = number of individuals in the queue at time t. If $X(t) = 0$, then there are arrivals (births) but no departures (deaths). If λ and μ are constant, then $X(t)$ is a birth and death process as described in Example 6.2 ($\mu = q$ and $\lambda = p$). The probabilities $\text{Prob}\{X(t) = i\} = p_i(t)$ are solutions of the forward Kolmogorov equations, $dp/dt = Qp$, where the generator matrix

$$Q = \begin{pmatrix} -\lambda & \mu & 0 & \cdots \\ \lambda & -\lambda-\mu & \mu & \cdots \\ 0 & \lambda & -\lambda-\mu & \cdots \\ \vdots & \vdots & \vdots & \end{pmatrix}.$$

The transition matrix T of the embedded Markov chain satisfies

$$T = \begin{pmatrix} 0 & \dfrac{\mu}{\lambda+\mu} & 0 & \cdots \\ 1 & 0 & \dfrac{\mu}{\lambda+\mu} & \cdots \\ 0 & \dfrac{\lambda}{\lambda+\mu} & 0 & \cdots \\ \vdots & \vdots & \vdots & \end{pmatrix}.$$

For this queueing system, there exists a unique stationary probability distribution iff $\lambda < \mu$. The ratio λ/μ is referred to as the *traffic intensity*. The stationary probability distribution is a geometric probability distribution,

$$\pi_i = \left(1 - \frac{\lambda}{\mu}\right)\left(\frac{\lambda}{\mu}\right)^i, \quad i = 0, 1, 2, \ldots.$$

If $\lambda \geq \mu$, then the queue length will tend to infinity. The mean of the stationary probability distribution represents the average number of customers C in the system (at equilibrium),

$$C = \frac{\lambda/\mu}{1 - \lambda/\mu} = \frac{\lambda}{\mu - \lambda}.$$

The average amount of time W a customer spends in the system (at equilibrium) is the average number of customers divided by the average arrival rate λ:

$$W = \frac{C}{\lambda} = \frac{1}{\mu - \lambda}.$$

Example 6.4. Suppose in an $M/M/1$ queueing system customer arrival rate is 3 per minute, $\lambda = 3$. The goal is to find the average service time so that 95% of the queue contains less than 10 customers. First,

$$\text{Prob}\{X(t) < 10\} = \sum_{i=0}^{9} \pi_i = 1 - \sum_{i=10}^{\infty} \pi_i.$$

Then

$$\text{Prob}\{X(t) \geq 10\} = \sum_{i=10}^{\infty} \pi_i$$

$$= \left(1 - \frac{\lambda}{\mu}\right) \sum_{i=10}^{\infty} \left(\frac{\lambda}{\mu}\right)^i$$

$$= \left(\frac{\lambda}{\mu}\right)^{10}.$$

To achieve the goal $(\lambda/\mu)^{10} = 0.05$. Substituting the value of $\lambda = 3$ leads to $\mu = 4.048$ customers per minute. ∎

Next, consider a queueing system of the type $M/M/1/K$. The queue is limited to K customers. Therefore, there are no arrivals after the number of customers has reached K:

$$\lambda_i = \begin{cases} \lambda, & i = 0, 1, \ldots, K-1. \\ 0, & i = K, K+1, \ldots. \end{cases}$$

The mean departure or death rate is μ. For the $M/M/1/K$ queueing system there exists a unique stationary probability distribution given by

$$\pi_i = \left(\frac{\lambda}{\mu}\right)^i \frac{1 - \lambda/\mu}{1 - (\lambda/\mu)^{K+1}}, \quad i = 0, 1, \ldots, K \tag{6.11}$$

(Exercise 14).

For the last example, consider a queueing system of type $M/M/s$. The mean arrival or birth rate is λ. Each of the s servers has an exponential service time with parameter μ. Therefore, the mean departure or death rate is

$$\mu_i = \begin{cases} i\mu, & i = 1, 2, \ldots, s-1, \\ s\mu, & i = s, s+1, \ldots. \end{cases}$$

In this case, there exists a unique stationary probability distribution iff $\lambda < s\mu$. The stationary probability distribution is given by

$$\pi_i = \begin{cases} \dfrac{(\lambda/\mu)^i}{i!} \pi_0, & i = 0, 1, \ldots, s-1, \\ \dfrac{(\lambda/\mu)^i s^{s-i}}{s!} \pi_0, & i = s, s+1, \ldots, \end{cases}$$

where π_0 is determined from $\sum_{i=0}^{\infty} \pi_i = 1$.

6.6 Population Extinction

In the simple birth and death processes, $\lambda_0 = 0 = \mu_0$; the zero state is absorbing. For many realistic birth and death processes, the zero state is absorbing. Eventually, the distribution for the total population size is concentrated at zero, $\lim_{t \to \infty} p_0(t) = 1$. The following theorem gives conditions for total population extinction as $t \to \infty$ in a general birth and death process.

THEOREM 6.2

Let $\mu_0 = 0 = \lambda_0$ in a general birth and death chain with $X(0) = m \geq 1$.

(i) Suppose $\mu_i > 0$ and $\lambda_i > 0$ for $i = 1, 2 \ldots$. If

$$\sum_{i=1}^{\infty} \frac{\mu_1 \mu_2 \cdots \mu_i}{\lambda_1 \lambda_2 \cdots \lambda_i} = \infty, \tag{6.12}$$

then $\lim_{t \to \infty} p_0(t) = 1$. If

$$\sum_{i=1}^{\infty} \frac{\mu_1 \mu_2 \cdots \mu_i}{\lambda_1 \lambda_2 \cdots \lambda_i} < \infty \tag{6.13}$$

and the probability of extinction approaches zero as $m \to \infty$, then for finite m,

$$\lim_{t \to \infty} p_0(t) = \frac{\sum_{i=m}^{\infty} \frac{\mu_1 \mu_2 \cdots \mu_i}{\lambda_1 \lambda_2 \cdots \lambda_i}}{1 + \sum_{i=1}^{\infty} \frac{\mu_1 \mu_2 \cdots \mu_i}{\lambda_1 \lambda_2 \cdots \lambda_i}}. \tag{6.14}$$

(ii) Suppose $\mu_i > 0$ for $i = 1, 2, \ldots$, $\lambda_i > 0$ for $i = 1, 2, \ldots, N-1$ and $\lambda_i = 0$ for $i = N, N+1, N+2, \ldots$. Then $\lim_{t \to \infty} p_0(t) = 1$.

The proof is given in the Appendix. Case (ii) is stated separately since the ratio in (i) is undefined if $\lambda_i = 0$. However, formally, the sum in case (i) is infinite if $\lambda_i = 0$ for any $i \geq 1$. Case (i) of Theorem 6.2 applies to an infinite Markov chain with state space $\{0, 1, 2, \ldots\}$, and case (ii) applies to a finite Markov chain with state space $\{0, 1, \ldots, N\}$ when $X(0) = m \leq N$. Note that in case (ii), the birth rate, λ_i, is zero for states i greater than or equal to the maximum size N. If $X(0) = m > N$, there is a simple death process occurring until size N is reached. Theorem 6.2(ii) also holds if the death and birth rates are zero, $\mu_i = 0 = \lambda_i$ for $i > N$ and $X(0) = m \leq N$.

Example 6.5. In the simple birth and death process, $\lambda_i = \lambda i$ and $\mu_i = \mu i$. Then

$$\sum_{i=1}^{\infty} \frac{\mu_1 \mu_2 \cdots \mu_i}{\lambda_1 \lambda_2 \cdots \lambda_i} = \sum_{i=1}^{\infty} \left(\frac{\mu}{\lambda}\right)^i = \begin{cases} \infty, & \text{if } \mu \geq \lambda, \\ < \infty, & \text{if } \mu < \lambda. \end{cases}$$

According to Theorem 6.2, if $\mu \geq \lambda$, then $\lim_{t \to \infty} p_0(t) = 1$ and if $\mu < \lambda$, then from equation (6.9),

$$\lim_{t \to \infty} p_0(t) = \left(\frac{\mu}{\lambda}\right)^m.$$

The formula agrees with equation (6.14). ∎

6.7 First Passage Time

6.7.1 Definition and Computation

The time until the process reaches a certain stage for the first time is known as the *first passage time*. Suppose a population size is a and we want to find the time it takes until it reaches a size equal to b, where either $a < b$ or $b < a$. It will be shown that the first passage time problems are related to the backward Kolmogorov differential equation.

Let $T_{i+1,i}$ be the random variable for the time it takes to go from state i to $i+1$. From the derivation of the interevent time, we know the p.d.f. for the interevent time has an exponential distribution with parameter $\lambda_i + \mu_i$ $(X(0) = i)$:

$$f_i(t) = (\lambda_i + \mu_i)e^{-(\lambda_i + \mu_i)t}.$$

Thus, the expected time to go from state i to either $i+1$ or $i-1$ is the mean of the exponential distribution, $1/(\lambda_i + \mu_i)$. The process jumps to state $i+1$ if there is a birth (probability $\lambda_i/(\lambda_i + \mu_i)$) and jumps to state $i-1$ if there is a death (probability $\mu_i/(\lambda_i + \mu_i)$). Thus, if $a < b$, the time it takes to go from state a to b is

$$T_{b,a} = T_{a+1,a} + T_{a+2,a+1} + \cdots + T_{b,b-1}$$

and the expected time to go from state a to b is

$$E(T_{b,a}) = E(T_{a+1,a}) + E(T_{a+2,a+1}) + \cdots + E(T_{b,b-1}).$$

Similar expressions can be derived for $a > b$. These are *mean first passage times*.

General expressions for $E(T_{i,i+1})$ and $E(T_{i+1,i})$ are derived for a birth and death process. Suppose the process is in state i. After an exponential amount of time the process jumps from i to $i+1$ with probability $\lambda_i/(\lambda_i + \mu_i)$ and to state $i-1$ with probability $\mu_i/(\lambda_i + \mu_i)$. To find the expected time of going from state i to $i+1$, we must consider that the process may jump to $i-1$;

then the expected time it takes to go back to $i+1$ must be added to this time,

$$
\begin{aligned}
E(T_{i+1,i}) &= \frac{\lambda_i}{\lambda_i + \mu_i}\left(\frac{1}{\lambda_i + \mu_i}\right) \\
&\quad + \frac{\mu_i}{\lambda_i + \mu_i}\left(\frac{1}{\lambda_i + \mu_i} + E(T_{i,i-1}) + E(T_{i+1,i})\right) \\
&= \frac{1}{\lambda_i + \mu_i} + \frac{\mu_i}{\lambda_i + \mu_i}\left[E(T_{i,i-1}) + E(T_{i+1,i})\right]
\end{aligned}
$$

(see Schinazi, 1999; Renshaw, 1993). The preceding relation can be simplified to obtain

$$
E(T_{i+1,i}) = \frac{1}{\lambda_i} + \frac{\mu_i}{\lambda_i}E(T_{i,i-1}). \tag{6.15}
$$

A similar derivation can be obtained for $E(T_{i-1,i})$:

$$
\begin{aligned}
E(T_{i-1,i}) &= \frac{\mu_i}{\lambda_i + \mu_i}\left(\frac{1}{\lambda_i + \mu_i}\right) \\
&\quad + \frac{\lambda_i}{\lambda_i + \mu_i}\left(\frac{1}{\lambda_i + \mu_i} + E(T_{i,i+1}) + E(T_{i-1,i})\right).
\end{aligned}
$$

Simplifying,

$$
E(T_{i-1,i}) = \frac{1}{\mu_i} + \frac{\lambda_i}{\mu_i}E(T_{i,i+1}). \tag{6.16}
$$

The two identities (6.15) and (6.16) will be applied to the simple birth and simple death processes.

Example 6.6. Consider a simple birth process beginning in state a. Recall that $\lambda_i = \lambda i$ and $\mu_i = 0$. The expected time $E(T_{i+1,i}) = 1/\lambda_i$. Therefore, the expected time to go from state a to state b, $a < b$ is

$$
\frac{1}{\lambda}\left(\frac{1}{a} + \frac{1}{a+1} + \cdots + \frac{1}{b-1}\right). \tag{6.17}
$$

The summation can be bounded by logarithms as follows:

$$
\ln\left(\frac{b}{a}\right) \le \sum_{i=a}^{b-1}\frac{1}{i} \le \ln\left(\frac{b-1}{a-1}\right).
$$

In particular,

$$
\lim_{n\to\infty}\left[\sum_{i=1}^{n}\frac{1}{i} - \ln(n)\right] = \gamma = 0.57721566490\ldots,
$$

where the constant γ is known as *Euler's constant*. Thus, for large values of a and b, the expression in (6.17) can be approximated by

$$
E(T_{b,a}) \approx \frac{1}{\lambda}\ln\left(\frac{b}{a}\right).
$$

This estimate agrees with the one obtained from the exponential growth model, $n(t) = ae^{\lambda t}$. The time it takes to go from state a to state b is found by solving the equation $b = ae^{\lambda t}$,

$$t = \frac{1}{\lambda} \ln\left(\frac{b}{a}\right).$$

It is left as an exercise to show that for the simple death process the expected time it takes to go from state a to state b for $a > b$ is approximately $E(T_{b,a}) \approx (1/\mu)\ln(a/b)$. ∎

Example 6.7. Consider a general birth and death process, where $\lambda_0 > 0$, $\mu_0 = 0$, and $\lambda_i, \mu_i > 0$ for $i = 1, 2, \ldots$. Then the mean time it takes to go from state 0 to state 3 is calculated. Note that $E(T_{1,0}) = 1/\lambda_0$. Then

$$E(T_{2,1}) = \frac{1}{\lambda_1} + \frac{\mu_1}{\lambda_1\lambda_0},$$

$$E(T_{3,2}) = \frac{1}{\lambda_2} + \frac{\mu_2}{\lambda_2}E(T_{2,1}) = \frac{1}{\lambda_2} + \frac{\mu_2}{\lambda_1\lambda_2} + \frac{\mu_1\mu_2}{\lambda_0\lambda_1\lambda_2},$$

so that

$$E(T_{3,0}) = E(T_{1,0}) + E(T_{2,1}) + E(T_{3,2})$$
$$= \frac{1}{\lambda_0} + \frac{1}{\lambda_1} + \frac{1}{\lambda_2} + \frac{\mu_1}{\lambda_1\lambda_0} + \frac{\mu_2}{\lambda_2\lambda_1} + \frac{\mu_2\mu_1}{\lambda_2\lambda_1\lambda_0}. \tag{6.18}$$

∎

The expression for $E(T_{3,0})$, equation (6.18), can be extended to $E(T_{b,0})$ (Taylor and Karlin, 1998). This latter expression is related to a nonexplosive process. If the expected time to blow-up approaches infinity, as $b \to \infty$, then the process is nonexplosive. That is, $\lim_{b\to\infty} E(T_{b,0}) = \infty$.

Suppose $\lambda_0 = 0 = \mu_0$ and $\lim_{t\to\infty} p_0(t) = 1$, so that ultimate extinction is certain. In this case, it is possible to derive a formula for $E(T_{0,i})$, the *expected time to extinction* beginning from state i. Notice that $E(T_{0,0}) = 0$. For simplicity, denote $\tau_i = E(T_{0,i})$. For a small interval of time Δt,

$$\tau_i = \lambda_i\Delta t(\tau_{i+1} + \Delta t) + \mu_i\Delta t(\tau_{i-1} + \Delta t) + [1 - (\lambda_i + \mu_i)\Delta t](\tau_i + \Delta t) + o(\Delta t).$$

Simplifying this expression, dividing by Δt, and letting $\Delta t \to 0$ leads to the following difference equation:

$$\tau_i = \frac{1}{\lambda_i + \mu_i} + \frac{\lambda_i}{\lambda_i + \mu_i}\tau_{i+1} + \frac{\mu_i}{\lambda_i + \mu_i}\tau_{i-1}.$$

These equations can be expressed as

$$\mu_i\tau_{i-1} - (\lambda_i + \mu_i)\tau_i + \lambda_i\tau_{i+1} = -1, \quad i = 1, 2, \ldots. \tag{6.19}$$

If the Markov chain is finite, with states $\{0, 1, 2, \ldots, N\}$, so that $\lambda_N = 0$, then for $i = 1, 2, \ldots, N$, the system of equations given in (6.19) can be expressed in matrix form:

$$\boxed{\tau^{tr} \tilde{Q} = -\mathbf{1}^{tr},} \tag{6.20}$$

where $\mathbf{1}$ is a column vector of ones and matrix $\tilde{Q}$ is the generator matrix with the first row and first column deleted (because $\tau_0 = 0$), equation (6.3).

A recursive formula for higher-order moments can be derived. The rth moment of the extinction time, $E(T_{0,i}^r)$, $i = 1, 2, \ldots, N$, can be expressed in terms of the $(r-1)$st moments, $E(T_{0,i}^{r-1})$ (Goel and Richter-Dyn, 1974; Norden, 1982; Richter-Dyn and Goel, 1972). If τ^r denotes the n-column vector of rth moments for the extinction time and the Markov chain is finite, then

$$[\tau^r]^{tr} \tilde{Q} = -r[\tau^{r-1}]^{tr}, \tag{6.21}$$

where $\tilde{Q}$ is the generator matrix with the first row and first column deleted. An explicit solution for the first moment can be derived for a general birth and death chain. This result is stated next; a sketch of the proof is in the Appendix for Chapter 6. See Chapter 3, Theorem 3.1.

THEOREM 6.3

Suppose $\{X(t) : t \in [0, \infty)\}$ *is a continuous-time birth and death chain with* $X(0) = m \geq 1$ *satisfying* $\lambda_0 = 0 = \mu_0$ *and* $\lambda_i > 0$ *and* $\mu_i > 0$ *for* $i = 1, 2, \ldots$. *In addition, suppose* $\lim_{t \to \infty} p_0(t) = 1$. *The expected time until extinction,* $\tau_m = E(T_{0,m})$, *satisfies*

$$\tau_m = \begin{cases} \dfrac{1}{\mu_1} + \displaystyle\sum_{i=2}^{\infty} \dfrac{\lambda_1 \cdots \lambda_{i-1}}{\mu_1 \cdots \mu_i}, & m = 1 \\[2mm] \tau_1 + \displaystyle\sum_{s=1}^{m-1} \left[\dfrac{\mu_1 \cdots \mu_s}{\lambda_1 \cdots \lambda_s} \displaystyle\sum_{i=s+1}^{\infty} \dfrac{\lambda_1 \cdots \lambda_{i-1}}{\mu_1 \cdots \mu_i} \right], & m = 2, 3, \ldots. \end{cases} \tag{6.22}$$

The condition $\lim_{t \to \infty} p_0(t) = 1$ is required in Theorem 6.3 (i.e., the condition (6.12) in Theorem 6.2), since if $\lim_{t \to \infty} p_0(t) < 1$, there is a positive probability that the population size will approach infinity which implies $\tau_m = \infty$. There is no positive stationary distribution since $\lambda_0 = 0$. If

$$\sum_{i=2}^{\infty} \frac{\lambda_1 \cdots \lambda_{i-1}}{\mu_1 \cdots \mu_i} = \infty,$$

then $\tau_m = \infty$. When the Markov chain is finite with maximal population size N, then automatically $\tau_m < \infty$ and the solution for $m = 1, 2, \ldots, N$, $\tau = (\tau_1, \tau_2, \ldots, \tau_N)^{tr}$, can be calculated via equation (6.20). Equation (6.22) applies to the finite case as well, where the summation to ∞ is replaced by N.

The formulas in equations (6.18) and (6.22) show that the expressions $E(T_{0,m})$ and $E(T_{m,0})$ are very different. In the expression (6.22), $\lambda_0 = 0$, the origin is absorbing, and in expression (6.18), $\lambda_0 > 0$, the origin is not absorbing. For example, the expression in (6.22) for the case $m = 3$ is

$$E(T_{0,3}) = \left[\frac{1}{\mu_1} + \frac{\lambda_1}{\mu_1\mu_2} + \frac{\lambda_1\lambda_2}{\mu_1\mu_2\mu_3} + \cdots \right] + \left[\frac{1}{\mu_2} + \frac{\lambda_2}{\mu_2\mu_3} + \frac{\lambda_2\lambda_3}{\mu_2\mu_3\mu_4} + \cdots \right]$$
$$+ \left[\frac{1}{\mu_3} + \frac{\lambda_3}{\mu_3\mu_4} + \frac{\lambda_3\lambda_4}{\mu_3\mu_4\mu_5} + \cdots \right].$$

The reason for this difference is that before reaching 0, the chain may visit states i, where $1 \leq i < \infty$, but when moving from 0 to m, the chain may only visit states i, $0 \leq i < m$.

6.7.2 Summary of First Passage Time

The results in the preceding section on first passage time can be related to the backward Kolmogorov differential equation. Let $T(m) = T_{0,m}$ be the discrete random variable for the time until the process first reaches state 0 (extinction), given $X(0) = m \neq 0$. Let $R(m,t)$ be the probability that the process does not reach 0 during the time interval $[0, t]$, referred to as a *reliability function* in engineering applications. That is,

$$R(m,t) = \text{Prob}\{T(m) > t\}.$$

Then $1 - R(m,t) = \text{Prob}\{T(m) \leq t\}$. The probability distribution for $T(m)$ is

$$p_T(m,t) = -\frac{dR(m,t)}{dt}.$$

Also, it follows from the definition of $R(m,t)$ that

$$R(m,t) = \sum_{j=1}^{n} p_{jm}(t),$$

where n is the maximal population size. Summing on the index j in $p_{ji}(t)$ in the transition matrix from the backward Kolmogorov differential equation, $dP(t)/dt = P(t)Q$, leads to

$$\sum_{j=1}^{n} \frac{dp_{jm}(t)}{dt} = \sum_{j=1}^{n} \sum_{i=0}^{n} p_{ji}(t) q_{im} = \sum_{i=0}^{n} \left(\sum_{j=1}^{n} p_{ji}(t) \right) q_{im}.$$

From the definition of $R(m,t)$, the preceding expression equals

$$\frac{dR(m,t)}{dt} = \sum_{i=0}^{n} R(i,t) q_{im}. \qquad (6.23)$$

It follows that $R(t) = (R(0,t), R(1,t), \ldots, R(n,t))$ is a solution of the backward Kolmogorov differential equation,

$$\frac{dR(t)}{dt} = R(t)Q,$$

where the initial conditions are $R(i,0) = 1$, $i = 1, 2, \ldots, n$ and it is assumed that $R(0,t) = 0$.

To relate $R(m,t)$ to the expected time to extinction τ_m, note that

$$\tau_m = \int_0^\infty t p_T(m,t)\, dt = -\int_0^\infty t \frac{dR(m,t)}{dt}\, dt$$
$$= \int_0^\infty R(m,t)\, dt.$$

The latter identity follows from integration by parts and $\lim_{t\to\infty}[tR(m,t)] = 0$. Integrating equation (6.23) from $t = 0$ to ∞, the left side is

$$\lim_{t\to\infty} R(m,t) - R(m,0) = -1$$

and the right side is $\sum_{i=0}^n \tau_i q_{im}$. It follows that the expected time to extinction, $\tau^{tr} = (\tau_1, \tau_2, \ldots, \tau_n)$, is the solution of system (6.20). That is,

$$\tau^{tr}\tilde{Q} = -\mathbf{1}^{tr},$$

where the $n \times n$ matrix $\tilde{Q}$ is the generator matrix with the first row and first column deleted (because $\tau_0 = 0$ and $R(0,t) = 0$).

This method for computing first passage time probabilities also applies to multivariate processes. But the method can become quite complicated because the generator matrix Q is large and more complex. For multivariate processes, numerical approximation methods and simulations are useful for studying first passage time problems.

6.8 Logistic Growth Process

To formulate a stochastic logistic model, recall that the deterministic logistic model is a solution of the differential equation

$$\frac{dn}{dt} = rn\left(1 - \frac{n}{K}\right), \quad n(0) > 0,$$

where $n = n(t)$ equals the population size at time t, r is the intrinsic growth rate, and K is the carrying capacity. Solutions $n(t)$ to this differential equation approach the carrying capacity K, $\lim_{t\to\infty} n(t) = K$. The derivative, dn/dt, equals the birth rate minus the death rate. Thus, in a stochastic logistic model, the birth and death rates, λ_n and μ_n, should satisfy

$$\lambda_n - \mu_n = rn - \frac{r}{K}n^2,$$

$\lambda_0 = 0 = \mu_0$, and $\lambda_K = \mu_K$. It is reasonable to assume that the birth and death rates are quadratic functions of the population size. The state space could be finite or infinite. In the case of an infinite state space $\{0, 1, \ldots\}$, assume that for each state $i \in \{0, 1, \ldots\}$, the birth and death rates are

$$\lambda_i = b_1 i + b_2 i^2 > 0 \quad \text{and} \quad \mu_i = d_1 i + d_2 i^2 > 0, \tag{6.24}$$

where the coefficients b_j and d_j, $j = 1, 2$ are constants. In the case of a finite state space $\{0, 1, 2 \ldots, N\}$, assume for each state $i \in \{0, 1, 2, \ldots, N\}$,

$$\lambda_i = \begin{cases} b_1 i + b_2 i^2 > 0, & \text{if } i = 1, 2, \ldots, N - 1 \\ 0, & \text{if } i = N \end{cases} \tag{6.25}$$

and

$$\mu_i = d_1 i + d_2 i^2 > 0.$$

The initial population size $X(0)$ can be greater than N, but the process acts as a death process until state N is reached and then the process remains in the state space $\{0, 1, 2, \ldots, N\}$. Putting the birth and death parameters in the differential equation leads to

$$\frac{dn}{dt} = (b_1 - d_1)n + (b_2 - d_2)n^2 = rn - \frac{r}{K}n^2,$$

so that

$$r = b_1 - d_1 > 0 \quad \text{and} \quad K = \frac{b_1 - d_1}{d_2 - b_2} > 0. \tag{6.26}$$

There are four constants in the stochastic logistic model, b_1, b_2, d_1, and d_2, but only two constants in the deterministic model, r and K. There is an infinite number of choices for the four constants in the stochastic logistic model that give the same values for r and K. Suppose, for example, $b_1 = r$, $b_2 = c$, $d_1 = 0$, and $d_2 = c + r/K$, where c is a constant. Then the relations (6.26) hold for an infinite number of possible choices for the constant c:

$$\lambda_n - \mu_n = (rn + cn^2) - \left(cn^2 + \frac{r}{K}n^2\right) = rn - \frac{r}{K}n^2.$$

If the coefficients b_2 and d_2 are not zero, then the per capita rates of birth and death depend on the population size (i.e., λ_n/n and μ_n/n depend on n). For example, if $b_2 < 0$ (or $d_2 < 0$), the number of births (deaths) decreases as the population size increases, but if the reverse inequality holds, $b_2 > 0$ ($d_2 > 0$), the number of births (deaths) increases as the population size increases. A reasonable assumption for many populations is that $d_2 > 0$, meaning that the death rate is density-dependent. Ultimately, the choice of the coefficients b_j and d_j depends on the dynamics of the particular population being modeled.

Example 6.8. Two stochastic logistic models are defined that have the same deterministic logistic model. Define the birth and death rates, λ_i and μ_i, as follows:

(a) $\lambda_i = i$ and $\mu_i = \dfrac{i^2}{10}$, $i = 0, 1, 2, \ldots$

(b) $\lambda_i = \begin{cases} i - \dfrac{i^2}{20}, & i = 0, 1, \ldots, 20 \\ 0, & i > 20 \end{cases}$ and $\mu_i = \dfrac{i^2}{20}$, $i = 0, 1, 2, \ldots$

In both cases the deterministic model is

$$\frac{dn}{dt} = n\left(1 - \frac{n}{10}\right),$$

where $r = 1$ and $K = 10$. In the deterministic model, solutions approach the carrying capacity $K = 10$. Three sample paths for models (a) and (b) are graphed in Figure 6.8. ∎

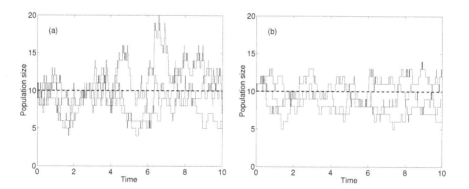

FIGURE 6.8: Three sample paths of the stochastic logistic model for cases (a) and (b) with $X(0) = 10$.

The assumptions (6.24) or (6.25) and application of Theorems 6.2 and 6.3 imply that in the stochastic logistic model, extinction occurs with probability 1, $\lim_{t \to \infty} p_0(t) = 1$, and the expected time to extinction is finite.

COROLLARY 6.1

Assume the stochastic logistic model satisfies (6.26) and either (6.24) or (6.25). Then

$$\lim_{t \to \infty} p_0(t) = 1 \tag{6.27}$$

and the expected time until extinction from state m is $\tau_m = E(T_{0,m}) < \infty$.

Proof. Suppose (6.25) is satisfied. Then (6.27) follows directly from Theorem 6.2 part (ii) and $\tau_m < \infty$.

Suppose (6.24) is satisfied. We will show that (6.12) is satisfied. The ratio of two successive terms, a_i/a_{i-1} in the summation of (6.12), equals μ_i/λ_i, where

$$\frac{\mu_i}{\lambda_i} = \frac{d_1 + d_2 i}{b_1 + b_2 i}.$$

Because λ_i and μ_i are positive for all i and from the assumptions in (6.26), it follows that $d_2 > b_2 \geq 0$. Thus,

$$\lim_{i \to \infty} \frac{\mu_i}{\lambda_i} = \begin{cases} \dfrac{d_2}{b_2} > 1, & \text{if } b_2 > 0 \\ \infty, & \text{if } b_2 = 0. \end{cases}$$

In either case, by the ratio test, it follows that (6.12) is satisfied and Theorem 6.2 (i) implies $\lim_{t \to \infty} p_0(t) = 1$. In addition, it is easy to show by the ratio test that $\sum_{i=1}^{\infty} \dfrac{\lambda_1 \cdots \lambda_{i-1}}{\mu_1 \cdots \mu_i} < \infty$:

$$\lim_{i \to \infty} \frac{\lambda_i}{\mu_{i+1}} = \begin{cases} \dfrac{b_2}{d_2} < 1, & \text{if } b_2 > 0 \\ 0, & \text{if } b_2 = 0. \end{cases}$$

Hence, by Theorem 6.3, $\tau_m < \infty$. $\square$

The expected time to extinction is calculated using the techniques from the previous section in the next two examples.

Example 6.9. Suppose the stochastic logistic model has the following birth and death rates:

$$\lambda_i = \begin{cases} i - \dfrac{i^2}{N}, & i = 0, 1, \ldots, N \\ 0, & i > N \end{cases} \quad \text{and} \quad \mu_i = \frac{i^2}{N}, \quad i = 0, 1, 2, \ldots.$$

The intrinsic growth rate $r = 1$ and the carrying capacity $K = N/2$. The expected time to extinction can be calculated for $X(0) = m$, $m = 1, 2, \ldots, N$ by solving the linear system (6.20), $\tau^{tr}\tilde{Q} = -1^{tr}$. When $N = 10$, the carrying capacity is $K = 5$, and when $N = 20$, the carrying capacity is $K = 10$. The expected time to extinction for a carrying capacity of $K = 5$ ranges from approximately 234 ($= \tau_1$) to 269 ($= \tau_{10}$), whereas for $K = 10$, the expected time to extinction is on the order of 10^5. The time units depend on the particular problem. If the time is measured in days, then for the first example, extinction occurs, on the average, in less than one year but for the second example, the mean time to extinction is over 300 years. ∎

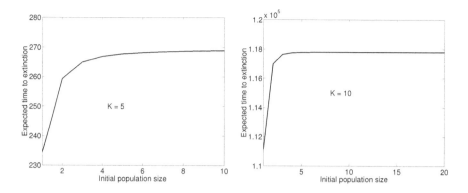

FIGURE 6.9: Expected time until extinction in the stochastic logistic model with $K = 5$ and $K = 10$.

Example 6.10. Assume the stochastic logistic model satisfies

$$\lambda_i = i \ \text{ and } \ \mu_i = \frac{i^2}{10}, \quad i = 0, 1, 2, \ldots.$$

In this case, the carrying capacity is $K = 10$. Formula (6.22) can be used to compute $\tau_1 = E(T_{0,1})$, a lower bound for the expected time to extinction, $\tau_m \geq \tau_1$, $m \geq 1$,

$$\tau_1 = \sum_{i=1}^{\infty} \frac{\lambda_1 \cdots \lambda_{i-1}}{\mu_1 \cdots \mu_i}.$$

An estimate for τ_1 is $\tau_1 \approx 2489$. Again, if time is measured in days, then the mean time to extinction is about seven years, much smaller than the estimate of over 300 years for the previous example. ∎

Applying the generating function technique to the forward Kolmogorov equations, it can be shown that the partial differential equation for the m.g.f. of the stochastic logistic model is

$$\frac{\partial M}{\partial t} = \left[b_1(e^{\theta} - 1) + d_1(e^{-\theta} - 1) \right] \frac{\partial M}{\partial \theta}$$
$$+ \left[b_2(e^{\theta} - 1) + d_2(e^{-\theta} - 1) \right] \frac{\partial^2 M}{\partial \theta^2}, \tag{6.28}$$

where $M(\theta, 0) = e^{N\theta}$ if $X(0) = N$. This differential equation can be used to derive differential equations satisfied by the mean and higher-order moments. Differentiating (6.28) with respect to θ and interchanging the order of the differentiation yields

$$\frac{\partial^2 M}{\partial t\, \partial \theta} = [b_1 e^\theta - d_1 e^{-\theta}]\frac{\partial M}{\partial \theta} + [b_1(e^\theta - 1) + d_1(e^{-\theta} - 1)]\frac{\partial^2 M}{\partial \theta^2}$$
$$+ [b_2 e^\theta - d_2 e^{-\theta}]\frac{\partial^2 M}{\partial \theta^2} + [b_2(e^\theta - 1) + d_2(e^{-\theta} - 1)]\frac{\partial^3 M}{\partial \theta^3}.$$

Evaluating this differential equation at $\theta = 0$ and using the fact that $\partial^k M/\partial \theta^k$ evaluated at $\theta = 0$ is $E(X^k(t))$ gives the identity

$$\frac{dm(t)}{dt} = [b_1 - d_1]m(t) + [b_2 - d_2]E(X^2(t))$$
$$= rm(t) - \frac{r}{K}E(X^2(t)). \tag{6.29}$$

The differential equation cannot be solved for $m(t)$ since it also depends on $E(X^2(t))$. A differential equation for $E(X^2(t))$ is also required. But then the differential equations for the second-order moments also depend on higher-order moments and so on, forming an infinite system of differential equations. The system is not "closed". Assumptions about the distribution lead to simplifications for the higher-order moments. For example, the distribution can be assumed to be approximately normal or lognormal, referred to as *moment closure assumptions* (e.g., Chan and Isham, 1998; Lloyd, 2004). In the simple birth and death processes, the equations for the mean do not depend on higher-order moments. For these simple models, the mean agrees with the solution of the deterministic model. This is not the case for nonlinear models. In particular, for the stochastic logistic model, it will be shown that the mean of the stochastic process is less than the solution of the deterministic model (Tognetti and Winley, 1980).

The variance $\sigma^2(t) = E(X^2(t)) - m^2(t) > 0$ so that $E(X^2(t)) > m^2(t)$. Because $m(t) \geq 0$, it follows from equation (6.29) that

$$\frac{dm(t)}{dt} < rm(t)\left[1 - \frac{m(t)}{K}\right], \quad t \in [0, \infty).$$

Suppose $n(t)$ is the solution of the logistic differential equation $dn/dt = rn(1 - n/K)$ with $0 < n(0) = m(0)$. By comparing the differential equations for $m(t)$ and $n(t)$, it can be shown that

$$m(t) \leq n(t) \quad \text{for} \quad t \in [0, \infty)$$

(Comparison theorem in the Appendix for Chapter 6). The solution of the deterministic logistic differential equation is greater than the mean of the stochastic logistic model. This result seems plausible because ultimate extinction is certain in the stochastic logistic model, $\lim_{t\to\infty} p_0(t) = 1$. As $t \to \infty$, the probability distribution becomes concentrated at zero, so that $\lim_{t\to\infty} m(t) = 0$.

A differential equation for $E(X^2(t))$ can be derived in a manner similar to the mean. Use the partial differential equation for the m.g.f. and differentiate twice with respect to θ; then evaluate at $\theta = 0$:

$$\frac{dE(X^2(t))}{dt} = [b_1 + d_1]m(t) + [2(b_1 - d_1) + (b_2 + d_2)]E(X^2(t))$$
$$+ 2[b_2 - d_2]E(X^3(t)). \tag{6.30}$$

The variance increases if $b_1 + d_1$ and $b_2 + d_2$ are large and positive. This can be seen in the stochastic simulations of the logistic model, Figure 6.8. In (a), $b_1 + d_1 = 1$ and $b_2 + d_2 = 1/10$ and in (b) $b_1 + d_1 = 1$ and $b_2 + d_2 = 0$ for $i \leq 20$ and $1/20$ for $i > 20$. Both models have the same deterministic logistic model, but model (b) has a smaller variance and the population modeled by (b) persists longer.

6.9 Quasistationary Probability Distribution

In birth and death models, when the origin is absorbing, there is no stationary probability distribution. However, even when $\lim_{t \to \infty} p_0(t) = 1$, prior to extinction, the probability distribution of $X(t)$ can be approximately stationary for a long period of time. This is especially true if the expected time to extinction is very long. This approximate stationary distribution is known as the *quasistationary probability distribution* or *quasiequilibrium probability distribution*.

Denote the probability distribution associated with $X(t)$ conditioned on nonextinction as $q_i(t)$. Then

$$q_i(t) = \frac{p_i(t)}{1 - p_0(t)}, \quad i = 1, 2, \ldots.$$

The quasistationary probabilities are solutions of a system of differential equations similar to the forward Kolmogorov differential equations,

$$\frac{dq_i}{dt} = \frac{dp_i/dt}{1 - p_0} + \frac{p_i}{(1 - p_0)}\frac{dp_0/dt}{(1 - p_0)}$$
$$= \lambda_{i-1}q_{i-1} - (\lambda_i + \mu_i)q_i + \mu_{i+1}q_{i+1} + q_i\mu_1 q_1,$$

where it is assumed that $\lambda_0 = 0 = \mu_0$.

The quasistationary probability distribution can be approximated by making the assumption $\mu_1 = 0$. Then $dq/dt = \tilde{Q}q$, where

$$\tilde{Q} = \begin{pmatrix} -\lambda_1 & \mu_2 & 0 & \cdots \\ \lambda_1 & -\lambda_2 - \mu_2 & \mu_3 & \cdots \\ 0 & \lambda_2 & -\lambda_3 - \mu_3 & \cdots \\ \vdots & \vdots & \vdots & \end{pmatrix}.$$

This system has unique positive stationary probability distribution given by $\tilde{\pi} = (\tilde{\pi}_1, \tilde{\pi}_2, \dots)^{tr}$ if the assumptions of Theorem 6.1 are satisfied. The stationary probability distribution is

$$\tilde{\pi}_i = \frac{\lambda_1 \lambda_2 \cdots \lambda_{i-1}}{\mu_2 \mu_3 \cdots \mu_i} \tilde{\pi}_1, \quad \text{and} \quad \sum_{i=1}^{\infty} \tilde{\pi}_i = 1.$$

Therefore, a unique positive stationary probability distribution exists to the system $dq/dt = \tilde{Q}q$ if

$$\sum_{i=2}^{\infty} \frac{\lambda_1 \lambda_2 \cdots \lambda_{i-1}}{\mu_2 \mu_3 \cdots \mu_i} < \infty.$$

The solution $\tilde{\pi}$ approximates the quasistationary probability distribution.

Example 6.11. Consider the two stochastic logistic models discussed previously, where the birth and death rates, λ_i and μ_i, are defined as follows:

(a) $\lambda_i = i$ and $\mu_i = \dfrac{i^2}{10}$, $i = 0, 1, 2, \dots$

(b) $\lambda_i = \begin{cases} i - \dfrac{i^2}{20}, & i = 0, 1, \dots, 20, \\ 0, & i > 20, \end{cases}$ and $\mu_i = \dfrac{i^2}{20}$, $i = 0, 1, 2 \dots$

In both cases the deterministic model is

$$\frac{dn}{dt} = n\left(1 - \frac{n}{10}\right),$$

so that the intrinsic growth rate is $r = 1$ and the carrying capacity is $K = 10$. The approximate quasistationary probabilities $\tilde{\pi}_n$ are calculated for each model and graphed in Figure 6.10. ∎

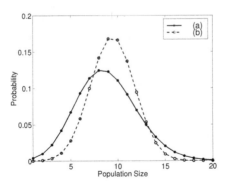

FIGURE 6.10: Approximate quasistationary probability distribution $\tilde{\pi}$ for cases (a) and (b). The mean and standard deviation for case (a) are $\tilde{m} \approx 8.85$, $\tilde{\sigma} \approx 3.19$ and for case (b), $\tilde{m} \approx 9.44$ and $\tilde{\sigma} \approx 2.31$.

Notice that the means of the quasistationary distributions are less than the equilibrium of the deterministic model, $K = 10$, and that the variance in model (a) is greater than model (b). It was shown in the last section that the mean $m(t)$ of the stochastic logistic model is always less than the solution of the deterministic model, $m(t) < n(t)$. In the deterministic model, $n(t) \to K$, and in the stochastic model, $m(t) \approx \tilde{m}$, so that it is reasonable to expect $\tilde{m} < K$. In addition, the variance is large when $b_1 + d_1$ and $b_2 + d_2$ are large. These parameter values are larger for case (a) than for case (b), implying that the variance is larger for case (a) than for case (b). This is evident in Figure 6.10.

For a discussion of generalized stochastic logistic models with immigration, see Matis and Kiffe (1999). In their models, a stationary probability distribution exists and the cumulant generating function is used in the analysis. In addition, it is shown that the stationary probability distribution is approximately normal. See Nåsell (2001) for a discussion of the quasistationary probability distribution in stochastic logistic models.

6.10   An Explosive Birth Process

Norris (1997) defines an explosive process in terms of the jump times or waiting times $\{W_i\}_{i=0}^{\infty}$ and the interevent times $\{T_i\}_{i=0}^{\infty}$. Recall that W_i is the time at which the process jumps to a new state and $T_i = W_{i+1} - W_i$. Let $W = \sup_i\{W_i\}$, $i \in \{0, 1, 2, \ldots\}$, and $T = \sum_{i=0}^{\infty} T_i$. If, for some state i,

$$\text{Prob}\{W < \infty | X(0) = i\} > 0 \quad \text{or} \quad \text{Prob}\{T < \infty | X(0) = i\} > 0, \quad (6.31)$$

then the process is said to be *explosive*; otherwise it is *nonexplosive*. In an explosive process, it follows that given $X(0) = i$, the probability distribution $\{p_i(t)\}_{i=0}^{\infty}$ corresponding to $X(t)$ has the property that there exists a time $t^* < \infty$, where $T = t^*$ such that

$$\sum_{i=0}^{\infty} p_i(t^*) < 1.$$

Therefore, a birth process is not explosive if

$$\sum_{i=0}^{\infty} p_i(t) = 1 \quad \text{for all } t \geq 0.$$

Necessary and sufficient conditions for a birth process to be explosive are given in the next theorem (Feller, 1968; Norris, 1997). Recall that in a birth process, $\lambda_i > 0$ and $\mu_i = 0$ for $i = 1, 2, \ldots$.

THEOREM 6.4

A birth process is explosive [equation (6.31)] iff

$$\sum_{i=1}^{\infty} \frac{1}{\lambda_i} < \infty. \tag{6.32}$$

Proof. Suppose $X(0) = m$ or $p_j(0) = \delta_{mj}$. Let $S_k(t) = \sum_{i=0}^{k} p_i(t)$. The forward Kolmogorov differential equations $dp/dt = Qp$ for a birth process are

$$\frac{dp_i}{dt} = \lambda_{i-1} p_{i-1} - \lambda_i p_i$$

$$\frac{dp_0}{dt} = -\lambda_0 p_0.$$

From the forward Kolmogorov differential equations it follows that

$$\frac{dS_k(t)}{dt} = -\lambda_k p_k(t).$$

If $k \geq m$, then $S_k(0) = 1$. Integrating from 0 to t leads to

$$S_k(t) - 1 = -\int_0^t \lambda_k p_k(\tau) \, d\tau, \quad k \geq m.$$

The sequence $\{S_k(t)\}_{k=m}^{\infty}$ is increasing. Therefore, the sequence $\{1 - S_k(t)\}_{k=m}^{\infty}$ is decreasing for $k \geq m$ and is bounded below by zero. The sequence must have a limit. Call this limit $L(t)$:

$$\lim_{k \to \infty} \lambda_k \int_0^t p_k(\tau) \, d\tau = L(t) \geq 0.$$

Also, $\int_0^t p_k(\tau) \, d\tau \geq L(t)/\lambda_k$ for $k \geq m$. It follows from the definition of $S_k(t)$ that for $k \geq m$,

$$\int_0^t S_k(\tau) \, d\tau \geq \sum_{i=m}^{k} \int_0^t p_i(\tau) \, d\tau \geq L(t) \sum_{i=m}^{k} \frac{1}{\lambda_i}. \tag{6.33}$$

Suppose the process is explosive given $X(0) = m$. Then there exists a time t^* such that

$$\lim_{k \to \infty} S_k(t^*) < 1 \quad \text{and} \quad \lim_{k \to \infty} \lambda_k \int_0^{t^*} p_k(\tau) \, d\tau > 0,$$

so that $L(t^*) > 0$. This fact, together with $t^* \geq \int_0^{t^*} S_k(\tau) \, d\tau$ in (6.33), implies that the summation (6.32) is convergent.

To show the converse, concepts presented by Norris (1997) are applied. Suppose the summation (6.32) is convergent and $X(0) = m$. The probability of a birth given the population size is i has an exponential distribution with parameter λ_i. Hence, the expectation

$$E\left(\sum_{i=0}^{k} T_i\right) = \sum_{i=0}^{k} E(T_i) = \sum_{i=m}^{m+k+1} \frac{1}{\lambda_i} < \sum_{i=1}^{\infty} \frac{1}{\lambda_i} < \infty.$$

Because $\sum_{i=0}^{\infty} E(T_i) < \infty$ it follows from the dominated convergence theorem (Rudin, 1987) that the limit of the left side as $k \to \infty$ is also finite, $E(T|X(0) = m) = E(\sum_{i=0}^{\infty} T_i|X(0) = m) < \infty$. Because m is arbitrary and the expectation of T is finite,

$$\text{Prob}\,\{T = \infty|X(0) = m\} = 0.$$

Thus, for any initial state m, $\text{Prob}\{T < \infty|X(0) = m\} = 1$; the process is explosive. $\square$

The following example describes a birth process that is explosive.

Example 6.12. Suppose a birth process satisfies $\lambda_i = bi^k > 0$, $i = 1, 2, \ldots,$ where $k > 1$. Note that

$$\sum_{i=1}^{\infty} \frac{1}{bi^k} = \frac{1}{b} \sum_{i=1}^{\infty} \frac{1}{i^k}$$

is a multiple of a convergent p-series with $p = k > 1$. According to Theorem 6.4, the birth process is explosive. A deterministic analogue of this model is the differential equation

$$\frac{dn}{dt} = bn^k.$$

Integration of n with initial condition $n(0) = N$ leads to the solution

$$n(t) = \left[N^{1-k} - (k-1)bt\right]^{-1/(k-1)}.$$

As $t \to N^{1-k}/[b(k-1)]$, then $n(t) \to \infty$. Hence, the deterministic solution approaches infinity or "explodes" at a finite time. ∎

6.11 Nonhomogeneous Birth and Death Process

In all of the birth and death processes discussed thus far the transition probabilities have been homogeneous with respect to time; that is, time-independent. Suppose the birth and death rate parameters λ_i and μ_i are time-dependent,

$$\lambda_i \equiv \lambda(i, t) \quad \text{and} \quad \mu_i \equiv \mu(i, t).$$

Then the transition probabilities are nonhomogeneous. The forward Kolmogorov equation is

$$\frac{dp_i(t)}{dt} = \lambda(i-1,t)p_{i-1}(t) + \mu(i+1,t)p_{i+1}(t) - (\lambda(i,t) + \mu(i,t))p_i(t).$$

Multiply by z^i and sum from $i = 0$ to ∞ to obtain a partial differential equation for the p.g.f. $\mathcal{P}(z,t)$:

$$\frac{\partial \mathcal{P}}{\partial t} = \sum_{i=0}^{\infty} \lambda(i-1,t)p_{i-1}(t)z^i + \sum_{i=0}^{\infty} \mu(i+1,t)p_{i+1}(t)z^i$$

$$- \sum_{i=0}^{\infty} [\lambda(i,t) + \mu(i,t)]p_i(t)z^i.$$

The right-hand side depends on the form of $\lambda(i,t)$ and $\mu(i,t)$. One example is discussed next.

Example 6.13. Suppose $\lambda(i,t) = \lambda(t)i$ and $\mu(i,t) = \mu(t)i$ (Bailey, 1990). Then the partial differential equation for the p.g.f. is

$$\frac{\partial \mathcal{P}}{\partial t} = [\lambda(t)(z^2 - z) + \mu(t)(1 - z)]\frac{\partial \mathcal{P}}{\partial z}, \quad \mathcal{P}(z,0) = z^N.$$

Along characteristic curves, τ and s,

$$\frac{dt}{d\tau} = 0, \quad \frac{dz}{d\tau} = (1-z)[\lambda(t)z - \mu(t)], \quad \text{and} \quad \frac{d\mathcal{P}}{d\tau} = 0$$

with initial conditions

$$t(s,0) = 0, \quad z(s,0) = s, \quad \text{and} \quad \mathcal{P}(s,0) = s^N.$$

The differential equation for z can be solved by letting $t = \tau$ and making the change of variable $1 - z = 1/y$ so that $y^2 = 1/(1-z)^2$ and $dz/d\tau = (1/y^2)\,dy/d\tau$. The differential equation for z can be transformed into a linear differential equation in y:

$$\frac{dy}{d\tau} = (\lambda(\tau) - \mu(\tau))y - \lambda(\tau).$$

Use of an integrating factor,

$$e^{\int_0^\tau [\mu(\alpha) - \lambda(\alpha)]\,d\alpha} = e^{\rho(\tau)},$$

leads to

$$ye^{\rho(\tau)} - y(s,0) = -\int_0^\tau \lambda(\beta)e^{\rho(\beta)}\,d\beta.$$

Now, $y(s,0) = 1/(1-s)$ and $y = 1/(1-z)$, so that

$$\frac{1}{s-1} = \frac{e^{\rho(\tau)}}{z-1} - \int_0^\tau \lambda(\beta)e^{\rho(\beta)}\,d\beta.$$

Solving for s,

$$s = 1 + \frac{1}{e^{\rho(\tau)}z - 1 - \int_0^\tau \lambda(\beta)e^{\rho(\beta)}\,d\beta}.$$

The solution of the p.g.f. is

$$\mathcal{P}(z,t) = \left[1 + \frac{1}{\frac{e^{\rho(t)}}{z-1} - \int_0^t \lambda(\tau)e^{\rho(\tau)}\,d\tau}\right]^N,$$

where $\rho(t) = \int_0^t [\mu(\tau) - \lambda(\tau)]\,d\tau$ and τ represents a dummy variable.

An expression for $p_0(t)$, the probability of extinction, can be found by evaluating $\mathcal{P}(z,t)$ at $z = 0$:

$$p_0(t) = \left[1 - \frac{1}{e^{\rho(t)} + \int_0^t \lambda(\tau)e^{\rho(\tau)}\,d\tau}\right]^N.$$

Because

$$e^{\rho(t)} + \int_0^t \lambda(\tau)e^{\rho(\tau)}\,d\tau = e^{\rho(t)} - \int_0^t \rho(\tau)e^{\rho(\tau)}\,d\tau + \int_0^t \mu(\tau)e^{\rho(\tau)}\,d\tau$$

$$= e^{\rho(t)} - e^{\rho(\tau)}|_0^t + \int_0^t \mu(\tau)e^{\rho(\tau)}\,d\tau$$

$$= 1 + \int_0^t \mu(\tau)e^{\rho(\tau)}\,d\tau,$$

the probability of extinction simplifies to

$$p_0(t) = \left[\frac{\int_0^t \mu(\tau)e^{\rho(\tau)}\,d\tau}{1 + \int_0^t \mu(\tau)e^{\rho(\tau)}\,d\tau}\right]^N.$$

Note that if $\lim_{t\to\infty} \int_0^t \mu(\tau)e^{\rho(\tau)}\,d\tau = \infty$, then $\lim_{t\to\infty} p_0(t) = 1$. But

$$\int_0^t \mu(\tau)e^{\rho(\tau)}\,d\tau = \int_0^t \rho(\tau)e^{\rho(\tau)}\,d\tau + \int_0^t \lambda(\tau)e^{\rho(\tau)}\,d\tau$$

$$\geq e^{\rho(t)} - 1.$$

Thus, if $\lim_{t\to\infty} \rho(t) = \infty$, it follows that $\lim_{t\to\infty} p_0(t) = 1$.

The analogous deterministic model satisfies the differential equation,

$$\frac{dn}{dt} = [\lambda(t) - \mu(t)]n, \quad n(0) > 0.$$

The solution of this differential equation is

$$n(t) = n(0)e^{-\int_0^t [\mu(\tau) - \lambda(\tau)]\, d\tau} = n(0)e^{-\rho(t)}.$$

For the deterministic model population extinction occurs, $\lim_{t\to\infty} n(t) = 0$, if and only if $\lim_{t\to\infty} \rho(t) = \infty$. This result differs from the stochastic model in that population extinction is still possible in the stochastic model even if the limit of $\rho(t)$ is not infinite. ∎

6.12 Exercises for Chapter 6

1. Suppose a CTMC $\{X(t) : t \in [0, \infty)\}$ is defined by the following infinitesimal transition probabilities:

$$p_{i+j,i}(\Delta t) = \text{Prob}\{\Delta X(t) = j | X(t) = i\}$$

$$= \begin{cases} \mu i \Delta t + o(\Delta t), & j = -1 \\ \lambda i \Delta t + o(\Delta t), & j = 1 \\ \nu \Delta t + o(\Delta t), & j = 2 \\ 1 - (\nu + \lambda i + \mu i)\Delta t + o(\Delta t), & j = 0 \\ o(\Delta t), & j \neq -1, 0, 1, 2. \end{cases}$$

The initial distribution is $\text{Prob}\{X(0) = N\} = 1$.

(a) Write the differential equations satisfied by each of the state probabilities

$$p_i(t) = \text{Prob}\{X(t) = i\} \quad \text{for} \quad i = 0, 1, 2, \ldots.$$

(b) Write the form of the generator matrix Q.

2. Let $X(t)$ be the random variable for the total population size for a birth-death-emigration process and $p_i(t) = \text{Prob}\{X(t) = i\}$. The infinitesimal transition probabilities are given by

$$p_{i+j,i}(\Delta t) = \text{Prob}\{\Delta X(t) = j | X(t) = i\}$$

$$= \begin{cases} (\mu i + \nu)\Delta t + o(\Delta t), & j = -1 \\ \lambda i^2 \Delta t + o(\Delta t), & j = 1 \\ 1 - (\lambda i^2 + \mu i + \nu)\Delta t + o(\Delta t), & j = 0 \\ o(\Delta t), & j \neq -1, 0, 1. \end{cases}$$

The initial probability distribution is $\text{Prob}\{X(0) = N\} = 1$.

(a) Show that the differential equations satisfied by $p_i(t)$ for $i = 1, 2, \ldots$ are

$$\frac{dp_i}{dt} = p_{i-1}[\lambda(i-1)^2] + p_{i+1}[\mu(i+1) + \nu] - p_i[\lambda i^2 + \mu i + \nu].$$

(b) Assume $\lambda = 0$. In addition, assume $\nu = 0$ when $i = 0$ (i.e., there is no emigration when the population size is zero). Notice that the state space consists of $\{0, 1, \ldots, N\}$, since it is a death-emigration process. Write the first-order partial differential equation satisfied by the p.g.f., $\mathcal{P}(z, t) = \sum_{i=0}^{N} p_i(t) z^i$.

3. Suppose a general birth and death process has birth and death rates given by $\lambda_i = b(i + 1)$ and $\mu_i = di$ for $i = 0, 1, 2, \ldots$, $b, d > 0$.

   (a) Determine conditions on b and d such that a unique positive stationary probability distribution exists.

   (b) Write the form of the unique stationary probability distribution.

4. Consider the simple birth process.

   (a) Write the form of the generator matrix Q and the transition matrix T for the embedded Markov chain based on the state space $\{N, N+1, \ldots\}$.

   (b) Show that all states are transient and there does not exist a stationary probability distribution.

5. Consider the simple death process.

   (a) Write the form of the generator matrix Q and the transition matrix T for the embedded Markov chain based on the state space $\{0, 1, \ldots, N\}$.

   (b) Show that zero is an absorbing state, the remaining states are transient, and the unique stationary probability distribution is $\pi = (1, 0, 0, \ldots, 0)^{tr}$.

6. Consider the simple birth and death process with immigration such that $\nu = 1 = \lambda$. Compute the stationary probability distribution π when $\mu > 1$.

7. Consider the simple birth and death process with immigration.

   (a) Use the m.g.f. $M(\theta, t)$ to find an expression for $\sigma^2(t)$.

   (b) Find the limit, $\lim_{t \to \infty} \sigma^2(t)$, when $\lambda < \mu$. Does $\sigma^2(\infty)$ agree with the variance of the stationary distribution in Example 6.3?

8. Suppose a general birth and death process has birth and death rates given by

$$\lambda_i = b_0 + b_1 i + b_2 i^2, \quad \text{and} \quad \mu_i = d_1 i + d_2 i^2, \quad \text{for} \quad i = 0, 1, 2, \ldots.$$

(a) Write the forward Kolmogorov equations. Then use the generating function technique to find the differential equations satisfied by the p.g.f. and the m.g.f.

(b) Write the partial differential equation for the m.g.f. in the more general case, where $\lambda_i = \sum_{k=0}^{n} b_k i^k$ and $\mu_i = \sum_{k=1}^{n} d_k i^k$.

9. Consider a death and immigration process, where $\lambda_i = \nu$ and $\mu_i = \mu i$.

(a) Show that the m.g.f. has the following form:

$$\frac{\partial M}{\partial t} = \mu(e^{-\theta} - 1)\frac{\partial M}{\partial \theta} + \nu(e^{\theta} - 1)M.$$

(b) Use the method of characteristics to solve for $M(\theta, t)$.

(c) Use the expression for $M(\theta, t)$ to find the mean, $m(t)$, and variance, $\sigma^2(t)$, of the process.

10. For the simple death process, show that the expected time to go from state a to state b can be approximated if a and b are large. Show that

$$E(T_{b,a}) \approx \frac{1}{\mu} \ln\left(\frac{a}{b}\right).$$

11. For the birth and death process with immigration, show that there exists a unique stationary probability distribution iff $\lambda < \mu$.

12. Consider the simple birth and death process with immigration.

(a) Assume $\lambda = 0.5$ and $\nu = 1 = \mu$. Show that the stationary probability distribution is

$$\pi_i = \frac{i+1}{2^{i+2}}, \quad i = 0, 1, 2, \ldots.$$

Calculate the mean of the distribution graphed in Figure 6.5.

(b) Assume $\lambda = 0.5$, $\mu = 1$ and $\nu = 1.5$. Show that the stationary probability distribution is

$$\pi_i = \frac{(i+2)(i+1)}{2^{i+4}}, \quad i = 0, 1, 2, \ldots.$$

Calculate the mean. This probability distribution is graphed in Figure 6.5.

13. In the simple birth and death process with immigration, verify the following statements about the mean.

(a) Verify the mean satisfies (6.10).

(b) Verify that the solution $n(t)$ of the deterministic model

$$\frac{dn}{dt} = (\lambda - \mu)n + \nu, \quad \nu(0) = N$$

equals the mean of the stochastic model, equation (6.10). That is, $n(t) = m(t)$.

14. In a queueing system of type $M/M/1/K$, derive the stationary probability distribution given in equation (6.11):

$$\pi_i = \left(\frac{\lambda}{\mu}\right)^i \frac{1 - \lambda/\mu}{1 - (\lambda/\mu)^{K+1}}, \quad i = 0, 1, \ldots, K.$$

Then find the average number of customers in the system, $C = \sum_{i=1}^{K} i\pi_i$.

15. In a queueing system of type $M/M/\infty$, there is an infinite number of servers. The arrival and departure rates are

$$\lambda_i = \lambda \text{ and } \mu_i = i\mu, \quad i = 0, 1, 2, \ldots.$$

(a) Calculate the stationary probability distribution.

(b) What is the average number of customers C in the system?

(c) What is the average amount of time W each customer spends in the system?

16. Assume the random variable $X(t)$ represents the total population size for the stochastic logistic model, where the birth and death rates have one of two different forms:

(i) $\lambda_i = \begin{cases} i - \dfrac{i^2}{100}, & i = 0, 1, \ldots, 100, \\ 0, & i > 100 \end{cases}$ and $\mu_i = \dfrac{i^2}{100}, i = 1, 2, \ldots$

(ii) $\lambda_i = i$ and $\mu_i = \dfrac{i^2}{50}, i = 0, 1, 2, \ldots$

Note that in the deterministic model, $r = 1$ and $K = 50$. Assume $X(0) = 5$.

(a) Write a computer program to simulate sample paths for the stochastic logistic models in (i) and (ii) up to time $t = 8$. Graph three sample paths for (i) and (ii). See Figure 6.11. Compare this figure with the DTMC logistic growth in Chapter 3, Figure 3.5.

(b) Calculate the mean and variance of 1000 sample paths at $t = 10$, $m(10)$, and $\sigma^2(10)$, for (i) and (ii).

(c) Calculate the quasistationary distributions $\{\tilde{\pi}_i\}_{i=1}^{\infty}$ for the logistic models (i) and (ii) and graph them.

(d) Calculate the mean and variance of the quasistationary distributions in (i) and (ii). How do they compare?

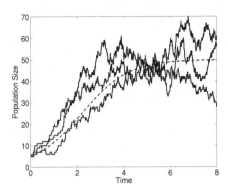

FIGURE 6.11: Three sample paths in case (ii) of Exercise 16.

17. Consider a general birth and death process with

$$\lambda_i = b_1 i + b_2 i + b_3 i^2 \quad \text{and} \quad \mu_i = d_1 + d_2 i + d_3 i^2$$

for $i = 0, 1, 2, \ldots$, where $d_3 \neq 0$ and $d_3 > b_3$. Assume $\lambda_i, \mu_i > 0$ for $i = 1, 2, \ldots$.

(a) Show that $d_3 > 0$ and $\lim_{t \to \infty} p_0(t) = 1$.

(b) Show that the expected time until population extinction is finite.

18. Consider the stochastic logistic model with m.g.f. $M(\theta, t)$ of the form given by equation (6.28).

(a) Write the differential equation for the c.g.f., $K(\theta, t)$.

(b) Use the differential equation for $K(\theta, t)$ in part (a) to find a differential equation for the mean $m(t)$ and variance $\sigma^2(t)$.

19. The m.g.f. $M(\theta, t)$ for a simple birth process is a solution of the following first order partial differential equation:

$$\frac{\partial M}{\partial t} = \lambda(e^\theta - 1)\frac{\partial M}{\partial \theta}, \quad M(\theta, 0) = e^{N\theta}.$$

(a) Differentiate the partial differential equation with respect to θ and evaluate at $\theta = 0$ to write a differential equation for the mean, $m(t) = E(X(t))$. Solve the differential equation for $m(t)$.

(b) Differentiate the partial differential equation twice with respect to θ and evaluate at $\theta = 0$ to write a differential equation for the second moment $E(X^2(t))$. Solve the differential equation for $E(X^2(t))$.

(c) Use parts (a) and (b) to find the variance $\sigma^2(t)$ of the process.

20. The m.g.f. $M(\theta, t)$ for a birth and death process $\{X(t) : t \in [0, \infty)\}$ is a solution of the following partial differential equation:

$$\frac{\partial M}{\partial t} = b(e^\theta - 1)\frac{\partial M}{\partial \theta} + d(e^{-\theta} - 1)\frac{\partial^2 M}{\partial \theta^2}.$$

(a) Differentiate the partial differential equation with respect to θ and evaluate at $\theta = 0$ to write a differential equation for the mean, $m(t) = E(X(t))$.

(b) Differentiate the partial differential equation twice with respect to θ and evaluate at $\theta = 0$ to write a differential equation for the second moment $E(X^2(t))$.

(c) Use parts (a) and (b) to write a differential equation for the variance $\sigma^2(t)$ of the process. Note that $dm^2(t)/dt = 2m(t)dm(t)/dt$. The system of differential equations for $m(t)$ and $\sigma^2(t)$ cannot be solved because it also depends on $E(X^3(t))$. The system is "not closed".

21. The following birth and death process is known as a *Prendiville process* (named after B. J. Prendiville) (Iosifescu and Tăutu, 1973). The birth and death rates are

$$\lambda_n = \alpha\left(\frac{n_2}{n} - 1\right), \quad 0 < n_1 \le n \le n_2$$

$$\mu_n = \beta\left(1 - \frac{n_1}{n}\right), \quad 0 < n_1 \le n \le n_2,$$

where α and β are positive constants. Outside of the interval $[n_1, n_2]$, the birth and death rates are zero, that is, $\lambda_n = 0 = \mu_n$ for $n < n_1$ or $n > n_2$. The p.g.f. $\mathcal{P}(z, t)$ for this process is a solution of the following partial differential equation:

$$\frac{\partial P}{\partial t} = (1 - z)(\alpha z + \beta)\frac{\partial P}{\partial z} + (z - 1)\left(\alpha n_2 + \frac{\beta n_1}{z}\right)P$$

with initial condition $\mathcal{P}(z, 0) = z^{n_0}$ (Iosifescu and Tăutu, 1973).

(a) Show that the solution of this first-order partial differential equation is

$$\mathcal{P}(z, t) = \frac{z^{n_1}[\alpha(1 - \rho(t))z + \alpha\rho(t) + \beta]^{n_2 - n_0}}{(\alpha + \beta)^{n_2 - n_1}[(\alpha + \beta\rho(t))z + \beta(1 - \rho(t))]^{n_1 - n_0}},$$

where $\rho(t) = e^{-(\alpha+\beta)t}$.

(b) Compute $\lim_{t\to\infty} \mathcal{P}(z,t)$ and denote this limit as $\mathcal{P}(z,\infty)$. This limit is the p.g.f. for the stable stationary probability distribution. Use $\mathcal{P}(z,\infty)$ to find the mean of this stationary distribution.

22. The deterministic model of the Prendiville process discussed in Exercise 21 is given by the differential equation

$$\frac{dn}{dt} = \lambda_n - \mu_n = \frac{\alpha n_2 + \beta n_1}{n} - (\alpha + \beta).$$

(a) Show that the equilibrium solution $\bar{n}$ (where $dn/dt = 0$) to this differential equation is

$$\bar{n} = \frac{\alpha n_2 + \beta n_1}{\alpha + \beta}.$$

Then show that $dn/dt > 0$ for $0 < n < \bar{n}$ and $dn/dt < 0$ for $n > \bar{n}$. Conclude that $\lim_{t\to\infty} n(t) = \bar{n}$.

(b) Compare the behavior of this deterministic model to the stochastic Prendiville process discussed in Exercise 21.

23. Consider the birth process with immigration, where $\lambda_i = b_0 + b_1 i^k$, $b_0 > 0$, $b_1 > 0$, for $i = 1, 2, \ldots$. Show that the process is not explosive if $k = 1$ and is explosive if $k = 2, 3, \ldots$.

24. Suppose a nonhomogeneous birth and death process satisfies $\lambda(i,t) = \lambda(t)i$ and $\mu(i,t) = \mu(t)i$ with $\mu(t) = t^2$ and $\lambda(t) = 2t^2$. Compute $\lim_{t\to\infty} p_0(t)$.

25. Consider a nonhomogeneous birth and death process, where $\mu(t)$ and $\lambda(t)$ are linear functions of t,

$$\mu(t) = \alpha t \quad \text{and} \quad \lambda(t) = \beta t.$$

(a) Show that if $\alpha > \beta > 0$, then $\lim_{t\to\infty} \rho(t) = \infty$, and if $\alpha = \beta > 0$, $\lim_{t\to\infty} \int_0^t \mu(\tau)e^{\rho(\tau)}\, d\tau = \lim_{t\to\infty} \int_0^t \alpha\tau\, d\tau = \infty$. Find $\lim_{t\to\infty} p_0(t)$.

(b) Show that if $0 < \alpha < \beta$ (births exceed deaths), then

$$\lim_{t\to\infty} \int_0^t \mu(\tau)e^{\rho(\tau)}\, d\tau = \frac{\alpha}{(\alpha-\beta)^2}.$$

Compute $\lim_{t\to\infty} p_0(t)$.

26. A birth and death process with immigration was developed by Alonso and McKane (2002) based on a spatially implicit patch model. Let $p(t)$ be the fraction of patches occupied by a particular species. Empty

patches can be colonized from occupied patches or via migration from the mainland. The deterministic model is expressed in terms of the following differential equation:

$$\frac{dp}{dt} = (m + cp)(1 - p) - ep, \quad 0 < p(0) < 1. \tag{6.34}$$

The positive constants m, c, and e are the rates of immigration from the mainland, colonization, and extinction, respectively. Model (6.34) was first discussed by Hanski (1999) in the context of metapopulation models. In the original model, studied by Levins (1969, 1970), there was no migration ($m = 0$).

(a) Show that model (6.34) has a unique positive equilibrium given by

$$E_1 = \frac{c - m - e + \sqrt{(c - m - e)^2 + 4cm}}{2c}$$

(when $dp/dt = 0$) and $\lim_{t \to \infty} p(t) = E_1$.

(b) In the stochastic formulation by Alonso and McKane (2002), it is assumed that $X(t)$ is the random variable for the number of patches occupied at time t. The maximal number of patches occupied is M, $X(t) \in \{0, 1, 2, \ldots, M\}$. The birth rate of new patches occupied is

$$\lambda_i = ci \left(1 - \frac{i}{M}\right) + m(M - i)$$

for $i = 0, 1, 2, \ldots, M$, where the first term in the preceding expression represents a colonization event and the second term represents an immigration event from the mainland. The death rate satisfies

$$\mu_i = ei, \quad i = 0, 1, 2, \ldots, M,$$

where death means population extinction on one of the patches. To relate this model to equation (6.34), note that if $p = i/M$, then

$$\frac{dp}{dt} = \frac{(\lambda_i - \mu_i)}{M}.$$

For the birth, death, and immigration process, find a formula for the stationary probability distribution.

(c) Compare the equilibrium value E_1 to the stationary probability distribution when $M = 40$, $c = 2$, $e = 1$, and $m = 0.1$.

6.13 References for Chapter 6

Alonso, D. and A. McKane. 2002. Extinction dynamics in mainland-island metapopulations: an N-patch stochastic model. *Bull. Math. Biol.* 64: 913–958.

Bailey, N. T. J. 1990. *The Elements of Stochastic Processes with Applications to the Natural Sciences.* John Wiley & Sons, New York.

Bharucha-Reid, A. T. 1997. *Elements of the Theory of Markov Processes and their Applications.* Dover Pub., New York.

Chan, M. S. and V. S. Isham. 1998. A stochastic model of schistosomiasis immuno-epidemiology. *Math. Biosci.* 151: 179–198.

Chao, X, M. Miyazawa, and M. Pinedo. 1999. *Queueing Networks.* John Wiley & Sons, Chichester, New York.

Corduneanu, C. 1977. *Principles of Differential and Integral Equations.* Chelsea Pub. Co., The Bronx, New York.

Feller, W. 1968. *An Introduction to Probability Theory and Its Applications.* Vol. 1. 3rd ed. John Wiley & Sons, New York.

Goel, N. S. and N. Richter-Dyn. 1974. *Stochastic Models in Biology.* Academic Press, New York.

Hanski, I. 1999. *Metapopulation Ecology.* Oxford Univ. Press, New York.

Hsu, H. P. 1997. *Schaum's Outline of Theory and Problems of Probability, Random Variables, and Random Processes.* McGraw-Hill, New York.

Iosifescu, M. and P. Tăutu. 1973. *Stochastic Processes and Applications in Biology and Medicine II. Models.* Springer-Verlag, Berlin, Heidelberg, New York.

Karlin, S. and H. Taylor. 1975. *A First Course in Stochastic Processes.* 2nd ed. Academic Press, New York.

Kleinrock, L. 1975. *Queueing Systems, Vol. 1, Theory.* John Wiley & Sons, New York.

Levins, R. 1969. Some demographic and genetic consequences of environmental heterogeneity for biological control. *Bull. Entomol. Soc. Am.* 15: 227–240.

Levins, R. 1970. Extinction. *Lecture Notes Math.* 2: 75–107.

Lloyd, A. L. 2004. Estimating variability in models for recurrent epidemics: assessing the use of moment closure techniques. *Theor. Pop. Biol.* 65: 49–65.

Matis, J. H. and T. R. Kiffe. 1999. Effects of immigration on some stochastic logistic models: a cumulant truncation analysis. *Theor. Pop. Biol.* 56: 139–161.

Nåsell, I. 2001. Extinction and quasi-stationarity in the Verhulst logistic model. *J. Theor. Biol.* 211: 11–27.

Nisbet, R. M. and W. S. C. Gurney. 1982. *Modelling Fluctuating Populations.* John Wiley & Sons, Chichester and New York.

Norden, R. H. 1982. On the distribution of the time to extinction in the stochastic logistic population model. *Adv. Appl. Prob.* 14: 687–708.

Norris, J. R. 1997. *Markov Chains.* Cambridge Series in Statistical and Probabilistic Mathematics, Cambridge Univ. Press, Cambridge, U. K.

Ortega, J. M. 1987. *Matrix Theory A Second Course.* Plenum Press, New York.

Renshaw, E. 1993. *Modelling Biological Populations in Space and Time.* Cambridge Univ. Press, Cambridge, U. K.

Richter-Dyn, N. and N. S. Goel. 1972. On the extinction of colonizing species. *Theor. Pop. Biol.* 3: 406–433.

Rudin, W. 1987. *Real and Complex Analysis.* 3rd ed. McGraw-Hill, New York.

Schinazi, R. B. 1999. *Classical and Spatial Stochastic Processes.* Birkhäuser, Boston.

Taylor, H. M. and S. Karlin. 1998. *An Introduction to Stochastic Modeling.* 3rd ed. Academic Press, New York.

Tognetti, K. and G. Winley. 1980. Stochastic growth models with logistic mean population. *J. Theor. Biol.* 82: 167–169.

6.14 Appendix for Chapter 6

6.14.1 Generating Functions for the Simple Birth and Death Process

The m.g.f. and the p.g.f. for the simple birth and death process are found by solving the following first-order partial differential equations. The p.g.f. is a solution of

$$\frac{\partial P}{\partial t} = [\mu(1 - z) + \lambda z(z - 1)]\frac{\partial P}{\partial z}, \quad P(z, 0) = z^N$$

and the m.g.f. is a solution of

$$\frac{\partial M}{\partial t} = [\mu(e^{-\theta} - 1) + \lambda(e^{\theta} - 1)]\frac{\partial M}{\partial \theta}, \quad M(\theta, 0) = e^{\theta N}.$$

Application of the method of characteristics to the m.g.f. equation leads to

$$\frac{dt}{d\tau} = 1, \quad \frac{d\theta}{\mu(e^{-\theta} - 1) + \lambda(e^{\theta} - 1)} = -d\tau, \quad \text{and} \quad \frac{dM}{d\tau} = 0,$$

with initial conditions

$$t(s, 0) = 0, \quad \theta(s, 0) = s, \quad \text{and} \quad M(s, 0) = e^{sN}.$$

The solutions satisfy

$$t = \tau \quad \text{and} \quad M(s, \tau) = e^{sN}.$$

Integrating the differential equation in θ (integration is made easier by a change of variable $x = e^{\theta}$),

$$\tau = \begin{cases} \dfrac{1}{\mu - \lambda}\left(\ln\left(\dfrac{e^{\theta} - 1}{\lambda e^{\theta} - \mu}\right) + \ln(\tau_1)\right), & \text{if } \lambda \neq \mu \\[3ex] \dfrac{1}{\lambda(e^{\theta} - 1)} + \tau_2, & \text{if } \lambda = \mu, \end{cases}$$

where τ_1 and τ_2 are constants. The two cases, $\lambda = \mu$ and $\lambda \neq \mu$, must be solved separately. The initial condition for θ is used to solve for the constants,

$$\tau = \begin{cases} \dfrac{1}{\mu - \lambda}\left[\ln\dfrac{(e^{\theta} - 1)(\lambda e^s - \mu)}{(\lambda e^{\theta} - \mu)(e^s - 1)}\right], & \text{if } \lambda \neq \mu \\[3ex] \dfrac{1}{\lambda(e^{\theta} - 1)} - \dfrac{1}{\lambda(e^s - 1)}, & \text{if } \lambda = \mu. \end{cases}$$

Because $M(s, \tau) = [e^s]^N$, these relations are solved for e^s,

$$
e^s = \begin{cases} \dfrac{e^{\tau(\mu-\lambda)}(\lambda e^\theta - \mu) - \mu(e^\theta - 1)}{e^{\tau(\mu-\lambda)}(\lambda e^\theta - \mu) - \lambda(e^\theta - 1)}, & \text{if } \lambda \neq \mu \\[4ex] \dfrac{1 - (\lambda\tau - 1)(e^\theta - 1)}{1 - \lambda\tau(e^\theta - 1)}, & \text{if } \lambda = \mu. \end{cases}
$$

Now the m.g.f. M can be expressed in terms of θ and t,

$$
M(\theta, t) = \begin{cases} \left(\dfrac{e^{t(\mu-\lambda)}(\lambda e^\theta - \mu) - \mu(e^\theta - 1)}{e^{t(\mu-\lambda)}(\lambda e^\theta - \mu) - \lambda(e^\theta - 1)} \right)^N, & \text{if } \lambda \neq \mu \\[4ex] \left(\dfrac{1 - (\lambda t - 1)(e^\theta - 1)}{1 - \lambda t(e^\theta - 1)} \right)^N, & \text{if } \lambda = \mu. \end{cases}
$$

Making the change of variable $\theta = \ln z$, the p.g.f. $\mathcal{P}$ is

$$
\mathcal{P}(z, t) = \begin{cases} \left(\dfrac{e^{t(\mu-\lambda)}(\lambda z - \mu) - \mu(z - 1)}{e^{t(\mu-\lambda)}(\lambda z - \mu) - \lambda(z - 1)} \right)^N, & \text{if } \lambda \neq \mu \\[4ex] \left(\dfrac{1 - (\lambda t - 1)(z - 1)}{1 - \lambda t(z - 1)} \right)^N, & \text{if } \lambda = \mu. \end{cases}
$$

6.14.2 Proofs of Theorems 6.2 and 6.3

Proof. (Proof of Theorem 6.2) Note that in a general birth and death chain

$$
\frac{dp_0(t)}{dt} = \mu_1 p_1(t).
$$

Since $0 \leq p_i(t) \leq 1$, it follows that $p_0(t)$ is an increasing function that is bounded above. Thus, $\lim_{t \to \infty} p_0(t)$ exists.

For part (i) of the theorem, let E_i be the probability of extinction given the population size is i (Karlin and Taylor, 1975; Renshaw, 1993). The ratio $\lambda_i/(\mu_i + \lambda_i)$ is the probability of a birth and the ratio $\mu_i/(\mu_i + \lambda_i)$ is the probability of a death given that an event has occurred (transition probabilities in the embedded Markov chain). Then

$$
E_i = \text{Prob}\{\text{first event is a birth}\}E_{i+1} + \text{Prob}\{\text{first event is a death}\}E_{i-1}
$$
$$
= \frac{\lambda_i}{\mu_i + \lambda_i}E_{i+1} + \frac{\mu_i}{\mu_i + \lambda_i}E_{i-1}.
$$

Because $\lambda_0 = 0 = \mu_0$, $E_0 = 1$. Rewriting the preceding expression,

$$
E_{i+1} = \frac{\mu_i + \lambda_i}{\lambda_i}\left[E_i - \frac{\mu_i}{\mu_i + \lambda_i}E_{i-1} \right]
$$
$$
= \left(1 + \frac{\mu_i}{\lambda_i} \right)E_i - \frac{\mu_i}{\lambda_i}E_{i-1}.
$$

For $i = 1$ and 2,

$$E_2 = \left(1 + \frac{\mu_1}{\lambda_1}\right) E_1 - \frac{\mu_1}{\lambda_1}$$
$$= E_1 + (E_1 - 1)\frac{\mu_1}{\lambda_1}.$$

$$E_3 = \left(1 + \frac{\mu_2}{\lambda_2}\right) E_2 - \frac{\mu_2}{\lambda_2} E_1$$
$$= \left(1 + \frac{\mu_2}{\lambda_2}\right) \left(E_1 + (E_1 - 1)\frac{\mu_1}{\lambda_1}\right) - \frac{\mu_2}{\lambda_2} E_1$$
$$= E_1 + (E_1 - 1)\left(\frac{\mu_1}{\lambda_1} + \frac{\mu_1\mu_2}{\lambda_1\lambda_2}\right).$$

By induction it follows that

$$E_n = E_1 + (E_1 - 1)\left(\sum_{i=1}^{n-1} \frac{\mu_1 \cdots \mu_i}{\lambda_1 \cdots \lambda_i}\right) \qquad (6.35)$$

for all $n = 2, 3, \ldots$. Also, $0 \le E_n \le 1$ for $n = 1, 2, \ldots$. Let $n \to \infty$. If the factor multiplied by $(E_1 - 1)$ approaches infinity, then $E_1 = 1$. But if $E_1 = 1$, then $E_n = E_1$, for $n > 1$, so that $E_n = 1$, the probability of extinction is 1, given the population size is n. Hence, if (6.12) holds, then

$$\lim_{t \to \infty} p_0(t) = 1.$$

Suppose (6.13) holds so that a solution $0 < E_1 < 1$ of (6.35) exists. Then E_n is a decreasing function of n. It is assumed that $\lim_{n\to\infty} E_n = 0$. According to Karlin and Taylor (1975), this assumption is not needed. By assuming the limit is not zero, a contradiction can be obtained via a probabilistic argument. Let $n \to \infty$ in (6.35). Then

$$E_1 = \frac{\displaystyle\sum_{i=1}^{\infty} \frac{\mu_1\mu_2 \cdots \mu_i}{\lambda_1\lambda_2 \cdots \lambda_i}}{1 + \displaystyle\sum_{i=1}^{\infty} \frac{\mu_1\mu_2 \cdots \mu_i}{\lambda_1\lambda_2 \cdots \lambda_i}}.$$

Substitution of E_1 into (6.35) with $n = m$ yields (6.14).

For part (ii) of the theorem, if the initial population size $m > N$, there are only deaths until the population size reaches N. The states $\{N+1, N+2, \ldots\}$ are transient. Once the population size reaches N, the population size will remain less than or equal to N for all time; $\{0, 1, 2, \ldots, N\}$ is a closed set. Thus, to find the probability of extinction, only consider population sizes $\{0, 1, \ldots, N\}$. Assume $p_n(t) = 0$ for $n > N$ and $t > T$, T sufficiently large. The system of differential equations satisfied by $p(t) = (p_0(t), \ldots, p_N(t))^{tr}$ for

$t > T$ is

$$\frac{dp_0(t)}{dt} = \mu_1 p_0(t)$$

$$\frac{dp_n(t)}{dt} = \lambda_{n-1}p_{n-1}(t) - (\lambda_n + \mu_n)p_n(t) + \mu_{n+1}p_{n+1}(t)$$

$$\frac{dp_N(t)}{dt} = \lambda_{N-1}p_{N-1}(t) - \mu_N p_N(t),$$

for $n = 1, 2, \ldots, N - 1$, where $0 \le p(T) \le 1$ and $\sum_{i=0}^{N} p_0(T) = 1$. Note that for this system of differential equations, $d[\sum_{i=0}^{N} p_i(t)]/dt = 0$, so that $\sum_{i=0}^{N} p_i(t) = $ constant. Because $\sum_{i=0}^{N} p_0(T) = 1$, it follows that $\sum_{i=0}^{N} p_i(t) = 1$. The forward Kolmogorov differential equations are $dp/dt = Qp$, where matrix Q is

$$Q = \begin{pmatrix} 0 & \mu_1 & 0 & \cdots & 0 & 0 \\ 0 & -\lambda_1 - \mu_1 & \mu_2 & \cdots & 0 & 0 \\ 0 & \lambda_1 & -\lambda_2 - \mu_2 & \cdots & 0 & 0 \\ \vdots & \vdots & \vdots & \cdots & \vdots & \vdots \\ 0 & 0 & 0 & \cdots & \lambda_{N-1} & -\mu_N \end{pmatrix}.$$

The solution is $p(t) = e^{Qt}p(0)$. The limit as $t \to \infty$ depends on the eigenvalues and eigenvectors of Q. Matrix Q has one zero eigenvalue with corresponding eigenvector $e_1 = (1, 0, 0, \ldots, 0)^{tr}$. Matrix Q also has N other eigenvalues that have negative real part. To show this latter assertion, we can apply Gershgorin's circle theorem and irreducible diagonal dominance to the submatrix of Q formed by deleting the first row and column (Ortega, 1987). Thus, the solution $p(t)$ satisfies $\lim_{t \to \infty} p(t) = c_0 e_1$. Because $\sum_{i=0}^{N} p_i(t) = 1$, it follows that $c_0 = 1$ and, hence,

$$\lim_{t \to \infty} p_0(t) = 1.$$

$\square$

Proof. (Proof of Theorem 6.3) Let $z_i = \tau_i - \tau_{i+1} \le 0$ and subtract τ_i from both sides of equation (6.19). Then

$$0 = \frac{1}{\lambda_i + \mu_i} + \frac{\lambda_i}{\lambda_i + \mu_i}(\tau_{i+1} - \tau_i) + \frac{\mu_i}{\lambda_i + \mu_i}(\tau_{i-1} - \tau_i)$$

$$\tau_i - \tau_{i+1} = \frac{1}{\lambda_i} + \frac{\mu_i}{\lambda_i}(\tau_{i-1} - \tau_i)$$

$$z_i = \frac{1}{\lambda_i} + \frac{\mu_i}{\lambda_i} z_{i-1}.$$

By induction, it follows that

$$z_m = \frac{1}{\lambda_m} + \frac{\mu_m}{\lambda_{m-1}\lambda_m} + \cdots + \frac{\mu_2 \cdots \mu_m}{\lambda_1 \lambda_2 \cdots \lambda_m} + \frac{\mu_1 \cdots \mu_m}{\lambda_1 \cdots \lambda_m} z_0$$

$$= \frac{\mu_1 \cdots \mu_m}{\lambda_1 \cdots \lambda_m} \left[\frac{1}{\mu_1} + \frac{\lambda_1}{\mu_1 \mu_2} + \cdots + \frac{\lambda_1 \cdots \lambda_{m-1}}{\mu_1 \cdots \mu_m} + z_0 \right]$$

$$= \frac{\mu_1 \cdots \mu_m}{\lambda_1 \cdots \lambda_m} \left[\sum_{i=1}^{m} \frac{\lambda_1 \cdots \lambda_{i-1}}{\mu_1 \cdots \mu_i} - \tau_1 \right],$$

since $z_0 = \tau_0 - \tau_1 = -\tau_1$. Then

$$\frac{\lambda_1 \cdots \lambda_m}{\mu_1 \cdots \mu_m} z_m = \frac{1}{\mu_1} + \sum_{i=2}^{m} \frac{\lambda_1 \cdots \lambda_{i-1}}{\mu_1 \cdots \mu_i} - \tau_1. \qquad (6.36)$$

Suppose

$$\sum_{i=2}^{\infty} \frac{\lambda_1 \cdots \lambda_{i-1}}{\mu_1 \cdots \mu_i} = \infty.$$

Then $\tau_1 = \infty$. But since $\{\tau_i\}_{i=1}^{\infty}$ is a nondecreasing sequence, it follows that $\tau_m = \infty$.

Suppose

$$\sum_{i=2}^{\infty} \frac{\lambda_1 \cdots \lambda_{i-1}}{\mu_1 \cdots \mu_i} < \infty.$$

For large m, deaths are much greater than births, so that $z_m \to \tau_m - \tau_{m+1} \approx 1/\mu_{m+1}$ as $m \to \infty$, which is the mean time for a death to occur when the population size is $m+1$ (Nisbet and Gurney, 1982). Then let $m \to \infty$ so that the left side of (6.36) is

$$\frac{\lambda_1 \cdots \lambda_m}{\mu_1 \cdots \mu_m} z_m \to \frac{\lambda_1 \cdots \lambda_m}{\mu_1 \cdots \mu_m \mu_{m+1}} \to 0$$

(Karlin and Taylor, 1975; Nisbet and Gurney, 1982). Since the expression on the left of (6.36) approaches zero,

$$\tau_1 = \frac{1}{\mu_1} + \sum_{i=2}^{\infty} \frac{\lambda_1 \cdots \lambda_{i-1}}{\mu_1 \cdots \mu_i}$$

and

$$z_m = -\frac{\mu_1 \cdots \mu_m}{\lambda_1 \cdots \lambda_m} \left[\sum_{i=m+1}^{\infty} \frac{\lambda_1 \cdots \lambda_{i-1}}{\mu_1 \cdots \mu_i} \right].$$

Now,

$$\tau_m - \tau_1 = -\sum_{s=1}^{m-1} z_s$$

$$= \sum_{s=1}^{m-1} \left[\frac{\mu_1 \cdots \mu_s}{\lambda_1 \cdots \lambda_s} \sum_{i=s+1}^{\infty} \frac{\lambda_1 \cdots \lambda_{i-1}}{\mu_1 \cdots \mu_i} \right].$$

The value of τ_m is given by (6.22). $\qquad\qquad\qquad\qquad\qquad\qquad\qquad\square$

6.14.3 Comparison Theorem

THEOREM 6.5 *Comparison theorem*
Suppose that f and g are continuous with continuous derivatives on $\mathbb{R}$ and $m(0) = n(0)$. If

$$\frac{dm(t)}{dt} = g(m(t)) < f(m(t)) \text{ and } \frac{dn(t)}{dt} = f(n(t)) \text{ for } t \in [0, \infty),$$

then

$$m(t) \leq n(t) \text{ for } t \in [0, \infty).$$

Proof. Let $y(t) = m(t) - n(t)$. At $t = 0$, $y(0) = 0$ and $dy(t)/dt < 0$. By continuity of y, there exists an interval $[0, T)$ such that $y(t) \leq 0$. We need to show that $T = \infty$. Suppose $T < \infty$ and $[0, T]$ is the largest interval such that $y(t) \leq 0$ for $t \in [0, T]$. Then $y(T) = 0$ and there exists a time $t_1 > T$ such that $y(t) > 0$ for $(T, t_1]$. But $y(T) = 0$ implies $m(T) = n(T)$ and $f(m(T)) = f(n(T))$. From the hypotheses, it follows that at $t = T$, $dy(t)/dt < 0$ which implies $y(t) < 0$ for some interval $(T, t_2]$, $T < t_2 < t_1$, a contradiction. $\qquad\qquad\qquad\qquad\qquad\qquad\qquad\square$

For more general comparison results, consult Corduneanu (1977).

Chapter 7

Biological Applications of Continuous-Time Markov Chains

7.1 Introduction

A variety of biological applications of CTMCs are discussed. Continuous-time branching processes are presented in the first section, an extension of discrete-time branching processes in Chapter 4. Applications to cellular and molecular biology and to epidemic models are presented for which the methods of Chapter 6 can be applied. For example, the stochastic SI epidemic model is a birth process, whereas the stochastic SIS epidemic model is a birth and death process. The remaining applications are to multivariate processes. Therefore, some notation and terminology associated with multivariate processes are reviewed. Then applications to enzyme kinetics, epidemic, competition, and predation processes are presented. First, the corresponding deterministic models are introduced, then the stochastic models. Additional biological applications are discussed in the exercises and others can be found in the references.

7.2 Continuous-Time Branching Processes

Continuous-time branching processes are the continuous analogues of the discrete-time branching processes in Chapter 4. If the interevent time is exponentially distributed, the continuous-time branching processes are CTMCs. In Galton-Watson processes, discussed in Chapter 4, an individual's lifetime is a fixed length of time, which for convenience, is denoted as one unit of time or one generation, $\tau = 1$. At the end of that time interval, the individual is replaced by his progeny (if we are speaking of male heirs). However, in the continuous-time process, an individual's lifetime is not fixed but may have an arbitrary distribution. In the case of an exponentially distributed lifetime, the branching process is *Markovian* and, if not, it is a general *age-dependent* process known as a *Bellman-Harris* branching process (Bharucha-Reid, 1997;

Harris, 1963; Kimmel and Axelrod, 2002).

Consider a Markov, continuous-time branching process, $\{X(t),\ t \in [0, \infty)\}$, where $X(t)$ is a discrete random variable for the total population size at time t. Suppose the random variable for the time between events, τ, has an exponential distribution with parameter λ. Then the cumulative distribution function for τ is

$$G(t) = \text{Prob}\{\tau \leq t\} = 1 - e^{-\lambda t}$$

and the p.d.f. is $g(t) = G'(t) = \lambda e^{-\lambda t}$. Assume the population starts with a single individual born at time $t = 0$. That individual either dies without giving birth, dies while giving birth, or survives and produces offspring. Let the "next generation" offspring p.g.f. be denoted $f(z)$, where

$$f(z) = \sum_{k=0}^{\infty} p_k z^k.$$

In this example, the probabilities p_k are independent of time. Each of the offspring behaves independently, lives for an exponential period of time, and serves as a parent for the next generation, producing a random number of offspring (same offspring distribution as the original individual). The process continues with subsequent generations behaving in the same manner. See Figure 7.1.

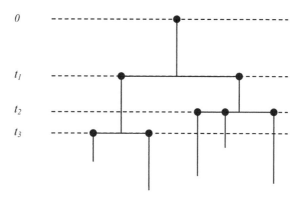

FIGURE 7.1: Sample path of a continuous-time branching process. At time t_1, an individual gives birth to two individuals and these individuals give birth to three and two individuals at times t_2 and t_3, respectively.

Let $\mathcal{P}(z, t)$ be the p.g.f. for $X(t)$. We shall derive a differential equation for $\mathcal{P}$. Let Δt be sufficiently small. Then, as in discrete-time branching processes, the p.g.f. $\mathcal{P}(z, t + \Delta t)$ is a composition of p.g.f.s; that is, it can be shown that

$$\mathcal{P}(z, t + \Delta t) = \mathcal{P}(\mathcal{P}(z, t), \Delta t). \tag{7.1}$$

(Kimmel and Axelrod, 2002). At time $t = 0$, $\mathcal{P}(z, 0) = z$. After a small period of time Δt,

$$
\begin{aligned}
\mathcal{P}(z, \Delta t) &= z\text{Prob}\{\tau > \Delta t\} + f(z)\text{Prob}\{\tau \leq \Delta t\} + o(\Delta t) \\
&= ze^{-\lambda \Delta t} + f(z)(1 - e^{-\lambda \Delta t}) + o(\Delta t). \tag{7.2}
\end{aligned}
$$

Subtracting $\mathcal{P}(z, t)$ from (7.1) and applying the identity in (7.2),

$$
\begin{aligned}
\mathcal{P}(z, t + \Delta t) - \mathcal{P}(z, t) &= \mathcal{P}(\mathcal{P}(z, t), \Delta t) - \mathcal{P}(z, t) \\
&= \mathcal{P}(z, t)e^{-\lambda \Delta t} + f(\mathcal{P}(z, t))(1 - e^{-\lambda \Delta t}) \\
&\quad - \mathcal{P}(z, t) + o(\Delta t) \\
&= [-\mathcal{P}(z, t) + f(\mathcal{P}(z, t))](1 - e^{-\lambda \Delta t}) + o(\Delta t).
\end{aligned}
$$

Dividing by Δt and letting $\Delta t \to 0$, the following differential equation is obtained:

$$\frac{\partial \mathcal{P}(z, t)}{\partial t} = -\lambda[\mathcal{P}(z, t) - f(\mathcal{P}(z, t))]. \tag{7.3}$$

The partial derivative can be treated as an ordinary derivative because there is no explicit dependence on z in the right-hand side of (7.3). Equation (7.3) with initial condition $\mathcal{P}(z, 0) = z$ has a unique solution provided $\lim_{z \to 1^-} \mathcal{P}(z, t) = 1$ for all times, i.e., the process does not explode (Kimmel and Axelrod, 2002). Equation (7.3) is known as the backward Kolmogorov differential equation, as opposed to the forward equation studied in Chapter 6.

As was the case for discrete-time branching processes, the mean number of births determines the asymptotic behavior. Let $m = f'(1)$. The asymptotic behavior depends on whether $m \leq 1$ or $m > 1$ Recall that $m < 1$ is subcritical, $m = 1$ is critical, and $m > 1$ is supercritical. Only for the supercritical case, $m > 1$, is there a positive probability of survival. Theorem 4.2 in Chapter 4 can be extended to the continuous-time process when the process is nonexplosive. Properties (1)-(5) in Section 4.4 are also assumed to hold for f, e.g., $f(0) = p_0 > 0$ and $p_0 + p_1 < 1$. The limiting result about the probability of extinction is stated in the next theorem. For a proof see Harris (1963).

THEOREM 7.1
Suppose $\{X(t) : t \in [0, \infty)\}$ is a nonexplosive, continuous-time Markov branching process with $X(0) = 1$. Assume f is the p.g.f. of the offspring distribution, where $m = f'(1)$ and $\mathcal{P}(z, t)$ is the p.g.f. of $X(t)$. If $m \leq 1$, then

$$\lim_{t \to \infty} \text{Prob}\{X(t) = 0\} = \lim_{t \to \infty} p_0(t) = 1$$

and if $m > 1$, then there exists a q satisfying $f(q) = q$ such that

$$\lim_{t \to \infty} \text{Prob}\{X(t) = 0\} = \lim_{t \to \infty} p_0(t) = q < 1.$$

The next example expresses the simple birth and death process as a continuous-time branching process.

Example 7.1. Suppose the cumulative distribution for the time between events is $G(t) = 1 - e^{-(\lambda + \mu)t}$, exponential with parameter $\lambda + \mu$ and the p.g.f. for the next generation

$$f(z) = \frac{\mu}{\lambda + \mu} + \frac{\lambda}{\lambda + \mu} z^2,$$

where $\lambda > 0$ is the birth rate and $\mu > 0$ is the death rate. An individual either dies or survives and gives birth. There is an important difference in this offspring p.g.f. and the one for the discrete-time model. For this p.g.f., a single offspring does not replace the parent but adds to the population size if the parent survives. Also, $m = f'(1) = 2\lambda/(\mu + \lambda) > 1$ if $\lambda > \mu$. In this case, the fixed point q of $f(q) = q$ is μ/λ. The differential equation for $\mathcal{P}(z, t)$ takes the form

$$\frac{d\mathcal{P}(z, t)}{dt} = -(\lambda + \mu)\mathcal{P}(z, t) + (\lambda + \mu)\left[\frac{\mu}{\lambda + \mu} + \frac{\lambda}{\lambda + \mu}\mathcal{P}^2(z, t)\right].$$

The notation for an ordinary derivative is used with respect to t, where z is treated as a constant. Simplifying,

$$\frac{d\mathcal{P}}{dt} = \mu - (\lambda + \mu)\mathcal{P} + \lambda\mathcal{P}^2, \tag{7.4}$$

with initial conditions $\mathcal{P}(z, 0) = z$ and $\lim_{z \to 1} \mathcal{P}(z, t) = 1$. This differential equation can be solved by separation of variables, yielding the solution

$$\mathcal{P}(z, t) = \begin{cases} \dfrac{e^{t(\mu - \lambda)}(\lambda z - \mu) - \mu(z - 1)}{e^{t(\mu - \lambda)}(\lambda z - \mu) - \lambda(z - 1)}, & \text{if } \lambda \neq \mu \\[2ex] \dfrac{\lambda t(z - 1) - z}{\lambda t(z - 1) - 1}, & \text{if } \lambda = \mu. \end{cases} \tag{7.5}$$

See Chapter 6, Section 6.4.3. It follows that

$$\lim_{t \to \infty} p_0(t) = \begin{cases} \dfrac{\mu}{\lambda}, & \text{if } \lambda > \mu \\ 1, & \text{if } \lambda \leq \mu \end{cases}$$

which agrees with the fixed point q derived from $f(z)$. ∎

In addition to the single type branching process, a multitype branching process can be defined. Assume there are k types of individuals, where $X_i(t)$ is the discrete random variable for the number of individuals of type i, $i = 1, \ldots, k$, at time t and $X(t) = (X_1(t), \ldots, X_k(t))^{tr}$. Let $f_i(z_1, \ldots, z_k)$ denote the next generation p.g.f. and $\mathcal{P}_i(z_1, \ldots, z_k, t)$ denote the p.g.f. for an individual of type i, $i = 1, \ldots, k$. Assume the waiting time between events is an exponential distribution with parameter λ, $G(t) = 1 - e^{-\lambda t}$. It can be shown that each of the $\mathcal{P}_i$ is a solution of a backward differential equation of the form,

$$\frac{d\mathcal{P}_i}{dt} = -\lambda \mathcal{P}_i + \lambda f_i(\mathcal{P}_1, \ldots, \mathcal{P}_k), \quad i = 1, \ldots, k. \tag{7.6}$$

A multitype process involving the development of drug resistance in cancer cells is discussed in the next example (Coldman and Goldie, 1987; Kimmel and Axelrod, 2002).

Example 7.2. Assume there are two types of cancer cells. Type 1 cells are sensitive to a particular drug and type 2 cells are resistant to that drug. Assume a population of cancer cells begins with a single type 1 cell. The time it takes for a cell to divide is a random variable τ with an exponential distribution having parameter λ. At each division of a type 1 sensitive cell, with probability p, one of the two daughter cells mutates and becomes resistant to the drug (type 2). A type 2 resistant cell produces two daughter cells that are both resistant (type 2) (i.e., mutations are irreversible). Then the birth p.g.f.s satisfy

$$f_1(z_1, z_2) = (1 - p)z_1^2 + pz_1z_2, \quad \text{and} \quad f_2(z_1, z_2) = z_2^2.$$

The p.g.f.s for each cell type are solutions of the following backward differential equations:

$$\frac{d\mathcal{P}_1}{dt} = -\lambda \mathcal{P}_1 + \lambda[(1 - p)\mathcal{P}_1^2 + p\mathcal{P}_1\mathcal{P}_2]$$

$$\frac{d\mathcal{P}_2}{dt} = -\lambda \mathcal{P}_2 + \lambda \mathcal{P}_2^2.$$

The initial conditions are $\mathcal{P}_1(z_1, z_2, 0) = z_1$ and $\mathcal{P}_2(z_1, z_2, 0) = z_2$. The differential equation for $\mathcal{P}_2$ can be solved by separation of variables (this is a logistic-type differential equation). Then the solution $\mathcal{P}_2$ can be substituted into the differential equation for $\mathcal{P}_1$ to find the solution for $\mathcal{P}_1$. It can be shown that the solutions are

$$\mathcal{P}_1(z_1, z_2, t) = \frac{z_1 e^{-\lambda t}[z_2 e^{-\lambda t} + 1 - z_2]^{-p}}{1 + z_1([e^{-\lambda t}z_2 + 1 - z_2]^{1-p} - 1)z_2^{-1}} \tag{7.7}$$

$$\mathcal{P}_2(z_1, z_2, t) = \frac{z_2}{z_2 + (1 - z_2)e^{\lambda t}} \tag{7.8}$$

(Kimmel and Axelrod, 2002). When the tumor is first identified, it is important to find out what proportion of the cells are resistant. Of course, it is hoped that there are no resistant cells. The probability there are no resistant cells at time t can be computed as follows:

$$\lim_{z_1 \to 1} \lim_{z_2 \to 0} \mathcal{P}_1(z_1, z_2, t) = \frac{1}{1 - p + pe^{\lambda t}}.$$

Eventually, as $t \to \infty$, all cells will be resistant. Therefore, it is important to discover the tumor very early, that is, when t is small. ∎

The general age-dependent branching process is known as a *Bellman-Harris process*, named after Richard Bellman and Theodore Harris, the first investigators of this type of process (Harris, 1963). In the Bellman-Harris process, the cumulative lifetime distribution is denoted as $G(t) = \text{Prob}\{\tau \leq t\}$, a general distribution. An integral equation can be derived for the p.g.f. $\mathcal{P}(z, t)$. Note that $\mathcal{P}(z, t)$ is

$$\mathcal{P}(z, t) = \begin{cases} z, & t < \tau \\ f(\mathcal{P}(z, t - \tau)), & t \geq \tau. \end{cases}$$

Then the integral equation for $\mathcal{P}(z, t)$ is

$$\mathcal{P}(z, t) = z(1 - G(t)) + \int_0^t f(\mathcal{P}(z, t - u)) \, dG(u). \tag{7.9}$$

In the special case where the lifetime distribution is exponential (Markov branching process), it can be shown that the integral equation (7.9) reduces to the differential equation (7.3). Let $G(t) = 1 - e^{-\lambda t}$. Then

$$\mathcal{P}(z, t) = ze^{-\lambda t} + \int_0^t f(\mathcal{P}(z, t - u))\lambda e^{-\lambda u} \, du.$$

Multiplying the preceding equation by $e^{\lambda t}$ and then making a change of variable $v = t - u$ in the integral leads to

$$e^{\lambda t}\mathcal{P}(z, t) = z + \lambda \int_0^t f(\mathcal{P}(z, v))e^{\lambda v} \, dv.$$

Differentiating the last equation with respect to t leads to

$$e^{\lambda t}\left[\lambda \mathcal{P}(z, t) + \frac{d\mathcal{P}}{dt}\right] = \lambda f(\mathcal{P}(z, t))e^{\lambda t}.$$

Finally, solving for $d\mathcal{P}/dt$ leads to the differential equation (7.3).

An extensive number of biological applications of continuous-time branching processes can be found in the books by Haccou, Jagers, and Vatutin (2005), Jagers (1975), Kimmel and Axelrod (2002), and Mode (1971).

7.3 SI and SIS Epidemic Processes

First, the dynamics of the deterministic SI and SIS epidemic models are reviewed. Recall that $S(t)$ is the number of susceptible individuals at time t, $I(t)$ is the number of infected individuals at time t, and N is the constant total population size, $N = S(t) + I(t)$. Infected individuals are also infectious (no latent period). The SI epidemic model has been applied to diseases such as influenza or the common cold, where generally no one is immune and over the course of the epidemic almost everyone eventually becomes infected. The SIS epidemic model has been applied to some sexually transmitted diseases, where there is recovery but no immunity; individuals can become infected immediately following recovery. (See the discussion of the SIS epidemic model in Chapter 3.)

The differential equations for the SI epidemic model are

$$\frac{dS}{dt} = -\frac{\beta}{N}SI$$
$$\frac{dI}{dt} = \frac{\beta}{N}SI,$$

$S(0) + I(0) = N$. The parameter β = transmission rate, the number of contacts per time that result in an infection of a susceptible individual. If there is recovery, then γI is included in the differential equation for I, where $1/\gamma$ is the average length of the infectious period. In addition, if there are births and deaths over the course of the epidemic, but the population size remains constant, then it will be assumed that the birth rate equals the death rate. The differential equations for the SIS epidemic model with recovery, births, and deaths are

$$\frac{dS}{dt} = -\frac{\beta}{N}SI + (\gamma + b)I$$
$$\frac{dI}{dt} = \frac{\beta}{N}SI - (\gamma + b)I,$$

$S(0) + I(0) = N$. It is easy to see that adding the differential equations gives $d(S + I)/dt = 0$ so that $S(t) + I(t) = N$.

For the SI epidemic model, substitution of $S = N - I$ into the differential equation for S yields a logistic differential equation for I. This differential equation can be solved directly via separation of variables; the solution $I(t)$ is

$$I(t) = \frac{NI(0)}{I(0) + (N - I(0))e^{-\beta t}}. \tag{7.10}$$

Eventually, everyone becomes infected, $\lim_{t \to \infty} I(t) = N$.

The SIS epidemic model has an endemic equilibrium given by

$$S^* = \frac{\gamma + b}{\beta} N, \quad I^* = \frac{N(\beta - \gamma - b)}{\beta}.$$

The endemic equilibrium is positive if the *basic reproduction number,*

$$\mathcal{R}_0 = \frac{\beta}{b + \gamma}, \tag{7.11}$$

satisfies $\mathcal{R}_0 > 1$. If $\mathcal{R}_0 > 1$, then solutions approach the endemic equilibrium and if $\mathcal{R}_0 \leq 1$ solutions approach the disease-free state.

The contact rate β may be a function of population size. If, for example, $\beta = cN$, then the expression $cNSI/N = cSI$ in the SI or SIS epidemic models is referred to as *mass action incidence rate*, and when β does not depend on N, the expression is referred to as *standard incidence rate*. In models where the population size is constant, this distinction makes little difference because β is constant in either case. However, the form of β can have a significant impact on the population dynamics when the population size is not constant (Allen and Cormier, 1996).

In the stochastic models, it will be shown that $I = N$ is the unique absorbing state for the SI epidemic model and $I = 0$ is the unique absorbing state for the SIS epidemic model. In addition, the expected duration until absorption is calculated. For comparison purposes, the duration of time until the infected population size reaches N is approximated in the deterministic model. The time T until $I(T) = N$ is actually infinite in the deterministic model because N is approached asymptotically. However, to obtain a realistic estimate of the time to absorption, solve for t in the identity (7.10), by letting $N - 1 = I(t)$. Then letting $t = T$,

$$T \approx \frac{\ln[(N - I(0))(N - 1)/I(0)]}{\beta}.$$

Table 7.1 gives the approximate duration until the entire population is infected for various values of N and β when the initial population size is one, $I(0) = 1$.

Table 7.1: Approximate duration (T) until infection in the deterministic SI epidemic model for population size N and contact rate β when $I(0) = 1$

β	N			
	10	100	1000	10000
1	4.605	9.210	13.816	18.421
10	0.461	0.921	1.382	1.842
100	0.046	0.092	0.138	0.184

7.3.1 Stochastic SI Model

The stochastic SI epidemic model is a birth process. Let $I(t)$ denote the random variable for the number of infected individuals at time t. The state space for $I(t)$ is $\{0, 1, 2 \ldots, N\}$. The transition probabilities are

$$\text{Prob}\{\Delta I(t) = j | I(t) = i\} = \begin{cases} \dfrac{\beta}{N} i(N-i)\Delta t + o(\Delta t), & j = 1 \\ 1 - \dfrac{\beta}{N} i(N-i)\Delta t + o(\Delta t), & j = 0 \\ o(\Delta t), & j \neq 0, 1, \end{cases}$$

leading to the forward Kolmogorov equations, $dp/dt = Qp$, with generator matrix

$$Q = \begin{pmatrix} -\beta(N-1)/N & 0 & \cdots & 0 & 0 \\ \beta(N-1)/N & -\beta 2(N-2)/N & \cdots & 0 & 0 \\ 0 & \beta 2(N-2)/N & \cdots & 0 & 0 \\ \vdots & \vdots & \cdots & \vdots & \vdots \\ 0 & 0 & \cdots & -\beta(N-1)/N & 0 \\ 0 & 0 & \cdots & \beta(N-1)/N & 0 \end{pmatrix}.$$

It is easy to see that the transition matrix of the embedded Markov chain is

$$T = \begin{pmatrix} 0 & 0 & 0 & \cdots & 0 & 0 \\ 1 & 0 & 0 & \cdots & 0 & 0 \\ 0 & 1 & 0 & \cdots & 0 & 0 \\ \vdots & \vdots & \vdots & \cdots & \vdots & \vdots \\ 0 & 0 & 0 & \cdots & 1 & 1 \end{pmatrix}.$$

State N is absorbing. The limiting stationary probability distribution is $\lim_{t\to\infty} p(t) = (0, 0, \ldots, 1)^{tr}$.

The time until absorption beginning from state 1 can be written as

$$T_{N,1} = \sum_{i=1}^{N-1} T_{i+1,i}.$$

Hence, the expected duration until absorption or until the entire population is infected is given by

$$E(T_{N,1}) = \sum_{i=1}^{N-1} E(T_{i+1,i}).$$

Since the time between events $i \to i + 1$ is exponentially distributed with parameter $\lambda = \beta i(N-i)/N$, the mean time between events i and $i + 1$ is

$E(T_{i+1,i}) = 1/\lambda = N/[\beta i(N - i)]$. Therefore, the expected time until the entire population is infected is

$$E(T_{N,1}) = \sum_{i=1}^{N-1} \frac{N}{\beta i(N - i)} = \frac{1}{\beta} \sum_{i=1}^{N-1} \left[\frac{1}{i} + \frac{1}{N - i}\right] = \frac{2}{\beta} \sum_{i=1}^{N-1} \frac{1}{i}.$$

(A partial fraction decomposition is used to simplify the summation.) In addition, the variance of the time to absorption has a simple expression,

$$\mathrm{Var}(T_{N,1}) = \sum_{i=1}^{N-1} \mathrm{Var}(T_{i+1,i}),$$

since the random variables $\{T_{i+1,i}\}_{i=1}^{N-1}$ are independent. Recall that the variance of an exponentially distributed random variable with parameter λ is $1/\lambda^2$. For the SI model, it follows that the variance for the interevent time is $Var(T_{i+1,i}) = 1/\lambda^2 = N^2/[\beta i(N - i)]^2$. Thus,

$$\mathrm{Var}(T_{N,1}) = \frac{1}{\beta^2} \sum_{i=1}^{N-1} \frac{N^2}{[i(N - i)]^2}$$

$$= \frac{1}{\beta^2} \sum_{i=1}^{N-1} \left[\frac{2/N}{i} + \frac{1}{i^2} + \frac{2/N}{N - i} + \frac{1}{(N - i)^2}\right]$$

$$= \frac{1}{\beta^2} \sum_{i=1}^{N-1} \left[\frac{4}{N}\frac{1}{i} + \frac{2}{i^2}\right].$$

(Again, a partial fraction decomposition is used to simplify the summation.) For large N, the mean and variance can be approximated using the following two identities:

$$\lim_{N \to \infty} \left[\sum_{i=1}^{N} \frac{1}{i} - \ln(N)\right] = \gamma \quad \text{and} \quad \lim_{N \to \infty} \sum_{i=1}^{N} \frac{1}{i^2} = \frac{\pi^2}{6},$$

where γ is Euler's constant, $\gamma \approx 0.5772156649\ldots$. Thus, for large N, approximations for the mean and variance are

$$E(T_{N,1}) \approx \frac{2}{\beta} [\gamma + \ln(N)] \approx \frac{2}{\beta} \ln(N - 1)$$

$$\mathrm{Var}(T_{N,1}) \approx \frac{1}{\beta^2} \left[\frac{4}{N} (\gamma + \ln(N - 1)) + \frac{\pi^2}{3}\right]$$

$$\approx \frac{\pi^2}{3\beta^2} \approx \frac{3.290}{\beta^2}.$$

Notice for a large population size, the variance does not depend on N and the approximate mean duration equals the approximate duration derived from the deterministic model when $I(0) = 1$. Table 7.2 gives the values for the approximate mean duration and the exact mean duration and also gives the exact variance for the distribution for the time until the entire population is infected. For the parameter $\beta = 1$, $\text{Var}(T_{N,1}) \approx 3.290$. It can be seen that the approximations approach the exact values as N increases.

Table 7.2: Approximate and exact mean durations $E(T_{N,1})$ and exact variance $\text{Var}(T_{N,1})$ for the time until absorption in the SI stochastic epidemic model when $I(0) = 1$ and $\beta = 1$

Mean and	N			
Variance	10	100	1000	10000
Approx. $E(T_{N,1})$	4.605	9.210	13.816	18.421
Exact $E(T_{N,1})$	5.658	10.355	14.969	19.579
Exact $Var(T_{N,1})$	4.211	3.477	3.318	3.294

The time units for an epidemic depend on the particular disease and the population being modeled. For humans, it depends on whether the time units are on the order of hours, days, weeks, or months. For example, if the time units are days, so that $\beta = 1$ successful contact/day, $I(0) = 1$, and $N = 1000$, the approximate duration until all 1000 individuals are infected would be, on the average, 15 days. It must be kept in mind that the SI epidemic model is oversimplified. For example, there is no latent period nor the possibility of recovery or immunity.

7.3.2 Stochastic SIS Model

In the SIS epidemic model, individuals can recover from the disease but do not develop immunity. They immediately become susceptible again. The SIS epidemic model discussed here is the continuous analogue of the discrete-time epidemic model discussed in Chapter 3. Let the transition probabilities equal

$$\text{Prob}\{\Delta I(t) = j | I(t) = i\}$$
$$= \begin{cases} \dfrac{\beta}{N} i(N - i)\,\Delta t + o(\Delta t), & j = 1 \\ (b + \gamma)i\,\Delta t + o(\Delta t), & j = -1 \\ 1 - \left[\dfrac{\beta}{N} i(N - i) + (b + \gamma)i\right]\Delta t \\ \quad + o(\Delta t), & j = 0 \\ o(\Delta t), & j \neq -1, 0, 1, \end{cases}$$

where $i \in \{0, 1, \ldots, N\}$. The SIS epidemic model is a birth and death process with

$$\lambda_i = \max\left\{0, \frac{\beta}{N}i(N - i)\right\} \quad \text{and} \quad \mu_i = (\gamma + b)i,$$

for $i = 0, 1, \ldots, N$. There is a single absorbing state at zero, $\lim_{t \to \infty} p_0(t) = 1$.

If $\mathcal{R}_0 > 1$, then the SIS epidemic model is a special case of the logistic model, studied in Chapter 6. In this case, $\lambda_n = b_1 n + b_2 n^2$ and $\mu_n = d_1 n + d_2 n^2$, where $b_1 = \beta$, $b_2 = -\beta/N$, $d_1 = b + \gamma$ and $d_2 = 0$. In order that the requirement $\lambda_n - \mu_n = rn(1 - n/K)$ in the logistic model be satisfied,

$$r = \beta - (b + \gamma) > 0 \quad \text{and} \quad -r/K = -\beta/N < 0.$$

The expression $r > 0$ is equivalent to $\mathcal{R}_0 > 1$. A generator matrix Q and transition matrix T can be easily defined. The expected duration until extinction, τ_i given $I(0) = i$, is a solution of

$$\tau^{tr}\tilde{Q} = -\mathbf{1}^{tr},$$

system (6.20), where $\tau^{tr} = (\tau_1, \ldots, \tau_N)$, $\mathbf{1}$ is a column vector of ones, and $\tilde{Q}$ is the generator matrix Q with the first row and first column deleted. See Theorem 6.3 for an explicit expression for τ. Higher-order moments can also be defined.

When $\mathcal{R}_0$ and N are large, the time until absorption can be very long. Therefore, prior to extinction, the infected population reaches a quasistationary probability distribution. Recall that the approximate quasistationary probability distribution, $\tilde{\pi} = (\tilde{\pi}_1, \ldots, \tilde{\pi}_N)^{tr}$, satisfies

$$\tilde{\pi}_{i+1} = \frac{\lambda_i}{\mu_{i+1}}\tilde{\pi}_i, \quad i = 1, 2, \ldots, N - 1.$$

Example 7.3. Let $\beta = 2$, $N = 100$, and $\gamma + b = 1$. Then $\mathcal{R}_0 = 2$. Graphs of τ and $\tilde{\pi}$ are given in Figure 7.2. The expected time until the infected population reaches zero, if $N = 100$ is on the order of 10^8. If time is measured in days, the epidemic ends, on the average, in 350,000 years. Thus, a quasistationary is a reasonable approximation to the dynamics. The mean and standard deviation of the approximate quasistationary probability distribution $\tilde{\pi}$ are approximately 48.9 and 7.2, respectively. Notice that the mean of $\tilde{\pi}$ is close to the value of the deterministic endemic equilibrium, $I^* = N(1 - 1/\mathcal{R}_0) = 50$. In addition, the distribution of $\tilde{\pi}$ is approximately normal. ∎

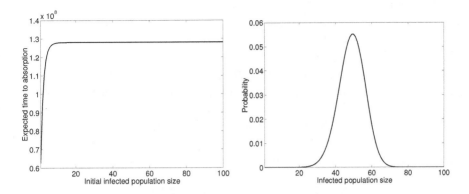

FIGURE 7.2: Expected duration until absorption at $I = 0$ and the quasistationary distribution of the stochastic SIS model when $\beta = 2$, $b + \gamma = 1$, and $N = 100$.

7.4   Multivariate Processes

In all of the CTMCs discussed thus far, the processes have been univariate, only a single random variable $X(t)$. In this section, the notation for multivariate processes will be introduced, processes for which there are two or more random variables.

For simplicity, consider a bivariate process. Let $\{(X(t), Y(t)) : t \geq 0\}$ denote a continuous-time, bivariate Markov process, where $X(t)$ and $Y(t) \in \{0, 1, 2, \ldots\}$. The *joint probability mass function (joint p.m.f.)* is

$$p_{(m,n)}(t) = \text{Prob}\{X(t) = m, Y(t) = n\}.$$

Be careful not to confuse this notation with the notation used in previous sections for the transition from state n to state m in a univariate process. The transition probability for the bivariate process is denoted as

$$p_{(m,n),(i,j)}(\Delta t) = \text{Prob}\{X(t + \Delta t) = m, Y(t + \Delta t) = n | X(t) = i, Y(t) = j\}.$$

The process is assumed to be homogeneous in time (i.e., the transition probabilities only depend on the length of time between transitions and do not depend on the time at which they occur).

The p.g.f. for this bivariate process is

$$P(w, z, t) = \sum_{m=0}^{\infty} \sum_{n=0}^{\infty} p_{(m,n)}(t) w^m z^n.$$

The m.g.f. and c.g.f. are $M(\theta, \phi, t) = P(e^\theta, e^\phi, t)$ and $K(\theta, \phi, t) = \ln M(\theta, \phi, t)$, respectively. The marginal probability distributions of $X(t)$ and $Y(t)$ are

$$\sum_{n=0}^{\infty} p_{(m,n)}(t) \quad \text{and} \quad \sum_{m=0}^{\infty} p_{(m,n)}(t),$$

respectively. Their means and variances are

$$\mu_X(t) = \sum_{m=0}^{\infty} \sum_{n=0}^{\infty} m p_{(m,n)}(t), \quad \mu_Y(t) = \sum_{n=0}^{\infty} \sum_{m=0}^{\infty} n p_{(m,n)}(t),$$

$$\sigma_X^2(t) = \sum_{m=0}^{\infty} \sum_{n=0}^{\infty} m^2 p_{(m.n)}(t) - \mu_X^2(t),$$

and

$$\sigma_Y^2(t) = \sum_{n=0}^{\infty} \sum_{m=0}^{\infty} n^2 p_{(m,n)}(t) - \mu_Y^2(t).$$

The means and higher-order moments can be obtained from the generating functions. For example,

$$\left.\frac{\partial M}{\partial \theta}\right|_{\theta=0=\phi} = \mu_X(t), \quad \left.\frac{\partial M}{\partial \phi}\right|_{\theta=0=\phi} = \mu_Y(t),$$

$$\left.\frac{\partial^2 M}{\partial \theta^2}\right|_{\theta=0=\phi} = E(X^2(t)), \quad \text{and} \quad \left.\frac{\partial^2 M}{\partial \phi^2}\right|_{\theta=0=\phi} = E(Y^2(t)).$$

Forward Kolmogorov differential equations can be derived from the transition probabilities in a manner similar to the univariate process. Let $\Delta X(t) = X(t + \Delta t) - X(t)$, $\Delta Y(t) = Y(t + \Delta t) - Y(t)$, and S be a finite subset of $Z \times Z$, where $Z = \{0, \pm 1, \pm 2, \ldots\}$ is the set of integers. Then

$$\text{Prob}\{\Delta X(t) = k, \Delta Y(t) = l | (X(t), Y(t))\}$$
$$= \begin{cases} h_{kl}(X(t), Y(t)) \Delta t + o(\Delta t), & (k, l) \in S \\ 1 - \sum_{(i,j) \in S} h_{ij}(X(t), Y(t)) \Delta t + o(\Delta t), & (k, l) \notin S. \end{cases}$$

For $(X(t), Y(t)) = (i, j)$, the preceding transition probability is denoted as $p_{(m,n),(i,j)}(\Delta t)$, where $m = i + k$ and $n = j + l$. Thus, the joint p.d.f. satisfies

$$p_{(m,n)}(t + \Delta t) = \sum_{(k,l) \in S} p_{(m-k,n-l)}(t) h_{kl}(m - k, n - l) \Delta t$$

$$+ p_{(m,n)}(t) \left[1 - \sum_{(k,l) \in S} h_{kl}(m, n) \Delta t \right] + o(\Delta t).$$

Subtracting $p_{(m,n)}(t)$ and dividing by Δt leads to the forward Kolmogorov differential equation for the bivariate process,

$$\frac{dp_{(m,n)}}{dt} = -p_{(m,n)} \sum_{(k,l) \in S} h_{kl}(m,n) + \sum_{(k,l) \in S} p_{(m-k,n-l)} h_{kl}(m-k, n-l).$$

Differential equations for the p.g.f. or m.g.f. can be derived using the generating function technique.

Example 7.4. Suppose $S = \{(1,0), (0,1), (-1,0), (0,-1)\}$; that is, for Δt small, the bivariate process can increase or decrease by one in each of its components. In addition, let $h_{10}(X,Y) = \lambda_1 X$, $h_{01}(X,Y) = \lambda_2 Y$, $h_{-1,0}(X,Y) = \mu_1 X$, and $h_{0,-1}(X,Y) = \mu_2 Y$ (Bailey, 1990). Then the forward Kolmogorov differential equation has the form

$$\begin{aligned}
\frac{dp_{(m,n)}}{dt} &= -p_{(m,n)} \left[\lambda_1 m + \lambda_2 n + \mu_1 m + \mu_2 n \right] \\
&\quad + p_{(m-1,n)} \lambda_1 (m-1) + p_{(m,n-1)} \lambda_2 (n-1) \\
&\quad + p_{(m+1,n)} \mu_1 (m+1) + p_{(m,n+1)} \mu_2 (n+1).
\end{aligned} \tag{7.12}$$

In addition, the m.g.f. has the form

$$\begin{aligned}
\frac{\partial M}{\partial t} &= \left[\lambda_1 (e^\theta - 1) + \mu_1 (e^{-\theta} - 1) \right] \frac{\partial M}{\partial \theta} \\
&\quad + \left[\lambda_2 (e^\phi - 1) + \mu_2 (e^{-\phi} - 1) \right] \frac{\partial M}{\partial \phi},
\end{aligned} \tag{7.13}$$

where $M(\theta, \phi, 0) = e^{N_1 \theta + N_2 \phi}$, $X(0) = N_1$, and $Y(0) = N_2$. Applying the differential equation for the m.g.f., it can be shown that the mean of $X(t)$ and $Y(t)$ are solutions of the following differential equations:

$$\frac{d\mu_X(t)}{dt} = (\lambda_1 - \mu_1)\mu_X(t), \quad \text{and} \quad \frac{d\mu_Y(t)}{dt} = (\lambda_2 - \mu_2)\mu_Y(t). \tag{7.14}$$

See Exercise 14. ∎

Numerical simulations of multivariate Markov chains can be performed in a manner similar to univariate processes. Suppose the process is in state (i,j) at time t; then, assuming the process can jump to at most a finite number of states, the time until the next event has an exponential distribution with parameter $\sum_{(k,l) \in S} h_{k,l}(i,j)$. For example, the bivariate process in Example 7.4 has an exponential interevent time distribution with parameter $(\lambda_1 + \mu_1)i + (\lambda_2 + \mu_2)j$ when the process is in state (i,j). The preceding methods can be easily extended to more than three random variables.

7.5   Enzyme Kinetics

Intracellular dynamics are governed by enzymes or catalysts. Suppose a nutrient or substrate is taken up by the cell and converted into some product. The model for this uptake is referred to as *Michaelis-Menten enzyme kinetics,* named for a German biochemist, Leonor Michaelis and a Canadian medical scientist, Maud Menten, whose research in the early part of the 20th century contributed much to the theory of enzyme kinetics. Nutrient molecules N (substrate, $N \equiv N(t)$) enter the cell membrane by attaching to membrane-bound receptors or enzymes E. After the nutrient binds to an enzyme, a complex formed from the enzyme and nutrient molecule is denoted as B. Then when the nutrient-bound molecule B enters the cell a product P is formed and the enzyme is released. The chemical process is illustrated by the following diagram:

$$N + E \xrightarrow{k_1} B, \quad B \xrightarrow{k_{-1}} N + E, \quad B \xrightarrow{k_2} P + E,$$

where the constants k_i, $i = -1, 1, 2$ denote reaction rates. The first reaction is reversible. The units of N and B are generally in terms of concentration, in particular, moles per liter (M), and time is measured in seconds (s). A deterministic model, a system of ordinary differential equations, and a CTMC model are formulated to describe the intracellular dynamics.

7.5.1   Deterministic Model

Based on these chemical reactions, differential equations can be derived for the temporal dynamics of the four variables:

$$\frac{dN}{dt} = -k_1 N E + k_{-1} B$$
$$\frac{dE}{dt} = -k_1 N E + (k_{-1} + k_2) B$$
$$\frac{dB}{dt} = k_1 N E - (k_{-1} + k_2) B$$
$$\frac{dP}{dt} = k_2 B,$$

where the units of the variables are in concentration (moles per liter, M). See Edelstein-Keshet (1988) or Allen (2007) for more details on the deterministic model formulation.

The enzyme is conserved and the nutrient is converted into a product:

$$E + B = e_\infty \quad \text{and} \quad N + B + P = p_\infty,$$

where e_∞ and p_∞ are positive constants. The preceding identities follow directly from the differential equations. That is, the sum of the corresponding differential equations is zero, e.g., $d(E+B)/dt = 0$ which implies $E+B$ is constant. Thus, the system of four differential equations simplifies to two differential equations in N and B:

$$\frac{dN}{dt} = -k_1 N(e_\infty - B) + k_{-1} B \tag{7.15}$$

$$\frac{dB}{dt} = k_1 N(e_\infty - B) - (k_{-1} + k_2) B, \tag{7.16}$$

where $N(0) = p_\infty$ and $B(0) = 0$. It is easy to show that solutions for N and B are nonnegative and bounded $(0 \le B(t) \le e_\infty)$ and there is only one equilibrium, namely, $\bar{N} = 0$ and $\bar{B} = 0$. It follows from the Poincaré-Bendixson theory of differential equations that solutions must tend to the equilibrium (Coddington and Levinson, 1955). The equilibrium $(\bar{N}, \bar{B}) = (0,0)$ is globally asymptotically stable, that is, $(N(t), B(t)) \to (0,0)$ and

$$\lim_{t \to \infty} (P(t), E(t)) = (p_\infty, e_\infty).$$

The dynamics of $N(t)$ and $B(t)$ are graphed in Figure 7.3 for the initial conditions $N(0) = p_\infty$ and $B(0) = 0$, where the following parameter values are assumed (Wilkinson, 2006):

$$p_\infty = 5 \times 10^{-7} M, \quad e_\infty = 2 \times 10^{-7} M, \quad k_1 = 10^6 M^{-1} s^{-1}, \tag{7.17}$$

$$k_{-1} = 10^{-4} s^{-1}, \quad \text{and} \quad k_2 = 0.1 s^{-1}. \tag{7.18}$$

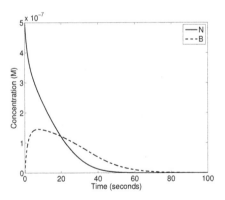

FIGURE 7.3: Concentration of $N(t)$ and $B(t)$ for $t \in [0, 100]$ for initial conditions $N(0) = 5 \times 10^{-7}$M and $B(0) = 0$.

If the nutrient concentration is much greater than the receptor concentration, then receptors are working at maximum capacity and their occupancy rate is approximately constant. Therefore, a reasonable assumption is that the nutrient-bound complex is at steady-state: $dB/dt = 0$. This assumption is often valid when there is a steady supply of nutrient (e.g., chemostat model in Exercise 12). However, we consider the simplifying assumption on the dynamics of N and P. It follows that

$$B = \frac{k_1 e_\infty N}{k_{-1} + k_2 + k_1 N}.$$

Then differential equations for the nutrient and the product are

$$\frac{dN}{dt} = -k_2 B = -\frac{dP}{dt}$$

or

$$\frac{dN}{dt} = -\frac{k_{max} N}{k_m + N} = -\frac{dP}{dt}, \qquad (7.19)$$

where $k_{max} = k_2 e_\infty$ and $k_m = (k_2 + k_{-1})/k_1$. The constant k_m is known as the Michaelis-Menten constant, i.e., the nutrient concentration $N = k_m$ at which the growth rate of P is half the maximum rate, $dP/dt = k_{\max}/2$. The rate of production of the product, $\dfrac{k_{max} N}{k_m + N}$, is generally referred to as Michaelis-Menten kinetics (Alon, 2007). The solutions for the differential equations for N and P are graphed in Figure 7.4. The limiting solution is $\lim_{t \to \infty} (N(t), P(t)) = (0, p_\infty)$. Note the difference in the dynamics for the two models. In this second model, the dynamics of the enzyme-nutrient complex are neglected, $dB/dt = 0$ so that the product is formed instantaneously from the nutrient.

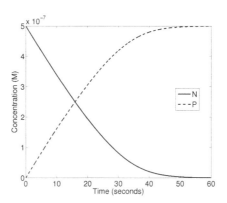

FIGURE 7.4: Solutions $N(t)$ and $P(t)$ for Michaelis-Menten enzyme kinetics model (7.19) when $N(0) = p_\infty$ and $P(0) = 0$.

7.5.2 Stochastic Model

How do the dynamics for the stochastic model differ from the deterministic model? In the stochastic model, there is finite-time extinction; the process stops, $(N(T), B(T)) = (0,0)$ for some random time T, the extinction time. The state (0,0) is absorbing. But finite time extinction does not occur in the deterministic model. For any time $t > 0$, $N(t) > 0$ and $B(t) > 0$ (although the values of $N(t)$ and $B(t)$ get very small.) We demonstrate this extinction process in the CTMC model for Michaelis-Menten enzyme kinetics, a bivariate process based on the reaction kinetics.

First, it is necessary to convert the units in the deterministic model to appropriate discrete units. The units for N, E, B, and P in the deterministic model are moles per liter (M). The units for the stochastic variables must be converted to number of molecules. If we let $[N]$ and $[B]$ be the concentration of N and B in moles per liter in equations (7.15) and (7.16), then to convert to number of molecules, it is necessary to multiply $[N]$ and $[B]$ by the volume V of a cell and the number of molecules in a mole, $n_A = 6.022 \times 10^{23}$, Avogadro's constant (Wilkinson, 2006).

The dynamics take place within the cell. Thus, the cell size is the volume V. For a bacterial cell, such as *Escherichia coli*, which has a cylindrical shape, its' volume is computed as $V = \pi r^2 l$, where r = radius and l = length (Wilkinson, 2006). The radius and length for a typical *E.coli* cell are $r = 0.5$ micrometers $= 0.5 \times 10^{-6}$ m and $l = 2$ micrometers $= 2 \times 10^{-6}$ m (one micrometer is one millionth of a meter, denoted as μm) (Wilkinson, 2006). Thus, the volume of an *E. coli* cell, expressed in terms of liters, is

$$V = (0.5\pi)10^{-18}\text{m}^3 = (0.5\pi)10^{-15}\text{L}.$$

The reaction rates k_1, k_{-1}, and k_2 in (7.15) and (7.16) are not necessarily the same as the reaction constants in the CTMC model. By changing the units in the differential equation, it will be straightforward to obtain the correct reaction constants for the CTMC model. Time is measured in seconds, the units on the right side of equations (7.15) and (7.16) are moles per liter per second (M/s). To convert the first-order reactions to molecules per second, multiply the terms $k_{-1}[B]$ and $k_2[B]$ in the differential equations (7.15) and (7.16) by $n_A V$. The units of k_{-1} and k_2 are per second (1/s), so they will be the same as in the stochastic model. But the terms with k_1 are second-order reactions and the units of k_1 are $M^{-1}s^{-1}$. Multiplying $k_1[N][e_0]$ or $k_1[N][B]$ by $(n_A V)^2$ changes the units of $[N][B]$ from M^2 to $(mol)^2$, mol = molecules. This means that k_1 must be changed to $mol^{-1}s^{-1}$ in the stochastic model. This yields a new reaction constant in the stochastic model to replace k_1:

$$\tilde{k}_1 = \frac{k_1}{n_A V}.$$

In the stochastic model, the same notation for the variables N and B are used as in the deterministic model. But note that the units are number of molecules (and not moles). The transition probabilities are

$$\text{Prob}\{\Delta N(t) = i, \Delta B(t) = j | (n, b)\}$$

$$= \begin{cases} \tilde{k}_1 n(\tilde{e}_\infty - b)\,\Delta t + o(\Delta t), & (i, j) = (-1, 1) \\ k_{-1} b\,\Delta t + o(\Delta t), & (i, j) = (1, -1) \\ k_2 b\,\Delta t + o(\Delta t), & (i, j) = (0, -1) \\ 1 - \left[\tilde{k}_1 n(\tilde{e}_\infty - n) + (k_{-1} + k_2)n\right]\Delta t \\ \quad + o(\Delta t), & (i, j) = (0, 0) \\ o(\Delta t), & \text{otherwise.} \end{cases}$$

Let $p_{(n,b)}(t) = \text{Prob}\{N(t) = n, B(t) = b\}$ be the joint probability distribution for N and P. The forward Kolmogorov differential equations corresponding to the bivariate process, given $(N(0), B(0)) = (p_\infty, 0)$, are

$$\frac{dp_{(n,b)}(t)}{dt} = \tilde{k}_1(n+1)(\tilde{e}_\infty - b - 1)p_{(n+1,b-1)}(t) + k_{-1}(b+1)p_{(n-1,b+1)}(t)$$

$$+ k_2(b+1)p_{(n,b+1)}(t) - [\tilde{k}_1 n(\tilde{e}_\infty - b) + (k_{-1} + k_2)b]p_{(n,b)}(t),$$

where $n = 0, 1, 2, \ldots, \tilde{p}_\infty$, $b = 0, 1, \ldots, \tilde{e}_\infty$ and $\text{Prob}\{N(0) = p_\infty, B(0) = 0\} = p_{(p_\infty, 0)}(0) = 1$. In addition, equations for the backward Kolmogorov differential equations can be derived. For a bivariate process containing a large number of states, the Kolmogorov differential equations become quite complex. Alternatives to applying the Kolmogorov differential equations are computational methods.

To perform numerical simulations, the parameter and initial values in (7.17) and (7.18) must be converted to number of molecules. The new units are $\tilde{p}_\infty = n_A V p_\infty$, $\tilde{e}_\infty = n_A V e_\infty$, and $\tilde{k}_1 = k_1/(n_A V)$, where $V = 10^{-15}\text{L}$. The remaining parameters have the same values as in the deterministic model (but the units are different). The parameter values for the stochastic model are:

$$\tilde{p}_\infty = 301 \; mol, \quad \tilde{e}_\infty = 120 \; mol, \quad \tilde{k}_1 = 1.66 \times 10^{-3} mol^{-1} s^{-1},$$

$$k_{-1} = 1 \times 10^{-4} s^{-1}, \quad \text{and} \quad k_2 = 0.1 s^{-1},$$

where mol = molecules (Wilkinson, 2006). One stochastic realization is graphed in Figure 7.5. In addition, 10,000 sample paths are generated to compute the approximate probability density for the duration T until $N(T) + B(T) = 0$, i.e., $(N(T), B(T)) = (0, 0)$, an absorbing state. The expected duration is $\mu_T \approx 85.0s$ and the standard deviation is $\sigma_T \approx 12.86s$. The stochastic model provides information about the variability in the process that cannot be obtained from the deterministic model.

A CTMC model for the (N, P) system simplifies if the dynamics of B are ignored. The CTMC model for N is a death process and for P it is a birth process. That is, the probability of a change $(\Delta N(t), \Delta P(t)) = (-1, 1)$ given $(N(t), P(t)) = (n, p)$ is $\mu_n + o(\Delta t) = \tilde{k}_{max} n\Delta t/(\tilde{k}_m + n) + o(\Delta t)$, where $\tilde{k}_{max} = k_2 \tilde{e}_\infty$ and $\tilde{k}_m = (k_{-1} + k_2)/\tilde{k}_1$. Three sample paths of this process are

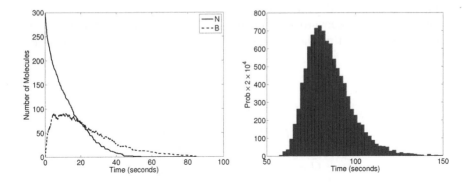

FIGURE 7.5: Sample path for the stochastic enzyme kinetics model, $N(t)$ and $B(t)$, and an approximate probability histogram for the time T until the number of molecules $N(T) + B(T) = 0$ given $N(0) = \tilde{p}_\infty$ and $B(0) = 0$.

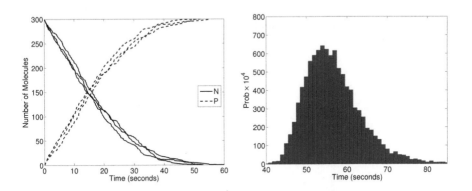

FIGURE 7.6: Three sample paths for the stochastic Michaelis-Menten enzyme kinetics model $N(t)$ and $P(t)$, and a probability histogram that approximates the probability density for the time T until $N(T) = 0$, given $N(0) = \tilde{p}_\infty$.

graphed in Figure 7.6 and the approximate probability distribution for the duration T of the process, that is, the time until $N(T) = 0$.

In the simpler CTMC model for (N, P), the expected duration can be calculated from the backward Kolmogorov differential equations for $N(t)$. (See Section 6.7.2.) Deleting the first row and first column of the generator matrix Q,

$$\tilde{Q} = \begin{pmatrix} -\mu_1 & \mu_2 & 0 & \cdots & 0 \\ 0 & -\mu_2 & 0 & \cdots & 0 \\ \vdots & \vdots & \vdots & \cdots & \vdots \\ 0 & 0 & 0 & \cdots & \mu_M \\ 0 & 0 & 0 & \cdots & -\mu_M \end{pmatrix},$$

where $\mu_n = \tilde{k}_{max} n / (\tilde{k}_m + n)$, $n = 1, 2, \ldots, M = \tilde{p}_\infty$. The expected time to extinction, $\tau^{tr} = (\tau_1, \ldots, \tau_M)$, is a solution of $\tau^{tr} \tilde{Q} = -1^{tr}$, where 1^{tr} is an row vector of size M. Applying the preceding method to the Michaelis-Menten kinetics model, where $N(0) = 301 = M$. The estimate for $\tau_{301} = 56.7s$. (See Figure 7.7 and Exercise 10.) The expected duration μ_T and standard deviation σ_T for $N(0) = 301$, calculated from 10,000 sample paths, gives an estimate of $\mu_T \approx 56.7s$ and standard deviation of $\sigma_T \approx 7.0s$, which agrees with τ_{301}.

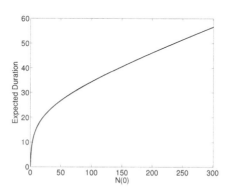

FIGURE 7.7: Expected duration for the stochastic enzyme kinetics model.

7.6   SIR Epidemic Process

In the SIR epidemic model, individuals recover and develop permanent immunity. The class R represents individuals that are permanently immune. Such types of models have been applied to childhood diseases such as measles, mumps, and chickenpox (e.g., Allen and Thrasher, 1998; Anderson and May, 1992; Hethcote, 2000 and references therein).

7.6.1   Deterministic Model

Ordinary differential equations describing the dynamics of an SIR epidemic are as follows:

$$\frac{dS}{dt} = -\frac{\beta}{N}SI$$
$$\frac{dI}{dt} = \frac{\beta}{N}SI - \gamma I$$
$$\frac{dR}{dt} = \gamma I,$$

where $S(0), I(0), R(0) \geq 0$, and $S(0) + I(0) + R(0) = N$. It can be shown that $\lim_{t \to \infty} I(t) = 0$. Also, $\lim_{t \to \infty} S(t) = S(\infty)$ and $\lim_{t \to \infty} R(t) = R(\infty)$ are finite but depend on initial conditions. The *effective reproduction number or replacement number* is defined as

$$\mathcal{R} = \frac{S(0)}{N}\frac{\beta}{\gamma}$$

(Anderson and May, 1992; Hethcote, 2000). Hethcote (2000) defines the replacement number, $\mathcal{R}$, as the average number of secondary infections produced by a typical infective during the entire period of infectiousness. Recall that $\mathcal{R}_0$ is the average number of secondary infections that occur when one infective is introduced into a completely susceptible population (Hethcote, 2000). Notice that $\mathcal{R} = \mathcal{R}_0 S(0)/N$. If $\mathcal{R} \leq 1$, then in the SIR model there is no epidemic; solutions $I(t)$ decrease monotonically to zero. But if $\mathcal{R} > 1$; then $I(t)$ increases first before decreasing to zero; an epidemic occurs.

Although the epidemic eventually ends, the severity of the epidemic can be determined by the total number of cases or the *final size* of the epidemic. If we assume $R(0) = 0$, then $R(\infty)$ represents the total number of cases or final size. Suppose $I(0) = 1$; then the number of cases equals the initial case plus all others infected during the course of the epidemic. The value of $R(\infty)$ can be obtained from the differential equations dI/dt and dS/dt,

$$\frac{dI}{dS} = \frac{(\beta/N)SI - \gamma I}{-(\beta/N)SI} = -1 + \frac{N\gamma}{\beta S}.$$

Separating variables, integrating, and applying the initial conditions, $S(0) = N - 1$ and $I(0) = 1$, the following solution is obtained:

$$I(t) + S(t) = \frac{N\gamma}{\beta} \ln S(t) + N - \frac{N\gamma}{\beta} \ln(N - 1).$$

Letting $t \to \infty$ and using the fact that $I(\infty) = 0$ yields

$$S(\infty) = \frac{N\gamma}{\beta} \ln \left(\frac{S(\infty)}{N - 1} \right) + N. \tag{7.20}$$

Equation (7.20) gives an implicit solution for $S(\infty)$, which can be used to find the value of $R(\infty) = N - S(\infty)$. The next example uses this formula to calculate the final size $R(\infty)$ for particular parameter values.

Example 7.5. Let $N = 100$, $\beta = 2$, and $\gamma = 1$. In addition, let $S(0) = 99$, $I(0) = 1$, and $R(0) = 0$. Then $S(\infty) = 50 \ln (S(\infty)/99) + 100$. The solution $S(\infty) \approx 19.98$ so that the final size is $R(\infty) \approx 80.02$. Table 7.3 gives the final size for the SIR epidemic model for various parameter values. ■

Table 7.3: Final size of an SIR epidemic, $R(\infty)$, when $\gamma = 1$, $S(0) = N - 1$, and $I(0) = 1$

	N		
β	20	100	1000
0.5	1.87	1.97	2.00
1	5.74	13.52	44.07
2	16.26	80.02	797.15
5	19.87	99.31	993.03
10	20.00	100.00	999.95

As the replacement number $\mathcal{R}$ increases, the number of cases increases until the entire population is infected. When $\mathcal{R} \leq 1$ ($\beta \leq 1$ in Table 7.3), there is no epidemic, so the total number of cases is relatively small. The data in Table 7.3 will be compared to the output of the stochastic SIR model.

7.6.2 Stochastic Model

Let $S(t)$, $I(t)$, and $R(t)$ denote random variables for the number of susceptible, infected, and immune individuals, respectively, where $S(t) + I(t) + R(t) = N$. There is no latent period so that infected individuals are also infectious.

Only two of the random variables are independent. Assume the transition probabilities are

$$\text{Prob}\{\Delta S(t) = i, \Delta I(t) = j | (S(t), I(t))\})$$

$$= \begin{cases} \dfrac{\beta}{N} S(t) I(t) \, \Delta t + o(\Delta t), & (i,j) = (-1,1) \\ \gamma I(t) \, \Delta t + o(\Delta t), & (i,j) = (0,-1) \\ 1 - \left[\dfrac{\beta}{N} S(t) I(t) + \gamma I(t) \right] \Delta t \\ \quad + o(\Delta t), & (i,j) = (0,0) \\ o(\Delta t), & \text{otherwise.} \end{cases}$$

When $\Delta I(t) = -1$, then $\Delta R(t) = 1$.

Assume the initial distribution is $(S(0), I(0)) = (s_0, i_0)$, where $s_0 + i_0 = N$, $s_0 \geq 0$ and $i_0 > 0$. Let $p_{(i,j)}(t) = \text{Prob}\{S(t) = i, I(t) = j\}$; then the state probabilities are solutions of the forward Kolmogorov equations:

$$\frac{dp_{(i,j)}(t)}{dt} = \frac{\beta}{N}(i+1)(j-1)p_{(i+1,j-1)}(t) + \gamma(j+1)p_{(i,j+1)}(t)$$

$$- \left[\frac{\beta}{N} ij + \gamma j \right] p_{(i,j)}(t),$$

where $i = 0, 1, 2, \ldots, N$, $j = 0, 1, 2, \ldots, N - i$, and $i + j \leq N$. If (i,j) lies outside of this range, the probabilities are assumed to be zero. For example, for $j = 0$,

$$\frac{dp_{(i,0)}}{dt} = \gamma p_{(i,1)}, \quad \text{and} \quad \frac{dp_{(N,0)}}{dt} = 0$$

for $i = 0, 1, 2, \ldots, N - 1$. The $N + 1$ states $(i,0)$ are closed, where $i = 0, 1, 2, \ldots, N$. No transitions occur out of any one of these states.

When $S(0) = N - j \approx N$ and the number of infected individuals $I(0) = j$ is small, the replacement number is approximately the basic reproduction number. That is, $\mathcal{R} = \frac{S(0)}{N} \frac{\beta}{\gamma} \approx \beta/\gamma = \mathcal{R}_0$. The SIR model can be related to the simple birth and death process. Death of an infected individual corresponds to recovery, $\mu = \gamma$, and birth of an infected individual corresponds to a new infection, $\lambda \approx \beta$. At the beginning of an epidemic, when $I(0) \approx j$ is small, the probability that the epidemic ends quickly or that there is no epidemic can be approximated by a simple birth and death process,

$$\text{probability no epidemic} = \begin{cases} 1, & \mathcal{R}_0 \leq 1 \\ \left(\dfrac{1}{\mathcal{R}_0} \right)^j, & \mathcal{R}_0 > 1. \end{cases}$$

[See also Daley and Gani (1999) and Whittle (1955).] If, for example, $N = 100$, $\mathcal{R}_0 = 2$, and $I(0) = 2$, an epidemic occurs with probability $3/4 = 1 - (1/\mathcal{R}_0)^2$ and no epidemic with probability $1/4 = (1/\mathcal{R}_0)^2$.

Even though the process is bivariate, an expression for the generator matrix Q and transition matrix T corresponding to the embedded Markov chain can be obtained. The form of matrices Q and T depends on how the states are ordered. There are $(N + 1)(N + 2)/2$ pairs of states in the SIR epidemic process. Suppose the pairs of of states are ordered as follows:

$$(N, 0), (N - 1, 0), \ldots, (0, 0), (N - 1, 1), (N - 2, 1), \ldots, (0, 1),$$
$$(N - 2, 2), (N - 3, 2), \ldots, (0, 2), \ldots, (0, N). \qquad (7.21)$$

Then $p(t) = (p_{(N,0)}, \ldots, p_{(0,N)})^{tr}$ and the generator matrix Q of $dp/dt = Qp$ depend on this particular ordering of the states. Matrix Q is a $(N + 1)(N + 2)/2 \times (N + 1)(N + 2)/2$ matrix with the elements in the first $N + 1$ columns zero because there are no transitions out of any states with zero infectives; they are absorbing states. This, in turn, means that the transition matrix T corresponding to the embedded Markov chain, $T = (t_{kl})$, satisfies $t_{ll} = 1$ for $l = 1, 2, \ldots, N + 1$, the first $N + 1$ states corresponding to $(i, 0)$, $i = 0, 1, 2, \ldots, N + 1$.

7.6.3 Final Size

The transition matrix T of the embedded Markov chain is useful in calculating the distribution for the final size of the epidemic. Explicit formulas are derived for the elements of the transition matrix T. In the embedded Markov chain, there are transitions from state (i, j) to either state $(i + 1, j - 1)$ representing a susceptible individual that becomes infected or to $(i, j - 1)$ representing a recovery of an infected individual. The probability of recovery is

$$p_i = \frac{\gamma j}{\gamma j + (\beta/N)ij} = \frac{\gamma}{\gamma + (\beta/N)i}. \qquad (7.22)$$

The probability that a susceptible individual becomes infected is

$$1 - p_i = \frac{(\beta/N)ij}{\gamma j + (\beta/N)ij} = \frac{(\beta/N)i}{\gamma + (\beta/N)i}.$$

In addition, it can be seen that the embedded Markov chain satisfies

$$p_{(i,j)} = \begin{cases} p_i p_{(i,j+1)}, & j = 0, 1 \\ p_i p_{(i,j+1)} + (1 - p_{i+1}) p_{(i+1,j-1)}, & j = 2, \ldots, N, \end{cases}$$

with the restriction that $0 \leq i + j \leq N$; otherwise the probabilities are zero (Daley and Gani, 1999).

Example 7.6. A stochastic SIR epidemic model with population size $N = 4$ has 15 pairs of states. The states and the directed digraph corresponding to the embedded Markov chain are graphed in Figure 7.8. Grouping the states into five sets according to (7.21) leads to

$$\text{I: } (4,0),\ (3,0),\ (2,0),\ (1,0),\ (0,0)$$
$$\text{II: } (3,1),\ (2,1),\ (1,1),\ (0,1)$$
$$\text{III: } (2,2),\ (1,2),\ (0,2)$$
$$\text{IV: } (1,3),\ (0,3)$$
$$\text{V: } (0,4).$$

The transition matrix of the embedded chain has the following block form:

$$T = \begin{pmatrix} I & A_1 & 0 & 0 & 0 \\ 0 & 0 & A_2 & 0 & 0 \\ 0 & B_1 & 0 & A_3 & 0 \\ 0 & 0 & B_2 & 0 & A_4 \\ 0 & 0 & 0 & B_3 & 0 \end{pmatrix},$$

which corresponds to the way the states are grouped into the five sets. The state probability vector can also be divided into the five sets $p = (p_{(i,j)}) = (p_I, p_{II}, p_{III}, p_{IV}, p_V)^{tr}$, where, for example,

$$p_I = (p_{(4,0)}, p_{(3,0)}, p_{(2,0)}, p_{(1,0)}, p_{(0,0)})^{tr}.$$

Each of the block matrices in T has different dimensions and represents different transitions between these sets. Matrix I is a 5×5 identity matrix, which means this set is absorbing. Matrix A_j represents recovery, transitions from j infected individuals to $j - 1$ infected individuals, $j = 1, 2, 3, 4$, and matrix B_j represents infection, transitions from j infected individuals to $j + 1$ infected individuals. For example, matrices A_1 and B_1 have the following forms:

$$A_1 = \begin{pmatrix} 0 & 0 & 0 & 0 \\ p_3 & 0 & 0 & 0 \\ 0 & p_2 & 0 & 0 \\ 0 & 0 & p_1 & 0 \\ 0 & 0 & 0 & p_0 \end{pmatrix}$$

and

$$B_1 = \begin{pmatrix} 1 - p_3 & 0 & 0 & 0 \\ 0 & 1 - p_2 & 0 & 0 \\ 0 & 0 & 1 - p_1 & 0 \end{pmatrix}.$$

If the initial state is $(3, 1)$, then the maximal number of transitions until absorption is seven. If we follow the path of the thick arrows represented in the digraph, it is easy to see that there are seven transitions. In general, for any population of size N, beginning with one infective or $p_{(N-1,1)}(0) = 1$, the maximal number of transitions until absorption is $2N - 1$. ■

Suppose, initially, the number of infected individuals is 1 and there are no immune individuals, $S(0) = N - 1$, $I(0) = 1$, and $R(0) = 0$. The probabilities associated with the final size of the epidemic $\{p_j^f\}$, $j = 1, 2, \ldots, N$, can be determined by computing the absorption probabilities. Because the states

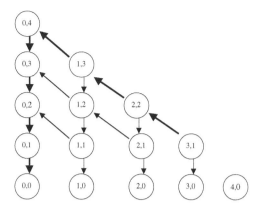

FIGURE 7.8: Directed graph of the embedded Markov chain of the SIR epidemic model when $N = 4$. The maximum path length beginning from state $(3, 1)$ is indicated by the thick arrows.

with no infected individuals are absorbing and $I(0) = 1$, $\displaystyle\lim_{t\to\infty} \sum_{i=0}^{N-1} p_{(i,0)}(t) = 1$. Thus, the probabilities for the final size distribution satisfy

$$\lim_{t\to\infty} p_{(i,0)}(t) = p^f_{N-i}$$

for $i = 0, 1, 2, \ldots, N-1$. If there are i susceptible individuals when the number of infected individuals has reached zero, the final size of the epidemic is $N-i$. Beginning with one infected individual, the maximal number of time steps until absorption is $2N - 1$. Hence, the absorption probabilities can be found using the transition matrix T of the embedded Markov chain (Daley and Gani, 1999). In particular,

$$\lim_{t\to\infty} p(t) = p(2N - 1) = T^{2N-1}p(0). \tag{7.23}$$

Example 7.7. Let $I(0) = 1$ and $S(0) = N - 1$. The distribution for the final size of the epidemic is computed for $N = 20$ and $N = 100$ for $\mathcal{R}_0 = 0.5$, 2, and 5, where $\gamma = 1$ using formula (7.23). A MATLAB® program for computing the final size is given in the Appendix for Chapter 7. The largest probabilities are confined to the tails of the distribution, either near 1 or N (see Figure 7.9). The largest part of the distribution is near 0 when $\mathcal{R}_0$ is less than 1 and near N when $\mathcal{R}_0$ is greater than 1 and sufficiently large. This distribution agrees with the conclusion derived from the preceding approximation; that is, when $\mathcal{R}_0 \leq 1$, there are no epidemics, so the final size of the epidemic should be small. When $\mathcal{R}_0 > 1$, the probability no epidemic occurs is approximately $1/\mathcal{R}_0$, so approximately $1 - 1/\mathcal{R}_0$ of the epidemics should be of large size. ∎

The expected final size distributions are given in Table 7.4. Notice that the expected values for the stochastic SIR epidemic model do not agree with the

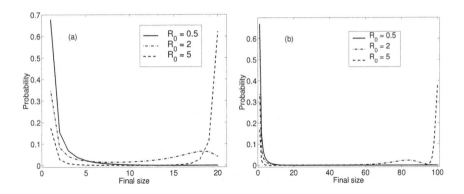

FIGURE 7.9: Probability distribution for the final size of a stochastic SIR epidemic model when $I(0) = 1$, $S(0) = N - 1$, $\gamma = 1$, and $\beta = 0.5$, 2, or 5 ($\mathcal{R}_0 = 0.5$, 2, or 5). In (a), $N = 20$ and in (b), $N = 100$.

final size calculated for the deterministic model in Example 7.5, Table 7.3, especially for $\mathcal{R}_0 > 1$. This difference is due to the fact that there is a positive probability of no epidemic $(1/\mathcal{R}_0)$ in the stochastic model, but, in the deterministic model, solutions always approach an endemic equilibrium.

Table 7.4: Expected final size of a stochastic SIR epidemic when $\gamma = 1$, $S(0) = N - 1$, and $I(0) = 1$

	N	
β	20	100
0.5	1.76	1.93
1	3.34	6.10
2	8.12	38.34
5	15.66	79.28
10	17.98	89.98

7.6.4 Duration

Formally, the duration of an SIR epidemic can be calculated using the first passage time method, introduced in Section 6.7. First, a generator matrix Q is formed for the bivariate process (S, I). Each element in the generator matrix

Q of a bivariate process is a transition rate between a pair of states $(s_1, i_1) \rightarrow (s_2, i_2)$. In addition, the vector of the expected time to extinction, τ, depends on the pairs (i, j). If the matrix $\tilde{Q}$ excludes the transitions to the absorbing states $(s, 0)$, then the solution τ is $\tau^{tr}\tilde{Q} = -\mathbf{1}^{tr}$. However, formation of the matrix Q depends on how the states are ordered. For example, using the notation from the previous section,

$$[\gamma j + (\beta/N)ij][p_i\tau_{(i,j-1)} - \tau_{(i,j)} + (1 - p_i)\tau_{(i-1,j+1)}] = -1.$$

Matrix $\tilde{Q}$ is of size $(N + 1)N/2 \times (N + 1)N/2$. For multivariate processes, it is often simpler to apply a computational method to approximate the distribution for the duration and the moments of this distribution for different parameter values and initial conditions.

Example 7.8. Let $N = 100$, $\beta = 2$, $\gamma = 1$, $S(0) = 99$, and $I(0) = 1$. Three sample paths are graphed in Figure 7.10. The MATLAB program that generated the three sample paths is included in the Appendix for Chapter 7. The durations of the three sample paths are 0.65, 8.49, and 12.49; two of the sample paths represent large epidemics. Approximately half of the sample paths will be large epidemics because the probability of no epidemic is approximately $(1/\mathcal{R}_0) = 1/2$. In Figure 7.11, the distribution for the duration of the epidemic is estimated from 1000 sample paths for two cases $I(0) = 1$ and $S(0) = 99$, and $I(0) = 5$ and $S(0) = 95$. Finally, in Figure 7.12, the expected duration and the corresponding standard deviation are graphed as a function of the initial number of infectives, $I(0) = i$ and $S(0) = 100 - i$, $i = 1, 2, \ldots, 100$. ∎

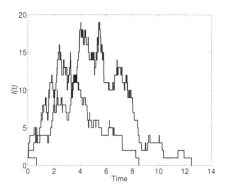

FIGURE 7.10: Three sample paths of a stochastic SIR epidemic model with $N = 100$, $\beta = 2$, $\gamma = 1$, $S(0) = 99$, and $I(0) = 1$; $\mathcal{R}_0 = 2$.

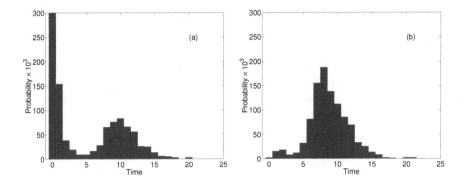

FIGURE 7.11: Probability distribution for the duration of an SIR epidemic, $N = 100$, $\beta = 2$, and $\gamma = 1$ (estimated from 1000 sample paths). In (a), $I(0) = 1$ and $S(0) = 99$ and in (b), $I(0) = 5$ and $S(0) = 95$.

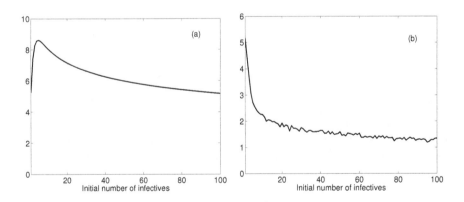

FIGURE 7.12: Mean and standard deviation, graphs (a) and (b), respectively, of the distribution for the duration of an SIR epidemic as a function of the initial number of infected individuals, $I(0) = i$ and $S(0) = N - i$, $i = 1, 2, \ldots, 100$, $N = 100$, when $\beta = 2$, and $\gamma = 1$ (estimated from 1000 sample paths).

The expected duration increases to a maximum, then decreases as the initial number of infected individuals increases. The standard deviation is a decreasing function of the initial number of infectives. This behavior is not unusual given the bimodal distribution of the duration and the fact that for a large number of infected individuals, the number of cases will not increase very much before the epidemic ends. But note that the mean duration of the SIR epidemic in Figure 7.12 is much less than that of the SIS epidemic. The SIR epidemic always ends quickly as compared to the SIS epidemic, where a quasistationary distribution may be established. See Example 7.3 and Figure 7.2.

7.7 Competition Process

7.7.1 Deterministic Model

The Lotka-Volterra competition model is one of the best known competition models. Interspecies competition results in a decrease in the population size or density of each species when its competitor is present. Let $x_i(t)$ be the population size of species i at time t. Then the deterministic model has the following form:

$$\frac{dx_1}{dt} = x_1(a_{10} - a_{11}x_1 - a_{12}x_2) \tag{7.24}$$

$$\frac{dx_2}{dt} = x_2(a_{20} - a_{21}x_1 - a_{22}x_2), \tag{7.25}$$

where $x_i(0) > 0$, $a_{ij} > 0$ for $i = 1, 2$ and $j = 0, 1, 2$. Solutions $(x_1(t), x_2(t))$ are nonnegative for $t \in [0, \infty)$. The parameters a_{i0} are the intrinsic growth rates for species i, a_{ii} are the intraspecific competition coefficients, and a_{ij}, $i \neq j$ are the interspecific competition coefficients, $i, j = 1, 2$. If the interspecific competition coefficients are zero, $a_{ij} = 0$, $i \neq j$, then the two equations (7.24) and (7.25) reduce to equations for logistic growth and each species grows to its carrying capacity,

$$\lim_{t \to \infty} x_i(t) = \frac{a_{i0}}{a_{ii}}, \quad i = 1, 2.$$

Competition between the species, $a_{ij} > 0$, $i \neq j$, changes the dynamics of one or both species.

The nullclines are the curves where $dx_i/dt = 0$, that is, where the rate of change in each species is zero. For equations (7.24) and (7.25), the nullclines are straight lines, $x_i = 0$, $a_{i1}x_1 + a_{i2}x_2 = a_{i0}$, $i = 1, 2$. The asymptotic dynamics of (7.24) and (7.25) depend on how the nullclines intersect. There are four cases.

I. If $\dfrac{a_{20}}{a_{22}} \leq \dfrac{a_{10}}{a_{12}}$ and $\dfrac{a_{20}}{a_{21}} \leq \dfrac{a_{10}}{a_{11}}$, then $\lim\limits_{t \to \infty} (x_1(t), x_2(t)) = (0, a_{10}/a_{11})$.

II. If $\dfrac{a_{20}}{a_{22}} \geq \dfrac{a_{10}}{a_{12}}$ and $\dfrac{a_{20}}{a_{21}} \geq \dfrac{a_{10}}{a_{11}}$, then $\lim\limits_{t \to \infty} (x_1(t), x_2(t)) = (a_{20}/a_{22}, 0)$.

III. If $\dfrac{a_{20}}{a_{22}} > \dfrac{a_{10}}{a_{12}}$ and $\dfrac{a_{20}}{a_{21}} < \dfrac{a_{10}}{a_{11}}$, then $\lim\limits_{t \to \infty} (x_1(t), x_2(t)) = (0, a_{10}/a_{11})$ or $\lim\limits_{t \to \infty} (x_1(t), x_2(t)) = (a_{20}/a_{22}, 0)$.

IV. If $\dfrac{a_{20}}{a_{22}} < \dfrac{a_{10}}{a_{12}}$ and $\dfrac{a_{20}}{a_{21}} > \dfrac{a_{10}}{a_{11}}$, then $\lim\limits_{t \to \infty} (x_1(t), x_2(t)) = (x_1^*, x_2^*)$.

The variables x_1^* and x_2^* represent the positive intersection points of the null-clines:

$$a_{10} = a_{11}x_1 + a_{12}x_2$$
$$a_{20} = a_{21}x_1 + a_{22}x_2.$$

At least one of the inequalities in cases I and II must be a strict inequality; otherwise there exists an infinite number of equilibria and the asymptotic behavior depends on initial conditions. Generally, survival of both species (case IV) requires that the interspecific competition coefficients, a_{ij}, $i \neq j$, be less than the intraspecific competition coefficients, a_{ii}.

7.7.2 Stochastic Model

Let $X_1(t)$ and $X_2(t)$ be random variables for the population size of two competing species at time t, where $X_1, X_2 \in \{0, 1, 2, \ldots\}$ and $t \in [0, \infty)$. Let $p_{(i,j)}(t) = \text{Prob}\{X_1(t) = i, X_2(t) = j\}$. The competition model is a birth and death process for two species in which births and deaths depend on the population sizes of one or both of the species. As in the case of logistic growth, there is a multitude of stochastic models corresponding to the one deterministic model.

Suppose for two competing species, the stochastic birth rates are denoted $\lambda_i(X_1, X_2)$ and death rates $\mu_i(X_1, X_2)$. A general stochastic Lotka-Volterra competition model birth and death rates equal to

$$\lambda_i(X_1, X_2) = \max\{0, X_i(b_{i0} + b_{i1}X_1 + b_{i2}X_2)\}$$

and

$$\mu_i(X_1, X_2) = \max\{0, X_i(d_{i0} + d_{i1}X_1 + d_{i2}X_2)\},$$

where

$$b_{i0} - d_{i0} = a_{i0}, \quad b_{i1} - d_{i1} = -a_{i1}, \quad \text{and} \quad b_{i2} - d_{i2} = -a_{i2},$$

for $i = 1, 2$. The max in the definitions of λ_i and μ_i is to ensure that the expressions are nonnegative, and the assumptions on the coefficients are to ensure that the deterministic model is of the form (7.24) and (7.25), $dx_i/dt = \lambda_i(x_1, x_2) - \mu_i(x_1, x_2)$, $i = 1, 2$. For example, consider birth and death rates given by

$$\lambda_i(X_1, X_2) = a_{i0}X_i \quad \text{and} \quad \mu_i(X_1, X_2) = X_i(a_{i1}X_1 + a_{i2}X_2).$$ (7.26)

Another example is discussed in the exercises. It follows that the transition probabilities are

$$\text{Prob}\{\Delta X_1(t) = i, \Delta X_2(t) = j | (X_1(t), X_2(t))\}$$
$$= \begin{cases} a_{10}X_1(t)\Delta t + o(\Delta t), & (i,j) = (1,0) \\ a_{20}X_2(t)\Delta t + o(\Delta t), & (i,j) = (0,1) \\ X_1(t)[a_{11}X_1(t) + a_{12}X_2(t)]\Delta t + o(\Delta t), & (i,j) = (-1,0) \\ X_2(t)[a_{21}X_1(t) + a_{22}X_2(t)]\Delta t + o(\Delta t), & (i,j) = (0,-1) \\ 1 - X_1(t)[a_{11}X_1(t) + a_{12}X_2(t)]\Delta t & \\ \quad - X_2(t)[a_{21}X_1(t) + a_{22}X_2(t)]\Delta t + o(\Delta t), & (i,j) = (0,0) \\ o(\Delta t), & \text{otherwise.} \end{cases}$$

The forward Kolmogorov equations can be derived from the infinitesimal transition probabilities. Assuming $X_i(0) = x_i(0)$, $i = 1, 2$,

$$\frac{dp_{(i,j)}}{dt} = \lambda_1(i-1,j)p_{(i-1,j)} + \lambda_2(i,j-1)p_{(i,j-1)}$$
$$+ \mu_1(i+1,j)p_{(i+1,j)} + \mu_2(i,j+1)p_{(i,j+1)}$$
$$- [\lambda_1(i,j) + \lambda_2(i,j) + \mu_1(i,j) + \mu_2(i,j)]\, p_{(i,j)}.$$

Applying the generating function technique, the partial differential equation for the m.g.f. can be obtained,

$$\frac{\partial M}{\partial t} = a_{10}(e^\theta - 1)\frac{\partial M}{\partial \theta} + a_{20}(e^\phi - 1)\frac{\partial M}{\partial \phi}$$
$$+ (e^{-\theta} - 1)\left[a_{11}\frac{\partial^2 M}{\partial \theta^2} + a_{12}\frac{\partial^2 M}{\partial \theta\, \partial \phi}\right]$$
$$+ (e^{-\phi} - 1)\left[a_{21}\frac{\partial^2 M}{\partial \theta\, \partial \phi} + a_{22}\frac{\partial^2 M}{\partial \phi^2}\right],$$ (7.27)

where $M(\theta, \phi, 0) = e^{(x_1\theta + x_2\phi)}$, $X_1(0) = x_1 \equiv x_1(0)$, and $X_2(0) = x_2 \equiv x_2(0)$. These equations can be used to derive differential equations for the the first- and higher-order moments of X_i, $i = 1, 2$. For simplicity, the notation of Bailey (1990) is used for the higher-order moments of the distribution,

$$m_{kl}(t) = E(X_1^k(t)X_2^l(t)).$$

Equations for the means are

$$\frac{dm_{10}(t)}{dt} = a_{10}m_{10}(t) - a_{11}m_{20}(t) - a_{12}m_{11}(t)$$ (7.28)

$$\frac{dm_{01}(t)}{dt} = a_{20}m_{01}(t) - a_{21}m_{02}(t) - a_{22}m_{11}(t).$$ (7.29)

These two differential equations depend on five variables, m_{ij}, $i = 1, 2, j = 0$ and m_{11}, and therefore, they do not form a "closed system". Note that the form of these equations is "similar" to the deterministic differential equations. In particular, if $E(X_i X_j) = E(X_i)E(X_j)$, then these differential equations are the same as the Lotka-Volterra competition equations. Of course, the random variables X_1 and X_2 are not independent and this assumption is not valid. Frequently, specific distributional assumptions are made, so that the expressions $E(X_1^k(t)X_2^l(t))$ can be simplified. For example, the joint distribution can be assumed to be approximately multivariate normal or multivariate lognormal, referred to as *moment closure assumptions*. (See, e.g., Chan and Isham, 1998; Keeling, 2000; Krishnarajah et al., 2005; Lloyd, 2004; Nåsell, 2003a, 2003b; Singh and Hespanha, 2007.)

Example 7.9. Let $a_{10} = 2$, $a_{20} = 1.5$, $a_{11} = 0.03$, $a_{12} = 0.02$, $a_{21} = 0.01$, and $a_{22} = 0.04$. Case IV holds; a positive equilibrium exists and is stable $(x_1^*, x_2^*) = (50, 25)$. A sample path is graphed in Figure 7.13 when the initial sizes are the equilibrium values, $X_1(0) = 50$ and $X_2(0) = 25$. At $t = 5$, the means and variances for each of the populations are calculated from 10,000 sample paths,

$$\mu_{X_1}(5) = m_{10}(5) \approx 49.5, \quad \mu_{X_2}(5) = m_{01}(5) \approx 23.4,$$

$$\sigma_{X_1}(5) \approx 9.3, \quad \text{and} \quad \sigma_{X_2}(5) = 6.8.$$

The expected values are close to their equilibrium values. The MATLAB programs that generated the graphs in Figure 7.13 are included in the Appendix for Chapter 7. ∎

If the population sizes are bounded, the expected duration until total population extinction or absorption can be found by solving a linear system. Suppose the random variables for two competing populations satisfy $X_1 \in \{0, 1, \ldots, N_1\}$ and $X_2 \in \{0, 1, \ldots, N_2\}$. Beginning from state (i, j), denote the expected duration as $\tau_{(i,j)}$. Then $\tau = (\tau_{(i,j)})$ is the unique nonnegative solution of the linear system $\tau^{tr}\tilde{Q} = -1^{tr}$ where matrix $\tilde{Q}$ is the truncated transition matrix of the process (delete row and column corresponding to the absorbing state $(X_1, X_2) = (0, 0)$) and 1 is a column vector of ones. See Section 6.7.2.

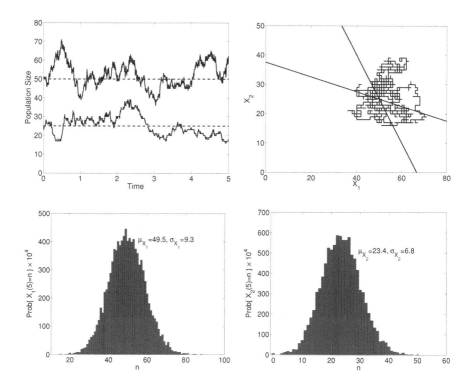

FIGURE 7.13: Sample paths of the CTMC competition model graphed as a function of time and in the X_1-X_2 phase plane (top two graphs). The dashed lines are the equilibrium values and the solid lines are the nullclines of the deterministic model. Approximate probability histograms for X_1 and X_2 at $t = 5$, based on 10,000 stochastic realizations (bottom two graphs). Birth and death rates are given in (7.26) with parameter values and initial conditions $a_{10} = 2$, $a_{20} = 1.5$, $a_{11} = 0.03$, $a_{12} = 0.02$, $a_{21} = 0.01$, $a_{22} = 0.04$, $X_1(0) = 50$, and $X_2(0) = 25$.

7.8 Predator-Prey Process

7.8.1 Deterministic Model

Let $x(t)$ and $y(t)$ denote the population sizes for the prey and predator at time t, respectively. The Lotka-Volterra predator-prey model has the form

$$\frac{dx}{dt} = x(a_{10} - a_{12}y) \tag{7.30}$$

$$\frac{dy}{dt} = y(a_{21}x - a_{20}), \tag{7.31}$$

where the parameters $a_{ij} > 0$, $x(0) > 0$, and $y(0) > 0$. The nullclines are lines and intersect at two equilibria, the origin and the equilibrium $(a_{20}/a_{21}, a_{10}/a_{12})$. The origin is unstable and the positive equilibrium is neutrally stable. For a given initial condition $(x(0), y(0))$, there exists a unique periodic solution $(x(t), y(t))$ in the x-y phase plane encircling the positive equilibrium. As the prey population size increases, the predator population follows suit until the predators consume too many prey and the prey population starts declining, then the predator population also starts to decline. However, each initial population size leads to a different cycle, an unrealistic behavior. Thus, the positive equilibrium is stable, but not asymptotically stable; it is "neutrally stable". Adding a term to the prey differential equation, accounting for intraspecific competition, leads either to damped oscillations in the predator-prey system or to predator extinction:

$$\frac{dx}{dt} = x(a_{10} - a_{11}x - a_{12}y),$$

The predator-prey dynamics with intraspecific competition in the prey population are summarized below.

I. If $\dfrac{a_{20}}{a_{21}} \geq \dfrac{a_{10}}{a_{11}}$, then $\lim_{t\to\infty} (x(t), y(t)) = (a_{10}/a_{11}, 0)$.

II. If $\dfrac{a_{20}}{a_{21}} < \dfrac{a_{10}}{a_{11}}$, then $\lim_{t\to\infty} (x(t), y(t)) = (a_{20}/a_{21}, [a_{10} - a_{11}a_{20}/a_{21}]/a_{12})$.

Numerous other predator-prey models have been studied in the literature (e.g., Edelstein-Keshet, 1988; Hassell, 1978; Kuang and Beretta, 1998; Murray, 1993).

7.8.2 Stochastic Model

Let $X(t)$ and $Y(t)$ denote random variables for the population sizes of the prey and predator at time t, respectively. Let the birth and death rates of the prey be $\lambda_1(X, Y) = a_{10}X$ and $\mu_1(X, Y) = a_{12}XY$ and for the predator be $\lambda_2(X, Y) = a_{21}XY$ and $\mu_2(X, Y) = a_{20}Y$ (no intraspecific competition in the prey). Then the transition probabilities are

$$
\begin{aligned}
&\text{Prob}\{\Delta X(t) = i, \Delta Y(t) = j | (X(t), Y(t))\} \\
&= \begin{cases}
a_{10}X(t)\Delta t + o(\Delta t), & (i, j) = (1, 0) \\
a_{21}X(t)Y(t)\Delta t + o(\Delta t), & (i, j) = (0, 1) \\
a_{12}X(t)Y(t)\Delta t + o(\Delta t), & (i, j) = (-1, 0) \\
a_{20}Y(t)\Delta t + o(\Delta t), & (i, j) = (0, -1) \\
1 - X(t)[a_{10} + a_{12}Y(t)]\Delta t & \\
\quad - Y(t)[a_{20} + a_{21}X(t)]\Delta t + o(\Delta t), & (i, j) = (0, 0) \\
o(\Delta t), & \text{otherwise.}
\end{cases}
\end{aligned}
$$

It is straightforward to write the forward Kolmogorov differential equations. Let $p_{(i,j)}(t) = \text{Prob}\{X(t) = i, Y(t) = j\}$. Then

$$
\begin{aligned}
\frac{dp_{(i,j)}}{dt} &= a_{10}(i-1)p_{(i-1,j)} + a_{21}i(j-1)p_{(i,j-1)} \\
&\quad + a_{12}(i+1)jp_{(i+1,j)} + a_{20}(j+1)p_{(i,j+1)} \\
&\quad - [a_{10}i + a_{21}ij + a_{12}ij + a_{20}j]p_{(i,j)}.
\end{aligned}
$$

From these equations, differential equations for higher-order moments or the moment generating function can be obtained. The next example illustrates the dynamics for the stochastic model.

Example 7.10. Let $a_{10} = 1$, $a_{20} = 1$, $a_{12} = 0.02$, $a_{21} = 0.01$ in the simple predator-prey model. For the initial conditions, $X(0) = 120$ and $Y(0) = 40$, graphs of the deterministic and stochastic models are compared in Figure 7.14. It can be seen that the stochastic model appears to jump from one neutrally stable cycle to another. ∎

CTMC models for predator and prey and competing species can be formulated in terms of queueing networks. Each species is represented by a node in the network. Arrivals and departures at each node can be births, deaths, and migration. A queueing network model for one predator and two prey is described by Chao, Miyazawa, and Pinedo (1999).

Numerous population, epidemic, cellular or genetics processes have been formulated and applied to problems biology. Population models with multiple competitors, predators or prey, with mutualistic interactions, population or epidemic models with spatial spread, and epidemic models with classes for latent individuals or individuals with maternal antibody protection provide additional realism. Applications of CTMC models to cellular kinetics and genetics processes are becoming increasingly important to accurately capture the complex dynamics. Please consult some of the references for the many variations on deterministic and stochastic models of competition, predation, mutualism, epidemics, cellular kinetics, and genetics processes (e.g., Allen, 2007, 2008; Alon, 2007; Brauer and Castillo-Chávez, 2001; Durrett, 1995,

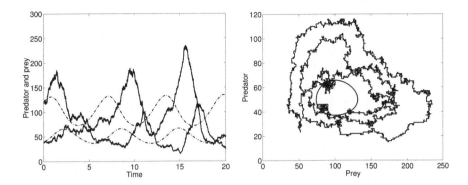

FIGURE 7.14: Sample path of the stochastic Lotka-Volterra predator-prey model with the solution of the deterministic model. Solutions are graphed over time and in the phase plane. The parameter values and initial conditions are $a_{10} = 1$, $a_{20} = 1$, $a_{12} = 0.02$, $a_{21} = 0.01$, $X(0) = 120$, and $Y(0) = 40$. Solutions with the smaller amplitude are the predator population.

1999; Gabriel et al., 1990; Goel and Richter-Dyn, 1974; Kot, 2001; Matis and Kiffe, 2000; Murray, 1993; Nisbet and Gurney, 1982; Renshaw, 1993; Szallasi et al., 2006; Wilkinson, 2006). Several examples are discussed in the exercises.

7.9 Exercises for Chapter 7

1. Consider the simple birth and death process discussed in Example 7.1.

   (a) Show that the solutions (7.5) satisfy the differential equation

   $$\frac{d\mathcal{P}}{dt} = \mu - (\lambda + \mu)\mathcal{P} + \lambda \mathcal{P}^2.$$

   (b) Show that the probability of extinction satisfies

   $$\lim_{t \to \infty} p_0(t) = \frac{\mu}{\lambda}.$$

2. An alternate method to find the limit of $p_0(t)$ in the simple birth and death process in Exercise 1 is to find the steady-state solutions of $\mathcal{P}(z, t)$. That is, find constant solutions, where $d\mathcal{P}/dt = 0$. For the simple birth and death process, show that there are two steady-state solutions, $\bar{\mathcal{P}} = 1$ and $\bar{\mathcal{P}} = \mu/\lambda$. Then show if $\mu < \lambda$ that solutions $\mathcal{P}(z, t)$ approach μ/λ for $z \neq 1$ (when $z = 1$, $\mathcal{P}(1, t) = 1$).

3. Consider the simple birth process in Chapter 6, Section 6.4.3.

   (a) Explain why the p.g.f. for the "offspring" or the "next generation" of each individual has the form $f(z) = z^2$ and why the probability of extinction is zero.

   (b) Write the ordinary differential equation satisfied by the p.g.f. of the total population size $\{X(t),\ t \geq 0\}$, that is, $\mathcal{P}(z,t)$ given $\mathcal{P}(z,0) = z$. (See Example 7.1.)

4. The simple birth and death process assumes at most one offspring is produced. Consider a model for a lytic viral population that reproduces by killing its host cell. That is, after successful attachment and entry into a host cell, the virus uses the host cell for its own reproduction, killing the cell to release the new virus particles or virions. The number of virions produced per host cell is referred to as the *burst size*. We describe a burst-death process studied by Hubbarde et al. (2007). Let μ be the death rate, λ be the birth rate, and β the burst size, $\mu > 0$, $\lambda > 0$, and $\beta \in \{1, 2, \ldots\}$. Then the p.g.f. for the next generation of virions is

$$f(z) = \frac{\mu}{\mu + \lambda} + \frac{\lambda}{\mu + \lambda} z^{\beta+1}.$$

   (a) For $\beta = 1$, this model is the same as the simple birth and death process and the probability of extinction is μ/λ. Show for $\beta > 1$ that the probability of extinction for the burst-death model is less than μ/λ.

   (b) Show that if $\beta \to \infty$ and μ and λ are fixed constants, then the probability of extinction approaches $\mu/(\mu + \lambda)$. This means the probability of survival is $\lambda/(\mu + \lambda)$. Even if the burst size is extremely large, the probability of viral survival is less than one (provided the birth and death rates remain constant).

5. Consider the simple burst process in Exercise 4, where the next generation or offspring p.g.f. is $f(z) = z^{\beta+1}$ and β is a positive integer greater than one (Hubbarde, et al., 2007).

   (a) Write the ordinary differential equation for the p.g.f. $\mathcal{P}(z,t)$ of the total population size $X(t)$ at time t. (See Example 7.1.)

   (b) Show that the solution $\mathcal{P}(z,t)$ of the differential equation with initial condition $\mathcal{P}(z,0) = z$ and $\lim_{z \to 1^-} \mathcal{P}(z,t) = 1$ is

$$\mathcal{P}(z,t) = \left(1 + e^{\beta \lambda t}(x^{-\beta} - 1)\right)^{-1/\beta}.$$

6. Consider the development of drug resistance in cancer cells discussed in Example 7.2.

(a) Show that the functions in (7.7) and (7.8) are solutions of the backward differential equations.

(b) Find the mean number of sensitive (type 1) and resistant (type 2) cells produced by type 1 cells; that is, $m_{11}(t) = \partial \mathcal{P}_1(1,1,t)/\partial z_1$ and $m_{21}(t) = \partial \mathcal{P}_1(1,1,t)/\partial z_2$ (Kimmel and Axelrod, 2002).

7. The cancer model with drug resistance in Example 7.2 can be extended to multi-drug resistance. Suppose there are four types of cells, $1, 2, 3, 4$. Cells of type 1 are sensitive to drug treatment. Cells of type 2 are resistant to drug A but sensitive to drug B and cells of type 3 are resistant to drug B but sensitive to drug A. Finally, cells of type 4 are resistant to both drugs. At cell division, a type 1 cell will produce two daughter cells with three possible outcomes: (1) one sensitive cell and one resistant to drug A with probability p_A; (2) one sensitive cell and one resistant to drug B with probability p_B; (2) two sensitive cells with probability $1 - p_A - p_B$. Division of a type 2 cell results in two possible outcomes: (1) one cell resistant to both drugs A and B and one resistant to only drug A with probability p_{AB}; (2) both cells resistant only to drug A with probability $1 - p_{AB}$. A similar outcome occurs for a type 2 cell. One daughter cell is resistant to both drugs and the other is resistant only to B with probability p_{BA} or both daughter cells are only resistant to drug B with probability $1 - p_{BA}$. The probabilities p_A, p_B, p_{AB}, and p_{BA} are the probabilities of a mutation. The multi-drug resistant cell type 4 only produces daughter cells of type 4; there are no mutations. The mutations are not reversible. The birth p.g.f.s f_i are

$$f_1(z_1, z_2, z_3, z_4) = (1 - p_A - p_B)z_1^2 + p_A z_1 z_2 + p_B z_1 z_3$$
$$f_2(z_1, z_2, z_3, z_4) = (1 - p_{AB})z_2^2 + p_{AB} z_2 z_4$$
$$f_3(z_1, z_2, z_3, z_4) = (1 - p_{BA})z_3^2 + p_{BA} z_3 z_4$$
$$f_4(z_1, z_2, z_3, z_4) = z_4^2.$$

(a) Assume an exponential distribution with parameter λ for the time for a cell to divide. Use the form of the differential equation (7.6) to write the differential equations for the p.g.f.s for each of the cell types, $\mathcal{P}_i(z_1, z_2, z_3, z_4, t)$, $i = 1, 2, 3, 4$.

(b) Show that if $p_B = 0 = p_{AB}$, then the differential equations for $\mathcal{P}_1$ and $\mathcal{P}_2$ reduce to those in Example 7.2.

(b) If $N_i(t)$ is the number of cells of type i at time t, then $\mathcal{P}_1(1,1,1,0,t) = \text{Prob}\{N_4(t) = 0\}$ is the probability that no cells of type 4 (resistant to both drugs) are present at time t, starting from one cell of type 1. For the special case that the mutation probabilities are equal, $p_A = p_B = p_{AB} = p_{BA} = p$, Kimmel and Axelrod (2002) derived the following formula for $\mathcal{P}_1(1,1,1,0,t)$:

$$\frac{e^{-\lambda t}[p + (1-p)e^{-\lambda t}]^{-2p/(1-p)}}{1 - [(1-2p)/(1-3p)]\{1 - [p + (1-p)e^{-\lambda t}]^{(1-3p)/(1-p)}\}}.$$

Graph this function for $\lambda = 1$ ($1/\lambda$ = average length of time for a cell to divide) and values of $p \in [10^{-5}, 10^{-3}]$. Discuss how soon multi-drug resistance develops.

8. This problem is based on a paper by Yakovlev et al. (1998). Brain cell development begins with precursor cells. A precursor cell divides and proliferates into daughter cells or transforms into another type of cell that does not divide and proliferate. Brain cell differentiation can be described simply by two types of cells in the central nervous system, the precursor cell, known as the progenitor cell (type 1 cell), which may be transformed into oligodendrocyte cells (type 2 cells). The progenitor cell is a stem cell, whereas an oligodendrocyte cell is a cell responsible for producing a fatty protein known as myelin, which insulates nerve cell axons. Myelinated axons are able to transmit nerve signals faster than unmyelinated ones. For example, in multiple sclerosis, a disease of the central nervous system characterized by neurological dysfunction, oligodendrocyte cells are often destroyed. In the model of Yakovlev et al. (1998), it is assumed that the process begins with a single type 1 cell. The type 1 cell divides and produces two daughter cells of the same type with probability p or transforms into a single type 2 cell with probability $1 - p$. A type 2 cell neither divides nor proliferates. Let $G(t)$ denote the cumulative lifetime distribution for a type 1 cell. Based on these assumptions, the birth p.g.f.s $f_1(z_1, z_2)$ and $f_2(z_1, z_2)$ can be expressed as follows:

$$f_1(z_1, z_2) = pz_1^2 + (1 - p)z_2 \quad \text{and} \quad f_2(z_1, z_2) = z_2.$$

The p.g.f.s for type i cells is an extension of the integral equation (7.9). That is,

$$P_i(z_1, z_2, t) = z_i(1 - G(t)) + \int_0^t f_i(P_1, P_2, t - u)dG(u).$$

(a) Use this latter expression and f_i for $i = 1, 2$ to write integral equations for P_i, $i = 1, 2$.

(b) Show that the solution P_2 simplifies to $P_2(z_1, z_2, t) = z_2$.

9. The following MATLAB program estimates the expected duration until the nutrient is depleted $N(T) = 0$ for the enzyme kinetics model. The expected duration μ_T given $N(0) = 301$ is computed based on 10,000 sample paths. An estimate for the standard deviation σ_T is also computed. Suppose the rate k_2 is changed to $k_2 = 0.2$ or to $k_2 = 0.05$. How do the dynamics change? Run the MATLAB program for these two cases. Graph several sample paths and estimates for μ_T and σ_T. Explain your results.

```
clear all
nsim=10000;
einf=120; pinf=301;
k1=1.66*10^(-3); k_1=10^(-4); k2=0.1;
kmax=(k2*einf); km=(k_1+k2)/k1;
nend=zeros(1,nsim);
for i=1:nsim
    j=1;
    t(1)=0.; n(1)=pinf; p(1)=0;
while n(j)>0
    ev=(kmax*n(j))/(km+n(j));
    rn1=rand;
    t(j+1)=t(j)-log(rn1)/ev;
    n(j+1)=n(j)-1;
    p(j+1)=p(j)+1;
    j=j+1;
end
nend(i)=t(j);
end
hist(nend,[30:1:150]);
xlabel('Time (seconds)');   ylabel('Prob \times 10^4');
mean(nend)
std(nend)
```

10. The expected duration for the simplified enzyme kinetics model, graphed in Figure 7.7, can be calculated by running the following MATLAB program.

    (a) Suppose the rate k_2 is changed to $k_2 = 0.2$ or to $k_2 = 0.05$. How do the dynamics change? Run the MATLAB program to compute the expected duration. Explain your results.

```
clear all
einf=120; pinf=301;
k1=1.66*10^(-3); k_1=10^(-4); k2=0.1;
kmax=(k2*einf); km=(k_1+k2)/k1;
d=-1*ones(1,pinf);
Q=zeros(pinf);
Q(1,1)=-kmax/(km+1);
for i=2:pinf
    Q(i,i)=-kmax*i/(km+i);
    Q(i-1,i)=-Q(i,i);
end
tau=d/Q
plot([1:1:pinf],tau,'k-','linewidth',2);
xlabel('N(0)'); ylabel('Expected Duration')
```

(b) Use the solution τ from part (a) and modify the MATLAB program to calculate the second moments for the time until extinction, $[\tau^2]^{tr}\tilde{Q} = -2\tau^{tr}$; see system (6.21). Then calculate the standard deviation. Show that for the original parameters with $N(0) = 301$ that $\sigma_T \approx 7.0s$. This can be accomplished by adding the following lines to the MATLAB program in the preceding exercise:

```
tau2=-2*tau/Q;
taustd=sqrt(tau2-tau.^2);
```

11. A model for intracellular viral kinetics follows the dynamics of three variables, T, G, and S, where T is the viral template, G is the viral genome, and S is structural proteins (Srivastava et al., 2002). Within a cell, the viral nucleic acid G can be (1) packaged within structural proteins S to form progeny viruses that leave the cell and infect other cells or can be (2) transcribed to form a template T. The template serves as a catalyst to promote the formation of structural proteins S and new genomic material G but is not consumed in the process. This model simplifies many of the complex intracellular pathways involved in viral production but still provides insight into viral synthesis and production. Also, it is assumed that the viral infection does not kill the cell, a nonlytic virus. Some of the relationships among the variables are described in the following diagram:

$$G \xrightarrow{k_1} T, \quad G + S \xrightarrow{k_4} [GS], \quad T \xrightarrow{k_2} \emptyset, \quad S \xrightarrow{k_6} \emptyset.$$

The rates of formation of G and S via the catalyst T are k_3 and k_5, respectively. The dynamics of the complex $[GS]$, which leaves the cell to infect other cells, is neglected. Hence, the underlying deterministic model is

$$\frac{dT}{dt} = k_1 G - k_2 T$$

$$\frac{dG}{dt} = k_3 T - k_1 G - k_4 GS$$

$$\frac{dS}{dt} = k_5 T - k_6 S - k_4 GS.$$

All rate constants k_i are in units of day^{-1}, except for k_4, which is molecules^{-1} day^{-1}. Because the units of T, G, and C are in number of molecules, the rate constants k_i in the deterministic model are the same as in the stochastic model. Let $k_1 = 0.025$, $k_2 = 0.25$, $k_3 = 1$, $k_4 = 7.5 \times 10^{-6}$, $k_5 = 1000$, and $k_6 = 2$ (Srivastava et al., 2002).

(a) Show that the positive equilibrium of the deterministic model is $(\bar{T}, \bar{G}, \bar{S}) \approx (20, 200, 10000)$. Calculate the Jacobian matrix at this equilibrium and show that the eigenvalues are negative meaning that the equilibrium is locally asymptotically stable.

(b) Formulate a CTMC model for this intracellular viral kinetics model.

(c) Write a computer program for the deterministic and stochastic models. Plot the deterministic solution when $T(0) = 1$, $G(0) = 0 = S(0)$ and when $T(0) = 5$, $G(0) = 5 = S(0)$. Then plot several sample paths of the stochastic model. Discuss how the two models differ.

12. An extension of the enzyme kinetics model is a model for bacterial growth in a chemostat. A chemostat is a laboratory device, where a constant supply of nutrient and a continuous harvest of the bacteria allows the volume in the chemostat to remain constant. An ODE model for nutrient n and bacteria concentration b is

$$\frac{db}{dt} = b\left(\frac{k_{max}n}{k_n + n} - D\right)$$

$$\frac{dn}{dt} = D(n_0 - n) - \beta\frac{k_{max}nb}{k_n + n}.$$

Consult Smith and Waltman (1995). Parameters k_n and k_{max} are Michaelis-Menten constants; k_n is the half-saturation constant and k_{max} is the maximum rate of bacterial growth. Parameter D is the dilution rate and n_0 is the constant rate of nutrient input. The dynamics of the ODE system depend on the two quantities: $m = k_{max}/D$ and BE$= (m - 1)/(k_n/n_0)$. If $m \leq 1$, then the high dilution rate $D \geq k_{max}$ causes the bacteria to be washed out, $b(t) \to 0$. Bacterial extinction also occurs if $m > 1$ and BE≤ 1. The value of BE$= 1$ is called the *break-even* value. If $m > 1$ and BE> 1, then there is sufficient amount of nutrient for the bacteria to survive.

First, the units of b and n need to be converted to number of molecules. In this example, the volume corresponds to the size of the chemostat device. The values of k_{max} and k_n change to $\tilde{k}_{max}$ and $\tilde{k}_n$, respectively, but the other parameter values remain unchanged because they are first-order processes. Table 7.5 gives the probabilities associated with the respective changes in a CTMC model $\{(b(t), n(t)) : t \in [0, \infty)\}$. The changes are integer-valued so that β is a positive integer.

Table 7.5: Probabilities associated with changes in the chemostat model

Change $(\Delta b, \Delta n)$	Transition Probabilities
$(1, -\beta)$	$\tilde{k}_{max}nb\Delta t/(\tilde{k}_n + n) + o(\Delta t)$
$(-1, 0)$	$bD\Delta t + o(\Delta t)$
$(0, -1)$	$nD\Delta t + o(\Delta t)$
$(0, 1)$	$n_0 D\Delta t + o(\Delta t)$

342

(a) Calculate the two equilibria for the ODE system and determine their local stability in terms of the parameters m and BE.

(b) Write the forward Kolmogorov differential equations for the joint p.m.f. $p_{(b,n)}(t)$.

13. Calculate the quasistationary probability distribution and the expected duration of an epidemic for an SIS epidemic model when $\beta = 1.5$, $b+\gamma = 1$, and $N = 100$. Sketch their graphs and compare them to Figure 7.2.

14. For the bivariate birth and death process in Example 7.4, show that the m.g.f. is a solution of (7.13). Then use the forward Kolmogorov differential equation (7.12) to show that the mean and variance of the process in Example 7.4 are solutions of (7.14).

15. Consider the transition matrix T corresponding to the embedded Markov chain of the SIR epidemic model in Example 7.6.

(a) Identify the remaining submatrices A_2, A_3, A_4, B_2, and B_3 of the transition matrix T.

(b) Let $N = 4$, $\gamma = 1$, and $\alpha = 2$. Then find $T^{2N-1} = T^7$ and show that the final size distribution when $I(0) = 1$ is

$$p_I = (0, 0.4, 0.15, 0.1556, 0.2944)^{tr}.$$

What is the final size distribution when $I(0) = 2$?

16. For the Lotka-Volterra competition model given in (7.24) and (7.25), assume the stochastic birth rates and death rates are

$$\lambda_i(X_1, X_2) = \max\left\{0, X_i\left(a_{i0} - \frac{a_{ii}}{2}X_i\right)\right\}, \quad i = 1, 2$$

and

$$\mu_i(X_1, X_2) = X_i\left(\frac{a_{ii}}{2}X_i + a_{ij}X_j\right), \quad i, j = 1, 2, \ i \neq j.$$

(a) Write the forward Kolmogorov differential equation for $p_{(i,j)}(t) = \text{Prob}\{X_1(t) = i, X_2(t) = j\}$ and use it to find the differential equations satisfied by the means, $m_{10}(t) = E(X_1(t))$ and $m_{01}(t) = E(X_2(t))$.

(b) For the same parameter values and initial conditions as in Example 7.9, $a_{10} = 2$, $a_{20} = 1.5$, $a_{11} = 0.03$, $a_{12} = 0.02$, $a_{21} = 0.01$, $a_{22} = 0.04$, $X_1(0) = 50$, and $X_2(0) = 25$, graph one sample path in the phase plane. Then generate 1000 sample paths and find the mean and variance at $t = 5$ and compare your answers with those given for Example 7.9.

17. The deterministic predator-prey system,

$$\frac{dx}{dt} = x\left[r\left(1 - \frac{x}{K}\right) - \frac{ay}{x+d}\right]$$

$$\frac{dy}{dt} = yb\left(1 - \frac{cy}{x}\right),$$

was shown by Murray (1993) to have a limit cycle for some parameter values.

(a) For the parameter values $r = 1$, $K = 100$, $a = 1$, $d = 20$, $b = 0.02$, and $c = 1$, show that a positive equilibrium exists and that it is unstable. It can be shown that a limit cycle exists.

(b) Assume for the prey x that the birth rate is rx and death rate is $x[r/K + ay/(x+d)]$ and for the predator y that the birth rate is by and the death rate is bcy^2/x. Write the transition probabilities for the stochastic model and the forward Kolmogorov differential equations for $p_{(i,j)}(t) = \text{Prob}\{X(t) = i, Y(t) = j\}$.

(c) For the parameter values in (a) and assumptions in (b), write a computer program to generate a sample path for the stochastic predator-prey system with initial conditions $X(0) = 20$ and $Y(0) = 30$ and graph the sample path.

(d) Due to the oscillations in the predator-prey model, it is very likely that either the predator or prey become extinct. What happens in the model if the predator y reaches zero first? Notice that there is a singularity in the differential equations when $x = 0$. When the prey hits zero, it is assumed that the predator also hits zero. An illustration of prey extinction is given in Figure 7.15.

18. For the three-species competition model,

$$\frac{dx_i}{dt} = x_i(a_{i0} - a_{i1}x_1 - a_{i2}x_2 - a_{i3}x_3), \quad i = 1, 2, 3,$$

define three random variables $X_i(t)$, $i = 1, 2, 3$, corresponding to the sizes of the three competing species. Define birth and death rates, $\lambda_i(X_1, X_2, X_3)$ and $\mu_i(X_1, X_2, X_3)$, and transition probabilities corresponding to a stochastic model. Then write the forward Kolmogorov differential equations in terms of $p_{(i,j,k)}(t) = \text{Prob}\{X_1(t) = i, X_2(t) = j, X_3(t) = k\}$. Assume the initial population sizes are fixed, $X_1(0) = a$, $X_2(0) = b$, and $X_3(0) = c$, where a, b and c are positive constants.

19. A Monte Carlo MATLAB simulation of an SIS epidemic is included in the Appendix for Chapter 7. Use the MATLAB program to approximate the probability the epidemic ends when $I(0) = 1, 2$ or 3 and to generate a probability histogram for the distribution of $I(t)$ at $t = 10$

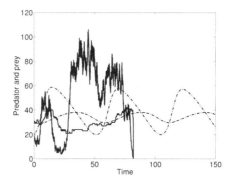

FIGURE 7.15: Sample path of the stochastic predator-prey model with the solution of the deterministic model. Parameter values and initial conditions are $r = 1$, $K = 100$, $a = 1$, $d = 20$, $b = 0.02$, $c = 1$, $X(0) = 20$, and $Y(0) = 30$. Solutions with the smaller amplitude are the predator population.

when $I(0) = 20$. Use the parameter values $\beta = 2$ and $\gamma = 1$ (same as $b + \gamma = 1$), and $N = 100$. Compute the mean and variance of $I(10)$. Compare your probability histogram with the quasistationary distribution in Example 7.3. The graphs should be very close.

20. Anderson and May (1986, 1992) applied the following SEIR model to measles outbreaks in various countries. The model includes births and deaths and an additional class of exposed or latent individuals, E, individuals that are not yet infectious. The differential equations for the deterministic model are as follows:

$$\frac{dS}{dt} = b(N - S) - \beta SI$$
$$\frac{dE}{dt} = \beta SI - \sigma E - bE$$
$$\frac{dI}{dt} = \sigma E - bI - \gamma I$$
$$\frac{dR}{dt} = \gamma I - bR,$$

where $S(t) + E(t) + I(t) + R(t) = N$. In this model, a mass action rate of incidence is assumed, βSI, rather than the standard incidence $\beta SI/N$. The birth rate b equals the death rate. The new parameter σ is the rate of becoming infectious or $1/\sigma$ is the average length of the latent period. The basic reproductive number for this model is

$$\mathcal{R}_0 = \left(\frac{\beta N}{\gamma + b} \right) \left(\frac{\sigma}{\sigma + b} \right).$$

The disease-free state, $S = N$ and $E = I = R = 0$, is locally asymptotically stable if $\mathcal{R}_0 < 1$. Anderson and May (1986, 1992) were interested in recurrent epidemics and added an immigration term Λ to the differential equation for the infected individuals,

$$\frac{dI}{dt} = \sigma E - bI - \gamma I + \Lambda.$$

With immigration, the disease-free equilibrium is not possible and the population size is not constant, $N(t) = S(t) + E(t) + I(t) + R(t)$. The dynamics of the SEIR immigration model are illustrated in Figure 7.16 for parameter values corresponding to measles in Iceland (Anderson and May, 1992):

$$\beta = \frac{0.008}{\text{year}}, \quad \frac{1}{\gamma} = 7 \text{ days}, \quad \frac{1}{\sigma} = 7 \text{ days},$$

$$\Lambda = \frac{7}{\text{year}}, \quad \text{and} \quad \frac{1}{b} = 70 \text{ years}. \tag{7.32}$$

The maximum of the first wave of the epidemic is not shown in Figure 7.16, but the first wave includes more than 90,000 infectives.

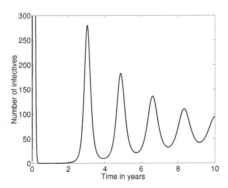

FIGURE 7.16: SEIR epidemic model with immigration of infectives; the system exhibits oscillations before convergence to an endemic equilibrium. Initial conditions are $S(0) = 249,995$, $E(0) = 0$, $I(0) = 5$, and $R(0) = 0$.

(a) Let $S(t)$, $E(t)$, $I(t)$, and $R(t)$ denote random variables for the number of susceptible, latent, infectious, and immune individuals. Write the equations for the infinitesimal transition probabilities for a CTMC model.

(b) Write a computer program to numerically simulate sample paths of this CTMC model. Consult Anderson and May (1992) for their stochastic simulations of this model with initial population sizes of $N = 230,000$ and $N = 100,000$. In both cases there are recurrent epidemics, but for the smaller population size, they are less frequent.

21. Consider the SEIR epidemic model with no immigration, $\Lambda = 0$.

    (a) Show that the disease-free equilibrium $S = N$, $E = I = R = 0$ is locally asymptotically stable if $\mathcal{R}_0 < 1$.

    (b) Convert the parameters given by Anderson and May (1992) to units of $(\text{year})^{-1}$. Then find the minimum value of N such that the disease-free equilibrium is no longer locally asymptotically stable (when $\mathcal{R}_0 = 1$). Anderson and May used the stochastic model with immigration to illustrate the importance of a sufficiently large population size to sustain an epidemic.

22. Consider a population genetics model, where a trait is determined by a single gene at a particular locus or site. Suppose there are only two different alleles for this gene, denoted as A and a, so that the genotype has one of three different forms, either

$$AA, \quad Aa, \quad \text{or} \quad aa.$$

Let $x_{AA}(t)$, $x_{Aa}(t)$, and $x_{aa}(t)$ denote the population sizes corresponding to the number of individuals with these particular genotypes at time t and let $N(t)$ be the total population size, $N(t) = x_{AA}(t) + x_{Aa}(t) + x_{aa}(t)$. Suppose b is the per capita population birth rate and d is the per capita population death rate. In random mating, a single gene from each parent forms the new gene pair in the next generation. A deterministic model for the change in the genotypic population sizes is

$$\frac{dx_i(t)}{dt} = bf_i - dx_i, \quad i \in \{AA, Aa, aa\}, \tag{7.33}$$

where under the assumption of random mating,

$$f_{AA}(x_{AA}, x_{Aa}) = \frac{(x_{AA} + x_{Aa}/2)^2}{N} \tag{7.34}$$

$$f_{Aa}(x_{AA}, x_{Aa}, x_{aa}) = \frac{2(x_{AA} + x_{Aa}/2)(x_{aa} + x_{Aa}/2)}{N} \tag{7.35}$$

$$f_{aa}(x_{aa}, x_{Aa}) = \frac{(x_{aa} + x_{Aa}/2)^2}{N}. \tag{7.36}$$

For example, the probability two A genes form an AA genotype is

$$\left(\frac{(x_{AA} + x_{Aa}/2)}{N} \right) \left(\frac{(x_{AA} + x_{Aa}/2)}{N} \right).$$

If the population birth rate is bN, the birth rate for genotype AA is

$$bN \left(\frac{(x_{AA} + x_{Aa}/2)}{N} \right) \left(\frac{(x_{AA} + x_{Aa}/2)}{N} \right) = bf_{AA}.$$

Since $f_{AA} + f_{Aa} + f_{aa} = N$, it follows that $dN/dt = (b - d)N$.

(a) For the deterministic model, let the frequency of the allele populations be denoted as $z_A = x_{AA} + x_{Aa}/2$ and $z_a = x_{aa} + x_{Aa}/2$, and their proportions be denoted as $p_A = z_A/N$ and $p_a = z_a/N$, where $p_A + p_a = 1$. Show that $dz_i/dt = (b - d)z_i$, $i = a, A$ and $dp_i/dt = 0$, $i = a, A$. These relationships imply that the population is at a *Hardy-Weinberg equilibrium*, an equilibrium where the proportions remain constant over time. If $b = d$, the population size N is constant but if $b \neq d$, the population size either increases or decreases exponentially.

(b) Let $X_{AA}(t)$, $X_{Aa}(t)$, and $X_{aa}(t)$ denote random variables for the genotypic population sizes and $N(t)$ denote the random variable for the total population size. Let $p_{(i,j,k)}(t) = \text{Prob}\{X_{AA}(t) = i, X_{Aa}(t) = j, X_{aa}(t) = k\}$, $i, j, k \in \{0, 1, 2, \ldots\}$ and

$$p_i^N(t) = \text{Prob}\{N(t) = i\}, \quad i \in \{0, 1, 2, \ldots\}$$

denote the joint probability function for the genotypes and the population size, respectively. Write equations for the transition probabilities for this multivariate stochastic process and the forward Kolmogorov differential equations for p_i^N.

(c) Write a computer program to numerically simulate several sample paths of the CTMC model with $X_{AA}(0) = 500$, $X_{Aa}(0) = 100$, $X_{aa}(0) = 200$ and $b = 0.1 = d$.

(d) For the parameter values in the previous example, calculate p_A, p_a, and

$$\lim_{t \to \infty} (x_{AA}(t), x_{Aa}(t), x_{aa}(t))$$

for the deterministic model. Compare the deterministic and the stochastic results.

23. Consider the stochastic population genetics model in the preceding exercise. Write the forward Kolmogorov differential equations for

$$p_{(i,j,k)} = \text{Prob}\{X_{AA}(t) = i, X_{Aa}(t) = j, X_{AA}(t) = k\}.$$

Assume the initial distribution is $X_{AA}(0) = a$, $X_{Aa}(0) = b$, $X_{aa}(0) = c$, where a, b, and c are positive constants.

24. A stochastic spatial predator-prey model was studied by Renshaw (1993), where the prey and predator move among a discrete set of spatial locations or patches, $i = 1, 2, \ldots, n$. For each spatial location there are random variables for the prey and predator, $X_i(t)$ and $Y_i(t)$, $i = 1, 2, \ldots, n$, respectively. In a small period of time Δt, the prey moves from location i to j with probability $u_{ji}X_i(t)\Delta t + o(\Delta t)$ and the predator moves from location i to j with probability $v_{ji}Y_i(t)\Delta t + o(\Delta t)$. The prey and predator dynamics within each spatial location or patch follow the simple Lotka-Volterra model, which has cyclic behavior. The model mimics some of the biological experiments performed by Huffaker in 1958 on mites. These experiments involved a predatory mite, *Typhlodromus occidentalis*, and another mite that served as the prey, *Eotetranychus sexmaculatus*. Oranges served as food for the prey, and the mites could move from one orange to another (Maynard Smith, 1974). The underlying deterministic model with two patches has the following form:

$$\frac{dx_i}{dt} = x_i(a_{10} - a_{12}y_i) + u_{ij}x_j - u_{ji}x_i, \quad j \neq i,$$
$$\frac{dy_i}{dt} = y_i(-a_{20} + a_{21}x_i) + v_{ij}y_j - v_{ji}y_i, \quad j \neq i$$

for $i, j = 1, 2$.

(a) Let X_i and Y_i, $i = 1, 2$ be the random variables for the prey and predator, respectively, in patches 1 and 2. Write equations for the infinitesimal transition probabilities for two patches. Assume the probabilities for a birth of a prey and a predator are $a_{10}X_i\Delta t$ and $a_{21}X_iY_i\Delta t$, respectively.

(b) Let $u = u_{ji}$ and $v = v_{ji}$ for i, j. Show that the deterministic two-patch model has exactly two nonnegative equilibria if $u, v > 0$:

$$(\bar{x}_1, \bar{y}_1, \bar{x}_2, \bar{y}_2) = (0, 0, 0, 0), \quad (a_{20}/a_{21}, a_{10}/a_{12}, a_{20}/a_{21}, a_{10}/a_{11}).$$

(c) If $u = 0 = v$ (no migration), show that the deterministic two-patch model has exactly four nonnegative equilibria.

(d) Let $a_{10} = 3$, $a_{12} = 0.1$, $a_{20} = 1$, $a_{21} = 0.01$ and $u = u_{ji} = 0.1 = v_{ji} = v$ for i, j. Calculate the two equilibria for the deterministic model. Let $X_1(0) = \bar{X}_i > 0$ and $X_2(0) = \bar{X}_2 > 0$. Write a computer program to simulate sample paths of the CTMC two-patch predator-prey model. Discuss your results.

25. Consider Renshaw's (1993) spatial predator-prey model with density dependence in the prey population and dispersal between two patches. Assume the deterministic model is

$$\frac{dx_i}{dt} = x_i(a_{10} - a_{11}x_i - a_{12}y_i) + u(x_j - x_i),$$
$$\frac{dy_i}{dt} = y_i(-a_{20} + a_{21}x_i) + v(y_j - y_i),$$

where $i, j = 1, 2$ and $j \neq i$. Let $X_i(t)$ and $Y_i(t)$ denote the random variables for the corresponding CTMC and let

$$[a_{11}X_i(t)X_i(t) + a_{12}X_i(t)Y_i(t)]\Delta t + o(\Delta t)$$

be the transition probability for death of a prey, $\Delta X_i(t) = -1$. Suppose the parameter values are $a_{10} = 1$, $a_{11} = 0.02$, $a_{12} = 0.1$, $a_{20} = 0.15$, and $a_{21} = 0.01$ and initial conditions are $X_1(0) = 15$, $Y_1(0) = 7$, and $X_2(0) = 0 = Y_2(0) = 0$.

(a) Show that the values of $x_i = 15$ and $y_i = 7$, $i = 1, 2$ are equilibrium values of the deterministic model (i.e., $dx_i/dt = 0 = dy_i/dt$).

(b) Let $u = 0 = v$ (no movement). Write a computer program to simulate 50 sample paths, and record the time when either (i) the prey population size or predator population size equals zero (extinction of one species) or (ii) time has reached $t = 200$ and neither population size is zero (coexistence). In what proportion of the sample paths are the prey extinct? predators extinct? both coexist?

(c) For two patches, consider three cases: (i) $u = 0.001 = v$ (prey and predator movement rates are equal), (ii) $u = 0.001$ and $v = 0.01$ (predator movement > prey movement), and (iii) $u = 0.01$ and $v = 0.001$ (prey movement > predator movement). Write a computer program to simulate 50 sample paths for each of these sets of parameter values and record the time when either the total prey population size or total predator population size equals zero (extinction of one species in both patches) or time has reached $t = 200$ and neither of the population sizes are zero (coexistence). In what proportion of the sample paths are the prey extinct? predators extinct? both coexist?

(d) Repeat part (c) with three patches, $X_1(0) = 15$, $Y_1(0) = 7$, $X_j(0) = 0$, and $Y_j(0) = 0$, $j = 2, 3$. Compare the results of (b) and (c), (b) and (d), and (c) and (d). When is extinction most likely to occur? in one, two, or three patches? When is coexistence most likely to occur? when prey movement is greater than predator movement or when predator movement is greater than prey movement?

26. Suppose one population is dispersing between two patches,

$$\frac{dx_i}{dt} = rx_i(1 - x_i/K) + u(x_j - x_i), \quad j \neq i$$

where $i, j = 1, 2$.

(a) Without migration, $u = 0$, the population grows logistically. What is the stable positive equilibrium when $u = 0$?

(b) Let $X_i(t)$ $i = 1, 2$ denote random variables for the stochastic process. Write the infinitesimal transition probabilities for a CTMC model based on the preceding deterministic model. Assume the birth and death rates are $\lambda_i = ri(1 - i^2/(2K))$ and $\mu_i = ri^2/(2K)$, $i = 0, 1, 2, \ldots, 2K$ (Section 6.8).

(c) Let $r = 2$, $K = 10$, and $u = 0$. The maximum population size is $N = 2K = 20$, with initial conditions $X_1(0) = 10$ and $X_2(0) = 0$ (a single population with logistic growth). Compute the expected time to extinction, τ, i.e., solve equation (6.20),

$$\tau^{tr} \tilde{Q} = -\mathbf{1}^{tr},$$

where $\tilde{Q}$ is the infinitesimal generator matrix with the first row and column deleted and $\tau^{tr} = (\tau_1, \ldots, \tau_N)$.

(d) For the same parameters and initial conditions in part (c), compute the second moment τ^2 for the time to extinction and use it to compute the standard deviation, i.e., $[\tau^2]^{tr} \tilde{Q} = -2\tau^{tr}$:

```
tau2=-2*tau/Q;
taustd=sqrt(tau2-tau.^2);
```

(e) For the same parameters and initial conditions as in part (c), except $u = 0.01$, write a computer program to simulate 1000 sample paths of a CTMC patch model. Fix a time $T = 2\tau_{10}$. Then record either (i) the time $t < T$ when either $X_1(t) = 0 = X_2(t)$ (species extinction) or (ii) the time as $t = T$ if both patches have a positive population by time $t = T$; the species persist. What proportion of the sample paths is there species extinction? species persistence? Compare the results with dispersal $u > 0$ to those without dispersal.

(f) Use the same parameters and initial conditions as in part (c), except $u = 0.1$, and answer the questions in part (e). Compare the results with those for $u = 0.01$.

7.10 References for Chapter 7

Allen, L. J. S. 2007. *An Introduction to Mathematical Biology.* Prentice Hall, Upper Saddle River, N.J.

Allen, L. J. S. 2008. An introduction to stochastic epidemic models. In: *Mathematical Epidemiology, Lecture Notes in Mathematics.* Brauer, F., P. van den Driessche, and J. Wu (eds.), Vol. 1945, pp. 81–130, Springer-Verlag, New York.

Allen, L. J. S. and P. J. Cormier. 1996. Environmentally-driven epizootics. *Math. Biosci.* 131: 51–80.

Allen, L. J. S. and D. Thrasher. 1998. The effects of vaccination in an age-dependent model for varicella and herpes zoster. *IEEE Trans. Aut. Control* 43: 779–789.

Alon, U. 2007. *An Introduction to Systems Biology.* Chapman & Hall/ CRC Mathematical and Computational Biology Series, London, UK.

Anderson, R. M. and R. May. 1986. The invasion, persistence and spread of infectious diseases within animal and plant communities. *Phil. Trans. Royal Soc.* B314: 533–570.

Anderson, R. M. and R. M. May. 1992. *Infectious Diseases of Humans: Dynamics and Control.* Oxford Univ. Press, Oxford.

Bailey, N. T. J. 1990. *The Elements of Stochastic Processes with Applications to the Natural Sciences.* John Wiley & Sons, New York.

Bharucha-Reid, A. T. 1997. *Elements of the Theory of Markov Processes and their Applications.* Dover Pub. Inc., New York.

Brauer, F. and C. Castillo-Chávez. 2001. *Mathematical Models in Population Biology and Epidemiology.* Springer-Verlag, New York.

Chan, M. S. and V. S. Isham. 1998. A stochastic model of schistosomiasis immuno-epidemiology. *Math. Biosci.* 151: 179–198.

Chao, X, M. Miyazawa, and M. Pinedo. 1999. *Queueing Networks.* John Wiley & Sons, Chichester, New York.

Coddington, E. A. and N. Levinson. 1955. *Theory of Ordinary Differential Equations.* McGraw-Hill Book Co. Inc., New York.

Coldman, A. J. and J. H. Goldie. 1987. Modeling resistance to cancer chemotherapeutic agents. In: *Cancer Modeling.* Thompson, J. R. and B. W. Brow (eds.), pp. 315-364, Marcel Dekker, Inc. N.Y.

Daley, D. J. and J. Gani. 1999. *Epidemic Modelling: An Introduction.* Cambridge Studies in Mathematical Biology, Vol. 15. Cambridge Univ. Press, Cambridge, U. K.

Durrett, R. 1995. Spatial epidemic models. In: *Epidemic Models: Their Structure and Relation to Data.* D. Mollison (ed.), pp. 187–201, Cambridge Univ. Press, Cambridge, U. K.

Durrett, R. 1999. Stochastic spatial models. *SIAM Review.* 41: 677–718.

Edelstein-Keshet, L. 1988. *Mathematical Models in Biology.* The Random House/Birkhäuser Mathematics Series, New York.

Gabriel, J. -P., C. Lefèvre, and P. Picard (eds.) 1990. *Stochastic Processes in Epidemic Theory.* Lecture Notes in Biomathematics, Springer-Verlag, New York.

Goel, N. S. and N. Richter-Dyn. 1974. *Stochastic Models in Biology.* Academic Press, New York.

Haccou, P., P. Jagers, and V. A Vatutin. 2005. *Branching Processes Variation, Growth, and Extinction of Populations.* Cambridge Studies in Adaptive Dynamics. Cambridge Univ. Press, Cambridge, U. K.

Harris, T. E. 1963. *The Theory of Branching Processes.* Prentice Hall, Englewood Cliffs, N. J.

Hassell, M. P. 1978. *The Dynamics of Arthropod Predator-Prey Systems.* Princeton Univ. Press, Princeton, N. J.

Hethcote, H. W. 2000. The mathematics of infectious diseases. *SIAM Review.* 42: 599–653.

Hubbarde, J. E., G. Wild, and L. M. Wahl. 2007. Fixation probabilities when generation times are variable: the burst-death model. *Genetics* 176: 1703-1712.

Huffaker, C. B. 1958. Experimental studies on predation: dispersion factors and predator-prey oscillations. *Hilgardia* 27: 343–383.

Jagers, P. 1975. *Branching Processes with Biological Applications.* John Wiley & Sons, London.

Keeling, M. J. 2000. Multiplicative moments and measure of persistence in ecology. *J. Theor. Biol.* 205: 269–281.

Kimmel, M. and D. Axelrod. 2002. *Branching Processes in Biology.* Springer-Verlag, New York.

Kot, M. 2001. *Elements of Mathematical Ecology.* Cambridge Univ. Press, Cambridge, U. K.

Krishnarajah, I., A. Cook A., G. Marion, and G. Gibson. 2005. Novel moment closure approximations in stochastic epidemics. *Bull. Math. Biol.* 67: 855–873.

Kuang, Y. and E. Beretta. 1998. Global qualitative analysis of a ratio-dependent predator-prey system. *J. Math. Biol.* 36: 389–406.

Lloyd, A. L. 2004. Estimating variability in models for recurrent epidemics: assessing the use of moment closure techniques. *Theor. Pop. Biol.* 65: 49–65.

Matis, J. H. and T. R. Kiffe. 2000. *Stochastic Population Models: A Compartmental Perspective.* Springer-Verlag, New York.

Maynard Smith, J. 1974. *Models in Ecology.* Cambridge Univ. Press, Cambridge, U. K.

Mode, C. J. 1971. *Multitype Branching Processes Theory and Applications.* American Elsevier Pub. Co., Inc., New York.

Murray, J. D. 1993. *Mathematical Biology.* 2nd ed. Spring-Verlag, Berlin, Heidelberg, New York.

Nåsell, I. 2003a. Moment closure and the stochastic logistic model. *Theor. Pop. Biol.* 63: 159–168.

Nåsell, I. 2003b. An extension of the moment closure method. *Theor. Pop. Biol.* 64: 233–239.

Nisbet, R. M. and W. S. C. Gurney. 1982. *Modelling Fluctuating Populations.* John Wiley & Sons, Chichester and New York.

Renshaw, E. 1993. *Modelling Biological Populations in Space and Time.* Cambridge Univ. Press, Cambridge, U. K.

Singh, A. and J. P. Hespanha. 2007. A derivative matching approach to moment closure for the stochastic logistic model. *Bull. Math. Biol.* 69: 1909–1925.

Smith, H. L. and P. Waltman. 1995. *The Theory of the Chemostat.* Cambridge Studies in Mathematical Biology. Cambridge Univ. Press, Cambridge, U. K.

Srivastava, R., L. You, J. Summers, and J. Yin. 2002. Stochastic vs. deterministic modeling of intracellular viral kinetics. *J. Theor. Biol.* 218: 309–321.

Szallasi, Z., J. Stelling, and V. Periwal. (eds.) 2006. *System Modeling in Cellular Biology From Concepts to Nuts and Bolts.* MIT Press, Cambridge, MA.

Whittle, P. 1955. The outcome of a stochastic epidemic–a note on Bailey's paper. *Biometrika* 42: 116–122.

Wilkinson, D. J. 2006. *Stochastic Modelling for Systems Biology.* Chapman & Hall/CRC Mathematical and Computational Biology Series, Boca Raton, FL.

Yakovlev, A. Y., M. Mayer-Proschel, and M. Noble. 1998. A stochastic model of brain cell differentiation in tissue culture. *J. Math. Biol.* 37: 49–60.

7.11 Appendix for Chapter 7

7.11.1 MATLAB® Programs

Programs 1-4 are MATLAB programs for the final size distribution of the SIR epidemic process (Program 1), three sample paths of the SIR epidemic process (Program 2), sample paths of the stochastic competition model (Program 3), and a Monte Carlo simulation of an SIS epidemic process (Program 4).

```
%MatLaB Program 1, Final Size
clear
N=input('Size of Population');
beta=input('beta value');
siz=(N+1)*(N+2)/2;
T=zeros(siz);
 gama=1;  e(1)=0;   T(1,1)=1;
 for i=1:N
       e(i+1)=e(i)+N+2-i;
       p(i)=gama*N/beta/(i-1+gama*N/beta);
       T(i+1,i+1)=1;
end
%Upper Blocks of T
for i=1:N
   for j=0:N-i
       T(e(i)+j+2,e(i+1)+j+1)=p(N-i-j+1);
     end
end
%Lower Blocks of  T
for i=1:N-1
    for j=1:N-i
        T(e(i+2)+j,e(i+1)+j)=1-p(N-j-i+2);
    end
end
 pa=zeros(siz,1);
 pa(N+2)=1; % Initial infectives, element N+2
 S=sparse(T);
 plim=S^(2*N-1)*pa;
```

```
 for i=1:N
     absorp(i)=plim(i+1);
 end
 meanN=dot(absorp,num)
 plot([1:N:N],absorp,'b-','LineWidth',2);
 xlabel('Size'); ylabel('Probability');
```

Note: Other more efficient numerical methods for computing T^{2N-1} could be employed. Here, we used the fact that T was a sparse matrix to make the MATLAB code more efficient.

```
%Matlab Program 2,  Three sample paths for the SIR epidemic.
 clear all
 beta=2;  g=1; N=100;
 for k=1:3
    clear  t s i
    t(1)=0;   i(1)=1; s(1)=N-i(1);
    j=1;
    while i(j)>0 & j<=1000
        y1=rand; y2=rand;
        t(j+1)=-log(y1)/((beta/N)*i(j)*s(j)+g*i(j))+t(j);
        if (y2<=(beta/N)*s(j)/(beta/N*s(j)+g))
    i(j+1)=i(j)+1;
            s(j+1)=s(j)-1;
        else
            i(j+1)=i(j)-1;
            s(j+1)=s(j);
        end
        j=j+1;
    end
    tend(k)=t(j) % Time  epidemic ends.
    l1=stairs(t,i); set(l1,'LineWidth',2);
    hold on
 end
 ylabel('I(t)'); xlabel('Time');
 hold off


% MatLaB Program 3, Stochastic Competition
 clear all
 a10=2; a11=.03; a12=.02;
 a20=1.5; a21=.01; a22=.04;
 t(1)=0; x1(1)=50; x2(1)=25;
 i=1;
 while (x1(i)>0 & x2(i)>0 & t(i)<5);
     b1=a10*x1(i);  b2=a20*x2(i);
```

```
       d1=x1(i)*(a11*x1(i)+a12*x2(i));
       d2=x2(i)*(a21*x1(i)+a22*x2(i));
       tot=b1+b2+d1+d2;
       u1=rand;
       u2=rand;
       t(i+1)=-log(u1)/tot+t(i);
       x1(i+1)=x1(i); x2(i+1)=x2(i);
       if (u2<=b1/tot)
                   x1(i+1)=x1(i)+1;
       elseif(u2>b1/tot& u2<=(b1+b2)/tot)
                   x2(i+1)=x2(i)+1;
       elseif (u2>(b1+b2)/tot & u2<=(b1+b2+d1)/tot)
           x1(i+1)=x1(i)-1;
        else
           x2(i+1)=x2(i)-1;
        end
        i=i+1;
end
xx=linspace(0,100,11);
isox1=a10/a12-(a11/a12)*xx;
isox2=a20/a22-(a21/a22)*xx;
l1=stairs(x1,x2,'r'); set(l1,'LineWidth',2);
axis([0,80,0,50]);
hold on
plot(xx,isox1,'k',xx,isox2,'k','LineWidth',2);
xlabel('X_1'); ylabel('X_2');


% Matlab Program 4, Monte Carlo simulation of an SIS epidemic
clear all
dt=0.01;time=10;
N=100; beta=2; g=1;
I(1)=1.; S(1)=N-I(1);
sim=10000;
for k=1:sim
    for t=1:time/dt;
        nI=0; nS=0;
    for j=1:S(t)
       p1=beta*I(t)/N*dt;
       if rand<p1
          nI=nI+1; nS=nS-1;
       else
          nI=nI; nS=nS;
       end
     end
   end
```

```
    for j=1:I(t);
        p2=g*dt;
        if rand<p2;
            nI=nI-1; nS=nS+1;
        else
            nI=nI; nS=nS;
        end
    end
    I(t+1)=I(t)+nI; S(t+1)=S(t)+nS;
    end
    if k==1 plot([0:dt:time],I,'b-','linewidth',2); end
    hold on
    if k==2  plot([0:dt:time],I,'r-','linewidth',2);   end
    if k==3  plot([0:dt:time],I,'g-','linewidth',2); end
    II(k)=I(time/dt+1);
end
probext=sum(II>0)/sim
hold off
```

Chapter 8

Diffusion Processes and Stochastic Differential Equations

8.1 Introduction

Stochastic processes continuous in time and in state are presented in this chapter. In the first section, it is shown that the random walk model leads to a diffusion process known as Brownian motion, also referred to as the Wiener process. A diffusion process is defined in Section 8.4, a Markov process having continuous sample paths with the additional property that the infinitesimal mean and variance are finite. It is shown that the transition probability density function of a diffusion process is a solution of the forward and backward Kolmogorov differential equations, the continuous analogues of the Kolmogorov differential equations from Markov chain theory. A more general definition of the Wiener process is given in Section 8.6 that leads to the formulation of Itô stochastic integrals and differential equations, equations whose solutions are stochastic realizations of a diffusion process. Then numerical methods for solution of Itô stochastic differential equations are discussed. The chapter ends with an application of Itô stochastic differential equations to the problem of monitoring the drug concentration in a body.

More theoretical and detailed discussions about diffusion processes and stochastic differential equations can be found in the references, including *Modeling with Itô Stochastic Differential Equations* by E. Allen (2007); *Stochastic Differential Equations: Theory and Applications*, by Arnold (1974); *Elements of the Theory of Markov Processes and Their Applications*, by Bharucha-Reid (1997); *Introduction to Stochastic Differential Equations*, by Gard (1988); *Stochastic Differential Equations and Diffusion Processes*, by Ikeda and Watanabe (1989); *A Second Course in Stochastic Processes*, by Karlin and Taylor (1981); *Stochastic Differential Equations: An Introduction with Applications*, by Øksendal (2000); and *Stochastic Processes in Physics and Chemistry*, by van Kampen (2007).

8.2   Definitions and Notation

Assume $\{X(t) : t \in [0, \infty)\}$ is a stochastic process, where $X(t)$ is a continuous random variable with state space

$$(-\infty, \infty) \quad \text{or} \quad [0, \infty) \quad \text{or} \quad [0, M]. \tag{8.1}$$

Associated with the continuous random variable $X(t)$ is a *probability density function (p.d.f.)*, $p(x, t)$. To find a probability associated with the random variable $X(t)$ requires integration of $p(x, t)$. For example,

$$\text{Prob}\{X(t) \in [a, b]\} = \int_a^b p(x, t)\, dx.$$

Both the time and state space are continuous. The stochastic processes studied in this chapter are Markov processes.

DEFINITION 8.1   *Assume* $\{X(t) : t \in [0, \infty)\}$ *is a stochastic process that is continuous in time with the same state space as in (8.1). Then* $\{X(t) : t \in [0, \infty)\}$ *is a* Markov *process if, given any sequence of times,* $0 < t_1 < \cdots < t_{n-1} < t_n$,

$$\text{Prob}\{X(t_n) \leq y | X(0) = x_0, X(t_1) = x_1, \ldots, X(t_{n-1}) = x_{n-1}\}$$
$$= \text{Prob}\{X(t_n) \leq y | X(t_{n-1}) = x_{n-1}\}.$$

The future state of a Markov process only depends on the current state and not its past history. In addition, note that the probability in Definition 8.1 is stated in terms of the cumulative distribution. The condition in Definition 8.1 is known as the *Markov property*.

DEFINITION 8.2   *The* transition p.d.f. *for a Markov process that is continuous in time and state is the density function for a transition from state* x *at time* t *to state* y *at time* s, $t < s$. *The transition p.d.f. is denoted as* $p(y, s; x, t)$. *The transition p.d.f. is said to be* homogeneous *or* time-homogeneous *if*

$$p(y, s + \Delta t; x, t + \Delta t) = p(y, s; x, t),$$

where $0 \leq t < s$ *and* $\Delta t > 0$. *If not, it is said to be* nonhomogeneous.

If the transition p.d.f. is time homogeneous, then the transitions only depend on the length of time between states $s - t$, and the notation for the transition p.d.f. is denoted simply as

$$p(y, x, s - t) = p(y, s; x, t).$$

In this case, the process moves from state x to y during the time interval $s-t$. The *Chapman-Kolmogorov equations* hold for the transition p.d.f.:

$$p(y, s; x, t) = \int_{-\infty}^{\infty} p(y, s; z, u)p(z, u; x, t)\, dz, \quad t < u < s. \qquad (8.2)$$

Similar equations were shown to hold for DTMCs and CTMCs.

The dynamics of a stochastic process depend on the initial density of $X(0)$. Frequently, in applications, the initial density is concentrated at a point x_0. Technically, this means that the p.d.f. of $X(0)$ is a Dirac delta function, $\delta(x - x_0)$, where

$$\delta(x - x_0) = 0, \quad x \neq x_0 \text{ and } \int_{-\infty}^{\infty} \delta(x - x_0)\, dx = 1.$$

For simplicity, we often write $X(0) = x_0$ when the initial p.d.f. is $p(x, t_0) = \delta(x - x_0)$. In this case, the p.d.f. of the random variable $X(t)$, $p(x, t)$ is the same as the transition probability density function, $p(x, t; x_0, 0)$. Applying the Chapman-Kolmogorov equations, it follows that

$$p(x, t; x_0, 0) = \int_{-\infty}^{\infty} p(x, t; z, u)p(z, u; x_0, 0)\, dz, \quad 0 < u < t$$

or

$$p(x, t) = \int_{-\infty}^{\infty} p(x, t; z, u)p(z, u)\, dz, \quad 0 < u < t.$$

Although the initial p.d.f. for $X(0)$ may be concentrated at x_0, the density $p(x, t)$ for $X(t)$, where $t > 0$, will, in general, be a smooth function of x. Hence,

$$\text{Prob}\{X(t) \in [a, b]\} = \text{Prob}\{X(t) \in (a, b)\} = \int_a^b p(x, t)\, dx, \quad t > 0.$$

The notation $p(x, t)$ for the p.d.f. of $X(t)$ is consistent with the notation used for a DTMC and a CTMC. Recall that for a DTMC $\{X_n\}_{n=0}^{\infty}$, the probability mass function (p.m.f) is denoted as

$$p_i(n) = \text{Prob}\{X_n = i\},$$

and for a CTMC $\{X(t) : t \in [0, \infty)\}$, the p.m.f. is denoted as

$$p_i(t) = \text{Prob}\{X(t) = i\},$$

where $i \in \{0, 1, 2, \ldots\}$. The p.d.f. $p(x, t)$ should not be confused with the transition p.d.f. $p(x, t; y, s)$ or $p(x, y; t - s)$.

The Chapman-Kolmogorov equations will be used to show that the transition p.d.f. is a solution of the forward and backward Kolmogorov differential equations. If an explicit solution of these equations can be determined, then the dynamics of the stochastic process are also determined. However, in many cases, the equations are too difficult to solve directly, and in these cases, other methods must be used to study the dynamics of the stochastic process.

In the next section, a classical example of a diffusion process is derived and discussed. The diffusion process is derived from the discrete random walk model and is referred to as Brownian motion. This example provides motivation for the general definition of a diffusion process and illustrates the form of the Kolmogorov differential equations for this process.

8.3 Random Walk and Brownian Motion

Consider a simple random walk on the set $\{0, \pm\Delta x, \pm 2\Delta x, \ldots\}$. Let p be the probability of moving to the right and q be the probability of moving to the left so that $p + q = 1$. Let $X(t)$ be the DTMC for this random walk, where $t \in \{0, \Delta t, 2\Delta t, \ldots\}$ and $X(t) \in \{0, \pm\Delta x, \pm 2\Delta x, \ldots\}$. Let

$$p_x(t) = \text{Prob}\{X(t) = x\} = u(x, t).$$

Recall from the analysis of DTMCs that

$$u(x, t + \Delta t) = pu(x - \Delta x, t) + qu(x + \Delta x, t).$$

Assume $u(x, t)$ is defined for all $x \in (-\infty, \infty)$ and $t \in [0, \infty)$. Applying Taylor's formula to expand the right-hand side of the preceding equation about the point (x, t) leads to

$$u(x, t + \Delta t) = p \left[u(x, t) + \frac{\partial u(x, t)}{\partial x}(-\Delta x) + \frac{\partial^2 u(x, t)}{\partial x^2} \frac{(\Delta x)^2}{2} + O((\Delta x)^3) \right]$$

$$+ q \left[u(x, t) + \frac{\partial u(x, t)}{\partial x} \Delta x + \frac{\partial^2 u(x, t)}{\partial x^2} \frac{(\Delta x)^2}{2} + O((\Delta x)^3) \right]$$

$$= u(x, t) + (q - p) \frac{\partial u(x, t)}{\partial x} \Delta x + \frac{\partial^2 u(x, t)}{\partial x^2} \frac{(\Delta x)^2}{2} + O((\Delta x)^3).$$

Subtracting $u(x, t)$ and dividing by Δt,

$$\frac{u(x, t + \Delta t) - u(x, t)}{\Delta t} = (q - p) \frac{\partial u(x, t)}{\partial x} \frac{\Delta x}{\Delta t}$$

$$+ \frac{1}{2} \frac{\partial^2 u(x, t)}{\partial x^2} \frac{(\Delta x)^2}{\Delta t} + O\left(\frac{(\Delta x)^3}{\Delta t} \right). \quad (8.3)$$

Assume

$$
\lim_{\Delta t, \Delta x \to 0} (p - q) \frac{\Delta x}{\Delta t} = c,
$$

$$
\lim_{\Delta t, \Delta x \to 0} \frac{(\Delta x)^2}{\Delta t} = D,
$$

$$
\lim_{\Delta t, \Delta x \to 0} \frac{(\Delta x)^3}{\Delta t} = 0.
$$

(8.4)

Letting Δt and Δx approach zero, the probability $u(x, t)$ represents the p.d.f. of a continuous-time and continuous-state process $X(t)$, which is a solution of the following partial differential equation:

$$
\frac{\partial u}{\partial t} = -c \frac{\partial u}{\partial x} + \frac{D}{2} \frac{\partial^2 u}{\partial x^2}, \quad x \in (-\infty, \infty).
$$

(8.5)

The partial differential equation (8.5) is known as the *diffusion equation with drift*, where D is the diffusion coefficient and c is the drift coefficient. This equation is also known as the *forward Kolmogorov differential equation* for this process. When $p = 1/2 = q$, so that movement is unbiased and symmetric, the limiting stochastic process is known as *Brownian motion*. In Brownian motion, $c = 0$, so that

$$
\frac{\partial u}{\partial t} = \frac{D}{2} \frac{\partial^2 u}{\partial x^2}, \quad x \in (-\infty, \infty).
$$

To solve the forward Kolmogorov differential equation, an assumption about the initial density $X(0)$ is required. The differential equation and the initial density form an initial value problem whose solution is well known. Assume that the initial density is concentrated at x_0. That is, $u(x, 0) = \delta(x - x_0)$, where $\delta(x)$ is the Dirac delta function [i.e., $X(0) = x_0$]. Then the solution of the initial value problem (8.5) is the normal density function with mean $x_0 + ct$ and variance Dt. That is,

$$
u(x, t) = \frac{1}{\sqrt{2\pi Dt}} \exp\left(-\frac{(x - x_0 - ct)^2}{2Dt} \right).
$$

(8.6)

In Figure 8.1, the solution $u(x, t)$ is graphed as a function of x for several values of t, where the initial mass is concentrated at zero, $x_0 = 0$. There is drift to the right since $c > 0$ implying $p > q$. Brownian motion is the case in which $c = 0$, where there is no drift to the right. *Standard Brownian motion* is Brownian motion with $X(0) = 0$ and $D = 1$. Standard Brownian motion is also referred to as the *Wiener process*. The Wiener process is discussed in detail in Section 8.6.

The assumptions on the limits in (8.4) were necessary to obtain the diffusion equation with drift. These assumptions are very important in the derivation

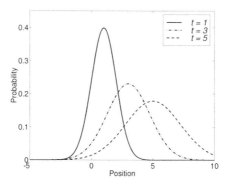

FIGURE 8.1: Solution of the diffusion equation with drift with $c = 1$ and $D = 1$ at $t = 1, 3, 5$.

of the Kolmogorov differential equations. It will be shown that the limits are related to the infinitesimal mean and variance of a diffusion process.

In a time period of $[0, t]$, there is a total of $t/\Delta t$ time steps of size Δt. Also, beginning at the origin, a particle at position x has moved spatially, in step sizes Δx, either to the right or to the left. It is assumed that the movements are independent of each other so that in a time interval $[0, t]$, the total displacement is the sum of $t/\Delta t$ movements, having values of either Δx or $-\Delta x$ with probabilities p and q, respectively (Bailey, 1990). In particular, the mean displacement in a time period of Δt is

$$p\Delta x + q(-\Delta x) = (p - q)\Delta x$$

and the variance of the displacement $(\sigma_Y^2 = E(Y^2) - \mu_Y^2)$ in time Δt is

$$
\begin{aligned}
p(\Delta x)^2 + q(-\Delta x)^2 - (p - q)^2(\Delta x)^2 &= (\Delta x)^2[p + q - p^2 + 2pq - q^2] \\
&= (\Delta x)^2[p(1 - p) + 2pq + q(1 - q)) \\
&= (\Delta x)^2(4pq).
\end{aligned}
$$

The total movement or displacement in time period $[0, t]$ is the sum of $t/\Delta t$ independent displacements. Thus, the mean displacement in time period t is

$$\frac{t}{\Delta t}(p - q)\Delta x = (p - q)t\left(\frac{\Delta x}{\Delta t}\right). \tag{8.7}$$

The variance of the displacement in time period t is

$$\frac{t}{\Delta t}(\Delta x)^2(4pq) = 4pqt\frac{(\Delta x)^2}{\Delta t}. \tag{8.8}$$

If the mean and variance of the process are finite, then as Δx and Δt approach zero, the expressions (8.7) and (8.8) must be finite in the limit. Hence, the assumptions (8.4) must hold. The variance of the process is $4pqtD$. However, for $(p-q)\Delta x/\Delta t$ to be finite, it must be the case that $(p-q)/\Delta x$ is bounded, and, thus,

$$\lim_{\Delta t, \Delta x \to 0} (p-q) = \lim_{\Delta x \to 0} \Delta x = 0.$$

The mean and variance of the process, as Δt and Δx approach zero, are ct and Dt, respectively. In the next section, these types of assumptions define a general diffusion process.

8.4 Diffusion Process

A diffusion process is a Markov process with additional properties regarding the infinitesimal mean and variance.

DEFINITION 8.3 *Let $\{X(t) : t \in [0, \infty)\}$ be a Markov process with state space $(-\infty, \infty)$, having continuous sample paths and a transition p.d.f. given by $p(y, s; x, t)$, $t < s$. Then $\{X(t) : t \in [0, \infty)\}$ is a diffusion process if its p.d.f. satisfies the following three assumptions for any $\epsilon > 0$ and $x \in (-\infty, \infty)$:*

(i) $\displaystyle \lim_{\Delta t \to 0+} \frac{1}{\Delta t} \int_{|y-x|>\epsilon} p(y, t+\Delta t; x, t)\, dy = 0.$

(ii) $\displaystyle \lim_{\Delta t \to 0+} \frac{1}{\Delta t} \int_{|y-x|\le\epsilon} (y-x) p(y, t+\Delta t; x, t)\, dy = a(x, t).$

(iii) $\displaystyle \lim_{\Delta t \to 0+} \frac{1}{\Delta t} \int_{|y-x|\le\epsilon} (y-x)^2 p(y, t+\Delta t; x, t)\, dy = b(x, t).$

The first condition states that the probability of a transition over a small time interval is negligible. The second and third conditions are restrictions on the infinitesimal mean and variance. Note that $b(x, t) \ge 0$.

It can be shown that the three preceding conditions follow from the stronger conditions:

(i)$'$ $\displaystyle \lim_{\Delta t \to 0+} \frac{1}{\Delta t} \int_{-\infty}^{\infty} |y-x|^\delta p(y, t+\Delta t; x, t)\, dy = 0$ for some $\delta > 2.$

(ii)$'$ $\displaystyle \lim_{\Delta t \to 0+} \frac{1}{\Delta t} \int_{-\infty}^{\infty} (y-x) p(y, t+\Delta t; x, t)\, dy = a(x, t).$

(iii)$'$ $\displaystyle \lim_{\Delta t \to 0+} \frac{1}{\Delta t} \int_{-\infty}^{\infty} (y-x)^2 p(y, t+\Delta t; x, t)\, dy = b(x, t).$

For $k = 0, 1, 2$, and $|y - x| > \epsilon$, it follows from assumption (i)' that

$$\int_{|y-x|>\epsilon} |y - x|^k p(y, t + \Delta t; x, t)\, dy$$

$$\leq \frac{1}{\epsilon^{\delta-k}} \int_{-\infty}^{\infty} |y - x|^\delta p(y, t + \Delta t; x, t)\, dy. \qquad (8.9)$$

The expression on the right-hand side is $o(\Delta t)$. Conditions (i)–(iii) follow from inequality (8.9) and conditions (ii)' and (iii)' (see Exercise 2). The conditions (i)'–(iii)', are sometimes easier to verify than (i)–(iii).

Conditions (i)'–(iii)' are generalizations of the conditions given in (8.4). These latter conditions can be expressed in terms of the expectation:

(i)' $\displaystyle \lim_{\Delta t \to 0+} \frac{1}{\Delta t} E(|\Delta X(t)|^\delta | X(t) = x) = 0$ for some $\delta > 2$.

(ii)' $\displaystyle \lim_{\Delta t \to 0+} \frac{1}{\Delta t} E(\Delta X(t) | X(t) = x) = a(x, t)$.

(iii)' $\displaystyle \lim_{\Delta t \to 0+} \frac{1}{\Delta t} E([\Delta X(t)]^2 | X(t) = x) = b(x, t)$,

where $\Delta X(t) = X(t + \Delta t) - X(t) = y - x$.

In the next section, the Chapman-Kolmogorov equations (8.2) and the assumptions (i)'–(iii)' will be used to show that the transition p.d.f. $p(y, s; x, t)$ of a diffusion process satisfies certain relationships known as the Kolmogorov differential equations.

8.5 Kolmogorov Differential Equations

The transition p.d.f. $p(y, s; x, t)$ of a diffusion process is a solution of the forward and backward Kolmogorov differential equations. These equations are derived based on the assumptions (i)'–(iii)' and the Chapman-Kolmogorov equation (8.2). It will be shown that the backward Kolmogorov differential equation corresponding to a nonhomogeneous diffusion process is

$$\frac{\partial p(y, s; x, t)}{\partial t} = -a(x, t) \frac{\partial p(y, s; x, t)}{\partial x} - \frac{1}{2} b(x, t) \frac{\partial^2 p(y, s; x, t)}{\partial x^2}, \qquad (8.10)$$

where a and b are defined in (ii) and (iii). It is a backward equation because the current position and time are (y, s), but the differential equation describes the dynamics of all the positions and times (x, t) prior to the current one or going "backward" in time.

If $a(x, t) = a(x)$ and $b(x, t) = b(x)$, then the process is time homogeneous, and $p(y, s; x, t) = p(y, x, s - t)$. Letting $\tau = s - t$, it follows that

$$\frac{\partial p(y, s; x, t)}{\partial t} = -\frac{\partial p(y, x, \tau)}{\partial \tau}.$$

Thus, the *backward Kolmogorov differential equation* for a *time-homogeneous process* is

$$\frac{\partial p(y,x,t)}{\partial t} = a(x)\frac{\partial p(y,x,t)}{\partial x} + \frac{1}{2}b(x)\frac{\partial^2 p(y,x,t)}{\partial x^2}. \tag{8.11}$$

To derive the backward Kolmogorov differential equation (8.10), the Chapman-Kolmogorov equations are applied and the following identity:

$$1 = \int_{-\infty}^{\infty} p(z,t;x,t-\Delta t)\,dz, \quad \Delta t > 0. \tag{8.12}$$

It follows that $p(y,s;x,t-\Delta t) - p(y,s;x,t)$ can be expressed as

$$\int_{-\infty}^{\infty} p(z,t;x,t-\Delta t)[p(y,s;z,t) - p(y,s;x,t)]\,dz.$$

Expanding $p(y,s;z,t)$ in the variable z using Taylor's formula about x,

$$
\begin{aligned}
p(y,s;x,&t-\Delta t) - p(y,s;x,t)\\
&= \int_{-\infty}^{\infty} p(z,t;x,t-\Delta t)\left[(z-x)\frac{\partial p(y,s;x,t)}{\partial x}\right.\\
&\quad \left. + \frac{(z-x)^2}{2}\frac{\partial^2 p(y,s;x,t)}{\partial x^2} + \frac{(z-x)^3}{6}\frac{\partial^3 p(y,s;\xi,t)}{\partial x^3}\right]dz\\
&= \frac{\partial p(y,s;x,t)}{\partial x}\int_{-\infty}^{\infty} p(z,t;x,t-\Delta t)(z-x)\,dz\\
&\quad + \frac{1}{2}\frac{\partial^2 p(y,s;x,t)}{\partial x^2}\int_{-\infty}^{\infty} p(z,t;x,t-\Delta t)(z-x)^2\,dz\\
&\quad + \frac{\partial^3 p(y,s;\xi,t)}{\partial x^3}\int_{-\infty}^{\infty} p(z,t;x,t-\Delta t)\frac{(z-x)^3}{6}\,dz,
\end{aligned}
$$

where ξ is between x and z. Dividing by Δt, letting $\Delta t \to 0$, and using the identities in (i)'–(iii)' with $\delta = 3$, we obtain the backward Kolmogorov differential equation (8.10).

The forward Kolmogorov differential equation can be derived from the assumptions (i)'–(iii)' and the identities (8.2) and (8.12). (This equation is derived for the time-homogeneous case in the Appendix for Chapter 8.) The *forward Kolmogorov differential equation* for a *nonhomogeneous diffusion process* is

$$\frac{\partial p(y,s;x,t)}{\partial s} = -\frac{\partial\,[a(y,s)p(y,s;x,t)]}{\partial y} + \frac{1}{2}\frac{\partial^2\,[b(y,s)p(y,s;x,t)]}{\partial y^2}. \tag{8.13}$$

Given the initial position and time (x,t), the forward Kolmogorov differential equation describes the dynamics of the process going "forward" in time. Using operator notation, the right side of the backward Kolmogorov differential equation (8.10) can be written as a differential operator, $\partial p/\partial t + Ap = 0$, where A is the following differential operator acting on p:

$$Ap = a(x,t)\frac{\partial p}{\partial x} + \frac{1}{2}b(x,t)\frac{\partial^2 p}{\partial x^2}.$$

Then the forward Kolmorogov differential equation (8.13) is $\partial p/\partial s + A^*p$, where A^* is known as the *adjoint operator* of A (Gard, 1988):

$$A^*p = \frac{\partial\,[a(y,s)p]}{\partial y} - \frac{1}{2}\frac{\partial^2\,[b(y,s)p]}{\partial y^2}.$$

If $a(x,t) = a(x)$ and $b(x,t) = b(x)$, the process is time homogeneous and $p(y,s;x,t) = p(y,x,s-t)$. Letting $\tau = s - t$, then

$$\frac{\partial p(y,s;x,t)}{\partial s} = \frac{\partial p(y,x,\tau)}{\partial \tau}.$$

The *forward Kolmogorov differential equation* for the *time-homogeneous process* is

$$\frac{\partial p(y,x,t)}{\partial t} = -\frac{\partial\,[a(y)p(y,x,t)]}{\partial y} + \frac{1}{2}\frac{\partial^2\,[b(y)p(y,x,t)]}{\partial y^2}. \tag{8.14}$$

The forward Kolmogorov differential equation is also referred to as the *Fokker-Planck equation* in physical applications (van Kampen, 2007).

The p.d.f. $p(x,t)$ of $X(t)$ is a solution of the forward Kolmogorov differential equation for the initial condition $X(0) = x_0$. Recall in this case $p(x,t) = p(x,t;x_0,0)$. Then the forward Kolmogorov differential equation (8.14) can be expressed as

$$\frac{\partial p(x,t)}{\partial t} = -\frac{\partial\,[a(x)p(x,t)]}{\partial x} + \frac{1}{2}\frac{\partial^2\,[b(x)p(x,t)]}{\partial x^2}.$$

Example 8.1. Brownian motion is a diffusion process, where $a(x) = 0$ and $b(x) = D$. It has already been shown in Section 8.3 that the p.d.f. for Brownian motion is a solution of the following forward Kolmogorov differential equation:

$$\frac{\partial p}{\partial t} = \frac{D}{2}\frac{\partial^2 p}{\partial x^2}, \quad x \in (-\infty, \infty).$$

Given the initial condition $X(0) = x_0$, this differential equation can be solved for $p(x,t)$:

$$p(x,t) = \frac{1}{\sqrt{2\pi Dt}}\exp\left(-\frac{(x-x_0)^2}{2Dt}\right).$$

It can be seen that the p.d.f. has a normal distribution with mean x_0 and variance Dt. In *standard* Brownian motion, $x_0 = 0$ and $D = 1$, so that the p.d.f is the normal distribution, $N(0,t)$. The forward and backward Kolmogorov differential equations for Brownian motion are the same. ∎

We mention briefly some of the underlying theory associated with forward and backward Kolmogorov differential equations. This theory shows some of the relationships between diffusion processes and DTMCs and CTMCs, and provides some background on the importance of the underlying assumptions to the solution of the Kolmogorov differential equations. For a more advanced treatment of this theory, please consult the references.

Whether or not the Kolmogorov differential equations have a solution depends on the coefficients and the domain. If, for example, $b(x) = 0$ at a particular value of x, $-\infty < x < \infty$, the differential equation becomes *singular*; the differential equation is no longer second-order and it cannot be solved on the entire spatial domain $(-\infty, \infty)$. In order to solve the Kolmogorov differential equations, it is required that on the interior of the spatial domain, the equation is nonsingular, $b(x) > 0$. Let the spatial domain be denoted as (r_1, r_2), $-\infty \leq r_1 < r_2 \leq \infty$. If every point in the interior of the interval may be reached from every other point with positive probability, the process is referred to as a *regular process* (Karlin and Taylor, 1981). A regular diffusion process is analogous to an irreducible Markov chain. The diffusion processes that we consider are regular.

We shall give a short description of the types of boundary classifications. A detailed classification scheme is based on the infinitesimal mean and variance $a(x)$ and $b(x)$, which shall not be given here [see, e.g., Bharucha-Reid (1997), Karlin and Taylor (1981), or Ricciardi (1986)]. We follow the discussion given by Ricciardi (1986). A boundary r_i may be either *accessible* or *inaccessible*. These boundaries are further classified as follows. An accessible boundary can be either a *regular* or an *exit* boundary, and an inaccessible boundary can be either a *natural* or an *entrance* boundary. If the boundary is never attained for finite time t, then the boundary is said to be *inaccessible*; otherwise it is *accessible*. An *exit* boundary is similar to an absorbing boundary; once the process reaches an exit boundary, it stays there (Ricciardi, 1986). At a *regular* boundary the process continues. For example, a regular boundary could be reflecting. At an *entrance* boundary, if the process begins there, it can enter the interior of the interval, but it can never return to an entrance boundary (Ricciardi, 1986). At a *natural boundary*, there is no probability flow. For example, in standard Brownian motion, the boundaries $r_1 = -\infty$ and $r_2 = \infty$ are natural boundaries. The forward and backward Kolmogorov equations can be solved using the initial condition. Often in population processes, where $X(t) \in [0, \infty)$ or $X(t) \in [0, M]$, the zero boundary is an exit boundary and the boundary $r_2 = \infty$ is a natural boundary or $r_2 = M$ is a regular boundary.

The coefficients in (ii) and (iii) require knowledge about the transition p.d.f. In Chapter 9, Section 9.9, these coefficients are derived from basic assumptions about population genetics processes. However, generally, they are difficult to compute directly from the model assumptions. We will show an alternate method for determining the coefficients in the Kolmogorov equations. It will be shown that they are related to the coefficients in a stochastic differential equation, which, in turn, are related to the coefficients in a deterministic model. Even if the coefficients a and b are known, analytical solutions to the forward and backward Kolmogorov equations are often difficult to obtain, especially if a and b are nonlinear. If the state space is $[0, \infty)$ or $[0, M]$, then, in addition to the initial condition, boundary conditions at 0 and M need to be specified in order to solve for the p.d.f. As a final example in this section, a simple forward Kolmogorov differential equation, where a and b are linear functions of x, is derived and then used to compute the mean and variance of the corresponding diffusion process.

Example 8.2. Assume that the infinitesimal mean and variance of a diffusion process are

$$a(x) \equiv \alpha x \quad \text{and} \quad b(x) \equiv \beta x. \tag{8.15}$$

It will be shown that the underlying stochastic process with these coefficients corresponds to an exponential growth or decay process. The forward Kolmogorov differential equation for this process takes the form

$$\frac{\partial p}{\partial t} = -\alpha \frac{\partial (xp)}{\partial x} + \frac{\beta}{2} \frac{\partial^2 (xp)}{\partial x^2}, \quad x \in (0, \infty). \tag{8.16}$$

The boundary at zero is assumed to be an exit or absorbing boundary, and the boundary at ∞ is a natural boundary. Assume the initial condition of the stochastic process is $X(0) = x_0$, so that $p(x_0, 0) = \delta(x - x_0)$. Bailey (1990) gives a solution of this partial differential equation, a Bessel function of the first kind. ∎

In the next example, expressions for the mean and variance, $\mu_X(t)$ and $\sigma_X^2(t)$, are derived for the preceding diffusion process.

Example 8.3. The mean and variance can be written as

$$\mu_X(t) = \int_0^\infty x p(x, t) \, dx$$

and

$$\sigma_X^2(t) = \int_0^\infty x^2 p(x, t) \, dx - m_X^2(t),$$

with initial conditions $\mu_X(0) = \int_{-\infty}^\infty x \delta(x - x_0) \, dx = x_0$ and $\sigma_X^2(0) = 0$.

Applying the forward Kolmogorov equation (8.16), then

$$\frac{d\mu_X}{dt} = \frac{d}{dt}\int_0^\infty xp(x,t)\,dx = \int_0^\infty x\frac{\partial p}{\partial t}\,dx$$

$$= \int_0^\infty \left[-\alpha x\frac{\partial(xp)}{\partial x} + \frac{1}{2}\beta x\frac{\partial^2(xp)}{\partial x^2}\right]dx$$

$$= -\alpha x^2 p(x,t)\big|_0^\infty + \alpha\int_0^\infty xp(x,t)\,dx$$

$$+ \frac{1}{2}\beta\, x\frac{\partial(xp)}{\partial x}\bigg|_0^\infty - \frac{1}{2}\beta\int_0^\infty \frac{\partial(xp)}{\partial x}\,dx$$

$$= \alpha\int_0^\infty xp(x,t)\,dx.$$

Thus,

$$\frac{d\mu_X}{dt} = \alpha\mu_X(t).$$

In the preceding derivation, it was assumed that

$$\lim_{x\to\infty} x^2\frac{\partial p(x,t)}{\partial x} = \lim_{x\to\infty} x^2 p(x,t) = \lim_{x\to\infty} xp(x,t) = 0,$$

which follows from the existence of the mean and variance of $X(t)$. The solution of the differential equation for the mean $\mu_X(t)$ with $\mu_X(0) = x_0$ is

$$\mu_X(t) = x_0 e^{\alpha t}.$$

The mean grows exponentially if $\alpha > 0$. The variance is equal to

$$\sigma_X^2(t) = \begin{cases} x_0\frac{\beta}{\alpha}e^{\alpha t}\left[e^{\alpha t} - 1\right], & \alpha \neq 0 \\ x_0\beta t, & \alpha = 0. \end{cases}$$

Higher-order moments can be obtained in a similar manner. ∎

The forward Kolmogorov differential equation in Example 8.2 corresponds to a continuous analogue of a simple birth and death Markov chain, where $\alpha = \lambda - \mu$ and $\beta = \lambda + \mu$, λ is the birth rate, and μ is the death rate. This can be seen by comparing the mean and variance derived previously with that of the simple birth and death Markov chain discussed in Section 6.4.3. In this case, the mean is a solution of an ordinary differential equation, where the ordinary differential equation only has the drift term from the stochastic model.

Stochastic realizations of diffusion processes are based on the Wiener process or standard Brownian motion, introduced in Section 8.3. The Wiener process is defined more formally in the next section.

8.6 Wiener Process

The Wiener process is also referred to as *standard Brownian motion* or as the *Wiener-Einstein process* (Taylor and Karlin, 1998). The term *Brownian motion* acknowledges the contributions made by Robert Brown, an English botanist, to the physical description of the process. In 1827, Brown noticed that pollen grains suspended in water moved constantly in a zigzag motion under the lens of the microscope. In 1905, Albert Einstein stated laws governing Brownian motion using principles from the kinetic theory of heat. Einstein received the Nobel prize for his work. However, it was Norbert Wiener's (1894–1964) research in 1923 that laid the mathematical foundation for Brownian motion.

Melsa and Sage (1973) relate the Wiener process to the physical process observed by Brown and the concept of Brownian motion. Suppose $W(t)$ is the displacement from the origin at time t of a small particle. The displacement of a particle over the time interval t_1 to t_2 is long compared to the time between impacts. The central limit theorem can be applied to the sum of a large number of these small disturbances so that it can be assumed $W(t_2) - W(t_1)$ has a normal density. The density of the particles' displacement depends on the length of the time interval and not on the time of observation; therefore, the probability density of the displacement from time t_1 to t_2 is the same as from time $t_1 + t$ to time $t_2 + t$ (Melsa and Sage, 1973). These concepts are made more rigorous in the following definition.

DEFINITION 8.4   *The stochastic process $\{W(t) : t \in [0, \infty)\}$ is a* Wiener process *(standard Brownian motion) if $W(t)$ depends continuously on t, $W(t) \in (-\infty, \infty)$, and the following three conditions hold:*

(1) For $0 \leq t_1 < t_2 < \infty$, $W(t_2) - W(t_1)$ is normally distributed with mean zero and variance $t_2 - t_1$; that is, $W(t_2) - W(t_1) \sim N(0, t_2 - t_1)$.

(2) For $0 \leq t_0 < t_1 < t_2 < \infty$, the increments $W(t_1) - W(t_0)$ and $W(t_2) - W(t_1)$ are independent.

(3) $\mathrm{Prob}\{W(0) = 0\} = 1$.

Definition 8.4 implies that the Wiener process has *stationary* and *independent increments*. For the intervals $0 \leq t_0 < t_1 < t_2 < \cdots < t_{n-1} < t_n$, the n random variables $W(t_1) - W(t_0)$, $W(t_2) - W(t_1)$, $\cdots$, $W(t_n) - W(t_{n-1})$ are independent. Also, the increments $W(t_1 + \Delta t) - W(t_1)$ and $W(t_2 + \Delta t) - W(t_2)$ are *stationary*, meaning that they have the same normal density, $N(0, \Delta t)$, for any $t_1, t_2 \in [0, \infty)$ and $\Delta t > 0$. To simplify the notation, define $\Delta t_i = t_{i+1} - t_i$ and $\Delta W(t_i) = W(t_{i+1}) - W(t_i)$. Applying the Law of Large Numbers (Theorem 1.2) to $\Delta W(i) \sim N(0,1)$ for $i = 0, 1, \ldots, t$, where

$$W(t) = \sum_{i=0}^{t} \Delta W(i) \sim N(0, t),$$

it follows that $W(t)/t \to 0$ with probability one:

$$\text{Prob} \left\{ \lim_{t \to \infty} \left| \frac{W(t)}{t} \right| = 0 \right\} = 1. \qquad (8.17)$$

It can be easily shown that the Wiener process $\{W(t) : t \in [0, \infty)\}$ is a diffusion process, and has the following transition p.d.f.:

$$p(y, x, t) = \frac{1}{\sqrt{2\pi t}} \exp\left(\frac{-(y-x)^2}{2t}\right).$$

Using the transition p.d.f. $p(y, x, \Delta t)$ in Definition 8.3 it can be shown directly that the three conditions (i)–(iii) are satisfied. We show that the stronger conditions (i)$'$–(iii)$'$ with $\delta = 4$ are also satisfied. Note that $\Delta W(t) = W(t + \Delta t) - W(t) \sim N(0, \Delta t)$. Hence,

$$E(\Delta W(t)) = 0, \quad E([\Delta W(t)]^2) = \Delta t, \quad \text{and} \quad E([\Delta W(t)]^4) = 3(\Delta t)^2.$$

Dividing these expressions by Δt and taking the limit as $\Delta t \to 0^+$ yields the conditions (i)$'$–(iii)$'$, where $a(x, t) = 0$ and $b(x, t) = 1$.

Sample paths of $W(t)$ are continuous functions of t but they do not have bounded variation and they are almost everywhere nondifferentiable. If $W(t)$ is a Wiener process, $dW(t)/dt$ has no meaning in the usual sense. In particular, the Riemann integral

$$\int_0^t g(\tau) \frac{dW(\tau)}{d\tau} \, d\tau$$

has no meaning since $dW(\tau)/d\tau$ is not defined. In addition, the Riemann-Stieltjes integral

$$\int_0^t g(\tau) \, dW(\tau)$$

has no meaning since $W(\tau)$ does not have bounded variation. Thus, a new definition of a stochastic integral is needed.

8.7 Itô Stochastic Integral

There are a variety of ways to define a stochastic integral. The two most well-known definitions of a stochastic integral are *Itô* and *Stratonovich*. The name Itô refers to the Japanese mathematician Kiyoshi Itô (1915-2008) who developed much of the basic theory. The name Stratonovich refers to the Russian physicist, Ruslan Stratonovich (1930-1997) who defined an alternative to the Itô stochastic integral. Each definition leads to a different "stochastic calculus", and, therefore, when speaking of stochastic integrals, it is important to specify whether the calculus refers to the Itô or to the Stratonovich definition. We shall use the Itô definition, which is the one frequently used in biological examples (see, e.g., E. Allen, 2007; Gard, 1988; Øksendal, 2000; Turelli, 1977). Braumann (2007) shows that in population growth models, whether Itô or Stratonovich calculi are applied is a matter of semantics; it is more important to carefully define the terms according to the observed population process.

Let $f(t, X(t))$ be a function of the random variable $X(t)$, where $t \in [a, b]$ and $\{X(t) : t \in [0, \infty)\}$ is a stochastic process. For simplicity, the random function $f(t, X(t))$ is denoted as $f(t)$. The Itô stochastic integral is defined for random functions f satisfying the following condition:

$$\int_a^b E(f^2(t)) \, dt < \infty, \tag{8.18}$$

provided that $E(f^2(t))$ exists. The integral (8.18) is a Riemann integral. In more advanced treatments of stochastic differential equations based on measure theory, $E(f^2(t))$ is assumed to be Lebesgue integrable (e.g., Arnold, 1974; Gard, 1988; Øksendal, 2000).

DEFINITION 8.5 *Assume $f(t)$ is a random function satisfying (8.18). Let $a = t_1 < t_2 < \cdots < t_k < t_{k+1} = b$ be a partition of $[a, b]$, $\Delta t = t_{i+1} - t_i = (b-a)/k$, and $\Delta W(t_i) = W(t_{i+1}) - W(t_i)$, where $W(t)$ is the standard Wiener process. Then the Itô stochastic integral of f is defined as*

$$\int_a^b f(t) \, dW(t) = l.i.m._{k \to \infty} \sum_{i=1}^k f(t_i) \, \Delta W(t_i), \tag{8.19}$$

where l.i.m. denotes mean square convergence. If $F_k = \sum_{i=1}^k f(t_i) \, \Delta W(t_i)$ and $\mathcal{I} = \int_a^b f(t) \, dW(t)$, then $l.i.m._{k \to \infty} F_k = \mathcal{I}$ means

$$\lim_{k \to \infty} E[(F_k - \mathcal{I})^2] = 0.$$

The need for assumption (8.18) is not evident from the definition, but it is important for existence of the limit in (8.19). The notation $l.i.m.$ for mean square convergence has been used by others (Hsu, 1997; Melsa and Sage, 1973). Mean square convergence, $E[(F_k - \mathcal{I})^2] \to 0$, implies convergence in the mean, $E(|F_k - \mathcal{I}|) \to 0$. However, the converse is not true; convergence in the mean does not imply mean square convergence. Mean square convergence is L^2 convergence for real-valued, measurable functions F_k and $\mathcal{I}$: $\|F_k - \mathcal{I}\|_2 = \sqrt{E[(F_k - \mathcal{I})^2]} \to 0$. In the definition of the Itô stochastic integral, a partition of equal intervals is used. The definition can be made more general by using a general partition. In the limit, however, the norm of the partition must approach zero (i.e., the length of the largest subinterval must approach zero).

The *Stratonovich stochastic integral* differs from the Itô stochastic integral in that $f(t_i)$ in (8.19) is replaced by $f([t_i + t_{i+1}]/2)$ (Øksendal, 2000). The random function is evaluated at the midpoint of the interval, rather than the left endpoint. However, it is more common practice to define the Stratonovich integral as the following limit:

$$l.i.m._{k \to \infty} \sum_{i=1}^{k} \left[\frac{f(t_i) + f(t_{i+1})}{2} \right] \Delta W(t_i),$$

where the function is averaged over the interval (see Kloeden and Platen, 1992, p. 100). In the Stratonovich stochastic integral, it is necessary to know f "in the future" whereas in the Itô stochastic integral it is only necessary to know f at the beginning of the time interval. Thus, the integrand in the Itô stochastic integral is referred to as *nonanticipative*. A function $f(t)$ is called *nonanticipating* if $f(t)$ is statistically independent of the Wiener increment $W(t + s) - W(t)$ for $s > 0$. There are certain advantages in applying the Itô definition, such as the Itô isometry and Itô's formula which are defined in this section. In addition, it will be shown that the Stratonovich integral can be converted to an Itô integral and vice versa. Therefore, we shall limit our discussion to Itô rather than Stratonovich stochastic integrals.

Some simple Itô stochastic integrals can be verified directly from the definition. For example, it is straightforward to verify that

$$\int_a^b dW(t) = W(b) - W(a),$$

and, in general, for any well-defined random function $F(W(t), t)$ satisfying (8.18),

$$\int_a^b dF(W(t), t) = F(W(b), b) - F(W(a), a). \tag{8.20}$$

The next theorem shows that the Itô stochastic integral is a linear operator on the set of functions f whose Itô stochastic integral exists [i.e., functions satisfying (8.18)].

THEOREM 8.1
Assume $f(t)$ and $g(t)$ are random functions satisfying the inequality (8.18) and α, a, b, and c are constants satisfying $a < c < b$. Then

$$\text{(i)} \quad \int_a^b \alpha f(t)\, dW(t) = \alpha \int_a^b f(t)\, dW(t)$$

$$\text{(ii)} \quad \int_a^b (f(t) + g(t))\, dW(t) = \int_a^b f(t)\, dW(t) + \int_a^b g(t)\, dW(t)$$

$$\text{(iii)} \quad \int_a^b f(t)\, dW(t) = \int_a^c f(t)\, dW(t) + \int_c^b f(t)\, dW(t).$$

A sketch of the proof for part (iii) is given. More rigorous proofs can be found in Arnold (1974), Gard (1988), and Øksendal (2000).

Proof. Let k_1 and k be two positive integers satisfying $1 < k_1 < k$. Let $F_{k_1} = \sum_{i=1}^{k_1} f(t_i)\,\Delta W(t_i)$ and $F_{k-k_1} = \sum_{i=k_1+1}^{k} f(t_i)\,\Delta W(t_i)$, where $\{t_i\}_{i=1}^{k_1+1}$ is a partition of $[a,c]$ and $\{t_i\}_{i=k_1+1}^{k+1}$ is a partition of $[c,b]$. Note that $t_1 = a$, $t_{k_1+1} = c$, and $t_{k+1} = b$. Then $\{t_i\}_{i=1}^{k+1}$ is a partition of $[a,b]$, so we can define $F_k = F_{k_1} + F_{k-k_1}$. In addition, let $\mathcal{I} = \int_a^b f(t)\, dW(t)$, $\mathcal{I}_1 = \int_a^c f(t)\, dW(t)$, and $\mathcal{I}_2 = \int_c^b f(t)\, dW(t)$.

For the proof of part (iii), let $\epsilon > 0$. There exists a positive integer K such that for k_1 and $k - k_1$ greater than K,

$$E\left[(F_{k_1} - \mathcal{I}_1)^2\right] < \epsilon/3 \quad \text{and} \quad E\left[(F_{k-k_1} - \mathcal{I}_2)^2\right] < \epsilon/3.$$

Then

$$\begin{aligned}
E\left[(F_k - \mathcal{I}_1 - \mathcal{I}_2)^2\right] &= E\left[(F_{k_1} + F_{k-k_1} - \mathcal{I}_1 - \mathcal{I}_2)^2\right] \\
&= E[(F_{k_1} - \mathcal{I}_1)^2] + 2E[(F_{k_1} - \mathcal{I}_1)(F_{k-k_1} - \mathcal{I}_2)] \\
&\quad + E[(F_{k-k_1} - \mathcal{I}_2)^2].
\end{aligned}$$

If f is a simple random function or an elementary function (Arnold, 1974; Gard, 1988; Øksendal, 2000), then $F_{k_1} - \mathcal{I}_1$ and $F_{k-k_1} - \mathcal{I}_2$ are independent because the intervals $[a,c]$ and $[c,b]$ are nonoverlapping. General random functions are limits of these simple functions, so it can be shown that

$$E[(F_{k_1} - \mathcal{I}_1)(F_{k-k_1} - \mathcal{I}_2)] = E(F_{k_1} - \mathcal{I}_1)E(F_{k-k_1} - \mathcal{I}_2)$$

(Øksendal, 2000). This latter expectation can be made sufficiently small ($< \epsilon/3$) for K sufficiently large. Part (iii) then follows from the Itô Definition 8.5 and the fact that the Itô integral is unique in the mean square sense. $\quad\square$

The next results show that the expectation of an Itô stochastic integral is zero and the expectation of the square of the integral is the integral of the expectation of the integrand squared.

THEOREM 8.2
Suppose $f(t)$ is a random function satisfying (8.18). Then

(i) $E\left[\int_a^b f(t)\,dW(t)\right] = 0$ and

(ii) $E\left[\left(\int_a^b f(t)\,dW(t)\right)^2\right] = \int_a^b E(f^2(t))\,dt.$

Property (ii) in Theorem 8.2 is known as the *Itô isometry property*. Properties (i) and (ii) are straightforward to verify for constant functions $f(t) = c$:

$$E\left[\int_a^b c\,dW(t)\right] = cE\left[W(b) - W(a)\right] = 0$$

because $W(b) - W(a) \sim N(0, b-a)$.

$$E\left[\left(\int_a^b c\,dW(t)\right)^2\right] = c^2 E\left[(W(b) - W(a))^2\right] = c^2(b-a)$$

because $E\left[(W(b) - W(a))^2\right] = \operatorname{Var}(W(b) - W(a)) = b - a$.

In the next example, the Itô stochastic integral is evaluated using Definition 8.5. As with Riemann integration, this method can be quite tedious. Some properties of Itô stochastic integrals will be stated in the next section that will help simplify the evaluation of some of these integrals.

Example 8.4. The following Itô stochastic integral will be shown to be equal to

$$\int_0^t W(\tau)\,dW(\tau) = \frac{1}{2}\left[W^2(t) - t\right]. \tag{8.21}$$

It is necessary to use the properties of the Wiener process, $\Delta W(t_i) = W(t_{i+1}) - W(t_i) \sim N(0, \Delta t)$, $W(0) = 0$, and $\Delta W(t_i)$ and $\Delta W(t_j)$ are independent for $i \neq j$ (since they are defined on nonoverlapping intervals). For simplicity of notation, let $\Delta W(t_i) = \Delta W_i$ and $W(t_i) = W_i$. Using the Itô stochastic integral definition,

$$\int_0^t W(\tau)\,dW(\tau) = l.i.m._{k\to\infty} \sum_{i=1}^k W_i\,\Delta W_i,$$

where the partition satisfies $0 = t_1 < t_2 < \cdots < t_{k+1} = t$ and $\Delta t = t/k$. Note that

$$\begin{aligned}
\Delta(W_i^2) &= W_{i+1}^2 - W_i^2 \\
&= (W_{i+1} - W_i)^2 + 2W_i(W_{i+1} - W_i) \\
&= (\Delta W_i)^2 + 2W_i\Delta W_i.
\end{aligned}$$

Applying the preceding identity,

$$\sum_{i=1}^{k} W_i \Delta W_i = \frac{1}{2} \sum_{i=1}^{k} [\Delta(W_i^2) - (\Delta W_i)^2]$$

$$= \frac{1}{2} W_{k+1}^2 - \frac{1}{2} \sum_{i=1}^{k} (\Delta W_i)^2$$

$$= \frac{1}{2} W^2(t) - \frac{1}{2} \sum_{i=1}^{k} (\Delta W_i)^2$$

since $W_{k+1} = W(t)$ and $W(t_1) = W(0) = 0$. To show the stochastic integral converges in the mean square sense to the limit given in (8.21), we need to show that $\sum_{i=1}^{k} (\Delta W_i)^2$ converges to t in the mean square sense. That is,

$$E\left[\left(t - \sum_{i=1}^{k} (\Delta W_i)^2\right)^2\right] \to 0 \quad \text{as} \quad k \to \infty.$$

The mean square is

$$E\left[t^2 - 2t \sum_{i=1}^{k} (\Delta W_i)^2 + \sum_{i=1}^{k} \sum_{j=1}^{k} (\Delta W_i)^2 (\Delta W_j)^2\right].$$

Because $E[(\Delta W_i)^2] = \Delta t$, it follows that the preceding expectation equals

$$t^2 - 2t(t) + E\left[\sum_{i=1}^{k} \sum_{j=1}^{k} (\Delta W_i)^2 (\Delta W_j)^2\right]. \tag{8.22}$$

The expectation of $(\Delta W_i)^4$ is the fourth moment of the normal distribution $N(0, \Delta t)$, which can be shown to equal

$$E((\Delta W_i)^4) = 3(\Delta t)^2 = 3(t/k)^2.$$

(See Chapter 1.) Returning to the expectation in (8.22) and using the fact that ΔW_i is independent of ΔW_j for $i \neq j$,

$$E\left[\sum_{i=1}^{k} \sum_{j=1}^{k} (\Delta W_i)^2 (\Delta W_j)^2\right] = E\left[\sum_{i=1}^{k} (\Delta W_i)^2\right] E\left[\sum_{j=1, j \neq i}^{k} (\Delta W_j)^2\right]$$

$$+ E\left[\sum_{i=1}^{k} (\Delta W_i)^4\right]$$

$$= \sum_{i=1}^{k} \Delta t \sum_{j=1}^{k-1} \Delta t + k 3(\Delta t)^2$$

$$= t(k-1)\frac{t}{k} + 3k \left(\frac{t}{k}\right)^2,$$

since $\Delta t = t/k$. The expression on the right side approaches t^2 as $k \to \infty$. Thus, the expression given in (8.22) approaches zero as $k \to \infty$. ∎

Example 8.5. When Theorem 8.1 is applied to Example 8.4, it follows that the Itô stochastic integral of $W(t)$ from a to b is

$$\int_a^b W(t)\, dW(t) = \frac{1}{2}\left[W^2(b) - W^2(a)\right] - \frac{1}{2}[b - a].$$

This simple example shows that the Itô integral cannot be treated in the same way as a Riemann-Stieltjes integral. ∎

Example 8.6. From the Itô isometry property in Theorem 8.2, it follows that

$$E\left[\left(\int_a^b W(t)\, dW(t)\right)^2\right] = \int_a^b E(W^2(t))\, dt = \int_a^b t\, dt = \frac{1}{2}(b^2 - a^2).$$

In addition, this identity can be verified directly using the value for the integral of $W(t)$ given in Example 8.5:

$$E\left[\left(\int_a^b W(t)\, dW(t)\right)^2\right] = \frac{1}{4}E\left[\left(W^2(b) - W^2(a) - (b - a)\right)^2\right]$$

$$= \frac{1}{4}\left\{E\left[W^4(b) - 2W^2(b)W^2(a) + W^4(a)\right]\right.$$

$$\left. - E\left[2(W^2(b) - W^2(a))(b - a) - (b - a)^2\right]\right\}$$

$$= \frac{1}{4}\left[3b^2 - 2E(W^2(b)W^2(a)) + 3a^2\right.$$

$$\left. - 2(b - a)^2 + (b - a)^2\right]$$

$$= \frac{1}{4}\left[2b^2 + 2a^2 + 2ab - 2E(W^2(b)W^2(a))\right],$$

since $W(t) \sim N(0, t)$ implies $E(W^2(t)) = t$ and $E(W^4(t)) = 3t^2$. The random functions $W^2(a)$ and $W^2(b)$ are not independent (intervals $[0, a]$ and $[0, b]$ are overlapping). Denote $W^2(b) = W^2(b) - W^2(a) + W^2(a)$, so that $W^2(b) - W^2(a)$ and $W^2(a)$ are independent (intervals $[0, a]$ and $[a, b]$ are nonoverlapping if $0 < a < b$). Then $W^2(b)W^2(a) = (W^2(b) - W^2(a))W^2(a) + W^4(a)$. This identity is used to calculate the expectation given previously,

$$E\left[\left(\int_a^b W(t)\, dW(t)\right)^2\right] = \frac{1}{4}\left[2b^2 + 2a^2 + 2ab - 2\{(b - a)a + 3a^2\}\right]$$

$$= \frac{1}{4}\left[2b^2 - 2a^2\right] = \frac{1}{2}(b^2 - a^2).$$

∎

Evaluation of an Itô integral using the definition can be very difficult. This is also the case for evaluation of a Riemann integral via Riemann sums. Some additional theory will be developed in the next section that allows some Itô integrals to be evaluated. First, some notation is introduced. Denote the solution of the integral in Example 8.4 as $X(t)$,

$$X(t) = \int_0^t W(\tau) \, dW(\tau).$$

This integral equation is often written in differential form as

$$dX(t) = W(t) \, dW(t), \quad X(0) = 0.$$

This latter form is referred to as an *Itô stochastic differential equation*. This differential notation is used only for convenience because the Wiener process $W(t)$ is not differentiable. The expression $dW(t)$ has meaning only in the stochastic sense of Definition 8.5.

In the next section, Itô stochastic differential equations are introduced and conditions for existence and uniqueness of solutions are discussed. In addition, the relationship between Itô stochastic differential equations (or Itô stochastic integrals) and diffusion processes is discussed.

8.8   Itô Stochastic Differential Equation

First, the definition of a general Itô stochastic differential equation (SDE) is given. Then conditions for existence and uniqueness of a solution of an Itô SDE are stated.

DEFINITION 8.6   *A stochastic process $\{X(t) : t \in [0, \infty)\}$ is said to satisfy an* Itô stochastic differential equation (SDE), *written as,*

$$dX(t) = \alpha(X(t), t) \, dt + \beta(X(t), t) \, dW(t), \tag{8.23}$$

if for $t \geq 0$ it is a solution of the integral equation,

$$X(t) = X(0) + \int_0^t \alpha(X(\tau), \tau) \, d\tau + \int_0^t \beta(X(\tau), \tau) dW(\tau),$$

where the first integral is a Riemann integral and the second integral is an Itô stochastic integral.

The conditions needed for existence and uniqueness of SDEs are closely related to the conditions for existence and uniqueness of ordinary differential equations (ODEs). In the case of an ODE, let $y(t)$ be a continuous and differentiable real-valued function of t, the unique solution of the following differential equation:

$$\frac{dy}{dt} = f(y(t), t), \quad y(0) = y_0.$$

From the theory of ODEs, there exists a unique solution of this initial value problem if f is continuous in the region $R = \{(y, t) : |y - y_0| \leq c_1, |t| < c_2, \}$ and satisfies the following Lipschitz condition: there exists a $K > 0$ such that

$$|f(x, t) - f(y, t)| \leq K|x - y|$$

(Corduneanu, 1977). The differential equation can be written in terms of differentials, $dy = f(t, y(t)) \, dt$, or as an integral equation,

$$y(t) = \int_0^t f(\tau, y(\tau)) \, d\tau.$$

It can be shown that there exists a unique Markov process $\{X(t) : t \in [0, T]\}$, which is a solution of the Itô SDE (8.23) on $[0, T]$, if the following two conditions are satisfied (Arnold, 1974; Gard, 1988; Kloeden and Platen, 1992; Øksendal, 2000). There exists a constant $K > 0$ such that

(a) $|\alpha(x, t) - \alpha(y, t)| + |\beta(x, t) - \beta(y, t)| \leq K|x - y|$, for $x, y \in \mathbb{R}$ and $t \in [0, T]$

(b) $|\alpha(x, t)|^2 + |\beta(x, t)|^2 \leq K^2(1 + |x|^2)$, for $x \in \mathbb{R}$ and $t \in [0, T]$

Then the solution $X(t)$ to the Itô SDE (8.23) is continuous with probability one,

$$\sup_{t \in [0,T]} E(X^2(t)) < \infty,$$

[compare with condition (8.18)] and is *pathwise unique*, which means that if X and Y are two solutions,

$$\text{Prob} \left\{ \sup_{[0,T]} |X(t) - Y(t)| = 0 \right\} = 1$$

(Gard, 1988). Please consult the references for a proof of the existence and uniqueness result. Condition (a) ensures that solutions $X(t)$ do not explode (become infinite in finite time), and condition (b) ensures that solutions are pathwise unique. Condition (b) can be replaced with the following condition:

(b)′ $|\alpha(x, t)| + |\beta(x, t)| \leq K^*(1 + |x|)$, where K^* is some positive constant.

Condition (b) implies condition (b)′ as a consequence of the following inequalities:

$$(|\alpha| + |\beta|)^2 \leq 2|\alpha|^2 + 2|\beta|^2 \quad \text{and} \quad \sqrt{1 + |x|^2} \leq 1 + |x|.$$

It can be shown under suitable smoothness of the coefficients that an Itô solution $X(t)$ of (8.23) satisfying (a) and (b) is a *diffusion process* on $[0, T]$; that is, it is a solution of the Kolmogorov equations with drift coefficient α and diffusion coefficient β^2 (see, e.g., E. Allen, 2007; Arnold, 1974; Gard, 1988; Kloeden and Platen, 1992). For example, the forward Kolmogorov equation corresponding to the Itô SDE (8.23) is

$$\frac{\partial p}{\partial t} = -\frac{\partial(\alpha(x, t)p)}{\partial x} + \frac{1}{2}\frac{\partial^2(\beta^2(x, t)p)}{\partial x^2}$$

where $p(x, t)$ is the p.d.f. of the stochastic process $\{X(t) : t \in [0, T]\}$.

Example 8.7. Consider the diffusion equation with drift, equations (8.5) and (8.6). The forward Kolmogorov differential equation is

$$\frac{\partial p}{\partial t} = -c\frac{\partial p}{\partial x} + \frac{D}{2}\frac{\partial^2 p}{\partial x^2}, \quad x \in (-\infty, \infty),$$

$p(x, 0) = \delta(x - x_0)$ with solution

$$p(x, t) = \frac{1}{\sqrt{2\pi Dt}}\exp\left(-\frac{(x - x_0 - ct)^2}{2Dt}\right).$$

The SDE corresponding to this process is

$$dX(t) = c\,dt + \sqrt{D}\,dW(t), \quad X(0) = x_0$$

so that $X(t) \sim N(x_0 + ct, Dt)$.   ∎

Example 8.8. Suppose $X(t)$ is a solution of the following SDE:

$$dX(t) = (\lambda - \mu)X(t)\,dt + \sqrt{(\lambda + \mu)X(t)}\,dW(t), \quad X(0) = x_0 > 0$$

for $X(t) \in [0, \infty)$, where λ and μ are positive constants. Then $X(t)$ has a p.d.f. that is a solution of the following forward Kolmogorov differential equation:

$$\frac{\partial p}{\partial t} = -(\lambda - \mu)\frac{\partial(xp)}{\partial x} + \frac{(\lambda + \mu)}{2}\frac{\partial^2(xp)}{\partial x^2}, \quad x \in (0, \infty),$$

$p(x, 0) = \delta(x - x_0)$. The solution of the SDE must be restricted to $[0, \infty)$. For a population process, a reasonable assumption is that zero is an exit or absorbing boundary, where the stochastic process $X(t + \tau) = 0$ for $\tau > 0$ if $X(t) = 0$. This particular partial differential equation is the one given in Example 8.2, where the mean and variance correspond to an exponential growth model. This SDE is a stochastic representation of an exponential growth model, $dX/dt = (\lambda - \mu)X$.   ∎

The next result is the "chain rule" for Itô SDEs and is known as *Itô's formula*.

THEOREM 8.3 Itô's Formula

Suppose $X(t)$ is a solution of the following Itô SDE:

$$dX(t) = \alpha(X(t), t)\, dt + \beta(X(t), t)\, dW(t).$$

If $F(x,t)$ is a real-valued function defined for $x \in \mathbb{R}$ and $t \in [a,b]$, with continuous partial derivatives, $\partial F/\partial t$, $\partial F/\partial x$, and $\partial^2 F/\partial x^2$, then

$$dF(X(t), t) = f(X(t), t)\, dt + g(X(t), t)\, dW(t), \qquad (8.24)$$

where

$$f(x,t) = \frac{\partial F(x,t)}{\partial t} + \alpha(x,t)\frac{\partial F(x,t)}{\partial x} + \frac{1}{2}\beta^2(x,t)\frac{\partial^2 F(x,t)}{\partial x^2} \qquad (8.25)$$

$$g(x,t) = \beta(x,t)\frac{\partial F(x,t)}{\partial x}. \qquad (8.26)$$

Verification of Theorem 8.3 can be found in Arnold (1974), Gard (1988), and Øksendal (2000). It should be noted that Theorem 8.3 does *not* hold for Stratonovich stochastic integrals.

Itô's formula when applied to an Itô SDE gives an "integration by parts" formula. It can be seen that Itô's formula gives a much different integration by parts formula than when Riemann integration is used. It is the term $(1/2)\beta^2(\partial^2 F/\partial x^2)\, dt$ that makes the stochastic formula different from the Riemann integration formula. For example, if x and t are real variables, then

$$dF(x,t) = \frac{\partial F}{\partial t}dt + \frac{\partial F}{\partial x}\, dx.$$

Replacing dx by $\alpha\, dt + \beta\, dW(t)$ yields Itô's formula minus the additional term $(1/2)\beta^2(\partial^2 F/\partial x^2)\, dt$.

To motivate why Itô's formula has the form given by (8.25) and (8.26), apply a Taylor series expansion to F:

$$\begin{aligned} dF(x,t) &\approx \Delta F \\ &\approx \frac{\partial F}{\partial t}\Delta t + \frac{\partial F}{\partial x}\Delta x + \frac{1}{2}\frac{\partial^2 F}{\partial t^2}(\Delta t)^2 \\ &\quad + \frac{1}{2}\frac{\partial^2 F}{\partial t \partial x}\Delta t\, \Delta x + \frac{1}{2}\frac{\partial^2 F}{\partial x^2}(\Delta x)^2 + \cdots. \end{aligned}$$

Replacing Δx by $\alpha\, \Delta t + \beta\, \Delta W$ and $(\Delta x)^2$ by $\alpha^2(\Delta t)^2 + 2\alpha\beta\, \Delta t\, \Delta W + \beta^2(\Delta W)^2$, then

$$\begin{aligned} dF(x,t) &\approx \frac{\partial F}{\partial t}\Delta t + \frac{\partial F}{\partial x}(\alpha\, \Delta t + \beta\, \Delta W) \\ &\quad + \frac{1}{2}\frac{\partial^2 F}{\partial t^2}(\Delta t)^2 + \frac{1}{2}\frac{\partial^2 F}{\partial t \partial x}(\alpha(\Delta t)^2 + \beta\, \Delta t\, \Delta W) \\ &\quad + \frac{1}{2}\frac{\partial^2 F}{\partial x^2}\left[\alpha^2(\Delta t)^2 + 2\alpha\beta\Delta t\Delta W + \beta^2(\Delta W)^2\right] + \cdots. \end{aligned}$$

Noting that $E\left[(\Delta W)^2\right]$ is Δt and approximating $(\Delta W)^2$ by Δt, the preceding expression can be written simply as

$$dF(x,t) \approx \left[\frac{\partial F}{\partial t} + \alpha\frac{\partial F}{\partial x} + \frac{\beta^2}{2}\frac{\partial^2 F}{\partial x^2}\right]\Delta t + \beta\frac{\partial F}{\partial x}\Delta W + o(\Delta t).$$

All terms except those in Itô's formula are $o(\Delta t)$. Itô's formula follows by letting $\Delta t \to 0$.

The usefulness of Itô's formula in evaluation of Itô stochastic integrals is illustrated in the next examples.

Example 8.9. Itô's formula can be used to verify the Itô stochastic integral in Example 8.5:

$$\int_a^b W(t)\,dW(t) = \frac{1}{2}\left[W^2(b) - W^2(a) - (b - a)\right]. \qquad (8.27)$$

Let $X(t) = W(t)$ and $F(x,t) = x^2$. The SDE satisfied by $X(t)$ is $dX(t) = dW(t)$, $\alpha = 0$, and $\beta = 1$. Applying (8.24) and the identities (8.25) and (8.26) in Itô's formula,

$$dF(W(t),t) = d\left[W^2(t)\right] = dt + 2W(t)\,dW(t).$$

Integrating from a to b,

$$W^2(b) - W^2(a) = b - a + 2\int_a^b W(t)\,dW(t),$$

shows that equation (8.27) is satisfied. ∎

Example 8.10. Itô's formula is used to evaluate the following stochastic integral:

$$\int_a^b t\,dW(t).$$

Let $X(t) = W(t)$ and $F(x,t) = tx$. Then it follows from Itô's formula that

$$dF(W(t),t) = d[tW(t)] = W(t)\,dt + t\,dW(t).$$

Integrate from a to b and rearrange terms. The integral satisfies

$$\int_a^b t\,dW(t) = bW(b) - aW(a) - \int_a^b W(t)\,dt.$$

∎

Example 8.11. Consider the following SDE:

$$dX(t) = rX(t)\,dt + cX(t)dW(t), \qquad (8.28)$$

where $rX(t)\,dt$ represents exponential growth, $r > 0$. It is assumed that the term $cX(t)dW(t)/dt$ represents environmental variation which is proportional to the population size. The stochastic process corresponding to this SDE is also known as *geometric Brownian motion*. To solve the SDE, Itô's formula is used with $F(x,t) = \ln(x)$. Applying Itô's formula,

$$dF(X(t),t) = d\left[\ln(X(t))\right] = \left(r - \frac{1}{2}c^2\right)dt + c\,dW(t).$$

Integrating from 0 to t yields

$$\ln\left(\frac{X(t)}{X(0)}\right) = (r - c^2/2)t + cW(t),$$

and solving for $X(t)$ gives the solution of the Itô stochastic integral as

$$X(t) = X(0)\exp\left([r - c^2/2]t + cW(t)\right).$$

From property (i) of Theorem 8.2 it follows that

$$E(X(t)) = E\left[\int_0^t rX(\tau)\,d\tau\right] + cE\left[\int_0^t X(\tau)dW(\tau)\right].$$

Applying property (i) of Theorem 8.2 and interchanging expectation and integration,

$$E(X(t)) = r\int_0^t E(X(\tau))\,d\tau.$$

Expressed as an ordinary differential equation,

$$\frac{dE(X(t))}{dt} = rE(X(t)).$$

The solution is

$$E(X(t)) = E(X(0))e^{rt} = X(0)e^{rt}.$$

The expectation of the Itô solution agrees with the deterministic exponential growth model. It can be shown that if $r < c^2/2$, then

$$\text{Prob}\left\{\lim_{t\to\infty}|X(t)| = 0\right\} = 1,$$

and if $r > c^2/2$, then

$$\text{Prob}\left\{\lim_{t\to\infty}|X(t)| = \infty\right\} = 1,$$

where $X(t)$ is the Itô solution (Gard, 1988; Øksendal, 2000). Three sample paths are graphed in Figure 8.2 for the stochastic model in Example 8.11. ∎

It can be shown that the solution of the Stratonovich SDE in Example 8.11 is

$$X(t) = X(0) \exp(rt + cW(t))$$

and its expectation is

$$E(X(t)) = E(X(0)) \exp(rt + c^2 t/2)$$

(Gard, 1988; Øksendal, 2000). Therefore, unlike the expectation for the Itô solution, the expectation for the Stratonovich solution does not agree with the deterministic solution. In general, the Stratonovich SDE,

$$d_S X(t) = \alpha(X(t), t)\,dt + \beta(X(t), t)dW(t)$$

solves the Itô SDE,

$$dX(t) = \left[\alpha + \frac{1}{2}\beta\frac{\partial\beta}{\partial x}\right](X(t), t)\,dt + \beta(X(t), t)\,dW(t)$$

(see, e.g., Gard, 1988).

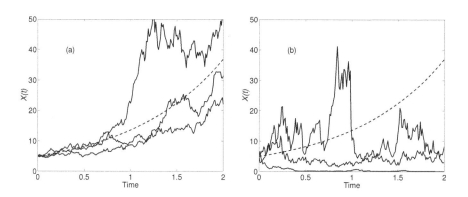

FIGURE 8.2: Three sample paths of the stochastic exponential growth model (8.28) with the deterministic solution, $X(0)\exp(rt)$, where $X(0) = 5$. In (a), $r = 1$ and $c = 0.5$. In (b), $r = 1$ and $c = 2$.

As a final example, Itô's formula is applied to a logistic-type SDE.

Example 8.12. The following SDE represents logistic growth with a hypothetical environmental variation affecting the population size,

$$dX(t) = rX(t)\left(1 - \frac{X(t)}{K}\right)dt + cX(t)\,dW(t).$$

It is assumed that the term $cX(t)\,dW(t)$ represents environmental variation which is proportional to the population size $X(t)$. Letting $F(x) = 1/x$ and applying Itô's formula,

$$dF(X(t), t) = d\left(\frac{1}{X(t)}\right) = \left[-\frac{r}{X(t)} + \frac{r}{K} + \frac{c^2}{X(t)}\right]dt - \frac{c}{X(t)}\,dW(t).$$

Letting $U = 1/X$, then

$$dU(t) = \left[(c^2 - r)U(t) + \frac{r}{K}\right]dt - cU(t)\,dW(t).$$

The solution $U(t)$ has the following form:

$$\exp\left([c^2/2 - r]t - cW(t)\right)\left[U(0) + \frac{r}{K}\int_0^t \exp\left([r - c^2/2]\tau + cW(\tau)\right)d\tau\right].$$

Since $X(t) = 1/U(t)$, the solution of the SDE is

$$X(t) = \frac{\exp\left([r - c^2/2]t + cW(t)\right)}{X^{-1}(0) + \frac{r}{K}\int_0^t \exp\left([r - c^2/2]\tau + cW(\tau)\right)d\tau}.$$

If $r/K = 0$, then the solution is the same as the one in Example 8.11. ∎

Except for relatively simple equations, such as the preceding examples, it is generally not possible to obtain explicit solutions to SDEs. In Section 8.10, numerical methods for obtaining approximations to sample paths of SDEs are discussed.

8.9 First Passage Time

The probability of extinction and the expected time until extinction can be calculated from the backward Kolmogorov differential equation. These types of problems are common in financial, physical, and engineering applications of stochastic processes, in addition to biological applications, and fit into the theory of first passage time problems (Roberts, 1986). For example, the probability of reaching state A before state B or the probability of reaching state A are known as *hitting probabilities*. In population processes, state A could be the zero state (extinction) and state B some other state such as carrying capacity of a population. Such types of problems are studied in Chapters 2 and 3 in connection with the random walk model on a bounded domain and in Chapter 6 in relation to birth and death chains.

Let $\{X(t) : t \in [0, \infty)\}$ be a stochastic process, a solution of the following Itô SDE:

$$dX(t) = a(X(t))dt + b(X(t))\,dW(t), \quad X(0) = x,$$

where $x \in (A, B)$ and $A < x < B$. The transition p.d.f. for this time-homogeneous process is a solution of the following backward Kolmogorov differential equation (equation (8.11)):

$$\frac{\partial p(y, x, t)}{\partial t} = a(x)\frac{\partial p(y, x, t)}{\partial x} + \frac{b^2(x)}{2}\frac{\partial^2 p(y, x, t)}{\partial x^2}. \tag{8.29}$$

Let $T(x)$ be the random variable for the time it takes for the stochastic process to reach either state A or B and let $p_T(x, t)$ be the corresponding p.d.f. of $T(x)$. In addition, let $Q(x, t)$ be the probability that the process does not reach states A or B in the time $[0, t]$, that is,

$$Q(x, t) = \text{Prob}\{T(x) > t\}.$$

Then $1 - Q(x, t) = \text{Prob}\{T(x) \leq t\}$. But $Q(x, t)$ also equals

$$Q(x, t) = \int_A^B p(y, x, t)\, dy, \tag{8.30}$$

where $p(y, x, t)$ is the transition p.d.f. of $X(t)$. Integration of $p(y, x, t)$ with respect to y from A to B in the backward Kolmorogov differential equation (8.29) shows that $Q(x, t)$ is also a solution of the backward Kolmogorov differential equation

$$\frac{\partial Q}{\partial t} = a(x)\frac{\partial Q}{\partial x} + \frac{b^2(x)}{2}\frac{\partial^2 Q}{\partial x^2}. \tag{8.31}$$

The boundary conditions determine whether the process reaches state A or B first. For example, to calculate the probability of reaching state B before A at time t, the partial differential equation is solved together with the boundary conditions

$$Q(A, t) = 0 \quad \text{and} \quad Q(B, t) = 1 \tag{8.32}$$

and initial condition

$$Q(x, 0) = 1 \tag{8.33}$$

(an initial boundary value problem). The p.d.f. $p_T(x, t)$ of $T(x)$ is

$$p_T(x, t) = \frac{\partial}{\partial t}(1 - Q(x, t)) = -\frac{\partial Q(x, t)}{\partial t}.$$

In addition, the backward Kolmogorov differential equation can be used to find the expected time to reach either states A or B. Denote this expectation by $m_T(x) = E(T(x))$. Then

$$m_T(x) = \int_0^\infty t p_T(x, t)\, dt$$
$$= -\int_0^\infty t\frac{\partial Q(x, t)}{\partial t}\, dt$$
$$= \int_0^\infty Q(x, t)\, dt.$$

The latter identity follows from integration by parts if $tQ(x,t) \to 0$ as $t \to \infty$. (One must be careful when applying this latter identity in practice because any terms $f(x)$ in the expression $Q(x,t) = f(x) + q(x,t)$ will vanish when differentiating with respect to t, then integrating.) This latter identity is useful in developing another method to find $m_T(x)$. Since $Q(x,t)$ is a solution of the backward Kolmogorov differential equation, integrate $Q(x,t)$ in equation (8.31) from $t = 0$ to infinity and assume $Q(x,t)$ approaches zero as $t \to \infty$. Since $Q(0,t) = 1$ it follows that the expected time to reach either A or B, as a function of the initial condition $X(0) = x$, is a solution of the boundary value problem:

$$-1 = a(x)\frac{dm_T(x)}{dx} + \frac{b^2(x)}{2}\frac{d^2 m_T(x)}{dx^2}, \quad m_T(A) = 0 = m_T(B).$$

If $Q(x,t)$ has a limiting stationary distribution, $Q(x)$, then this stationary distribution will give the probability of reaching state A or B first, beginning from state x (independent of time t). The solution $Q(x)$ is a stationary solution of the backward Kolmogorov differential equation (setting $\partial Q/\partial t = 0$). To find the probability of reaching state A before state B, the following boundary value problem is solved:

$$0 = a(x)\frac{dQ(x)}{dx} + \frac{b^2(x)}{2}\frac{d^2 Q(x)}{dx^2}, \quad Q(A) = 1, \quad Q(B) = 0. \tag{8.34}$$

Similarly, the probability of reaching B before A can be solved by changing the boundary conditions to $Q(A) = 0$ and $Q(B) = 1$.

In general, exact analytical solutions for first passage time problems cannot be found for many realistic problems. Numerical methods are usually needed to solve the initial boundary value problems and boundary value problems. For the simple case of Brownian motion, explicit solutions can be found.

Example 8.13. Recall for Brownian motion $X(t) \sim N(0, Dt)$, the SDE has the following form:

$$dX(t) = \sqrt{D}\,dW(t), \quad X(0) = x,$$

$D > 0$. The backward Kolmogorov differential equation is the same as the forward Kolmogorov differential equation for Brownian motion. The function $Q(x,t)$ is a solution of the backward Kolmogorov differential equation,

$$\frac{\partial Q}{\partial t} = \frac{D}{2}\frac{\partial^2 Q}{\partial x^2}.$$

If the boundary conditions (8.32) and initial condition (8.33) are applied, then the solution of the initial boundary value problem is the probability of reaching state B before A in time $[0,t]$. This initial boundary value problem can be solved by applying separation of variables and obtaining a Fourier series solution (see e.g., Farlow, 1993). For example, if $A = 0$ and $B = L$, the solution $Q(x,t)$ of the initial boundary value problem can be expressed as a Fourier sine series,

$$Q(x,t) = \frac{x}{L} + \sum_{n=1}^{\infty} a_n e^{-n^2\pi^2 Dt/(2L)} \sin\left(\frac{n\pi x}{L}\right),$$

where the Fourier sine coefficients are

$$a_n = \frac{2}{L} \int_0^L \frac{x}{L} \sin\left(\frac{n\pi x}{L}\right) dx = \frac{2}{n\pi}(-1)^n,$$

$n = 1, 2, \ldots$. The p.d.f. for the time to extinction is $p_T(x,t) = -\dfrac{\partial Q}{\partial t}$. Note that the limiting stationary distribution $Q(x) = x/L$ gives the probability of reaching state L before state zero. Alternately, $Q(x)$ is the solution of the simple boundary value problem:

$$\frac{d^2 Q}{dx^2} = 0, \ Q(A) = 0, \ Q(B) = 1$$

which yields the solution

$$Q(x) = \frac{x - A}{B - A} = \frac{x}{L}$$

for $A = 0$ and $B = L$.

Finally, the expected time to reach either state A or B is the solution of the following boundary value problem:

$$\frac{D}{2}\frac{d^2 m_T}{dx^2} = -1, \ m_T(A) = 0 = m_T(B).$$

The solution is

$$m_T(x) = \frac{1}{D}(B - x)(x - A).$$

Suppose the first passage time problem is to find the time it takes to reach A for the first time, given $X(0) = x \in (-\infty, \infty)$, $x \neq A$. An explicit solution for the p.d.f. $p_T(x,t)$ of this first passage time problem is given by

$$p_T(x,t) = \frac{|A - x|}{\sqrt{2\pi Dt^3}} \exp\left(\frac{-(A - x)^2}{2Dt}\right)$$

(Karlin and Taylor, 1981). ∎

Example 8.14. Suppose a population process has infinitesimal mean and variance

$$a(x) = 0 \ \text{and} \ b(x) = 2x,$$

respectively. These assumptions give rise to an exponential growth process with birth rate equal to death rate, where $\lambda = 1 = \mu$,

$$a(x) = \lambda x - \mu x = 0 \quad \text{and} \quad b(x) = \lambda x + \mu x = 2x.$$

The differential equation for the mean time to reach either a population size of 1 or of K, $K > 1$, is the solution of the following boundary value problem:

$$-1 = x \frac{d^2 m_T(x)}{dx^2}, \quad m_T(1) = 0 = m_T(K).$$

The solution of this boundary value problem is

$$m_T(x) = \frac{K \ln(K)}{K - 1}(x - 1) - x \ln(x), \quad x \in [1, K].$$

A graph of this solution is given in Figure 8.3. See Exercises 22 and 23. ∎

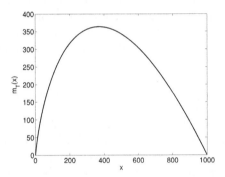

FIGURE 8.3: Mean time to reach either a population size of one or a population size of $K = 1000$ for the stochastic process in Example 8.14.

Of course, the study of population extinction should consider other stochastic factors that contribute to population extinction, including environmental stochasticity and catastrophes (Lande, 1993). The theory behind the analysis of populations at risk of extinction is known as population viability analysis (e.g. Beissinger and McCullough, 2002; Soulé, 1987).

8.10 Numerical Methods for SDEs

Numerical methods can be used to approximate the solution of SDEs. A single realization of a numerical solution approximates a sample path. There are many types of numerical methods, such as the Euler-Maruyama method, Milstein's method, other Taylor approximations, and Runge Kutta methods (Kloeden and Platen, 1992; Kloeden et al., 1997; Shurz, 2002). We shall discuss the Euler-Maruyama method because it is a simple and straightforward method to implement and converges in a mean-square sense.

We demonstrate the Euler-Maruyama method for the following Itô SDE on the interval $[0, T]$:

$$dX(t) = \alpha(X(t), t) \, dt + \beta(X(t), t) \, dW(t). \tag{8.35}$$

It is assumed that the initial condition $X(0) = X_0$ is a fixed constant and $X(t)$ is the solution on $[0, T]$. The interval $[0, T]$ is partitioned into k subintervals of equal length. Let $0 = t_0 < t_1 < \ldots < t_{k-1} < t_k = T$ be a partition of $[0, T]$, where the length of each subinterval is $\Delta t = t_{i+1} - t_i = T/k$. Then $t_{i+1} = t_i + \Delta t = i \, \Delta t$ and $\Delta W_i = \Delta W(t_i) = W(t_i + \Delta t) - W(t_i)$. In the Euler-Maruyama method, the differential $dX(t_i)$ is approximated by $\Delta X(t_i) = X(t_{i+1}) - X(t_i)$. For each sample path, the value of $X(t_{i+1})$ is approximated using only the value at the previous time step, $X(t_i)$. Let X_i denote the approximation to the solution at t_i, $X(t_i)$. Then the Euler-Maruyama method for (8.35) is given by the formula:

$$X_{i+1} = X_i + \alpha(X_i, t_i) \, \Delta t + \beta(X_i, t_i) \, \Delta W_i,$$

for $i = 0, 1, \ldots, k-1$. To write a computer program using the preceding method, we need to know how to compute ΔW_i. Because the partition is made up of equal intervals, the differences ΔW_i, $i = 0, 1, \ldots, k-1$ have the same distribution, $\Delta W_i \sim N(0, \Delta t)$. Let η be a random variable with a standard normal distribution, $\eta \sim N(0, 1)$. Then $\sqrt{\Delta t} \, \eta$ has a normal distribution with mean zero and variance Δt; that is, $\sqrt{\Delta t} \, \eta \sim N(0, \Delta t)$. To generate sample paths by numerical methods, random values from the standard normal distribution are needed. A computer program that applies the Euler-Maruyama method to (8.35) can be written using the following formula:

$$X_{i+1} = X_i + \alpha(X_i, t_i) \, \Delta t + \beta(X_i, t_i) \sqrt{\Delta t} \, \eta_i, \tag{8.36}$$

$i = 0, 1, \ldots, k-1$, and $X_0 = X(0)$.

The formula (8.36) also indicates why the sample paths corresponding to solutions of SDEs are not differentiable. Dividing equation (8.36) by Δt,

$$\frac{X_{i+1} - X_i}{\Delta t} = \alpha(X_i, t_i) + \beta(X_i, t_i) \frac{\eta_i}{\sqrt{\Delta t}}.$$

Then taking the limit as $\Delta t \to 0$, the limit on the left side approaches dX/dt, but the limit of the right side of the preceding equation does not exist.

It can be shown that the numerical solution generated by the Euler-Maruyama method in (8.36) has a distribution that is close to the distribution of the exact solution in the mean square sense on $[0, T]$. Sufficient conditions for convergence are the existence and uniqueness assumptions (a) and (b), discussed in Section 8.8, and one additional assumption, if the coefficients α and β depend on time. In particular, convergence is guaranteed with

(c) $|\alpha(x, t_1) - \alpha(x, t_2)| + |\beta(x, t_1) - \beta(x, t_2)| \leq K|t_1 - t_2|^{1/2}$, for all $t_1, t_2 \in [0, T]$ and $x \in \mathbb{R}$.

Then the numerical approximation given by the Euler-Maruyama method X_i and the exact solution $X(t_i)$ satisfy

$$E\left(|X(t_i) - X_i|^2 \mid X(t_{i-1}) = X_{i-1}\right) = O((\Delta t)^2) \qquad (8.37)$$

and

$$E\left(|X(t_i) - X_i|^2 \mid X(0) = X_0\right) = O(\Delta t) \qquad (8.38)$$

(Kloeden and Platen, 1992). Each step in the Euler-Maruyama method has a mean-square error that is of order $(\Delta t)^2$. The error over the entire interval $[0, T]$ is of order Δt. Hence, if Δt is small, then the approximation improves. However, if Δt is too small, there will be rounding error, which will accumulate due to the large number of calculations (Arciniega and Allen, 2003). The procedure and convergence results for the Euler-Maruyama method are summarized in the next theorem.

THEOREM 8.4
Assume that the Itô SDE (8.35) satisfies conditions (a), (b), and (c). Then the Euler-Maruyama method approximates a sample path $X(t)$ at fixed points t_i, $X(t_i) \approx X_i$, on the interval $[0, T]$, based on the formula:

$$X_{i+1} = X_i + \alpha(X_i, t_i)\,\Delta t + \beta(X_i, t_i)\sqrt{\Delta t}\,\eta_i \qquad (8.39)$$

for $i = 0, 1, \ldots, k - 1$, where $0 = t_0 < t_1 < \ldots < t_{k-1} < t_k = T$, $\Delta t = t_{i+1} - t_i = T/k$, $X_0 = X(0)$, and $\eta_i \sim N(0, 1)$. The error in the Euler-Maruyama method satisfies (8.37) and (8.38).

Example 8.15. The Euler-Maruyama method is applied to the following SDE:

$$dX(t) = dW(t),$$

where $X(0) = W(0) = 0$. The solution $X(t) = W(t)$ is the Wiener process. The Euler-Maruyama method for this SDE is

$$X_{i+1} = X_i + \sqrt{\Delta t}\,\eta_i,$$

$\eta_i \in N(0, 1)$. A MATLAB® program for the Euler-Maruyama method on $[0, 1]$ is given next. In MATLAB, the command **randn** generates pseudorandom numbers from a standard normal distribution. Two sample paths of the Wiener process are graphed in Figure 8.4. The sample paths in Example 8.11 were also generated using the Euler-Maruyama method. The Euler-Maruyama method can also be applied to systems of Itô SDEs.

```
clear
T=1; k=1000;
dt=T/k; %  dt=Deltat=0.001
for j=1:2 % Two sample paths
    W(1)=0;
    for i=1:k
         W(i+1)=W(i)+sqrt(dt)*randn;
    end
    plot([0:dt:T],W,'LineWidth',2);
     hold on
end
xlabel('t'); ylabel('W(t)');
```

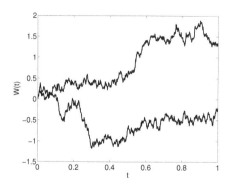

FIGURE 8.4: Two sample paths $W(t)$ of the Wiener process on $[0, 1]$.

If a programming language does not have a built-in standard normal random number generator but does have a built-in uniform random number generator, then standard normal random numbers can be generated using the Box-Muller method (Hogg and Craig, 1995). If u_1 and u_2 are two random values from a uniform distribution on $[0, 1]$, then the Box-Muller method calculates two standard normal random values x_1 and x_2 as follows:

$$x_1 = \sqrt{-2 \ln u_1} \cos(2\pi u_2) \text{ and } x_2 = \sqrt{-2 \ln u_1} \sin(2\pi u_2)$$

(Hogg and Craig, 1995).

There are many numerical methods that can be used to approximate solutions to SDEs (e.g., multi–step methods, Milstein's method). Consult E. Allen (2007), Kloeden and Platen (1992), Kloeden et al. (1997), or Schurz

(2002) for a discussion on various numerical methods for SDEs. An article by Higham (2001) gives a brief introduction to efficient numerical simulation of SDEs using MATLAB. In Higham's article, the Euler-Maruyama method is presented along with several MATLAB programs implementing this numerical method.

8.11 An Example: Drug Kinetics

The concentration C of a drug in the body is modeled by the following Itô SDE:

$$dC(t) = -kC(t)\,dt + \sigma C(t)\,dW(t), \quad C(0) = C_0$$

(Ramanathan, 1999). The term $-kC(t)$ is the mean rate at which the drug is eliminated from the body (drift) and the term $\sigma^2 C^2(t)$ is the variance rate for the change in the concentration (diffusion coefficient). The coefficient of variation (standard deviation/mean= $\sigma C/(-kC)$) is constant. The stochastic process corresponding to this SDE is known as *geometric Brownian motion.* (See Example 8.11.) Applying Itô's formula with $F(C) = \ln(C)$ it follows that

$$d(\ln C(t)) = -\left(k + \frac{\sigma^2}{2}\right)dt + \sigma\,dW(t).$$

Thus, the drug concentration is lognormally distributed. That is,

$$\ln(C(t)) \sim N\left(C_0 - \left[k + \frac{\sigma^2}{2}\right]t, \sigma^2 t\right)$$

(Example 8.7). This solution is only valid provided $C(t) \geq 0$.

Another model was formulated by Ramanathan (2000) to model the drug concentration assuming a strategy to maintain an approximate steady-state concentration C_s by continuous drug dosing at a rate α:

$$dC(t) = \alpha(C_s - C(t))\,dt + \sigma C(t)\,dW(t). \tag{8.40}$$

The rate constant α is related to the speed of drug dosing and $1/\alpha$ is the time scale over which the dose takes effect (Ramanathan, 2000). For example, based on the deterministic approximation, the time scale for the drug difference $C_s - C(t)$ to reach $1/2$ of its original level is $\ln(2)/\alpha$, i.e., the solution of

$$dC(t) = \alpha(C_s - C(t))\,dt$$

is $C_s - C(t) = (C_s - C(0))e^{-\alpha t}$ yielding the value for the time at half-concentration to equal $t = \ln(2)/\alpha$. The SDE equation (8.40) was shown to provide a good fit to data for the cardiac antiarrhythmic medication procainamide, given to patients to control irregular heartbeats, e.g., tachycardia (Ramanathan, 2000).

Applying Itô's formula and the method of Example 8.11, it follows that

$$\frac{dE(C(t))}{dt} = \alpha[C_s - E(C(t))], \quad E(C(0)) = C_0. \tag{8.41}$$

In addition, applying Itô's formula to $F(C) = C^2$, it can be shown that

$$\frac{dE(C^2(t))}{dt} = (\sigma^2 - 2\alpha)E(C^2(t)) + 2\alpha C_s E(C(t)), \quad E(C^2(0)) = C_0^2. \tag{8.42}$$

The solution of these two linear differential equations are

$$E(C(t)|C_0) = C_s + e^{-\alpha t}(C_0 - C_s) \tag{8.43}$$

$$E(C^2(t)|C_0) = \frac{2\alpha C_s^2}{2\alpha - \sigma^2} + \frac{2\alpha C_s(C_s - C_0)}{\sigma^2 - \alpha}e^{-\alpha t}$$

$$+ \left(C_0^2 + \frac{2\alpha C_s}{\sigma^2 - \alpha}\left[C_0 + \frac{\alpha C_s}{\sigma^2 - 2\alpha}\right]\right)e^{(\sigma^2 - 2\alpha)t}. \tag{8.44}$$

If the speed of the dosing rate is sufficiently large, $2\alpha > \sigma^2$, then

$$\lim_{t \to \infty} E(C(t)|C_0) = C_s$$

$$\lim_{t \to \infty} Var(C(t)|C_0) = \frac{\sigma^2 C_s^2}{2\alpha - \sigma^2}.$$

The mean concentration approaches a steady-state level with a constant variance. A graph of the mean and three sample paths $C(t)$ on the interval $t \in [0,1]$ are plotted in the case $2\alpha > \sigma^2$ in Figure 8.5. (See Exercise 26.) Parameters α and C_s are controlled by the caregiver. Maintaining a steady-state level of drug concentration that is of sufficient magnitude to be effective but not so large so as to become toxic can be accomplished through careful selection of these parameters. However, if there is a large variability in the drug kinetics $\sigma \gg 0$, then it may be difficult to maintain this steady-state level, since $2\alpha < \sigma^2$ implies $Var(C(t)|C_0) \to \infty$ as $t \to \infty$.

8.12 Exercises for Chapter 8

1. Verify that

$$u(x,t) = \frac{1}{\sqrt{2\pi Dt}} \exp\left(-\frac{(x - x_0 - ct)^2}{2Dt}\right)$$

is a solution of the diffusion equation with drift, equation (8.5), for $t \in (0, \infty)$ and $x \in (-\infty, \infty)$.

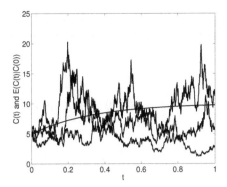

FIGURE 8.5: Three sample paths of $C(t)$ and the mean $E(C(t)|C(0))$ (smooth curve) for the drug kinetics example with parameter values and initial condition $\alpha = 3$, $\sigma = 1$, $C_s = 10$, and $C(0) = 5$ for $t \in [0,1]$, $2\alpha > \sigma^2$.

2. Show that the conditions (i)–(iii) in Definition 8.3 of a diffusion process follow from inequality (8.9) and conditions (i)′–(iii)′.

3. Use the method of Example 8.3 to derive a formula for the variance $\sigma_X^2(t)$ of the diffusion process, discussed in Example 8.2.

   (a) Let $g(t) = \int_{-\infty}^{\infty} x^2 p(x,t)\, dx$. Then show that $dg/dt = 2\alpha g + \beta \mu_X(t)$.

   (b) Solve for $g(t)$. Then show that $\sigma_X^2(t)$ equals

   $$\sigma_X^2(t) = \begin{cases} x_0 \dfrac{\beta}{\alpha} e^{\alpha t}\left[e^{\alpha t} - 1\right], & \alpha \neq 0 \\ x_0 \beta t, & \alpha = 0. \end{cases}$$

   (c) Compare the formulas for the mean and variance in Example 8.3 to those for a simple birth and death Markov chain in Section 6.4.3 (see Table 6.1), where $\alpha = \lambda - \mu$ and $\beta = \lambda + \mu$.

4. Use the transition p.d.f. of the Wiener process $W(t)$,

   $$p(y, x, t) = \frac{1}{\sqrt{2\pi t}} \exp\left(\frac{-(y-x)^2}{2t}\right),$$

   to show directly that the three conditions (i)′ (for $\delta = 3$), (ii)′, and (iii)′ for a diffusion process are satisfied (listed after Definition 8.3). The process is time-homogeneous so that $p(y, t + \Delta t; x, t) = p(y, x, \Delta t)$.

5. The transition p.d.f corresponding to the diffusion equation with drift is

   $$p(y, x, t) = \frac{1}{\sqrt{2\pi D t}} \exp\left(\frac{-(y - x - ct)^2}{2Dt}\right).$$

The process is time-homogeneous (Exercise 4). The transition p.d.f. is the p.d.f. of a normal random variable Y, where $p(y, x, t) = p(y)$ when x and t are fixed. The mean of Y is $x + ct$ and variance of Y is Dt. That is,

$$\int_{-\infty}^{\infty} p(y)\, dy = 1, \quad E(Y) = \int_{-\infty}^{\infty} yp(y)\, dy = x + ct,$$

and

$$E((Y - x - ct)^2) = \int_{-\infty}^{\infty} (y - x - ct)^2 p(y)\, dy = Dt.$$

6. Use Definition 8.5 to verify the Itô stochastic integral given in (8.20):

$$\int_a^b dF(W(t), t) = F(W(b), b) - F(W(a), a).$$

7. Verify the following:

   (a) The Itô stochastic integral in Example 8.4 converges in the mean; that is,

   $$\lim_{k \to \infty} E\left[t - \sum_{i=1}^{k} (\Delta W_i)^2 \right] = 0,$$

   where $\Delta W_i = W(t_{i+1}) - W(t_i)$ and $\Delta t = t_{i+1} - t_i = t/k$.

   (b) The expectations $E((\Delta W_i)^3) = 0$ and $E((\Delta W_i)^4) = 3(\Delta t)^2$.

8. Use the Itô isometry property to evaluate the following expectations:

   (a) $E\left[\left(\int_a^b W^2(t)\, dW(t) \right)^2 \right]$

   (b) $E\left[\left(\int_a^b [W(t)]^{3/2}\, dW(t) \right)^2 \right]$

   (c) $E\left[\left(\int_a^b [W(t)]^{(2n-1)/2}\, dW(t) \right)^2 \right], \ n = 3, 4 \ldots$

9. Suppose the forward Kolmogorov equation of a diffusion process has the form

   $$\frac{\partial p}{\partial t} = -\frac{\partial}{\partial x} \left[\{(b_1 - d_1)x - (b_2 + d_2)x^2\} p \right]$$
   $$+ \frac{1}{2} \frac{\partial^2}{\partial x^2} \left[\{(b_1 - d_1)x + (b_2 + d_2)x^2\} p \right],$$

   $x \in (0, \infty)$, where $b_i, d_i > 0$ for $i = 1, 2$ and $b_1 > d_1$.

   (a) Write a corresponding SDE for this process.

(b) Write a corresponding ODE that is related to this process.

10. Consider the SDE

$$dX(t) = a\,dt + b\,dW(t), \quad X(0) = x_0,$$

where $t \in [0, \infty)$, a and b are constants, and $b > 0$.

(a) Write the SDE as an Itô stochastic integral equation. Solve for $X(t)$.

(b) Find the p.d.f. $p(x, t)$ of $X(t)$, a solution of the forward Kolmogorov differential equation with $x \in (-\infty, \infty)$, $t \in (0, \infty)$, and $p(x, t) = \delta(x - x_0)$.

(c) The p.d.f. has a normal density. What are the mean and variance of this process, $\mu_X(t)$ and $\sigma_X^2(t)$?

11. Consider the simple birth and death process,

$$dX(t) = (\lambda - \mu)X(t)\,dt + \sqrt{(\lambda + \mu)X(t)}\,dW(t),$$

$X(0) = x_0$ and $X(t) \in [0, \infty)$, where λ and μ are positive constants.

(a) Use property (i) in Theorem 8.2 and Itô's formula in Theorem 8.3 to show that

$$\frac{dE(X^2(t))}{dt} = 2\alpha E(X^2(t)) + \beta^2 E(X(t)),$$

where $\alpha = \lambda - \mu$ and $\beta = \sqrt{\lambda + \mu}$. Compare this differential equation to the one in Example 8.3, where $g(t) = E(X^2(t))$.

(b) Use the fact that $E(X(t)) = x_0 e^{\alpha t}$; then solve the differential equation in (a) to show that

$$\text{Var}(X(t)) = \begin{cases} x_0 \frac{\beta^2}{\alpha} e^{\alpha t}[e^{\alpha t} - 1], & \alpha \neq 0 \\ 2x_0\lambda t, & \alpha = 0. \end{cases}$$

This is the same solution as in Example 8.3.

12. Consider the equation

$$dX(t) = \mu X(t)\,dt + \sigma\,dW(t), \quad X(0) = x_0, \tag{8.45}$$

where μ, σ, and x_0 are constants. The corresponding stochastic process is known as the *Ornstein Uhlenbeck process*. Use Itô's formula to find the solution.

13. For the Ornstein Uhlenbeck process modeled by equation (8.45) in Exercise 12, use the solution $X(t)$, the square of the solution $X^2(t)$, and properties (i) and (ii) in Theorem 8.2 to find the mean $E(X(t))$ and the variance $Var(X(t))$.

14. Consider the Itô SDE for geometric Brownian motion (Example 8.11), where $X(0) = x_0$.

   (a) Use property (i) in Theorem 8.2 and Itô's formula in Theorem 8.3 to show that

$$\frac{dE(X(t))}{dt} = aE(X(t)) \text{ and}$$

$$\frac{dE(X^2(t))}{dt} = (2a + b^2)E(X^2(t)).$$

   (b) Solve the differential equations in (a) to find the mean $E(X(t))$ and the variance $E(X^2(t))$.

15. Solve the following Itô SDE:

$$dX(t) = r(X(t) - E)\, dt + c(X(t) - E)\, dW(t), \quad X(0) = x_0,$$

   where r, c and E are positive constants.

16. Use Itô's formula to verify the following Itô stochastic integrals.

   (a) $\int_0^t W^2(\tau)\, dW(\tau) = \frac{1}{3} W^3(t) - \int_0^t W(\tau)\, d\tau$

   (b) $\int_0^t W^n(\tau)\, dW(\tau) = \frac{1}{n+1} W^{n+1}(t) - \frac{n}{2} \int_0^t W^{n-1}(\tau)\, d\tau$, $n = 1, 2, \ldots$

   (c) $\int_0^t \tau^n\, dW(\tau) = t^n W(t) - n \int_0^t \tau^{n-1} W(\tau)\, d\tau$, $n = 1, 2, \ldots$

17. Use Itô's formula to show that $X(t) = e^{aW(t)}$ is the solution of $dX(t) = \frac{a^2}{2} X(t)dt + aX(t)dW(t)$, $X(0) = 1$.

18. The following stochastic process is known as the *Brownian bridge* from a to b. The solution begins at $X(0) = a$ and at $t = 1$ reaches the value $X(1) = b$. Let $dX(t) = \frac{b - X(t)}{1 - t} dt + dW(t)$, $X(0) = a$ for $0 \leq t < 1$.

   (a) Verify that the solution is

$$X(t) = a(1 - t) + bt + (1 - t) \int_0^t \frac{1}{1 - s} dW(s).$$

   (b) Write a computer program to simulate three sample paths of this SDE.

19. For the diffusion equation with drift and $X(t) = x \in (A, B)$, the probability of reaching either state A or B first can be solved in a manner similar to Example 8.13. It follows from Example 8.7 that the SDE for diffusion with drift is

$$dX(t) = cdt + \sqrt{D}\, dW(t), \quad X(0) = x, \tag{8.46}$$

   $c, D > 0$.

(a) Find the probability of reaching state A before state B by solving the following boundary value problem:

$$\frac{D}{2}\frac{d^2Q}{dx^2} + c\frac{dQ}{dx} = 0, \quad Q(A) = 1, \quad Q(B) = 0.$$

(b) Find the probability of reaching state B first.

20. For the diffusion equation with drift, $X(0) = x \in (A, B)$, equation (8.46), find the expected time $m_T(x)$ to reach either state A or B. See Example 8.13.

21. For the diffusion equation with drift, the probability distribution $Q(x, t)$ for the time $T(x)$ until the process, $\{X(t) : t \in [0, \infty)\}$ with $X(0) = x \in (A, B)$, reaches state B before state A is a solution of the backward Kolmogorov differential equation, equation (8.31):

$$\frac{\partial Q(x, t)}{\partial t} = c\frac{\partial Q(x, t)}{\partial x} + \frac{D}{2}\frac{\partial^2 Q(x, t)}{\partial x^2}.$$

(a) Set up an initial boundary value problem for $Q(x, t) = \text{Prob}\{T(x) > t\}$ so that at time t the process reaches state B before A, equations (8.32), and (8.33).

(b) Show that the solution $Q(x, t)$ of the initial boundary value problem is $Q(x, t) = e^{ct}\bar{Q}(x, t)$, where $\bar{Q}(x, t)$ is the corresponding solution for $c = 0$ (given in Example 8.13).

22. Suppose $\{X(t) : t \in [0, \infty)\}$ is a population process, where $X(t) \in [0, M]$ and $X(0) = x$. Consider the backward Kolmogorov equation for the population process, with the birth rate equal to the death rate, $b(x) = d(x)$,

$$\frac{\partial p(y, x, t)}{\partial t} = \frac{(b(x) + d(x))x}{2}\frac{\partial^2 p(y, x, t)}{\partial x^2}$$

$$= b(x)x\frac{\partial^2 p(y, x, t)}{\partial x^2}, \tag{8.47}$$

$y \in [0, M]$. The function $p(y, x, t)$ is the transition p.d.f. Let $Q(x, t) = \text{Prob}\{T(x, t) > t\} = \int_0^M p(y, x, t)\, dy$, where $T(x, t)$ is the random variable for the time to extinction given $X(0) = x$.

(a) Show that Q is a solution of the backward Kolmogorov equation (8.47).

(b) Suppose $Q(0, t) = 0$ and $\partial Q/\partial x = 0$ at $x = M$, $\lim_{t\to\infty} Q(x, t) = 0$, and $Q(x, 0) = 1$ for $x > 0$. Denote the associated p.d.f. of $T(x, t)$ by $p_T(x, t)$; then show that

$$p_T(x, t) = -\frac{\partial Q}{\partial t}.$$

(c) Let the expectation of T be denoted $m_T(x) = \int_0^\infty t p_T(x,t)\, dt$. Show that

$$-1 = xb(x)\frac{d^2 m_T(x)}{dx^2}.$$

The boundary conditions are $m_T(0^+) = 0 = dm_T(M)/dx$, where $m_T(0^+) = \lim_{x \to 0^+} m_T(x)$.

(d) Show that

$$m_T(M) = \int_0^M \frac{dy}{b(y)},$$

which is the expected time to extinction from the maximum population size M.

23. In the preceding Exercise 22, let $b(y) = 1$.

   (a) Show that the expected time to extinction from the maximum population size M is M: $m_T(M) = M$.

   (b) Show that the solution of

$$-1 = x\frac{d^2 m_T(x)}{dx^2}$$

   with boundary conditions $m_T(0^+) = 0 = dm_T(M)/dx$ is

$$m_T(x) = x(\ln(M) + 1) - x\ln(x), \quad x \in (0, M).$$

   Graph $m_T(x)$ for $M = 1000$ and compare it with the graph in Figure 8.3.

24. For the SDE model corresponding to the drug kinetics example, equation (8.40), use Itô's formula to derive differential equations for the first and second moments, equations (8.41) and (8.42). Then show that the solutions $E(C(t)|C_0)$ and $E(C^2(t)|C_0)$ are given by equations (8.43) and (8.44).

25. The drug kinetics example with the steady state solution C_s, equation (8.40), differs from the following *mean-reverting Ornstein Uhlenbeck process*. In this SDE, the noise is additive:

$$dX(t) = \alpha(X_e - X(t))dt + \sigma dW(t),$$

X_e is constant. Find the solution $X(t)$, then use the solution to find the mean $E(X(t)|X(0))$ and variance $Var(X(t)|X(0))$ for the mean-reverting Ornstein Uhlenbeck process.

26. Use the Euler-Maruyama numerical method (program in the Appendix for Chapter 8) to generate $10,000$ sample paths for the solution of $dX(t) = \alpha(X_e - X(t))dt + \sigma X(t)dW(t)$, where $\alpha = 3$, $\sigma = 1$, $X_e = 10$, and $X(0) = 5$.

(a) Find the first two moments at $t = 1$,

$$E(X(1)|X(0)) \quad \text{and} \quad E(X^2(1)|X(0)),$$

and compare your answers with the drug kinetics example (equations (8.43) and (8.44)).

(b) Graph three sample paths $X(t)$ and the expectation $E(X(t)|X(0))$ on the interval $[0, 1]$.

8.13   References for Chapter 8

Allen, E. 2007. *Modeling with Itô Stochastic Differential Equations.* Springer, Dordrecht, The Netherlands.

Arciniega, A. and E. J. Allen. 2003. Rounding error in numerical solution of stochastic differential equations. *Stoch. Anal. Appl.* 21: 281–300.

Arnold, L. 1974. *Stochastic Differential Equations: Theory and Applications.* John Wiley & Sons, New York.

Bailey, N. T. J. 1990. *The Elements of Stochastic Processes with Applications to the Natural Sciences.* John Wiley & Sons, New York.

Beissinger, S. R. and D. R. McCullough. 2002. *Population Viability Analysis.* The University of Chicago Press, Chicago and London.

Bharucha-Reid, A. T. 1997. *Elements of the Theory of Markov Processes and Their Applications.* Dover Pub., Inc., New York.

Braumann, C. A. 2007. Itô versus Stratonovich calculus in random population growth. *Math. Biosci.* 206: 81–107.

Corduneanu, C. 1977. *Principles of Differential and Integral Equations.* Chelsea Pub. Co., The Bronx, New York.

Farlow, 1993. *Partial Differential Equations for Scientists and Engineers.* Dover Pub., Inc, New York.

Gard, T. C. 1988. *Introduction to Stochastic Differential Equations.* Marcel Dekker, Inc., New York and Basel.

Higham, D. J. 2001. An algorithmic introduction to numerical simulation of stochastic differential equations. *SIAM Review* 43: 525–546.

Hogg, R. V. and A. T. Craig. 1995. *Introduction to Mathematical Statistics.* 5th ed. Prentice Hall, Upper Saddle River, N. J.

Hsu, H. P. 1997. *Schaum's Outline of Theory and Problems of Probability, Random Variables, and Random Processes.* McGraw-Hill, New York.

Ikeda, N. and S. Watanabe. 1989. *Stochastic Differential Equations and Diffusion Processes.* 2nd ed., North-Holland.

Karlin, S. and H. Taylor. 1981. *A Second Course in Stochastic Processes.* Academic Press, New York.

Kloeden, P. E. and E. Platen. 1992. *Numerical Solution of Stochastic Differential Equations.* Springer-Verlag, New York.

Kloeden, P. E., E. Platen, and H. Schurz. 1997. *Numerical Solution of SDE through Computer Experiments.* Springer-Verlag, Berlin.

Lande, R. 1993. Risks of population extinction from demographic and environmental stochasticity and random catastrophes. *Am. Nat.* 142: 911–927.

Melsa, J. L. and A. P. Sage. 1973. *An Introduction to Probability and Stochastic Processes.* Prentice Hall, Englewood Cliffs, N. J.

Øksendal, B. 2000. *Stochastic Differential Equations: An Introduction with Applications.* 5th ed. Springer-Verlag, Berlin, Heidelberg, New York.

Ramanathan, M. 1999. An application of Ito's lemma in population pharmacokinetics and pharmacodynamics. *Pharmaceutical Research* 16: 584–586.

Ramanathan, M. 2000. Pharmacokinetic variability and therapeutic drug monitoring actions at steady state. *Pharmaceutical Research* 17: 589–592.

Ricciardi, L. M. 1986. Stochastic population theory: diffusion processes. In: *Mathematical Ecology.* Biomathematics, Vol. 17, Hallam, T. G. and S. A. Levin (eds.), pp. 191–238, Springer-Verlag, Berlin, Heidelberg, New York.

Roberts, J. B. 1986. First passage probabilities for randomly excited systems: diffusion methods. *Probabilistic Engineering Mechanics* 1: 66–81.

Schurz, H. 2002. Numerical analysis of stochastic differential equations without tears. In: *Handbook of Stochastic Analysis and Applications.* Kannan, D. and V. Lakshmikantham (eds.), pp. 237–359, Marcel-Dekker, New York.

Soulé, M. E. (ed.) 1987. *Viable Populations for Conservation.* Cambridge Univ. Press, Cambridge, U. K.

Taylor, H. M. and S. Karlin. 1998. *An Introduction to Stochastic Modeling.* 3rd ed. Academic Press, New York.

Turelli, M. 1977. Random environments and stochastic calculus. *Theor. Pop. Biol.* 12: 140–178.

van Kampen, N. G. 2007. *Stochastic Processes in Physics and Chemistry.* 3rd ed. Elsevier, Amsterdam, Boston, and Heidelberg.

8.14 Appendix for Chapter 8

8.14.1 Derivation of Kolmogorov Equations

To derive the *forward Kolmogorov differential equation* (8.14) for the time-homogeneous process, a test function h is used, where $h \in C^3((-\infty, \infty))$, $h(x) = 0$ for $x \notin [A, B]$, $h(A) = 0 = h(B)$ and $h'(A) = 0 = h'(B)$ for any A and B satisfying $-\infty < A < B < \infty$. Then

$$\int_{-\infty}^{\infty} h(y) \frac{\partial p(y, x, t)}{\partial t} dy = \frac{\partial}{\partial t} \int_{-\infty}^{\infty} h(y) p(y, x, t) \, dy.$$

This latter expression can be written as

$$\lim_{\Delta t \to 0} \frac{1}{\Delta t} \int_{-\infty}^{\infty} h(y) [p(y, x, t + \Delta t) - p(y, x, t)] \, dy.$$

Applying the Chapman-Kolmogorov equations,

$$\lim_{\Delta t \to 0} \frac{1}{\Delta t} \left[\int_{-\infty}^{\infty} h(y) \int_{-\infty}^{\infty} p(z, x, t) p(y, z, \Delta t) \, dz \, dy - \int_{-\infty}^{\infty} h(z) p(z, x, t) \, dz \right].$$

Interchanging the limits of integration and applying (8.12),

$$\lim_{\Delta t \to 0} \frac{1}{\Delta t} \left[\int_{-\infty}^{\infty} p(z, x, t) \int_{-\infty}^{\infty} p(y, z, \Delta t) [h(y) - h(z)] \, dy \, dz \right].$$

Expand $h(y)$ in the preceding expression using Taylor's formula about the value z:

$$\lim_{\Delta t \to 0} \frac{1}{\Delta t} \left[\int_{-\infty}^{\infty} p(z,x,t) \int_{-\infty}^{\infty} p(y,z,\Delta t) \left[(y-z)h'(z) + \frac{(y-z)^2}{2}h''(z) \right. \right.$$
$$\left. \left. + \frac{(y-z)^3}{6}h'''(\xi) \right] dy \, dz \right],$$

where ξ is between y and z. Applying the diffusion assumptions (i)–(iii) yields

$$\int_{-\infty}^{\infty} p(z,x,t) \left[a(z)h'(z) + \frac{1}{2}b(z)h''(z) \right] dz. \tag{8.48}$$

Next each of the terms on the right-hand side of (8.48) is integrated by parts and the assumptions on h are applied:

$$\int_{-\infty}^{\infty} p(z,s,t)a(z)h'(z) \, dz = -\int_{-\infty}^{\infty} h(z) \frac{\partial \left[a(z)p(x,z,t) \right]}{\partial z} \, dz$$

and

$$\int_{-\infty}^{\infty} p(z,s,t)b(z)h''(z) \, dz = \int_{-\infty}^{\infty} h(z) \frac{\partial^2 \left[b(z)p(z,s,t) \right]}{\partial z^2} \, dz.$$

Substituting the preceding expressions into (8.48) yields

$$\int_{-\infty}^{\infty} h(z) \left[\frac{\partial p(z,x,t)}{\partial t} + \frac{\partial \left[a(z)p(z,x,t) \right]}{\partial z} - \frac{1}{2} \frac{\partial^2 \left[b(z)p(z,x,t) \right]}{\partial z^2} \right] dz.$$

This expression holds for any test function h. Therefore, the forward Kolmogorov differential equation (8.14) holds.

8.14.2 MATLAB® Program

The following MATLAB program generates 10,000 sample paths using the Euler-Maruyama numerical method on the time interval $[0,1]$ for the SDE model $dX(t) = \alpha(X_e - X(t))dt + \sigma X(t)dW(t)$. The parameter values are $\alpha = 3$, $\sigma = 1$, $X(0) = 5$, and $X_e = 10$. See Exercise 26 (a).

```
clear all % this program uses vector processing
nsim=10000;
Xe=10; X0=5;
alpha=3; sigma=1;
dt=0.001; time=1;
randn('state',1);
X=ones(nsim,1)*X0;
for t=1:time/dt
    r=randn(nsim,1);
```

```
    X=X+alpha*(Xe-X)*dt+sigma*X*sqrt(dt).*r;
end
meanX=sum(X)/nsim
meanx2=sum(X.*X)/nsim
EX=Xe+exp(-alpha)*(X0-Xe)
d1=sigma^2-alpha;
d2=sigma^2-2*alpha;
hlp=(X0^2+Xe/d2)*exp(d2)-2*alpha*Xe^2/d2;
EX2=2*alpha*Xe*(X0-Xe)/d1*(exp(d2)-exp(-alpha))+hlp
```

Chapter 9

Biological Applications of Stochastic Differential Equations

9.1 Introduction

The number and variety of modeling applications in biology that employ SDEs have increased rapidly in the 21st century. Only a few applications are discussed in this chapter with the intent of illustrating some methods for derivation, analysis, and numerical simulation of SDEs and of providing a background in some classical biological applications. First, some of the theoretical results for scalar SDEs are extended to systems of SDEs, that is, multivariate SDEs. Then a method for formulating scalar or multivariate Itô SDE models from first principles is described. Two population processes that give rise to scalar Itô SDEs are presented: a birth, death, and immigration process and a logistic growth process. Applications to multivariate Itô SDE processes are discussed in the remaining sections of this chapter (enzyme kinetics, competition, predation, and population genetics). Additional biological examples can be found in the exercises.

9.2 Multivariate Processes

The randomness associated with interacting populations, the spread of epidemics, or population genetics often requires more than one random variable to describe the process. Suppose the biological process is described by a vector of random variables $X = (X_1, X_2, \ldots, X_n)^{tr}$. In addition, suppose this process can be modeled by a system of Itô SDEs, where X_i is a solution of

$$dX_i(t) = \alpha_i(X(t), t)\, dt + \sum_{j=1}^{m} \beta_{ij}(X(t), t)\, dW_j(t), \quad i = 1, \ldots, n \qquad (9.1)$$

and $W = (W_1, W_2, \ldots, W_m)^{tr}$ is a vector of m independent Wiener processes. As in the scalar case, the Itô SDEs (9.1) can be written as Itô integral equations:

$$X_i(t) = X_i(0) + \int_0^t \alpha_i(X(\tau),\tau)\, d\tau + \int_0^t \sum_{j=1}^m \beta_{ij}(X(\tau),\tau)\, dW_j(\tau), \quad i = 1,\ldots,n.$$

Writing the system of SDEs (9.1) in vector form leads to

$$dX(t) = a(X(t),t)\, dt + B(X(t),t)\, dW(t), \tag{9.2}$$

where $a = (\alpha_1, \alpha_2, \ldots, \alpha_n)^{tr}$ and $B = (\beta_{ij})$ is an $n \times m$ matrix. The first term $a(X(t),t)$ is referred to as the *drift* vector and the second term $B(X(t),t)$ is referred to as the *diffusion* matrix. We will only consider equations of the Itô type in this chapter.

Many of the properties of scalar Itô integral equations also apply to systems. In particular, Theorems 8.1 and 8.2 in Chapter 8 apply to systems of Itô integral equations. For example, a useful result in Theorem 8.2 is property (i):

$$E\left[\int_a^b f(t)\, dW(t) \right] = 0.$$

A requirement for application of these theorems to a random function $f(t)$ on the interval $[a,b]$ is given in (8.18):

$$\int_a^b E(f^2(t))\, dt < \infty.$$

In addition, Itô's formula in Theorem 8.3, Chapter 8, can be extended to a vector random variable.

THEOREM 9.1 Multivariate Itô's formula

Suppose $X(t)$ is a solution of (9.2). If $F(x,t)$ is a real-valued function defined for $x = (x_1, x_2, \ldots, x_n) \in \mathbb{R}^n$ and $t \in [a,b]$, $F : \mathbb{R}^n \times [a,b] \to \mathbb{R}$, with continuous partial derivatives in t and x_i, $\partial F/\partial t$, $\partial F/\partial x_i$, $\partial^2 F/\partial x_i \partial x_j$, then

$$dF(X(t),t) = f(X(t),t)\, dt + g(X(t),t) \cdot dW(t), \tag{9.3}$$

where

$$f(x,t) = \frac{\partial F}{\partial t} + \sum_{i=1}^n \frac{\partial F}{\partial x_i}\alpha_i + \sum_{i=1}^n \sum_{j=1}^n \sum_{k=1}^m \frac{1}{2}\frac{\partial^2 F}{\partial x_i \partial x_j}\beta_{ik}\beta_{jk}$$

and

$$g(x,t) \cdot dW(t) = \sum_{j=1}^m \sum_{i=1}^n \frac{\partial F}{\partial x_i}\beta_{ij}\, dW_j(t).$$

The expression for $g(x,t)$ is an m-vector $(g_1, \ldots, g_m)^{tr}$, where

$$g_j = \sum_{i=1}^n \frac{\partial F}{\partial x_i}\beta_{ij}$$

so that $g(x,t) \cdot dW(t)$ is the dot product of two vectors, $g \cdot dW$. Itô's formula (9.3) is a scalar SDE. This formula is useful, for example, in calculating the moments of the random variables X_i, as will be demonstrated in the examples.

Existence and uniqueness for scalar SDEs carry over to systems of SDEs (Øksendal, 2000). Existence and uniqueness of a vector solution $X(t)$ to (9.1) for $t \in [0,T]$, with initial condition $X(0) = Z$, $E(|Z|^2) < \infty$, follow from the conditions (a) and (b) in Section 8.8, where the notation $| \cdot |$ means

$$|a|^2 = \sum_{i=1}^{n} |\alpha_i|^2 \text{ and } |B|^2 = \sum_{i=1}^{n} \sum_{j=1}^{m} |\beta_{ij}|^2.$$

The transition p.d.f, denoted as $p(y,s;x,t)$, where $x = (x_1, x_2, \ldots, x_n)$ and $y = (y_1, y_2, \ldots, y_n)$, is a solution of the *multivariate forward Kolmogorov differential equation*, an extension of equation (8.13) to vector random variables,

$$\frac{\partial p(y,s;x,t)}{\partial s} = \frac{1}{2} \sum_{i=1}^{n} \sum_{j=1}^{n} \frac{\partial^2}{\partial y_i \partial y_j} \left[p(y,s;x,t) \sum_{l=1}^{m} \beta_{il}(y,s) \beta_{jl}(y,s) \right]$$
$$- \sum_{i=1}^{n} \frac{\partial [p(y,s;x,t)\alpha_i(y,s)]}{\partial y_i}. \tag{9.4}$$

The preceding equation is also known as the *multivariate Fokker-Planck equation*. The *multivariate backward Kolmogorov differential equation* is an extension of equation (8.10) to vector random variables,

$$\frac{\partial p(y,s;x,t)}{\partial t} = -\frac{1}{2} \sum_{i=1}^{n} \sum_{j=1}^{n} \sum_{l=1}^{m} \beta_{il}(x,t) \beta_{jl}(x,t) \frac{\partial^2}{\partial x_i \partial x_j} [p(y,s;x,t)]$$
$$- \sum_{i=1}^{n} \alpha_i(x,t) \frac{\partial [p(y,s;x,t)]}{\partial x_i}. \tag{9.5}$$

In the case of time-homogeneous processes, $p(y,s;x,t) = p(y,x,\tau)$, where $\tau = s - t$, then $\partial p/\partial t = -\partial p/\partial \tau$ and $\partial p/\partial s = \partial p/\partial \tau$. The joint p.d.f., $p(x,t)$ of $X(t)$ is a solution of the forward Kolmogorov differential equation, where $p(x,0) = \delta(x - x_0)$. In this case, the forward Kolmogorov differential equation (9.4) is generally written as

$$\frac{\partial p(x,t)}{\partial t} = \frac{1}{2} \sum_{i=1}^{n} \sum_{j=1}^{n} \frac{\partial^2}{\partial x_i \partial x_j} \left[p(x,t) \sum_{l=1}^{m} \beta_{il}(x,t) \beta_{jl}(x,t) \right]$$
$$- \sum_{i=1}^{n} \frac{\partial [p(x,t)\alpha_i(x,t)]}{\partial x_i}. \tag{9.6}$$

These partial differential equations are generally too difficult to solve except in very special cases. But they are useful equations for computing, for example, the time-independent stationary solutions, $p(x,t) \equiv p(x)$, and in computing the expected time to extinction. In both of these cases, the problems are reduced to boundary value problems. The next example is an application of the multivariate Itô's formula.

Example 9.1. Consider a linear system of SDEs with multiplicative noise:

$$dX_i(t) = \sum_{j=1}^{n} a_{ij} X_j(t)\, dt + \sigma_i X_i(t) dW_i(t)$$

for $i = 1, \ldots, n$. Relating the preceding system to the general system (9.1), the coefficients are $\alpha_i = \sum_{j=1}^{n} a_{ij} X_j$ and $\beta_{ii} = \sigma_i X_i$ so that $B = \text{diag}(\beta_{11}, ..., \beta_{nn})$. Taking the expectation of the corresponding stochastic integral equation, applying property (i) in Theorem 8.2, then differentiating leads to a linear differential equation for $E(X_i(t))$:

$$\frac{dE(X_i(t))}{dt} = \sum_{j=1}^{n} a_{ij} E(X_j(t)).$$

The expectation will be equal to the solution corresponding to the deterministic equation (when $\sigma_i = 0$).

Applying the multivariate Itô's formula, an SDE for the product of two random variables can be obtained. Let $F_i(x, t) = x_i^2$, where $x = (x_1, x_2, \ldots, x_n)$. Then $\partial F_i/\partial t = 0$, $\partial F_i/\partial x_i = 2x_i$, and $\partial^2 F_i/\partial x_i^2 = 2$. The functions f and g (a scalar function) in Itô's formula are $f(x, t) = \alpha_i 2x_i + \frac{1}{2} 2\beta_{ii}^2$ and $g(x, t) = 2x_i \beta_{ii}$. Thus,

$$dX_i^2(t) = \left[(2a_{ii} + \sigma_i^2) X_i^2(t) + \sum_{j=1, j \neq i}^{n} 2a_{ij} X_i(t) X_j(t) \right] dt + 2\sigma_i X_i^2(t) dW_i(t),$$

for $i = 1, \ldots, n$. Let $F_{ik}(x, t) = x_i x_k$, where $i \neq k$. In this case, $\partial F_{ik}/\partial t = 0$, $\partial F_{ik}/\partial x_i = x_k$, $\partial F_{ik}/\partial x_k = x_i$, $\partial^2 F_{ik}/\partial x_i \partial x_k = 1$, and $\partial^2 F_{ik}/\partial x_l^2 = 0$ for $l = i, k$. Applying Itô's formula to $d(F_{ik})$ yields $f(x, t) = \alpha_i x_k + \alpha_k x_i$ and $g = (g_i, g_k)$, where $g_i(x, t) = x_k \beta_{ii}$ and $g_k(x, t) = x_i \beta_{kk}$ so that

$$d(X_i(t) X_k(t)) = \left[a_{ik} X_k^2(t) + a_{ki} X_i^2(t) + \sum_{j=1, j \neq k}^{n} a_{ij} X_j(t) X_k(t) \right.$$
$$\left. + \sum_{j=1, j \neq i} a_{kj} X_j(t) X_i(t) \right] dt$$
$$+ \sigma_i X_i(t) X_k(t) dW_i(t) + \sigma_k X_i(t) X_k(t) dW_k(t).$$

Consider the case $n = 2$. Applying property (i) in Theorem 8.2 to the stochastic integral equations for $X_i^2(t)$, $i = 1, 2$, and $X_1(t) X_2(t)$, and then differentiating, leads to ODEs for $E(X_i^2(t))$ and $E(X_1(t) X_2(t))$:

$$\frac{dE(X_i^2(t))}{dt} = (2a_{ii} + \sigma_i^2) E(X_i^2(t)) + 2a_{ij} E(X_i(t) X_j(t)), \quad i = 1, 2, i \neq j$$

and

$$\frac{dE(X_1(t)X_2(t))}{dt} = (a_{11} + a_{22})E(X_1(t)X_2(t)) + a_{12}E(X_2^2(t)) + a_{21}E(X_1^2(t)).$$

The differential equations for the first- and second-order moments are linear systems, $dY/dt = AY$, that can be easily solved, given the initial conditions $E(X_i(0))$, $E(X_i^2(0))$, ..., $E(X_1(0)X_2(0))$. (See Exercise 2.)

In the special case $n = 1$, the system reduces to a linear scalar SDE for $X = X_1$ ($a_{11} = a$, $\sigma_1 = \sigma$), studied previously in Example 8.11 (geometric Brownian motion). Applying property (i) in Theorem 8.2 to the stochastic integral equation, and then differentiating, leads to an ODE for $E(X^2(t)|X(0))$:

$$\frac{dE(X^2(t)|X(0))}{dt} = (2a + \sigma^2)E(X^2(t)).$$

For $2a + \sigma^2 < 0$, $\lim_{t \to \infty} E(X^2(t)) = 0$. But if $2a + \sigma^2 > 0$, then

$$\lim_{t \to \infty} E(X^2(t)) = \infty.$$

If $a < 0$ and $\sigma = 0$, the deterministic system approaches zero. But if $\sigma > 0$ is sufficiently large, the expectation approaches zero and the variance approaches infinity. The noise terms destabilize the system. ∎

9.3   Derivation of Itô SDEs

Itô SDEs for more than one interacting population are derived from a DTMC model. From this derivation, the relationship between the underlying deterministic ODE model, Markov chain model, and Itô SDE model can be observed. The variation over time is due to variation within the population process such as that due to births, deaths, immigration, emigration, or transitions between states. These types of variations are often referred to as *demographic variations*. However, variations from outside the population, such as external or *environmental variations* can be easily modeled using the same method, if it is known how the external variations affect the population (Allen et al., 2005).

We follow the derivation procedure presented in E. Allen (1999, 2007) and Allen et al. (2008). The derivation procedure shows the close relationship between DTMC and CTMC processes and the derived system of SDEs. This relationship has been known for a long time. In 1970, Kurtz gave conditions for the CTMC process to converge to the underlying system of ODEs. In a second paper in 1971, Kurtz remarked that density-dependent CTMC population processes with sufficiently large population densities are "close" to the solution of a corresponding system of SDEs i.e., weak convergence to a diffusion process under suitable conditions. Gardiner (1985), Gillespie (1992), and van Kampen (2007) derive the Fokker-Planck equation or the forward Kolmogorov differential equation of the SDE system, from the master equation or the forward Kolmogorov differential equation of the CTMC process. The Fokker-Planck equation is an approximation of another equation known as the Kramers-Moyal expansion which is identical to the master equation (van Kampen, 2007). In physical and biochemical applications, the derived SDE system is often referred to as the *Langevin* or *chemical Langevin equation*, an approximation to the underlying CTMC (Gillespie, 2000, 2002). As an approximation to the CTMC process, the system of SDEs must be justified. That is, the population densities must be sufficiently large. But even for small population densities, the agreement between the jump process and the continuous state processes can be very good, as will be illustrated in some of the examples. Nice summaries of the derivation of the chemical Langevin equation and its relationship to the underlying deterministic model, the reaction-rate equations, can be found in Gillespie (2000), Gillespie and Petzold (2006), Higham (2008), and Manninen et al. (2006).

The derivation procedure is applied to two interacting populations. Let X_1 and X_2 be the random variables for each of the two populations. The two populations may represent two chemical species or predator and prey, or may be two subpopulations or stages within a single population, such as susceptible and infectious individuals.

The first step in the derivation of an SDE model is to identify all of the possible interactions that lead to a change in one or more of the populations (or subpopulations) and the probability associated with this particular change in a given interval of time Δt. This first step is closely related to derivation of a Markov chain model. Thus, as will be seen, the dynamics of the SDE model are related to that of a corresponding Markov chain model.

Let $\Delta X = (\Delta X_1, \Delta X_2)^{tr}$, where $\Delta X_i = X_i(t + \Delta t) - X_i(t)$, $i = 1, 2$. Recall that a birth or a death in a population modeled via a DTMC model causes a change in state in discrete units, generally an increase by one or a decrease by one unit (but it need not be only a change of one unit). Again, for simplicity, assume a birth in time Δt for the first population causes an increase in the population size by one, $X_1(t) \to X_1(t) + 1$, denoted as $\Delta X = (\Delta X_1, \Delta X_2)^{tr} = (1, 0)^{tr}$ with probability $b_1 \Delta t$, where $b_1 \equiv b_1(X_1(t), X_2(t), t)$ is the birth rate of population X_1, which may depend on both populations and on the current time t. In addition to a birth in the first population, there are probabilities of a birth in the second population, deaths in the first or second population, and transitions between the two populations (which may be migration between the two populations or a susceptible individual that becomes infectious). Suppose there are a total of six possible changes and associated probabilities, denoted as $(\Delta X)_i$ and $p_i \equiv p_i(X_1, X_2, t)$, $i = 1, 2, \ldots, 6$, respectively, defined in Table 9.1.

Table 9.1: Probabilities associated with changes in the interacting population model, $0 \leq p_i \leq 1$

i	Change, $(\Delta X)_i$	Probability, p_i
1	$(1,0)^{tr}$	$b_1 \Delta t$
2	$(0,1)^{tr}$	$b_2 \Delta t$
3	$(-1,0)^{tr}$	$d_1 \Delta t$
4	$(0,-1)^{tr}$	$d_2 \Delta t$
5	$(-1,1)^{tr}$	$m_{21} \Delta t$
6	$(1,-1)^{tr}$	$m_{12} \Delta t$

In addition to these six changes in Table 9.1, there is a probability of no change, $\Delta X = (0,0)^{tr}$, whose probability of occurrence is one minus the sum of these six probabilities, $p_7 = 1 - \sum_{i=1}^{6} p_i$. The seven probabilities account for all of the possible changes in time Δt: $\sum_{i=1}^{7} p_i = 1$. It should be noted that the changes $(\Delta X)_i$ do not have to be integer-valued. See Exercise 13.

In the second step of the derivation, the expectation and covariance for the change in the two populations are computed based on the probabilities in Table 9.1. The expectation, $E(\Delta X) = \sum_{i=1}^{7} p_i (\Delta X)_i$ is a 2×1 vector. To order Δt, the expectation can be expressed as follows:

$$E(\Delta X) = \begin{pmatrix} b_1 - d_1 - m_{21} + m_{12} \\ b_2 - d_2 + m_{21} - m_{12} \end{pmatrix} \Delta t.$$

In addition, the covariance matrix associated with these changes is a 2×2 matrix, $\Sigma(\Delta X) = E([\Delta X][\Delta X]^{tr}) - E(\Delta X)[E(\Delta X)]^{tr}$, which to order Δt is approximately $\Sigma(\Delta X) = E([\Delta X][\Delta X]^{tr}) = \sum_{i=1}^{7} p_i (\Delta X)_i (\Delta X)_i^{tr}$:

$$\Sigma(\Delta X) = E \begin{pmatrix} (\Delta X_1)^2 & (\Delta X_1)(\Delta X_2) \\ (\Delta X_1)(\Delta X_2) & (\Delta X_2)^2 \end{pmatrix}$$

$$= \begin{pmatrix} b_1 + d_1 + m_{21} + m_{12} & -m_{21} - m_{12} \\ -m_{21} - m_{12} & b_2 + d_2 + m_{21} + m_{12} \end{pmatrix} \Delta t = V \Delta t.$$

That is, $V = \Sigma(\Delta X)/\Delta t$.

In the third step, it is shown that there are alternate, but equivalent, ways to write a system of Itô SDEs that model this stochastic process. Although the SDEs have different forms, they each have the same mean vector and covariance matrix and their corresponding p.d.f. is a solution of the same Kolmogorov differential equations. For example, two equivalent formulations are

$$dX(t) = \mu(X(t), t)\,dt + S(X(t), t)dW(t) \tag{9.7}$$

$$dX(t) = \mu(X(t), t)\,dt + B(X(t), t)dW^*(t) \tag{9.8}$$

where S is a 2×2 matrix, B is a 2×6 matrix, W is a 2×1 vector and W^* is a 6×1 vector of independent Wiener processes. The equations agree in the drift term but differ in the diffusion matrix, either S or B. It will be shown that $S^2 = V = BB^{tr}$, a necessary condition for equivalence of the SDEs (E. Allen, 2007; Allen et al., 2008).

In the first derivation procedure, note that the covariance matrix $\Sigma(\Delta X) = V \Delta t$ is diagonally dominant. Matrix V has a nonnegative trace, determinant, and discriminant, $\operatorname{Tr} V \geq 0$, $\det V \geq 0$, and $(\operatorname{Tr} V)^2 - 4\det V \geq 0$, which implies the eigenvalues, λ_1 and λ_2, of V are real and nonnegative ($\operatorname{Tr} V = \lambda_1 + \lambda_2$ and $\det V = \lambda_1 \lambda_2$). Matrix V is symmetric and positive semidefinite. Therefore, it can be shown that $V = P^{tr} D P$, where P is an orthogonal matrix $P^{tr} P = I$ and $D = \operatorname{diag}(\lambda_1, \lambda_2)$ (Ortega, 1987). Matrix $\Sigma(\Delta X)$ has a unique positive semidefinite square root given by $S = P^{tr} \sqrt{D} P$, $S^2 = V$. Therefore,

$$S = \sqrt{V} = \sqrt{\Sigma/\Delta t}.$$

Assuming that $X(t)$ is sufficiently large, so that $\Delta X(t)$ has an approximate normal distribution with mean vector $\mu \Delta t$ and covariance matrix $\Sigma(\Delta X) = V \Delta t$ [i.e., $\Delta X(t) \sim N(\mu \Delta t, V \Delta t)$]. The assumption that $\Delta X(t)$ is normally distributed follows from the Central Limit Theorem. Let Δt be subdivided into k equal subintervals, $\Delta t = k(\Delta t/k)$, and let $\Delta_j X(t) = X(t + j(\Delta t/k)) - X(t + (j-1)\Delta t/k)$. Then suppose that Δt is sufficiently small and k and $X(t)$ are sufficiently large, so that the $\Delta_j X(t)$ have *approximately* the same distribution, which only depends on $X(t)$. The Central Limit Theorem can be applied to $\Delta X(t) = \sum_{j=1}^{k} \Delta_j X(t)$ to show that $\Delta X(t)$ has an approximate normal distribution. The approximation improves for large k and small Δt. See Serfling (1980) for a generalization of the Central Limit Theorem to vector random variables. Also, an argument can be applied using the normal approximation to Poisson random variables.

Let $\eta = (\eta_1, \eta_2)^{tr} \sim N(0, I)$ be a two-dimensional standard Wiener process, then $\Delta X(t) = \mu \Delta t + S\sqrt{\Delta t}\, \eta \sim N(\mu \Delta t, V \Delta t)$. Thus $X(t+\Delta t)$ can be written in vector form as follows:

$$X(t + \Delta t) = X(t) + \Delta X(t)$$
$$= X(t) + \mu \Delta t + S\sqrt{\Delta t}\, \eta.$$

Expressing the preceding equation in terms of X_1 and X_2 and denoting $S \equiv S(t) = (S_{ij}(t))$ yields

$$X_1(t + \Delta t) = X_1(t) + \mu_1(t) \Delta t + S_{11}(t)\eta_1\sqrt{\Delta t} + S_{12}(t)\eta_2\sqrt{\Delta t} \quad (9.9)$$
$$X_2(t + \Delta t) = X_2(t) + \mu_2(t) \Delta t + S_{21}(t)\eta_1\sqrt{\Delta t} + S_{22}(t)\eta_2\sqrt{\Delta t}. \quad (9.10)$$

Equations (9.9) and (9.10) represent one iteration of the Euler-Maruyama method applied to a system of Itô SDEs (nonanticipative). If $\Delta t \to 0$ and assuming the stochastic integral exists and is unique, then $\eta_i\sqrt{\Delta t} \to dW_i(t)$, where $W_i(t)$ is a Wiener process. The system converges in the mean square sense to the following system of Itô SDEs:

$$dX_1(t) = \mu_1\, dt + S_{11}\, dW_1(t) + S_{12}\, dW_2(t)$$
$$dX_2(t) = \mu_2\, dt + S_{21}\, dW_1(t) + S_{22}\, dW_2(t),$$

where $W_1(t)$ and $W_2(t)$ are two independent Wiener processes (E. Allen, 1999, 2007). In vector form, the preceding system can be expressed as

$$dX(t) = \mu(X(t), t)\, dt + S(X(t), t)\, dW(t) \tag{9.11}$$

where

$$\mu = \begin{pmatrix} \mu_1 \\ \mu_2 \end{pmatrix} = \begin{pmatrix} b_1 X_1 - d_1 X_1 - m_{21} X_1 + m_{12} X_2 \\ b_2 X_2 - d_2 X_2 - m_{12} X_2 + m_{21} X_1 \end{pmatrix},$$

$S = \sqrt{V}$ and $W = (W_1, W_2)^{tr}$.

The joint p.d.f. $p(x, t)$ corresponding to the Itô SDE (9.11) is a solution of the multivariate forward Kolmogorov differential equation,

$$\frac{\partial p(x, t)}{\partial t} = \frac{1}{2} \sum_{i=1}^{2} \sum_{j=1}^{2} \frac{\partial^2}{\partial x_i \partial x_j} \left[p(x, t) \sum_{i=1}^{2} S_{il} S_{jl} \right] - \sum_{i=1}^{2} \frac{\partial [p(x, t) \mu_i]}{\partial x_i}, \tag{9.12}$$

where $S_{ij} \equiv S_{ij}(x, t)$ and $\mu_i \equiv \mu_i(x, t)$. Of interest are the terms in the first summation:

$$\sum_{i=1}^{2} S_{il} S_{jl} = (SS^{tr})_{ij} = (S^2)_{ij} = V_{ij}.$$

This expression also appears in the multivariate backward Kolmogorov differential equation. Thus, it is the matrix product $S^2 = V$ rather than S that determines the p.d.f. for the stochastic process. This fact is used in the next derivation.

In the second derivation procedure, a similar procedure to the first one is followed to obtain the mean and covariance matrices for the change ΔX. But instead of computing the square root of the covariance matrix to find the diffusion matrix, another matrix B is shown to have the property that $BB^{tr} = V$ or $BB^{tr} \Delta t = \Sigma(\Delta X)$. Denote the ith change in Table 9.1, $(\Delta X)_i$, $i = 1, 2, \ldots, 6$, as $(\Delta_{1i}, \Delta_{2i})^{tr}$, where each component reflects the amount and direction (sign) of the change in the first variable, Δ_{1i}, and the second variable, Δ_{2i}. Define the (i, j) entry in matrix B as follows:

$$\boxed{B_{ij} = \Delta_{ij} \sqrt{p_j / \Delta t.}} \tag{9.13}$$

Based on the entries in Table 9.1, matrix B is

$$B = \begin{pmatrix} \sqrt{b_1} & 0 & -\sqrt{d_1} & 0 & -\sqrt{m_{21}} & \sqrt{m_{12}} \\ 0 & \sqrt{b_2} & 0 & -\sqrt{d_2} & \sqrt{m_{21}} & -\sqrt{m_{12}} \end{pmatrix}.$$

It is straightforward to show that $BB^{tr} = V = S^2$ and $\sum_{l=1}^{2} S_{il}S_{jl} = V_{ij} = \sum_{l=1}^{6} B_{il}B_{jl}$. In particular, it can be shown that sample paths of the Itô SDE (9.11) are also sample paths of the following system of Itô SDEs:

$$dX(t) = \mu(X(t), t) + B(X(t), t)dW^*(t), \qquad (9.14)$$

where $W^* = (W_1^*, W_2^*, \ldots, W_6^*)^{tr}$ is a vector of six independent Wiener processes (E. Allen, 2007; Allen et al., 2008). The joint p.d.f. of the process $X(t)$ is a solution of (9.12). Equation (9.14) is referred to as the *Langevin equation* in physical applications or as the *chemical Langevin equation* in biochemical applications (Gillespie, 2000, 2002).

Other diffusion matrices G that account for the changes in the process can be derived. A necessary condition is $GG^{tr} = V$. For example, G can be taken as the lower triangular matrix in the Cholesky factorization of V (Ackleh et al. 2010). Another example of a 2×4 matrix is

$$G = \begin{pmatrix} \sqrt{b_1 + d_1} & 0 & -\sqrt{m_{21}} & \sqrt{m_{12}} \\ 0 & \sqrt{b_2 + d_2} & \sqrt{m_{21}} & -\sqrt{m_{12}} \end{pmatrix}.$$

Therefore, another equivalent system of SDEs is

$$dX(t) = \mu(X(t), t)\, dt + G(X(t), t)\, d\tilde{W}(t),$$

where $\tilde{W} = (\tilde{W}_1, \tilde{W}_2, \tilde{W}_3, \tilde{W}_4)^{tr}$ is a vector of four independent Wiener processes.

Systems of SDEs based on three or more interacting populations can be derived in a similar manner. The covariance matrix is a symmetric positive semidefinite matrix of dimension $n \times n$, where $n \geq 3$. To apply the first derivation procedure, it is necessary to compute $\sqrt{V}$. Unfortunately, there are no analytical methods for obtaining the square root of a general $n \times n$ positive definite matrix V for $n > 2$, but there are computational methods.

In the special case of a 2×2 positive definite matrix, there is a simple formula for the square root of a matrix V and for the Cholesky factorization. E. J. Allen (1999) derived a formula for S, the square root of the 2×2 matrix V, $S^2 = V$,

$$S = \frac{1}{d} \begin{pmatrix} V_{11} + g & V_{12} \\ V_{21} & V_{22} + g \end{pmatrix}, \qquad (9.15)$$

where $V_{12} = V_{21}$, g is the square root of the determinant of matrix V,

$$g = \sqrt{V_{11}V_{22} - V_{12}^2},$$

and d is the square root of the sum of the entries along the diagonal,

$$d = \sqrt{V_{11} + V_{22} + 2g}.$$

It can be easily verified that $S^2 = V$. The Cholesky factorization of a 2×2 positive definite matrix is $SS^{tr} = V$, where

$$S = \begin{pmatrix} \sqrt{V_{11}} & 0 \\ V_{12}/\sqrt{V_{11}} & \sqrt{V_{22} - V_{12}^2/V_{11}} \end{pmatrix}.$$

See Ackleh et al. (2010).

Simple computational methods for calculating the square root of a positive definite square matrix of any size can be found in E. J. Allen et al. (2000). Software programs often have built-in commands for computation of a square root or of a Cholesky factorization of a matrix. In MATLAB®, the command for calculation of a square root of a matrix V is **sqrtm**(V) and the command for the Cholesky factorization of V is **chol**(V).

9.4 Scalar Itô SDEs for Populations

Scalar SDE models are derived for a simple birth, death, and immigration process and for the classical density-dependent logistic growth process.

9.4.1 Simple Birth and Death with Immigration

Assume $X(t)$ is the random variable for the total population size that experiences births, deaths, and a constant rate of immigration. Table 9.2 lists the changes and the associated probabilities for this population process, where λ, μ, and ν are positive constants. The parameter notation is the same as for the simple birth and death process with immigration in Chapter 6, Section 6.4.4.

Table 9.2: Probabilities associated with changes in the birth, death, and immigration model

i	Change, $(\Delta X)_i$	Probability, p_i
1	1	$\lambda X(t)\Delta t$
2	-1	$\mu X(t)\Delta t$
3	1	$\nu \Delta t$

The expectation and variance for the change, to order Δt, in the stochastic process are

$$E(\Delta X) = [(\lambda - \mu)X + \nu]\Delta t$$

and

$$\Sigma(\Delta X) = [(\lambda + \mu)X + \nu]\Delta t = V\Delta t.$$

An Itô SDE corresponding to this process is

$$dX(t) = [(\lambda - \mu)X(t) + \nu]\,dt + \sqrt{(\lambda + \mu)X(t) + \nu}\,dW(t). \qquad (9.16)$$

Alternately, considering each of the three possible changes in Table 9.2, $i = 1, 2, 3$, another SDE formulation is

$$dX(t) = [(\lambda - \mu)X(t) + \nu]\,dt + \sqrt{\lambda X(t)}\,dW_1(t) + \sqrt{\mu X(t)}\,dW_2(t) + \sqrt{\nu}\,dW_3(t). \qquad (9.17)$$

The 1×3 matrix $B = (\sqrt{\lambda X(t)},\ \sqrt{\mu X(t)},\ \sqrt{\nu})$ has the property that $BB^{tr} = V$. These two SDEs are equivalent in that they will generate equivalent sample paths and have the same p.d.f.

Dropping the terms with the Wiener processes, the underlying ODE model is

$$\frac{dx}{dt} = (\lambda - \mu)x + \nu, \quad x(0) > 0.$$

If $\lambda < \mu$, then solutions approach a positive equilibrium,

$$\lim_{t\to\infty} x(t) = \frac{\nu}{\mu - \lambda}.$$

Written as a stochastic integral equation, the SDE population model (9.16) is

$$X(t) = X(0) + \int_0^t [(\lambda - \mu)X(s) + \nu]\,ds + \int_0^t \sqrt{(\lambda + \mu)X(s) + \nu}\,dW(s).$$

Taking the expectation of both sides, applying property (i) in Theorem 8.2, and then differentiating, leads to a differential equation for the expectation,

$$\frac{dE(X(t))}{dt} = (\lambda - \mu)E(X(t)) + \nu,$$

whose solution is the same as the underlying deterministic model. For $\lambda < \mu$, $\lim_{t\to\infty} E(X(t)) = \dfrac{\nu}{\mu - \lambda}$. Applying Itô's formula to $F(x) = x^2$ with SDE (9.16) yields

$$dX^2(t) = [2(\lambda - \mu)X^2(t) + (\lambda + \mu + 2\nu)X(t) + \nu]dt$$
$$+ 2X(t)\sqrt{(\lambda + \mu)X(s) + \nu}\,dW(t).$$

A differential equation for the variance, $Var(X) = E(X^2) - E(X)^2$, is

$$\frac{dVar(X)}{dt} = 2(\lambda - \mu)Var(X) + (\lambda + \mu)E(X) + \nu.$$

The variance approaches a constant in the case $\lambda < \mu$:

$$\lim_{t \to \infty} Var(X(t)) = \frac{\nu\mu}{(\mu - \lambda)^2}.$$

The limiting moments agree with those for the simple birth and death process with immigration, as discussed in Chapter 6, Section 6.4.4. All the moments for this stochastic process can be easily computed because the differential equations are linear and do not depend on higher-order moments. The dynamics of this population process are explored further in Exercises 6 and 7.

9.4.2 Logistic Growth

The classical ODE model for logistic growth is

$$\frac{dx}{dt} = rx\left(1 - \frac{x}{K}\right),$$

where $x(t)$ is the population size at time t. For $x(0) > 0$, it is well known that solutions approach the carrying capacity K, $\lim_{t \to \infty} x(t) = K$. Let $X(t)$ denote the random variable for the total population size. Two different sets of assumptions, (a) and (b), are made regarding births and deaths that give rise to two distinct, nonequivalent SDEs. See Chapter 6, Section 6.8. The assumptions are summarized in Table 9.3.

Table 9.3: Probabilities associated with changes in the logistic growth model

i	Change, $(\Delta X)_i$	(a) Probability, p_i	(b) Probability, p_i
1	1	$rX\Delta t$	$rX[1 - X/(2K)]\Delta t$
2	-1	$(rX^2/K)\Delta t$	$rX^2/(2K)\Delta t$

In case (a), an SDE for logistic growth is

$$dX(t) = rX(t)\left(1 - \frac{X(t)}{K}\right) dt + \sqrt{rX(t)\left(1 + \frac{X(t)}{K}\right)}\, dW(t), \qquad (9.18)$$

$X(t) \in [0, \infty)$. In case (b), an SDE for logistic growth is

$$dX(t) = rX(t)\left(1 - \frac{X(t)}{K}\right) dt + \sqrt{rX(t)}\, dW(t), \qquad (9.19)$$

$X(t) \in [0, 2K]$. Consequently, the forward Kolmogorov differential equations corresponding to the SDEs (9.18) and (9.19) are

$$\frac{\partial p}{\partial t} = -\frac{\partial}{\partial x}\left[rx\left(1 - \frac{x}{K}\right)p\right] + \frac{1}{2}\frac{\partial^2}{\partial x^2}\left[rx\left(1 + \frac{x}{K}\right)p\right], \qquad (9.20)$$

$x \in (0, \infty)$ and

$$\frac{\partial p}{\partial t} = -\frac{\partial}{\partial x}\left[rx\left(1 - \frac{x}{K}\right)p\right] + \frac{1}{2}\frac{\partial^2}{\partial x^2}\left[rxp\right], \qquad (9.21)$$

$x \in (0, 2K)$, respectively. It is easy to see that the infinitesimal variance is larger in (9.18) than in (9.19).

The Euler-Maruyama method applied to (9.18) and (9.19) yields the following iterative schemes:

$$X_{i+1} = X_i + rX_i\left(1 - \frac{X_i}{K}\right)\Delta t + \sqrt{rX_i\left(1 + \frac{X_i}{K}\right)}\sqrt{\Delta t}\,\eta_i,$$

$X_i \in [0, \infty)$ and

$$X_{i+1} = X_i + rX_i\left(1 - \frac{X_i}{K}\right)\Delta t + \sqrt{rX_i}\sqrt{\Delta t}\,\eta_i,$$

$X_i \in [0, 2K]$, respectively, for $i = 0, 1, 2, \ldots, k - 1$, where $\eta_i \sim N(0, 1)$ is the standard Wiener process. Three samples paths for the SDEs (9.18) and (9.19) on the interval $[0,10]$ are graphed in Figure 9.1. The Euler-Maruyama method is applied with $\Delta t = 0.01$, parameter values $K = 10$ and $r = 1$, and initial condition $X(0) = K = 10$. Note that there is larger variation in the solutions for (9.18) than there is for (9.19). These figures should be compared with those of the CTMC logistic model in Example 6.8 in Chapter 6.

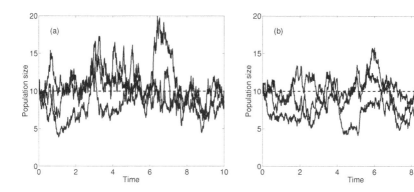

FIGURE 9.1: Three stochastic realizations of the logistic model for case (a), equation (9.18), and case (b), equation (9.19), with $r = 1$, $K = 10$, and $X(0) = 10$.

9.4.3 Quasistationary Density Function

For logistic growth, the boundary $x = 0$ is an exit boundary, so that if $X(t) = 0$, then $X(t + \tau) = 0$ for $\tau > 0$. The boundary at infinity is a natural boundary and the boundary at $x = 2K$ is reflecting. The population size can never go beyond $2K$. Thus, the probability of extinction will approach one as $t \to \infty$. However, it may take a long time for extinction to occur. Therefore, a probability density function conditioned on nonextinction, a quasistationary p.d.f., can be approximated by finding a time-independent solution of the forward Kolmogorov differential equations (subject to certain restrictions on the boundary).

The forward Kolmogorov differential equation for logistic growth is

$$\frac{\partial p(x,t)}{\partial t} = -\frac{\partial [\alpha(x)p(x,t)]}{\partial x} + \frac{1}{2}\frac{\partial^2 [\beta(x)p(x,t)]}{\partial x^2},$$

where, for example, in case (a), $\alpha(x) = rx(1 - x/K)$ and $\beta(x) = rx(1 + x/K)$, and in case (b), $\beta(x) = rx$, equations (9.20) and (9.21). To compute a quasistationary p.d.f., let $p(x,t) \equiv p(x)$ so that the derivative $\partial p(x)/\partial t = 0$. Assume $\beta(0)p(0) = 0$ and $d\beta(x)p(x)/dx = 0$ at $x = 0$. Then the partial derivatives in the Kolmogorov differential equation are ordinary derivatives, the left side equals zero, and the right side can be integrated from zero to x to obtain the following first-order differential equation:

$$\frac{d[\beta(x)p(x)]}{dx} - 2\frac{\alpha(x)}{\beta(x)}[\beta(x)p(x)] = 0, \tag{9.22}$$

subject to the restriction $\int_1^\infty p(x)\,dx = 1$ (Nisbet and Gurney, 1982). Solving the differential equation requires the following integrating factor:

$$\exp\left(-2\int_0^x \frac{\alpha(x)}{\beta(x)}\,dx\right).$$

Multiplying the differential equation by the integrating factor, then integrating and solving for $p(x)$ leads to the solution

$$p(x) = c[\beta(x)]^{-1}\exp\left(2\int_0^x \frac{\alpha(x)}{\beta(x)}\,dx\right),$$

where the constant c is calculated from the condition $\int_1^\infty p(x)\,dx = 1$.

An approximation to the quasistationary p.d.f. for the stochastic logistic growth model can be calculated from the preceding formula. In case (a),

$$p(x) = \frac{c}{rx}e^{-2x}(1 + x/K)^{4K-1}$$

and in case (b),

$$p(x) = \frac{c}{rx}e^K \exp(-[x - K]^2/K). \tag{9.23}$$

Example 9.2. Let the parameter values be the same as in the CTMC logistic model in Chapter 6, Example 6.11: $r = 1$ and $K = 10$. Then the approximate quasistationary p.d.f. from cases (a) and (b) are graphed in Figure 9.2. Compare the graphs in Figure 9.2 with those in Figure 6.10 in Chapter 6. The approximate quasistationary p.d.f., obtained from the solution of the forward Kolmogorov differential equation, is often referred to as the "diffusion approximation" for the CTMC model (Nisbet and Gurney, 1982). ■

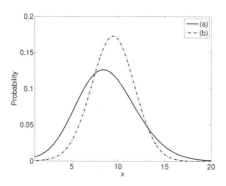

FIGURE 9.2: Approximate quasistationary p.d.f. for logistic growth in cases (a) and (b) when $r = 1$ and $K = 10$. In case (a), the mean and standard deviation are $m = 8.86$ and $\sigma = 3.19$ and in case (b), they are $m = 9.43$ and $\sigma = 2.32$.

The forward Kolmogorov differential equations can be used to set up a differential equation for the stationary probability distribution associated with the simple birth and death process with immigration, studied in Section 9.4.1. In particular for the case that the birth rate is less than the death rate, $\lambda < \mu$, there is a stationary probability distribution. Unfortunately, the stationary probability distribution $p(x)$ is not a solution of equation (9.22) because $p(0) \neq 0$ and $dp(x)/dx \neq 0$ at $x = 0$. Thus, the right-hand side of the equation is a nonzero constant that depends on $p(0)$ and the value of $dp(x)/dx$ at $x = 0$.

The mean persistence time and higher-order moments for the persistence of a population are first passage time problems. They are solutions of boundary value problems, as shown in Chapter 8, Section 8.9. Allen and Allen (2003) computed the mean and variance of the persistence time in SDE logistic models and compared them to the persistence time calculations from the corresponding CTMC and DTMC formulations. In most cases, the agreement between the three models was very good.

9.5 Enzyme Kinetics

A system of SDEs is derived based on the CTMC enzyme kinetics model in Chapter 7, Section 7.5.2. The underlying ODE model is

$$\frac{dN}{dt} = -\tilde{k}_1 N(\tilde{e}_\infty - B) + k_{-1} B$$

$$\frac{dB}{dt} = \tilde{k}_1 N(\tilde{e}_\infty - B) - (k_{-1} + k_2)B,$$

where N is the number of nutrient molecules and B is the number of molecules formed when the nutrient binds to an enzyme E. The product that is formed within the cell is denoted by P. The following quantities are conserved:

$$E + B = \tilde{e}_\infty \quad \text{and} \quad N + B + P = \tilde{p}_\infty,$$

so that $E = \tilde{e}_\infty - B$. Thus, only N and B are modeled. To formulate an SDE model, let N and B be continuous random variables for the nutrient and the complex that is formed and $\Delta X = (\Delta N, \Delta B)^{tr}$. Based on the infinitesimal transition probabilities, a table of probabilities can be constructed.

Table 9.4: Probabilities associated with changes in the enzyme kinetics model

i	Change, $(\Delta X)_i$	Probability, p_i
1	$(-1, 1)^{tr}$	$\tilde{k}_1 N(\tilde{e}_\infty - B)\,\Delta t$
2	$(1, -1)^{tr}$	$k_{-1} B\,\Delta t$
3	$(0, -1)^{tr}$	$k_2 B\,\Delta t$

The expectation vector and covariance matrix, to order Δt, are

$$E(\Delta X) = \mu \Delta t = \begin{pmatrix} -\tilde{k}_1 N(\tilde{e}_\infty - B) + k_{-1} B \\ \tilde{k}_1 N(\tilde{e}_\infty - B) - (k_{-1} + k_2)B \end{pmatrix} \Delta t$$

and

$$\Sigma(\Delta X) = \begin{pmatrix} \tilde{k}_1 N(\tilde{e}_\infty - B) + k_{-1} B & -\tilde{k}_1 N(\tilde{e}_\infty - B) - k_{-1} B \\ -\tilde{k}_1 N(\tilde{e}_\infty - B) - k_{-1} B & \tilde{k}_1 N(\tilde{e}_\infty - B) + (k_{-1} + k_2)B \end{pmatrix} \Delta t.$$

Equivalent SDEs can be formulated based on the derivation procedures in the preceding sections. Since the process involves only two random variables, the square root of matrix $V = \Sigma(\Delta X)/\Delta t$ can be computed (Exercise 12). Alternately, accounting for each of the changes, a diffusion matrix G of dimension 2×3 is

$$G = \begin{pmatrix} -\sqrt{\tilde{k}_1 N(\tilde{e}_\infty - B)} & \sqrt{k_{-1}B} & 0 \\ \sqrt{\tilde{k}_1 N(\tilde{e}_\infty - B)} & -\sqrt{k_{-1}B} & -\sqrt{k_2 B} \end{pmatrix},$$

where $GG^{tr}\Delta t = \Sigma$. A system of SDEs for the enzyme kinetics model is

$$dX(t) = \mu(X(t), t)\, dt + G(X(t), t)\, dW(t), \tag{9.24}$$

where $W = (W_1, W_2, W_3)^{tr}$ and $X(0) = (N(0), B(0))^{tr}$. The same parameter values as in the CTMC model, Chapter 7, Section 7.5.2, are used to compute numerical simulations for the preceding stochastic enzyme kinetics model. Let

$$\tilde{p}_\infty = 301\ mol, \quad \tilde{e}_\infty = 120\ mol, \quad \tilde{k}_1 = 1.66 \times 10^{-3} mol^{-1}s^{-1},$$

$$k_{-1} = 1 \times 10^{-4}s^{-1}, \quad \text{and} \quad k_2 = 0.1s^{-1},$$

with initial conditions, $N(0) = \tilde{p}_\infty$ and $B(0) = 0$ (Wilkinson, 2006). One sample path and the average of 10,000 sample paths are graphed in Figure 9.3 for the SDE enzyme kinetics model (Euler-Maruyama numerical method with $\Delta t = 0.01$).

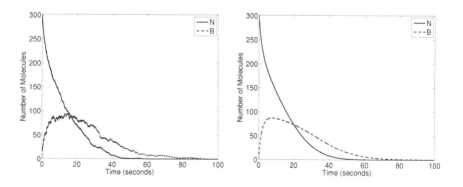

FIGURE 9.3: Sample path of the stochastic enzyme kinetics model, $N(t)$ and $B(t)$, and the average of 10,000 sample paths when $N(0) = \tilde{p}_\infty$ and $B(0) = 0$.

The multivariate Itô's formula and Theorem 8.2 can be applied to obtain differential equations for the moments of the process. For example, taking expectations of the integral form of the equations and applying Theorem 8.2

leads to the following differential equations:

$$\frac{dE(N)}{dt} = -\tilde{k}_1 \tilde{e}_\infty E(N) + \tilde{k}_1 E(NB) + k_{-1} E(B)$$

$$\frac{dE(B)}{dt} = \tilde{k}_1 \tilde{e}_\infty E(N) - \tilde{k}_1 E(NB) - (k_{-1} + k_2) E(B)$$

with initial conditions $E(N(0)) = \tilde{p}_\infty$ and $E(B(0)) = 0$. However, these equations cannot be solved for $E(N)$ and $E(B)$ because they depend on $E(NB)$. Thus, it is necessary to use Itô's formula and Theorem 8.2 to obtain an SDE for $E(NB)$. However, this equation will, in turn, depend on $E(N^2)$ and $E(B^2)$ which also depend on higher-order terms. Thus, given a nonlinear system of SDEs, differential equations for the moments depend on higher-order moments; the equations are not closed. Generally, assumptions are made regarding the distributions of $N(t)$ and $B(t)$, normal or lognormal, so that higher-order moments can be defined in terms of lower-order moments, i.e., moment closure assumptions (Ekanayake and Allen, 2010). These methods have been applied frequently to CTMC models. (See the discussion in Section 7.7.2).

In the next sections, the derivation procedures for Itô SDEs are applied to epidemic, competition, predation, and population genetics processes.

9.6 SIR Epidemic Process

Recall the deterministic SIR epidemic model formulated in Chapter 7, Section 7.6, where $S =$ susceptible individuals, $I =$ infected and infectious individuals, and $R =$ recovered and immune individuals. The ODE model has the form

$$\frac{dS}{dt} = -\frac{\beta}{N} SI$$

$$\frac{dI}{dt} = \frac{\beta}{N} SI - \gamma I$$

$$\frac{dR}{dt} = \gamma I,$$

where β is the contact rate, γ is the recovery rate and N is the total population size, $S(t) + I(t) + R(t) = N$. Solutions satisfy $S(t) \to \bar{S}_0$, $I(t) \to 0$, and $R(t) \to 0$. The limiting value $\bar{S}_0$ depends on the initial conditions. Whether there is an increase in the number of infected individuals (an epidemic) depends on the effective reproduction number given by

$$\mathcal{R} = \frac{\beta}{\gamma} \frac{S(0)}{N}.$$

When $\mathcal{R} > 1$, there is an epidemic, and when $\mathcal{R} \leq 1$, there is no epidemic.

For the SDE epidemic model, let S and I denote continuous random variables for the susceptible and infectious individuals and let $X = (S, I)^{tr}$. There are no births in this model. Infected individuals recover at a rate γI which can be considered a death rate for I since the immune individuals are not modeled. There is a transfer of a susceptible individual to an infected individual after successful contact, $\beta SI/N$. Table 9.5 lists the probabilities for the two possible changes in the SIR epidemic model for a small time interval Δt.

Table 9.5: Probabilities associated with changes in the SIR model

i	Change, $(\Delta X)_i$	Probability, p_i
1	$(-1, 1)^{tr}$	$\beta SI/N \, \Delta t$
2	$(0, -1)^{tr}$	$\gamma I \, \Delta t$

The expectation vector and the covariance matrix, to order Δt, have the following form:

$$E(\Delta X) = \begin{pmatrix} -\beta SI/N \\ \beta SI/N + \gamma I \end{pmatrix} \Delta t$$

and

$$\Sigma(\Delta X) = \begin{pmatrix} \beta SI/N & -\beta SI/N \\ -\beta SI/N & \beta SI/N + \gamma I \end{pmatrix} \Delta t = V \Delta t.$$

Applying formula (9.15), the square root of V can be computed, $G = V^{1/2}$. The SDE SIR epidemic model has the following form:

$$dS = -\frac{\beta}{N} SI \, dt + G_{11} dW_1 + G_{12} \, dW_2 \tag{9.25}$$

$$dI = \left(\frac{\beta}{N} SI - \gamma I \right) dt + G_{21} dW_1 + G_{22} \, dW_2, \tag{9.26}$$

where $S \in [0, N - I]$, $I \in [0, N - S]$, and the diffusion matrix $G = (G_{ij})$. The state space for (S, I) is the triangular region, $\{(S, I) : 0 \leq S, 0 \leq I, S + I \leq N\}$. The boundary $I = 0$ is an exit boundary; once the infected individuals reach zero, they remain at zero.

Alternately, an equivalent Itô SDE model for the SIR epidemic process is $dX = \mu \, dt + B dW^*$, where matrix B, a 2×2 matrix, is computed from the entries in Table 9.5 and formula (9.13),

$$B = \begin{pmatrix} -\sqrt{\beta SI/N} & 0 \\ \sqrt{\beta SI/N} & -\sqrt{\gamma I} \end{pmatrix}.$$

The explicit form for the SDE epidemic model, in this case, is

$$dS = -\frac{\beta}{N} SI \, dt - \sqrt{\beta SI/N} \, dW_1^* \tag{9.27}$$

$$dI = \left(\frac{\beta}{N} SI - \gamma I\right) dt + \sqrt{\beta SI/N} dW_1^* - \sqrt{\gamma I} \, dW_2^*, \tag{9.28}$$

where $BB^{tr} = V$.

Stochastic differential equations for an SIS epidemic model can be derived in a similar manner. If $S + I = N = $ constant, then the model can be reduced to a single variable I. A MATLAB program for numerical simulation of an SIS epidemic model is included in the Appendix for Chapter 9.

9.7   Competition Process

Consider the Lotka-Volterra competition model discussed in Chapter 7, Section 7.7. The system of ODEs that models two competing populations, x_1 and x_2, is

$$\frac{dx_1}{dt} = x_1(a_{10} - a_{11}x_1 - a_{12}x_2)$$

$$\frac{dx_2}{dt} = x_2(a_{20} - a_{21}x_1 - a_{22}x_2),$$

where $a_{ij} > 0$, $x_1(0)$, and $x_2(0) > 0$. The dynamics of the ODE model are summarized in Section 7.7. There are four cases that depend on the value of the coefficients, a_{ij}, and the initial conditions:

   I. Species 1 outcompetes species 2 for all positive initial values.

  II. Species 2 outcompetes species 1 for all positive initial values.

 III. Either species 1 or species 2 wins the competition. The outcome depends on the initial values.

 IV. Both species survive for all positive initial values.

Let X_i, $i = 1, 2$ denote continuous random variables for the two competing populations and let $X = (X_1, X_2)^{tr}$. Derivation of the SDE model depends on how the births and deaths are defined in this model. As noted for logistic growth, there is an infinite number of possibilities for the SDE model that give rise to the same deterministic model. For illustration purposes, assume the birth and death rates are $b_1 = a_{10}X_1$, $d_1 = X_1[a_{11}X_1 + a_{12}X_2]$, $b_2 = a_{20}X_2$, and $d_2 = X_2[a_{21}X_1 + a_{22}X_2]$, the same assumptions that were made in constructing the CTMC competition model in Section 7.7. Table 9.6 summarizes the changes in the competition process for a small time interval Δt.

Table 9.6: Probabilities associated with changes in the competition model

i	Change, $(\Delta X)_i$	Probability, p_i
1	$(1,0)^{tr}$	$a_{10}X_1 \Delta t$
2	$(-1,0)^{tr}$	$X_1[a_{11}X_1 + a_{12}X_2]\Delta t$
3	$(0,1)^{tr}$	$a_{20}X_2 \Delta t$
4	$(0,-1)^{tr}$	$X_2[a_{21}X_1 + a_{22}X_2]\Delta t$

The expectation vector and covariance matrix, to order Δt, are

$$E(\Delta X) = \begin{pmatrix} X_1(a_{10} + a_{11}X_1 + a_{12}X_2) \\ X_2(a_{20} + a_{21}X_1 + a_{22}X_2) \end{pmatrix} \Delta t = \mu \Delta t$$

and

$$\Sigma(\Delta X) = \begin{pmatrix} X_1(a_{10} + a_{11}X_1 + a_{12}X_2) & 0 \\ 0 & X_2(a_{20} + a_{21}X_1 + a_{22}X_2) \end{pmatrix} \Delta t$$
$$= V\Delta t.$$

The square root of a diagonal matrix is just the diagonal matrix with the square roots of the corresponding diagonal entries. That is,

$$S = \sqrt{V} = \begin{pmatrix} \sqrt{X_1(a_{10} + a_{11}X_1 + a_{12}X_2)} & 0 \\ 0 & \sqrt{X_2(a_{20} - a_{21}X_1 - a_{22}X_2)} \end{pmatrix}.$$

In this case, the system of Itô SDEs is $dX = \mu\, dt + S dW$, where μ is basically the right-hand side of the ODE model. The restriction, $X_1(t) \in [0, \infty)$ and $X_2(t) \in [0, \infty)$ is also imposed. If $X_i(t) = 0$, then $X_i(t+\tau) = 0$ for any $\tau > 0$; $X_i = 0$ are exit boundaries, $i = 1, 2$. Also, $X_i = \infty$ is a natural boundary, $i = 1, 2$. If either of the competing species hits zero (extinction), they cannot survive because there is no immigration.

An alternate but equivalent form for the Itô SDE competition model is $dX = \mu dt + B dW^*$, where $W^* = (W_1^*, W_2^*, W_3^*, W_4^*)^{tr}$ and the entries of matrix B can be calculated directly from (9.13). That is, an equivalent SDE competition model is

$$dX_1 = X_1(a_{10} - a_{11}X_1 - a_{12}X_2)dt + \sqrt{a_{10}X_1}\, dW_1^*$$
$$- \sqrt{X_1(a_{11}X_1 + a_{12}X_2)}\, dW_2^*$$
$$dX_2 = X_2(a_{20} - a_{21}X_1 - a_{22}X_2)dt + \sqrt{a_{20}X_2}\, dW_3^*$$
$$- \sqrt{X_2(a_{21}X_1 + a_{22}X_2)}\, dW_4^*.$$

One stochastic realization and the approximate probability histogram for the SDE two-species competition model (based on 10,000 stochastic realizations) are graphed in Figure 9.4. The MATLAB program that generated the graphs in Figure 9.4 is included in the Appendix for Chapter 9. Compare the dynamics of the SDE competition model with the CTMC competition model in Section 7.7, Figure 7.13. The mean and variance for the CTMC and SDE models are close. The mean and standard deviations for the probability histograms generated from 10,000 stochastic realizations of the SDE competition model at $t = 5$ are

$$\mu_{X_1} = 49.3, \quad \mu_{X_2} = 23.6$$

and

$$\sigma_{X_1} = 9.2, \quad \sigma_{X_2} = 6.8.$$

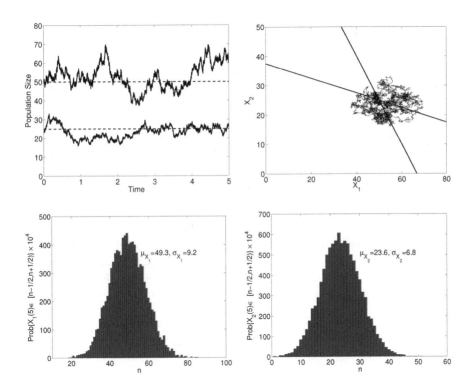

FIGURE 9.4: Sample paths of the SDE competition model as a function of time and in the X_1-X_2 phase plane (top two graphs). The dashed lines are the equilibrium values and the solid lines are the nullclines of the deterministic model. Approximate probability histograms for X_1 and X_2 for $t = 5$, based on 10,000 stochastic realizations (bottom two figures). The parameter values and initial conditions are $a_{10} = 2$, $a_{20} = 1.5$, $a_{11} = 0.03$, $a_{12} = 0.02$, $a_{21} = 0.01$, $a_{22} = 0.04$, $X_1(0) = 50$, and $X_2(0) = 25$.

9.8 Predator-Prey Process

Recall the ODE model for a Lotka-Volterra predator-prey system, as described in Chapter 7, Section 7.8. The variable x is the prey population size and y is the predator population size. The ODE model is

$$\frac{dx}{dt} = x(a_{10} - a_{12}y)$$
$$\frac{dy}{dt} = y(a_{21}x - a_{20}),$$

where $a_{ij} > 0$, $x(0) > 0$, and $y(0) > 0$. For each initial condition, there is a unique periodic solution encircling the positive equilibrium

$$(\bar{x}, \bar{y}) = (a_{20}/a_{21}, a_{10}/a_{12})$$

in the x-y phase plane, representing predator-prey cycles.

To formulate a stochastic model, it is important to know the birth and death rates for each of the species. Let X and Y be the random variables for the prey and the predator, respectively, and let $Z = (X, Y)^{tr}$. For illustration purposes, let the birth and death rates for the prey be $a_{10}X$ and $a_{12}XY$, respectively, and the birth and death rates for the predator be $a_{21}XY$ and $a_{20}Y$, respectively. Table 9.7 lists the probabilities associated with the four possible changes in a small period of time Δt.

Table 9.7: Probabilities associated with changes in the predator-prey model

i	Change, $(\Delta Z)_i$	Probability, p_i
1	$(1, 0)^{tr}$	$a_{10}X\,\Delta t$
2	$(-1, 0)^{tr}$	$a_{12}XY\,\Delta t$
3	$(0, 1)^{tr}$	$a_{21}XY\,\Delta t$
4	$(0, -1)^{tr}$	$a_{20}Y\,\Delta t$

The covariance matrix for the change ΔZ, to order Δt, has the following form:

$$\Sigma(\Delta Z) \approx \begin{pmatrix} a_{10}X + a_{12}XY & 0 \\ 0 & a_{21}XY + a_{20}Y \end{pmatrix} \Delta t.$$

Based on the probabilities in Table 9.7, an Itô SDE model has the form $dZ(t) = \mu(Z(t), t) + S(Z(t), t)\,dW(t)$ or an equivalent form $dZ(t) = \mu(Z(t), t)\,dt + B(Z(t), t)\,dW^*(t)$. The diffusion matrix S is of size 2×2 and W is a vector of two independent Wiener processes, whereas the diffusion matrix B is of size 2×4 and W^* is a vector of four independent Wiener processes. The two equivalent systems are

$$dX = X(a_{10} - a_{12}Y)\,dt + \sqrt{X(a_{10} + a_{12}Y)}\,dW_1$$
$$dY = Y(a_{21}X - a_{20})\,dt + \sqrt{Y(a_{21}X + a_{20})}\,dW_2,$$

and

$$dX = X(a_{10} - a_{12}Y)\,dt + \sqrt{a_{10}X}\,dW_1^* - \sqrt{a_{12}Y}\,dW_2^*$$
$$dY = Y(a_{21}X - a_{20})\,dt + \sqrt{a_{21}XY}\,dW_3^* - \sqrt{a_{20}Y}\,dW_4^*,$$

where it is assumed that $X(t) \in [0, \infty)$ and $Y(t) \in [0, \infty)$. The zero boundaries are exit boundaries and the boundaries at infinity are natural boundaries.

The Euler-Maruyama method for approximating a sample path to the first system of SDEs, $(X(t_i), Y(t_i)) = (X_i, Y_i)$ for $t_i = 0, \Delta t, 2\Delta t, \ldots, T$ and $i = 0, 1, 2, \ldots, k - 1$ is

$$X_{i+1} = X_i + X_i(a_{10} - a_{12}Y_i)\,\Delta t + \sqrt{X_i(a_{10} + a_{12}Y_i)}\sqrt{\Delta t}\,\eta_{1i},$$
$$Y_{i+1} = Y_i + Y_i(a_{21}X_i - a_{20})\,\Delta t + \sqrt{Y_i(a_{21}X_i + a_{20})}\sqrt{\Delta t}\,\eta_{2i},$$

where η_{1i} and η_{2i} are two random numbers sampled from a normal distribution with zero mean and unit variance. A single sample path for the stochastic Lotka-Volterra predator-prey models is graphed in Figure 9.5 as a function of time and in the phase plane. The sample path is compared to the solution of the deterministic model. There is a neutrally stable positive equilibrium at $(\bar{X}, \bar{Y}) = (100, 50)$. The deterministic solution is periodic. Compare Figure 9.5 with the CTMC predator-prey model in Figure 7.14. See also Gard and Kannan (1976) and Abundo (1991) for examples of other SDE predator-prey systems.

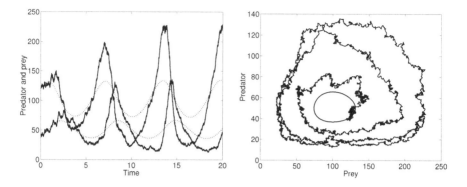

FIGURE 9.5: Sample path of the stochastic Lotka-Volterra predator-prey model with the solution of the deterministic model. Solutions are graphed over time as well as in the phase plane. The parameter values and initial conditions are $a_{10} = 1$, $a_{20} = 1$, $a_{12} = 0.02$, $a_{21} = 0.01$, $X(0) = 120$, and $Y(0) = 40$. Solutions with the smaller amplitude are the predator population.

9.9   Population Genetics Process

The Kolmogorov differential equations for the gene frequencies of a population are derived assuming random mating and no selection or mutation. This process in the population genetics literature is known as *random drift*. The continuous process is based on the Wright-Fisher model, a DTMC, discussed in the exercises in Chapter 3. The method of derivation for the SDEs is different from the preceding sections. The expectation and variance for the change in the dynamics are derived based on the assumptions in the Wright-Fisher model.

Assume that the population is diploid; each individual has two copies of the chromosomes. Assume that the gene is determined by a single locus in which there are only two alleles, A and a. There are three possible genotypes,

$$AA, \quad Aa, \quad \text{and} \quad aa.$$

In addition, assume the total population size is N. Because the population is diploid, the total number of alleles equals $2N$. Let $Y(t)$ denote the number of A alleles in the population in generation t and $X(t)$ denote the proportion of A alleles in the population, $X(t) = Y(t)/(2N)$. Suppose individuals mate randomly and generations are nonoverlapping. The number of genes in generation $t + 1$ is derived by sampling with replacement from the genes in generation t (Ewens, 1979). Given $X(t) = x$, then $Y(t + 1)$ has a binomial distribution, $b(2N, x)$; that is,

$$Y(t + 1) \sim b(2N, x).$$

Under these assumptions the infinitesimal mean and variance can be derived; that is, the coefficients $a(x)$ and $b(x)$ in the Kolmogorov differential equations (Ludwig, 1974). Let $\Delta Y(t) = Y(t + 1) - Y(t)$ and $\Delta X(t) = X(t + 1) - X(t)$. Then given $X(t) = x$, $Y(t) = 2Nx$,

$$E(Y(t + 1)|X(t) = x) = 2Nx$$

and

$$E(\Delta Y(t)|X(t) = x) = 2Nx - 2Nx = 0.$$

Then $E([\Delta Y(t)]^2|X(t) = x)$ equals

$$E(Y^2(t + 1) - 2Y(t + 1)Y(t) + Y^2(t)|X(t) = x)$$

which can be simplified to

$$E(Y^2(t + 1)|X(t) = x) - 2(2Nx)(2Nx) + 4N^2x^2$$
$$= E(Y^2(t + 1)|X(t) = x) - 4N^2x^2.$$

It can be seen that

$$E([\Delta Y(t)]^2|X(t) = x) = \text{Var}(Y(t + 1)|X(t) = x)$$
$$= 2Nx(1 - x).$$

Because $X(t) = Y(t)/(2N)$,

$$E(\Delta X(t)|X(t) = x) = \frac{1}{2N}E(\Delta Y(t)|X(t) = x) = 0 = a(x)$$

$$E([\Delta X(t)]^2|X(t) = x) = \frac{1}{(2N)^2}E([\Delta Y(t)]^2|X(t) = x)$$

$$= \frac{2Nx(1 - x)}{(2N)^2} = \frac{x(1 - x)}{2N} = b(x).$$

Hence, the forward Kolmogorov differential equation for random genetic drift, expressed in terms of the p.d.f. $p(x, t)$, is

$$\frac{\partial p}{\partial t} = \frac{1}{4N}\frac{\partial^2(x(1 - x)p)}{\partial x^2}, \quad 0 < x < 1, \tag{9.29}$$

where $p(x, 0) = \delta(x - x_0)$ or $X(0) = x_0$. Note that the forward Kolmogorov differential equation is singular at the boundaries, $x = 0$ and $x = 1$. Both boundaries are exit boundaries. At either of the states zero or one, there is fixation of allele a or A, respectively. The solution to the forward Kolmogorov equation was derived by Kimura (1955). See also Crow and Kimura (1970). The solution is a complicated expression depending on the hypergeometric function. The solution behavior of $p(x,t)$ is examined through the corresponding Itô SDE for this process.

The Itô SDE for random genetic drift has the form

$$dX(t) = \sqrt{\frac{X(t)(1 - X(t))}{2N}}\, dW(t), \quad X(t) \in [0,1],$$

where $X(0) = x_0$, $0 < x_0 < 1$. The boundaries 0 and 1 are absorbing [e.g., if $X(t) = 0$ (or 1), then $X(t + \tau) = 0$ (or 1) for $\tau > 0$]. The Euler-Maruyama method is used to calculate some numerical simulations of this SDE. Three sample paths for $X(t)$ are graphed in Figure 9.6 when $X(0) = 1/2$ and $N = 100$.

By numerically solving the SDE via Euler-Maruyama method, 10,000 sample paths were generated up to a fixed time t. A probability histogram was generated, an approximation to the p.d.f. corresponding to random genetic drift at times $t = 10$, $t = 50$, and $t = 200$, $p(x, 10)$, $p(x, 50)$, and $p(x, 200)$ in Figure 9.6. It was shown by Kimura (1955, 1994), for large t, that

$$p(x,t) \approx Ce^{-t/(2N)}, \quad 0 < x < 1.$$

The p.d.f. is approximately constant and very small when $0 < x < 1$ and t is large. This can be seen in the probability histogram in Figure 9.6 at $t = 200$. The probability of fixation at either $x = 0$ and $x = 1$ increases for large t. The value of the p.d.f. $p(x,t)$ tends to infinity at $x = 0$ and $x = 1$ and to zero for $0 < x < 1$ as t approaches infinity. When $X(0) = 1/2$, fixation at 0 or 1 is equally likely; the probability distribution is symmetric about $x = 1/2$. In particular, $E(X(t)) = 1/2$. This latter property is true in general for the population genetics process,

$$E(X(t)) = X(0).$$

This can be verified directly from the SDE (Exercise 21). In deterministic population genetics models, this property is known as a Hardy-Weinberg equilibrium. A Hardy-Weinberg equilibrium exists when there is random mating, no mutation, and no selection; the proportion of alleles stays constant in the population. Consult Ewens (1979), Goel and Richter-Dyn (1974), Kimura (1994), and Nagylaki (1992) for additional theory and models of population genetics processes.

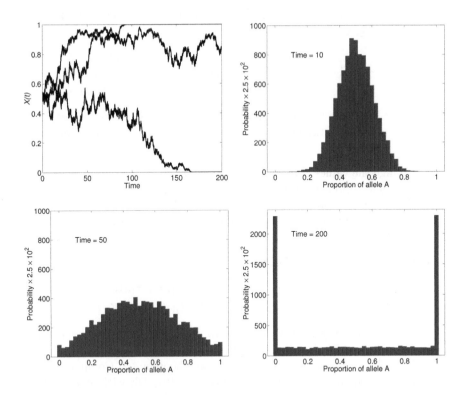

FIGURE 9.6: Three sample paths for the population genetics random drift model with $X(0) = 1/2$ and $N = 100$. The approximate probability histograms of $X(t)$ at $t = 10, 50, 200$.

9.10 Exercises for Chapter 9

1. For Example 9.1 with $n = 1$, let $X(0) = c$ =constant. Show that $Var(X(t)|X(0)) = c^2 e^{\sigma^2 t}$, which implies $\lim_{t \to \infty} Var(X(t)|X(0)) = \infty$. Compare this result with the variance for the drug kinetics example in Section 8.11.

2. For Example 9.1 with $n = 2$, assume $a_{11} = -2$, $a_{12} = 1 = a_{21}$, $a_{22} = -2$, and $\sigma_1 = 1 = \sigma_2$. Let matrix $A_1 = (a_{ij})$.

   (a) Show the expectations of X_i, $i = 1, 2$ can be found by solving $dY_1/dt = A_1 Y_1$, where $Y_1 = (E(X_1(t)), E(X_2(t)))^{tr}$.

(b) Show that the second-order moments are solutions of $dY_2/dt = A_2 Y_2$, where

$$A_2 = \begin{pmatrix} -3 & 0 & 2 \\ 0 & -3 & 2 \\ 1 & 1 & -4 \end{pmatrix} \quad \text{and} \quad Y_2 = \begin{pmatrix} E(X_1^2(t)) \\ E(X_2^2(t)) \\ E(X_1(t)X_2(t)) \end{pmatrix}.$$

(c) Show that the eigenvalues of A_1 and A_2 have negative real parts and hence, the mean and variance of the system of SDEs approach zero.

3. Change the values of σ_i in the preceding Exercise 2 so that the variance does not approach zero, even though the expectation approaches zero. This example shows that noise can destabilize the system. State some general conditions on the parameters a_{ij} and σ_i so that $E(X_i(t)) \to 0$ but $E(X_i^2(t)) \to \infty$.

4. Consider a linear Itô SDE with additive noise:

$$dX_1(t) = [a_{11}X_1(t) + a_{12}X_2(t)] \, dt + \sigma_1 dW_1(t)$$
$$dX_2(t) = [a_{21}X_1(t) + a_{22}X_2(t)] \, dt + \sigma_2 dW_2(t).$$

Use the multivariate Itô's formula to derive ODEs for the expectation $E(X_i(t))$ and for the second-order moments $E(X_i^2(t))$ and $E(X_1(t)X_2(t))$, $i = 1, 2$. Apply properties (i) and (ii) in Theorem 8.2. Compare the differential equations for the moments of the Itô SDE with additive noise to those with multiplicative noise (Example 9.1).

5. Show that $S^2 = V$, where S is defined in equation (9.15).

6. Consider the simple birth and death with immigration model discussed in Section 9.4.1.

(a) Solve the two differential equations for the mean, $E(X(t))$ and the variance, $Var(X(t))$, given the initial conditions $E(X(0)) = X_0$ and $Var(X(0)) = 0$.

(b) Apply the Itô SDE model (9.17) with Itô's formula to show that the same differential equations for the mean and variance are obtained as for the SDE (9.16).

(c) Use Itô's formula to write the differential equation for the third moment $E(X^3(t))$.

7. Apply the Euler-Maruyama method to compute three sample paths of the simple birth and death with immigration model (9.16). Then generate 10,000 sample paths to compute the mean and variance at $t = 10$ and compare your answers with part (a) of the preceding exercise. Use parameter values and initial values $\beta = 0.01$, $\gamma = 0.02$, $m = 1$, and $X(0) = 100$.

8. For the Itô SDE models for logistic growth, (9.18) and (9.19), formulate two alternate but equivalent SDE formulations based on the changes listed in Table 9.3.

9. For the stochastic logistic growth model, case (b), the approximate quasistationary p.d.f. (9.23) is related to a normal distribution:

$$f(x) = \frac{\exp(-(x-\mu)^2/[2\sigma^2])}{\sqrt{2\pi\sigma^2}},$$

where μ is the mean and σ^2 is the variance. Let $\mu = K$ and $\sigma^2 = K/2$. For the parameter values in Example 9.2, $r = 1$ and $K = 10$, graph the normal p.d.f. and the p.d.f. associated with (9.23).

10. In the derivation of the stochastic logistic model, let the birth and death probabilities be $p_1 = [r_1 X - rX^2/(2K)]\Delta t$ and $p_2 = [r_2 X + rX^2/(2K)]\Delta t$, where $r_1 - r_2 = r > 0$ and $X \equiv X(t)$.

   (a) Compute the approximate quasistationary p.d.f. That is, find the solution of equation (9.22) with the restriction $\int_1^\infty p(x)\, dx = 1$.

   (b) Show that the approximate quasistationary p.d.f. is related to a normal p.d.f. with $\mu = K$ and $\sigma^2 = K(r_1 + r_2)/(2r)$.

11. Consider the Itô SDE

$$dX(t) = [b(X) - d(X)]\, dt + \sqrt{b(X) + d(X)}\, dW(t),$$

$X(0) = x_0$ and $X(t) \in [0, \infty)$, where $b(X)$ is the birth rate and $d(X)$ is the death rate. The zero state is an absorbing boundary.

   (a) Let $b(X) = X$, $d(X) = X^2/15$, and $X(0) = x_0 = 15$. Apply the Euler-Maruyama method for $t \in [0, 2]$. Compute the mean and variance of 10,000 sample paths at $t = 1$ and $t = 2$. Graph three sample paths.

   (b) Let $b(X) = X - X^2/30$, $d(X) = X^2/30$, and $X(0) = x_0 = 15$. In this case, $X(t) \in [0, 30]$. Apply the Euler-Maruyama method for $t \in [0, 2]$. Compute the mean and variance of 10,000 sample paths at $t = 1$ and $t = 2$. Graph three sample paths.

   (c) Discuss the behavior of the SDE models in these two cases and relate their behavior to the CTMC models and the underlying deterministic logistic model,

$$\frac{dx}{dt} = b(x) - d(x).$$

12. For the stochastic enzyme kinetics model, formulate a system of SDEs equivalent to (9.24). In particular, formulate a system of the form $dX(t) = \mu\, dt + S dW(t)$, where S is the square root of matrix V, $\Sigma(\Delta X) = V \Delta t$.

13. An extension of the enzyme kinetics model is a model for bacterial growth in a chemostat. See Exercise 12 in Chapter 7. An ODE model for nutrient n and bacteria concentration b grown in a laboratory device known as a chemostat is

$$\frac{db}{dt} = b\left(\frac{k_{max}n}{k_n + n} - D\right)$$

$$\frac{dn}{dt} = D(n_0 - n) - \beta\frac{k_{max}nb}{k_n + n}.$$

See Smith and Waltman (1995). Parameter D is the dilution rate and n_0 is the constant rate of nutrient input. Parameters k_n and k_{max} are Michaelis-Menten constants; k_n is the half-saturation constant and k_{max} is the maximum rate of bacterial growth. It is not necessary to convert the units of b and n to number of molecules for SDEs. Use Table 9.8 to formulate a system of Itô SDEs for $X = (b, n)^{tr}$. Bacterial growth due to consumption of the nutrient is not a one-to-one relationship; the parameter β is not necessarily an integer.

Table 9.8: Probabilities associated with changes in the chemostat model

i	Change, $(\Delta X)_i$	Probability, p_i
1	$(1, -\beta)^{tr}$	$k_{max}nb\Delta t/(k_n + n)$
2	$(-1, 0)^{tr}$	$bD\Delta t$
3	$(0, -1)^{tr}$	$nD\Delta t$
4	$(0, 1)^{tr}$	$n_0 D\Delta t$

(a) Use the table of probabilities to set up a system of SDEs.

(b) Show that the same covariance matrix is obtained if the probability p_1 is changed to $p_1 = \beta k_{max}nb\Delta t/(k_n + n)$ with the corresponding change $(\Delta X)_1 = (1/\beta, -1)^{tr}$.

14. Recall that the SIS epidemic model can be expressed as a single differential equation,

$$\frac{dI}{dt} = \frac{\beta}{N}SI - (b + \gamma)I,$$

where $S(t) = N - I(t)$ and N is the constant total population size.

(a) Write an Itô SDE for the diffusion process.

(b) Write the forward Kolmogorov differential equation for the diffusion process.

15. For the Itô SDE SIS epidemic model in the preceding exercise, let $\beta = 2$, $b = 0.5 = \gamma$, and $N = 100$.

(a) Use the Euler-Maruyama method to compute three sample paths of the Itô SDE for $t \in [0, 10]$, $I(0) = 1$ and graph them. See the MATLAB program in the Appendix for Chapter 9.

(b) Estimate the probability that no epidemic occurs, $I(t) = 0$ before $t = 10$ for five different initial conditions, $I(0) = 1, 2, 3, 4, 5$. Use 10,000 sample paths for each initial condition. Compare these estimates with the approximation $(1/\mathcal{R}_0)^i$, $i = 1, 2, 3, 4, 5$ (probability of no epidemic), estimated from the CTMC and DTMC SIS epidemic models.

16. An approximation to the quasistationary p.d.f. for the SIS epidemic model can be obtained from the forward Kolmogorov differential equations, as was done for the stochastic logistic model in Section 9.4.3. Let $\beta = 2$, $\gamma + b = 1$ and $N = 100$, then find the approximate quasistationary p.d.f., that is, the solution of (9.22) with the restriction $\int_1^{100} p(x)\, dx = 1$. Compare the graph of the approximate quasistationary p.d.f. of the SDE SIS model with the quasistationary probability distribution of the CTMC SIS model, graphed in Figure 7.3.

17. Consider the stochastic SIR epidemic process.

(a) Use the entries in Table 9.5 and the formula for the square root of a 2×2 matrix, equation (9.15), to calculate $G = V^{1/2}$ for the SDE model (9.25)-(9.26).

(b) Use the Euler-Maruyama method to compute several sample paths of the SIR epidemic process, either model (9.25)-(9.26) or model (9.27)-(9.28) for $S(0) = 99$, $I(0) = 1$, $N = 100$, $\beta = 1$, and $\gamma = 0.5$. Calculate the effective reproduction number, $\mathcal{R}$. Plot the infected individuals I as a function of time, $t \in [0, 5]$. Do the sample paths indicate that there may be an epidemic? Estimate the probability of an epidemic.

(c) Change the computer program in the part (b) so that $I(0) = 2$ or $I(0) = 5$. Do the sample paths indicate that there may be an epidemic? Estimate the probability of an epidemic.

18. Suppose the stochastic predator-prey model in Section 9.8 is modified by assuming that the prey death rate depends on both prey and predator densities. In particular, suppose the SDE model is

$$dX = X(a_{10} - a_{11}X - a_{12}Y)\, dt + \sqrt{X(a_{10} + a_{11}X + a_{12}Y)}\, dW_1$$
$$dY = Y(a_{21}X - a_{20})\, dt + \sqrt{Y(a_{21}X + a_{20})}\, dW_2,$$

where $X(t)$ and $Y(t)$ are nonnegative with absorbing boundaries when $X = 0$ or $Y = 0$.

(a) Make a table of probabilities for births and deaths that leads to the preceding Itô SDE predator-prey model.

(b) Write an alternative set of SDEs that model this same process, $dZ = \mu\, dt + B\, dW^*$.

(c) Let $a_{10} = 1$, $a_{11} = 0.005$, $a_{12} = 0.02$, $a_{20} = 1$, and $a_{21} = 0.01$. Show that for the underlying deterministic model, $(x(t), y(t))$, there exists a positive equilibrium given by $(\bar{x}, \bar{y}) = (100, 25)$, which is locally asymptotically stable.

(d) Use the Euler-Maruyama method to compute several sample paths of the stochastic model for the parameter values in part (b) when $X(0) = 90$ and $Y(0) = 30$ on the interval $[0, 10]$. Plot two sample paths.

19. Suppose a deterministic model for a predator-prey system is

$$\frac{dx}{dt} = rx\left(1 - \frac{x}{K}\right) - \frac{xy}{x+c}$$

$$\frac{dy}{dt} = y\left(-b + \frac{x}{x+c}\right),$$

where r, K, c, and b are positive constants.

(a) Determine the possible equilibria of this system and the conditions for their local stability. An equilibrium is locally stable if the eigenvalues of the Jacobian matrix J, evaluated at the equilibrium, have negative real parts or if matrix J satisfies $\text{Tr } J < 0$ and $\det J > 0$.

(b) Set up two different but equivalent systems of SDEs that correspond to this predator-prey model.

(c) Select some parameter values for these models based on your results from part (a), then apply the Euler-Maruyama method to the SDEs, and plot some sample paths.

(d) Discuss the results obtained from parts (a)–(c).

20. The SDE for random genetic drift is

$$dX(t) = \sqrt{\frac{X(t)(1 - X(t))}{2N}}\, dW(t), \quad X(t) \in [0, 1].$$

The two boundaries, 0 and 1, are absorbing. Suppose $N = 100$. Use the Euler-Maruyama method with $\Delta t = 0.01$ and perform 1000 simulations until $T = 50$. Let $X(0) = 1/4$. Compute the mean value of the distribution at $T = 10, 20, 30, 40$, and 50.

21. Express the SDE for the population genetics model as an integral equation. Then use property (i) of Theorem 8.2 to show that $E(X(t)) = X(0)$. This is a consequence of random mating and is known as a Hardy-Weinberg equilibrium.

22. Consider the stochastic viral kinetics model formulated in Exercise 11 in Chapter 7. Variable T is the viral template, G is the viral genome, and S is structural proteins (Srivastava et al., 2002). The template serves as a catalyst to promote the formation of structural proteins S and new genomic material G but is not consumed in the process. Let T, G, and S be continuous random variables for this process and let $X = (T, G, S)^{tr}$. The units of T, G, and S are the number of molecules, and time is measured in days. The probabilities associated with the changes in the process are given in Table 9.9.

Table 9.9: Probabilities associated with changes in the viral kinetics model

i	Change, $(\Delta X)_i$	Probability, p_i
1	$(1, -1, 0)^{tr}$	$k_1 G \Delta t$
2	$(-1, 0, 0)^{tr}$	$k_2 T \Delta t$
3	$(0, 1, 0)^{tr}$	$k_3 T \Delta t$
4	$(0, -1, -1)^{tr}$	$k_4 GS \Delta t$
5	$(0, 0, 1)^{tr}$	$k_5 T \Delta t$
6	$(0, 0, -1)^{tr}$	$k_6 S \Delta t$

(a) Write a system of Itô SDEs and a system of Itô integral equations for this intracellular viral kinetics process.

(b) Derive a system of ODEs for the expectations, $E(T)$, $E(G)$ and $E(S)$. Apply property (i) in Theorem 8.2.

(c) Let $k_1 = 0.025$, $k_2 = 0.25$, $k_3 = 1$, $k_4 = 7.5 \times 10^{-6}$, $k_5 = 1000$, and $k_6 = 2$ (Srivastava et al., 2002). Use the Euler-Maruyama method and compute several sample paths for the SDE model when $T(0) = 1$, $G(0) = 0 = S(0)$ and when $T(0) = 5$, $G(0) = 5 = S(0)$. Compare your results with Exercise 11 in Chapter 7.

9.11 References for Chapter 9

Abundo, M. 1991. A stochastic model for predator-prey systems: basic properties, stability and computer simulation. *J. Math. Biol.* 29: 495–511.

Ackleh, A. S., E. J. Allen, R. B. Kearfott, and P. Seshaiyer. 2010. *Classical and Modern Numerical Analysis.* CRC Press, Boca Raton.

Allen, E. J. 1999. Stochastic differential equations and persistence time for two interacting populations. *Dyn. Cont., Discrete and Impulsive Systems* 5: 271–281.

Allen, E. 2007. *Modeling with Itô Stochastic Differential Equations.* Springer, Dordrecht, The Netherlands.

Allen, E. J., L. J. S. Allen, A. Arciniega, and P. Greenwood. 2008. Construction of equivalent stochastic differential equation models. *Stoch. Anal. Appl.* 26: 274–297.

Allen, E. J., L. J. S. Allen, and H. Schurz. 2005. A comparison of persistence-time estimation for discrete and continuous stochastic population models that include demographic and environmental variability. *Math. Biosci.* 196: 14–38.

Allen, E. J., J. Baglama, and S. K. Boyd. 2000. Numerical approximation of the product of the square root of a matrix and a vector. *Linear Algebra and its Applications.* 310: 167-181.

Allen, L. J. S. and E. J. Allen. 2003. A comparison of three different stochastic population models with regard to persistence time. *Theor. Pop. Biol.* 64: 439–449.

Crow, J. F. and M. Kimura. 1970. *An Introduction to Population Genetics Theory.* Harper & Row, New York.

Ekanayake, A. J. and L. J. S. Allen. 2010. Comparison of Markov chain and stochastic differential equation population models under higher-order moment closure approximations. *Stoch. Anal. Appl.* 28: 907-927.

Ewens, W. J. 1979. *Mathematical Population Genetics.* Springer-Verlag, Berlin, Heidelberg, New York.

Gard, T. C. and D. Kannan. 1976. On a stochastic differential equation modeling of prey-predator evolution. *J. Appl. Prob.* 13: 429–443.

Gardiner, C. W. 1985. *Handbook of Stochastic Methods for Physics, Chemistry and the Natural Sciences.* 2nd ed. Springer, Berlin and Heidelberg.

Gillespie, D. T. 1992. *Markov Processes. An Introduction for Physical Scientists.* Academic Press, Inc., San Diego, CA.

Gillespie, D. T. 2000. The chemical Langevin equation. *J. Chem. Phys.* 113: 297-306.

Gillespie, D. T. 2002. The chemical Langevin and Fokker-Planck equations for the reversible isomerization reaction. *J. Phys. Chem. A* 106: 5063-5071.

Gillespie, D. T. and L. Petzold. 2006. Numerical simulation for biochemical kinetics. In: *System Modeling in Cellular Biology From Concepts to Nuts and Bolts*. Szallasi, Z., J. Stelling, and V. Periwal (eds.), pp. 331–353, MIT Press, Cambridge, MA.

Higham, D. J. 2008. Modeling and simulating chemical reactions. *SIAM Review* 50: 347–368.

Goel, N. S. and N. Richter-Dyn. 1974. *Stochastic Models in Biology*. Academic Press, New York.

Kimura, M. 1955. Solution of a process of random genetic drift with a continuous model. *Proc. Nat. Acad. Sci.* 41: 144–150.

Kimura, M. 1994. *Population Genetics, Molecular Evolution, and the Neutral Theory, Selected Papers*. The University of Chicago Press, Chicago and London.

Kurtz, T. G. 1970. Solutions of ordinary differential equations as limits of pure jump Markov processes. *J. Appl. Prob.* 7: 49-58.

Kurtz, T. G. 1971. Limit theorems for sequences of jump Markov processes approximating ordinary differential processes. *J. Appl. Prob.* 8: 344-356.

Ludwig, D. 1974. *Stochastic Population Theories*. Springer-Verlag, New York.

Manninen, T., M.-L. Linne, and K. Ruohonen. 2006. Developing Itô stochastic differential equation models for neuronal signal transduction pathways. *Comp. Biol. and Chem.* 30: 280–291.

Nagylaki, T. 1992. *Introduction to Theoretical Population Genetics*. Springer-Verlag, New York.

Nisbet, R. M. and W. S. C. Gurney. 1982. *Modelling Fluctuating Populations*. John Wiley & Sons, Chichester, New York.

Øksendal, B. 2000. *Stochastic Differential Equations: An Introduction with Applications*. 5th ed. Springer-Verlag, Berlin, Heidelberg, New York.

Ortega, J. M. 1987. *Matrix Theory*. Plenum Press, New York and London.

Serfling, R. J. 1980. *Approximation Theorems of Mathematical Statistics*. John Wiley & Sons, New York.

Smith, H. L. and P. Waltman. 1995. *The Theory of the Chemostat.* Cambridge Studies in Mathematical Biology. Cambridge Univ. Press, Cambridge, U. K.

Srivastava, R., L. You, J. Summers, and J. Yin. 2002. Stochastic vs. deterministic modeling of intracellular viral kinetics. *J. Theor. Biol.* 218: 309-321.

van Kampen, N. G. 2007. *Stochastic Processes in Physics and Chemistry.* 3rd ed. Elsevier Science B. V., Amsterdam, The Netherlands.

Wilkinson, D. 2006. *Stochastic Modelling for Systems Biology.* Chapman & Hall/CRC Mathematical and Computational Biology Series. Boca Raton, London, New York.

9.12   Appendix for Chapter 9

9.12.1   MATLAB® Programs

The following MATLAB programs are for the SDE competition model and for the SDE SIS epidemic model.

```
% MatLab program for the SDE competition model
% One sample path is graphed in the phase plane
clear
set(0,'DefaultAxesFontSize',18);
a10=2; a20=1.5; a11=0.03;
a12=0.02; a21=0.01; a22=0.04;
x1(1)=50; x2(1)=25;
k=5000; T=5; dt=T/k;
for i=1:k
  rn=randn(2,1);
  f1=x1(i)*(a10-a11*x1(i)-a12*x2(i));
  f2=x2(i)*(a20-a21*x1(i)-a22*x2(i));
  g1=x1(i)*(a10+a11*x1(i)+a12*x2(i));
  g2=x2(i)*(a20+a21*x1(i)+a22*x2(i));
  x1(i+1)=x1(i)+f1*dt+sqrt(g1*dt)*rn(1);
  x2(i+1)=x2(i)+f2*dt+sqrt(g2*dt)*rn(2);
  x1p=[x1(i+1)>0];
  x2p=[x2(i+1)>0];
  x1(i+1)=x1(i+1)*x1p;
  x2(i+1)=x2(i+1)*x2p;
end
```

```
plot(x1,x2,'r-'); % One sample path in the phase plane
hold on
xlabel('X_1'); ylabel('X_2');
x=[0:5:80]; axis([0,80,0,50]);
y1=(a10-a11*x)/a12; % x_1 nullcline
y2=(a20-a21*x)/a22; % x_2 nullcline
plot(x,y1,'k-',x,y2,'k-','linewidth',2);


% MatLab program for the SDE SIS epidemic model.
% Three sample paths and probability of no epidemic.
clear
set(0,'DefaultAxesFontSize',18);
beta=2; b=0.5; g=0.5; N=100;
init=1; dt=0.01; sdt=sqrt(dt); time=10;
sim=1000; tot=0;
for k=1:sim
    clear i t
    j=1; i(j)=init; t(j)=0;
    while i(j)>0 & t(j)<time
        mu=beta*i(j)*(N-i(j))/N-(b+g)*i(j);
        sigma=sqrt(beta*i(j)*(N-i(j))/N+(b+g)*i(j));
        i(j+1)=i(j)+mu*dt+sigma*randn*sdt;
        t(j+1)=t(j)+dt;
        j=j+1;
    end
    if i(j)<=0
        tot=tot+1;
    end
    if k==1
        plot(t,i,'r-','Linewidth',2);
    elseif k==2
        plot(t,i,'b-','Linewidth',2);
    elseif k==3
        plot(t,i,'g-','Linewidth',2);
    end
    hold on
end
noepid=tot/sim
y(1)=init;
for k=1:time/dt
    f1=beta*(N-y(k))*y(k)/N-(b+g)*y(k);
    y(k+1)=y(k)+dt*f1;
end
plot([0:dt:time],y,'k--','LineWidth',2);
```

```
axis([0,time,0,80]);
xlabel('Time'); ylabel('Infectious Individuals');
hold off
```

Appendix A: Hints and Solutions to Selected Exercises

A.1 Chapter 1

3. (b) $\mu_X = \dfrac{1-p}{p}$, $\sigma_X^2 = \dfrac{1-p}{p^2}$

6. $\mu = r$, $\sigma^2 = 2r$.

8. (a) $f(x_1, x_2) = f_1(x_1)f_2(x_2) = e^{-x_1}e^{-x_2}$.

    (b) $M_{X_i}(t) = \dfrac{1}{1-t}$

12. Note that $\sum_x f(x) = e^{-\lambda}\sum_x \dfrac{\lambda^x}{x!} = e^{-\lambda}e^{\lambda} = 1$ so that $\sum_x f(x)t^x = e^{-\lambda}\sum_x \dfrac{(\lambda t)^x}{x!}$.

15. (b) $\mu_X(t) = \alpha\beta$, $\sigma_X^2 = \alpha\beta^2$.

16. (a) $M_X(t) = 1/(1-\theta t)$, $K_X(t) = -\ln(1-\theta t)$.

22. The p.d.f. is $f(x) = \exp(-x/10)/10$.

    (a) $\mathrm{Prob}\{5 < X < 15\} = \exp(-1/2) - \exp(-3/2)$.

29. A Poisson distribution with parameter λ has mean and variance equal to λ. For a relatively large sample size $n = 25$, the Central Limit Theorem can be applied: $\mathrm{Prob}\{\bar{X} - \mu \le 0\} \approx 0.5$ and $\mathrm{Prob}\left\{\dfrac{\bar{X} - \mu}{\sigma/\sqrt{25}} \le 1\right\} \approx 0.84$.

30. For a continuous and nonnegative random variable X,

$$\frac{\mu}{c} = \int_0^\infty \frac{x}{c}f(x)\,dx \ge \int_c^\infty \frac{x}{c}f(x)\,dx.$$

31. Let $c = k^2\sigma^2$ in Markov's inequality, $\mathrm{Prob}\{Y^2 \ge c\} \le \dfrac{\sigma^2}{c}$.

A.2   Chapter 2

3. $\sum_{i=1}^{\infty} p_{ij}^{(2)} = \sum_{i=1}^{\infty} \sum_{k=1}^{\infty} p_{ik}p_{kj} = \sum_{k=1}^{\infty} \left(p_{kj} \sum_{i=1}^{\infty} p_{ik} \right) = \sum_{k=1}^{\infty} p_{kj} = 1.$

5. Apply the relationship $P_{ji}(s) = F_{ji}(s)P_{jj}(s)$ when s equals 1.

9. (b) $\{1, 2, 3\}$, periodic of period 2, recurrent

(c) $f_{33}^{(2n)} = (1/2)^n$, $f_{33}^{(2n-1)} = 0$, $f_{11}^{(n)} = f_{33}^{(n)}$. $f_{22}^{(2)} = 1$ and $f_{22}^{(k)} = 0$, $k \neq 2$.

(d) $\mu_{11} = 4 = \mu_{33}$, $\mu_{22} = 2$.

10. (a) (i) reducible, (ii) reducible, (iii) reducible.

11. (a) (i) reducible, (ii) reducible, (iii) irreducible.

13. (a) P_1, P_2, P_3 reducible.

(b) P_1: Communicating classes all transient, $\{1\}$ aperiodic, $\{2, 3\}$ period 2, $\{4, 5, 6\}$ period 3, $\{7, 8, 9, 10\}$ period 4, ...
P_2: Communicating classes all transient with $d(i) = 0$, $\{1\}$, $\{2\}$, $\{3\}$, ...
P_3: Communicating classes all aperiodic, $\{1\}$ is positive recurrent, $\{2\}, \{3\}, \{4\}, \ldots$, are transient.

18. Note that $\sum_{j=1}^{N} \pi_{ij}/N = 1/N$.

19. (b) $P^{2n} = \begin{pmatrix} 1/2 & 1/2 & 0 \\ 1/2 & 1/2 & 0 \\ 0 & 0 & 1 \end{pmatrix}$

22. The even powers of P are

$$P^{2n} = (1)^{2n}\pi(1, 1, \ldots, 1)^{tr} + (-1)^{2n}x_2y_2^{tr} + \sum_{i=3}^{n} \lambda_i^{2n}x_iy_i^{tr}.$$

24. (a) $E((\nu_i + 1)^k) = E(\nu_i^k) + \sum_{r=1}^{k-1} \binom{k}{r} E(\nu_i^{k-r}) + 1$ and

$$[\sum_{j \in \mathcal{A}} + \sum_{j \in \mathcal{T}}]p_{ji} = 1.$$

26. (b) $E(X_{100}|X_0 = 0)$ with $p = 0.45$ and $q = 0.55$ can be found exactly. See the following *Maple* command:

```
sum(binomial(100,k)*0.45^(100-k)*0.55^k*(-100+2*k),k=0..100)
```

A.3 Chapter 3

2. (a) The first-order difference equation is

$$p_{11}^{(n)} = p_{11}^{(n-1)}(1-a) + p_{12}^{(n-1)}a.$$

Because P^{n-1} is a stochastic matrix the column sums equal one. Thus,

$$p_{12}^{(n-1)}(N-1) + p_{11}^{(n-1)} = 1.$$

Solve for $p_{12}^{(n-1)}$ and substitute into the first equation:

$$p_{11}^{(n)} = p_{11}^{(n-1)}\left(1 - \frac{aN}{N-1}\right) + \frac{a}{N-1}.$$

Find the general solution and apply the initial condition $p_{11}^{(0)} = 1$:

$$p_{11}^{(n)} = \frac{(N-1)}{N}\left(1 - \frac{aN}{N-1}\right)^n + \frac{1}{N}.$$

4. (c) Solve $d_{JC} = 0.051745 = -\frac{3}{4}\ln\left(1 - \frac{4q}{3}\right)$ for α, if

$$q = \frac{3}{4}\left[1 - \left(1 - \frac{4\alpha}{3}\right)^{100}\right].$$

6. (c) When $p = q$, $a_k = \frac{N-k}{N}$ but $\tau_k = \frac{k(N-k)}{1-r}$. The general solution for the mean time to extinction, $\tau_k = c_1 + c_2 k - \frac{k^2}{1-r}$. Apply boundary conditions $\tau_0 = 0$ and $\tau_N = 0$.

11. (a) $b_0 = pb_1 + qb_0$ which gives $b_0 = b_1$.

    (b) Solve $b_k = pb_{k+1} + qb_{k-1}$ with boundary conditions $b_0 = b_1$ and $b_N = 1$.

19. (a) Matrix $I - T$ is

$$\begin{pmatrix} \Pi_1 + (b+\gamma) & -2(b+\gamma) & 0 & \cdots & 0 \\ -\Pi_1 & \Pi_2 + 2(b+\gamma) & -3(b+\gamma) & \cdots & 0 \\ 0 & -\Pi_2 & \Pi_3 + 3(b+\gamma) & \cdots & 0 \\ \vdots & \vdots & \vdots & \cdots & \vdots \\ 0 & 0 & 0 & \cdots & N(b+\gamma) \end{pmatrix}.$$

Calculate $\tau^{tr} = 1^{tr}F$, where $F = (I - T)^{-1}$.

452

24. (d) Show that $a_i = i/2N$ is the solution of $a_k = \sum_{i=0}^{2N} p_{ik} a_i$.

25. (c) Show that $a_i = i/2N$ is the solution of $a_k = \sum_{i=k-1}^{k+1} p_{ik} a_i$.

A.4 Chapter 4

2. (b) $\lim_{n\to\infty} \text{Prob}\{X_n = 0\} = \dfrac{1 - (a + b)}{b(1 - b)}$.

6. (a) $p_0 = 0.1$ and $p_1 = 0.09$.

10. $m = R_0 = 1.5$. The probability the disease becomes endemic is $1 - q^k \approx 1 - 0.42^5 = 0.013$.

12. (a) $M = \begin{pmatrix} 1/3 & 1/2 & 1/2 \\ 2/3 & 1/2 & 1/2 \\ 1/3 & 1/2 & 1/2 \end{pmatrix}$, dominant eigenvalue $\lambda = 1.448 > 1$.

(b) $0.514^{r_1} 0.565^{r_2} 0.445^{r_3}$.

16. Applying Sykes criteria, it can be seen that M_1 is regular (g.c.d.$\{1,4\} = 1$) but that M_2 is not regular (g.c.d.$\{2,4\} = 2$).

21. (a) $\lambda_A = 2.12$, $\lambda_W = 1.02$, $\lambda_{Sp} = 0.831$, and $\lambda_{Su} = 0.759$. The product P of these matrices has a dominant eigenvalue equal to $\lambda_P = 0.913$ and corresponding eigenvector equal to

$$\bar{x} = c(0.415, 0.291, 0.205, 0.0885)^{tr},$$

where c is an arbitrary nonzero constant.

A.5 Chapter 5

1. The generator matrix for the simple birth process with states $\{1, 2, 3, \ldots\}$ is

$$Q = \begin{pmatrix} -b & 0 & 0 & 0 & \cdots \\ b & -2b & 0 & 0 & \cdots \\ 0 & 2b & -3b & 0 & \cdots \\ 0 & 0 & 3b & -4b & \cdots \\ \vdots & \vdots & \vdots & \vdots & \vdots \end{pmatrix}.$$

3. (a) $T = \begin{pmatrix} 0 & 1/3 & 0 & 0 \\ 1 & 0 & 2/5 & 0 \\ 0 & 2/3 & 0 & 1 \\ 0 & 0 & 3/5 & 0 \end{pmatrix}$.

4. (c) $\mu_{ii} = 3/2$, $i = 1, 2$.

5. The linear system for the coefficients a_i, $i = 1, 2, 3, 4$ is

$$a_1 + a_2 + a_3 + a_4 = 1$$
$$-a_2 - 2a_3 - 3a_4 = -3$$
$$a_2 + 4a_3 + 9a_4 = 9$$
$$-a_2 - 8a_3 - 27a_4 = -27.$$

7. (e) $\mu_{11} = 4/3$, $\mu_{22} = 2$, $\mu_{33} = 8/3$.

11. (a) $\tau_{101} = \dfrac{1}{d} \sum_{i=1}^{100} \dfrac{1}{i}$.

16. The transition matrix $P(\Delta t)$ for the simple birth and death process with states $\{0, 1, 2, 3, \ldots\}$ is

$$P(\Delta t) = \begin{pmatrix} 1 & d\Delta t & 0 & 0 & \cdots \\ 0 & 1 - (b+d)\Delta t & 2d\Delta t & 0 & \cdots \\ 0 & b\Delta t & 1 - 2(b+d)\Delta t & 3d\Delta t & \cdots \\ 0 & 0 & 2b\Delta t & 1 - 3(b+d)\Delta t & \cdots \\ \vdots & \vdots & \vdots & \vdots & \vdots \end{pmatrix} + o(\Delta t).$$

The corresponding generator matrix Q is

$$Q = \begin{pmatrix} 0 & d & 0 & 0 & \cdots \\ 0 & -b-d & 2d & 0 & \cdots \\ 0 & b & -2(b+d) & 3d & \cdots \\ 0 & 0 & 2b & -3(b+d) & \cdots \\ \vdots & \vdots & \vdots & \vdots & \vdots \end{pmatrix}.$$

17. (a) $\dfrac{dp_N(t)}{dt} = -d_N p_N(t)$.

A.6   Chapter 6

2. (b) $\dfrac{\partial P}{\partial t} = \mu(1 - z)\dfrac{\partial P}{\partial z} + \nu\left(\dfrac{1}{z} - 1\right)P(z, t)$.

4. (a) $Q = \begin{pmatrix} -\lambda N & 0 & 0 & 0 & \cdots \\ \lambda N & -\lambda(N+1) & 0 & 0 & \cdots \\ 0 & \lambda(N+1) & -\lambda(N+2) & 0 & \cdots \\ 0 & 0 & \lambda(N+2) & -\lambda(N+3) & \cdots \\ 0 & 0 & 0 & \lambda(N+3) & \cdots \\ \vdots & \vdots & \vdots & \vdots & \vdots \end{pmatrix}.$

8. (a) The equations for the p.g.f. and m.g.f. are given by

$$\frac{\partial P}{\partial t} = (z-1)\left[b_0 + b_1 z\frac{\partial P}{\partial z} + b_2\left(z^2\frac{\partial^2}{\partial z^2} + z\frac{\partial}{\partial z} \right)P \right]$$
$$+ \left(\frac{1}{z} - 1 \right)\left[d_1 z\frac{\partial P}{\partial z} + d_2\left(z^2\frac{\partial^2}{\partial z^2} + z\frac{\partial}{\partial z} \right)P \right]$$

and

$$\frac{\partial M}{\partial t} = (e^\theta - 1)\left[b_0 + b_1\frac{\partial}{\partial \theta} + b_2\frac{\partial^2}{\partial \theta^2} \right]M$$
$$+ (e^{-\theta} - 1)\left[d_1\frac{\partial}{\partial \theta} + d_2\frac{\partial^2}{\partial \theta^2} \right]M.$$

9. (b) Express the characteristic equation for M in terms of θ. Then solve $dM/d\theta$ to show that

$$M(\theta, t) = [1 + (e^\theta - 1)e^{-\mu t}]^N \exp\left(\frac{\nu}{\mu}[e^\theta - 1][1 - e^{-\mu t}] \right).$$

11. According to Theorem 6.1 a stationary probability distribution exists for the simple birth, death, and immigration process if

$$\sum_{i=1}^\infty \frac{\lambda_0 \lambda_1 \cdots \lambda_{i-1}}{\mu_1 \mu_2 \cdots \mu_i} = \sum_{i=1}^\infty \frac{\nu(\nu + \lambda)\cdots(\nu + (i-1)\lambda)}{i!\mu^i} < \infty.$$

Let $c > 0$ such that $\nu = c\lambda$ and simplify the preceding summation.

13. (a) In the case $\lambda = \mu$, apply L'Hopital's rule. Let $u = \lambda - \mu$ and take the limit as $u \to 0$:

$$m(t) = \lim_{u \to 0} \frac{e^{ut}(Nu + \nu) - \nu}{u} = \nu t + N.$$

16. The following MATLAB® program for case (ii) generates three sample paths and calculates the mean and standard deviation for 1000 sample paths at $t = 10$. See Figure 6.11.

```
clear all
hold off
r=1; K=50; sim=1000;
time=10; nn(1)=1; dt=.01;
set(0,'DefaultAxesFontSize',18)
for k=1:sim
    j=1;
    clear n t
    n(1)=1; t(1)=0;
    while t(j)<time & n(j)>0;
        lam=r*n(j);
        mu=r*n(j)^2/K;
        tot=lam+mu;
        u1=rand; u2=rand;
        t(j+1)=t(j)-log(u1)/(tot);
        if u2<lam/tot
            n(j+1)=n(j)+1;
        else
            n(j+1)=n(j)-1;
        end
        j=j+1;
    end
    nend(k)=n(j);
    if k==1 stairs(t,n,'g-','linewidth',2); hold on; end
    if k==2 stairs(t,n,'b-','linewidth',2); end
    if k==3 stairs(t,n,'r-','linewidth',2); end
end
for tt=1:time/dt
    nn(tt+1)=nn(tt)+dt*r*nn(tt)*(1-nn(tt)/K);
end
plot([0:dt:time],nn,'k--','linewidth',2)
axis([0,time,0,70]);
xlabel('Time'); ylabel('Population size');
m10=mean(nend)
std10=std(nend)
```

20. (a) $\dfrac{dm(t)}{dt} = bm(t) - dE(X^2(t)).$

   (b) $\dfrac{dE(X^2(t))}{dt} = bm(t) + (2b+d)E(X^2(t)) - 2dE(X^3(t)).$

A.7 Chapter 7

5. (a) $\dfrac{d\mathcal{P}}{dt} = -\lambda(\mathcal{P} - \mathcal{P}^{\beta+1})$.

    (b) $\displaystyle\int_z^{\mathcal{P}} \dfrac{du}{u^{\beta+1} - u} = \lambda t$ leads to $\dfrac{1}{\beta}\left[\ln\left(\dfrac{\mathcal{P}^\beta - 1}{\mathcal{P}^\beta}\right) - \ln\left(\dfrac{z^\beta - 1}{z^\beta}\right)\right] =$
    λt. Solve for $\mathcal{P}$.

8. (a)

$$\mathcal{P}_1(z_1, z_2, t) = z_1(1 - G(t)) + p\int_0^t \mathcal{P}_1^2(z_1, z_2, t - u)\,dG(u)$$

$$+ (1 - p)\int_0^t \mathcal{P}_2(z_1, z_2, t - u)\,dG(u)$$

$$\mathcal{P}_2(z_1, z_2, t) = z_2(1 - G(t)) + \int_0^t \mathcal{P}_2(z_1, z_2, t - u)\,dG(u).$$

   (b) Differentiate the expression for $\mathcal{P}_2$ and use the initial condition $\mathcal{P}_2(z_1, z_2, 0) = z_2$.

19. Add the following lines to the Monte Carlo simulation of the SIS epidemic model to graph the probability histogram, where the vertical axis is $\text{Prob}\{I(10) = i\} \times 10^4$ and the horizontal axis is i. Use a total of sim $= 10,000$ sample paths at time $= 10$, and $I(1) = 20$. Then II gives the value of $I(t)$ at $t = 10$.

```
hist(II,[0:1:100]);
mean(II)
std(II)
```

20. (a) The transition probabilities are

$$\text{Prob}\{(\Delta S, \Delta E, \Delta I, \Delta R) = (i, j, k, l)|(S(t), E(t), I(t), R(t))\}$$

$$= \begin{cases} bN(t)\Delta t + o(\Delta t), & (i, j, k, l) = (1, 0, 0, 0) \\ bS(t)\Delta t + o(\Delta t), & (i, j, k, l) = (-1, 0, 0, 0) \\ bE(t)\Delta t + o(\Delta t), & (i, j, k, l) = (0, -1, 0, 0) \\ bI(t)\Delta t + o(\Delta t), & (i, j, k, l) = (0, 0, -1, 0) \\ bR(t)\Delta t + o(\Delta t), & (i, j, k, l) = (0, 0, 0, -1) \\ \beta S(t)I(t)\Delta t + o(\Delta t), & (i, j, k, l) = (-1, 1, 0, 0) \\ \sigma E(t)\Delta t + o(\Delta t), & (i, j, k, l) = (0, -1, 1, 0) \\ \gamma I(t)\Delta t + o(\Delta t), & (i, j, k, l) = (0, 0, -1, 1) \\ \Lambda\Delta t + o(\Delta t), & (i, j, k, l) = (0, 0, 1, 0) \\ 1 - [\beta S(t)I(t) + \sigma E(t) + \gamma I(t)]\Delta t & \\ \quad - [\Lambda + 2bN(t)]\Delta t + o(\Delta t), & (i, j, k, l) = (0, 0, 0, 0) \\ o(\Delta t), & \text{otherwise.} \end{cases}$$

where $(\Delta S, \Delta E, \Delta I, \Delta R) = (\Delta S(t), \Delta E(t), \Delta I(t), \Delta R(t))$.

22. (b) The transition probabilities are

$$\text{Prob}\{(\Delta X_{AA}, \Delta X_{Aa}, \Delta X_{aa}) = (i, j, k)|(X_{AA}, X_{Aa}, X_{aa})\}$$

$$= \begin{cases} bf_{AA}(X_{AA}, X_{Aa})\Delta t + o(\Delta t), & (i, j, k) = (1, 0, 0) \\ bf_{Aa}(X_{AA}, X_{Aa}, X_{aa})\Delta t + o(\Delta t), & (i, j, k) = (0, 1, 0) \\ bf_{aa}(X_{aa}, X_{Aa})\Delta t + o(\Delta t), & (i, j, k) = (0, 0, 1) \\ dX_{AA}\Delta t + o(\Delta t), & (i, j, k) = (-1, 0, 0) \\ dX_{Aa}\Delta t + o(\Delta t), & (i, j, k) = (0, -1, 0) \\ dX_{aa}\Delta t + o(\Delta t), & (i, j, k) = (0, 0, -1) \\ 1 - (b + d)N(t)\Delta t + o(\Delta t), & (i, j, k) = (0, 0, 0) \\ o(\Delta t), & \text{otherwise.} \end{cases}$$

The stochastic process for the total population size, $\{N(t)\}$, $t \geq 0$, is a simple birth and death process provided $b = \lambda = $ constant and $d = \mu = $ constant. In this case, the transition probabilities are

$$\text{Prob}\{\Delta N(t) = i|N(t)\} = \begin{cases} bN(t)\Delta t + o(\Delta t), & i = 1 \\ dN(t)\Delta t + o(\Delta t), & i = -1 \\ 1 - [b + d]N(t)\Delta t + o(\Delta t), & i = 0 \\ o(\Delta t), & \text{otherwise.} \end{cases}$$

The forward Kolmogorov differential equation has the form

$$\frac{dp_i^N}{dt} = b(i - 1)p_{i-1}^N + d(i + 1)p_{i+1}^N - (b + d)ip_i^N.$$

24. (a) The transition probabilities are

$$\text{Prob}\{(\Delta X_i, \Delta Y_i) = (k_i, l_i), i = 1, 2|(X_1(t), Y_1(t), X_2(t), Y_2(t))\}$$

$$= \begin{cases} a_{10}X_1(t)\Delta t + o(\Delta t), & (k_1, l_1, k_2, l_2) = (1, 0, 0, 0) \\ a_{21}X_1(t)Y_1(t)\Delta t + o(\Delta t), & (k_1, l_1, k_2, l_2) = (0, 1, 0, 0) \\ a_{12}X_1(t)Y_1(t)\Delta t + o(\Delta t), & (k_1, l_1, k_2, l_2) = (-1, 0, 0, 0) \\ a_{20}Y_1(t)\Delta t + o(\Delta t), & (k_1, l_1, k_2, l_2) = (0, -1, 0, 0) \\ a_{10}X_2(t)\Delta t + o(\Delta t), & (k_1, l_1, k_2, l_2) = (0, 0, 1, 0) \\ a_{21}X_2(t)Y_2(t)\Delta t + o(\Delta t), & (k_1, l_1, k_2, l_2) = (0, 0, 0, 1) \\ a_{12}X_2(t)Y_2(t)\Delta t + o(\Delta t), & (k_1, l_1, k_2, l_2) = (0, 0, -1, 0) \\ a_{20}Y_2(t)\Delta t + o(\Delta t), & (k_1, l_1, k_2, l_2) = (0, 0, 0, -1) \\ u_{21}X_1(t)\Delta t + o(\Delta t), & (k_1, l_1, k_2, l_2) = (-1, 0, 1, 0) \\ u_{12}X_2(t)\Delta t + o(\Delta t), & (k_1, l_1, k_2, l_2) = (1, 0, -1, 0) \\ v_{21}Y_1(t)\Delta t + o(\Delta t), & (k_1, l_1, k_2, l_2) = (0, -1, 0, 1) \\ v_{12}Y_2(t)\Delta t + o(\Delta t), & (k_1, l_1, k_2, l_2) = (0, 1, 0, -1) \end{cases}$$

The probability of no change in the process is

$$1 - \sum_{i=1}^{2} X_i(t)[a_{10} + a_{12}Y_i(t)]\Delta t - \sum_{i=1}^{2} Y_i(t)[a_{20} + a_{21}X_i(t)]\Delta t$$

$$- \sum_{j=1, j\neq i}^{2} \sum_{i=1}^{2} [u_{ji}X_i(t) + v_{ji}Y_i(t)]\Delta t + o(\Delta t).$$

(b) Show that if $\bar{x}_1 < \dfrac{a_{20}}{a_{21}} < \bar{x}_2$, then $\bar{y}_1 < \dfrac{a_{10}}{a_{12}} < \bar{y}_2$.

A.8   Chapter 8

4. The transition p.d.f. is the p.d.f. of a normal random variable Y, where $p(y, x, t) \equiv p(y)$ when x and t are fixed. It follows that

$$\int_{\mathbb{R}} p(y)\, dy = 1, \quad E(Y) = \int_{\mathbb{R}} yp(y)\, dy = x,$$

and

$$E((Y - x)^2) = \int_{\mathbb{R}} (y - x)^2 p(y)\, dy = t.$$

5. Use the properties of Y to show that conditions (i)$'$ (for $\delta = 3$), (ii)$'$, and (iii)$'$ are satisfied (listed after Definition 8.3). For conditions (ii)$'$ and (iii)$'$ show that $a(x) = c$ and $b(x) = D$. For condition (i)$'$, make a change of variable for $u = y - x - c\Delta t$ so that

$$\int_{-\infty}^{\infty} |y - x|^3 p(y, x, \Delta t)\, dy = \int_{-\infty}^{\infty} |u + c\Delta t|^3 \phi(u)\, du$$

$$\leq 2 \int_{0}^{\infty} (u^3 + 3u^2|c|\Delta t + 3u(c\Delta t)^2$$

$$+ |c|^3 (\Delta t)^3)\phi(u)\, du,$$

where $\phi(u) = \dfrac{\exp(-u^2/[2D\Delta t])}{\sqrt{2\pi D\Delta t}}$. Then apply the following identities which follow from properties of the error function and the normal density:

$$\int_0^\infty u^3 \phi(u)\, du = \sqrt{\frac{2}{\pi}} (\Delta t)^{3/2}$$

$$\int_0^\infty u^2 \phi(u)\, du = \frac{1}{2} \Delta t$$

$$\int_0^\infty u \phi(u)\, du = \frac{1}{\sqrt{2\pi}} (\Delta t)^{1/2}$$

$$\int_0^\infty \phi(u)\, du = \frac{1}{2}.$$

8. (a) $b^3 - a^3$.

9. (a) An SDE is $dX(t) = [(b_1 - d_1)X(t) - (b_2 + d_2)X^2(t)]dt + [(b_1 - d_1)X(t) + (b_2 + d_2)X^2(t)]^{1/2}dW(t)$.

12. $X(t) = x_0 e^{\mu t} + \sigma e^{\mu t} \int_0^t e^{-\mu s}\, dW(s)$.

13. The first two moments are $E(X(t)) = x_0 e^{\mu t}$ and $E(X^2(t)) = x_0^2 e^{2\mu t} + \frac{\sigma^2}{2\mu}(e^{2\mu t} - 1)$.

14. The first two moments are $E(X(t)) = x_0 e^{at}$ and $E(X^2(t)) = x_0^2 e^{(2a+b^2)t}$.

17. Let $F(x) = \ln(ax)$ in Itô's formula.

19. (a) The solution is $Q(x) = \dfrac{e^{-2cB/D} - e^{-2cx/D}}{e^{-2cB/D} - e^{-2cA/D}}$.

22. (d) Rewrite the differential equation as $-1/b(x) = x d^2 m_T(x)/dx^2$, then integrate.

A.9 Chapter 9

4. The SDEs for the second-order moments are

$$dX_1^2(t) = [2a_{11}X_1^2(t) + 2a_{12}X_1(t)X_2(t)]dt + \sigma_1^2 dW_1(t)$$
$$dX_2^2(t) = [2a_{21}X_1(t)X_2(t) + 2a_{22}X_2^2(t)]dt + \sigma_2^2 dW_2(t)$$
$$d[X_1(t)X_2(t)] = [a_{11}X_1(t)X_2(t) + a_{12}X_2^2(t) + a_{21}X_1^2(t)$$
$$+ a_{22}X_1(t)X_2(t)]dt + \sigma_1 X_2(t)dW_1(t) + \sigma_2 X_1(t)dW_2(t).$$

10. $p(x) = \dfrac{c}{(r_1 + r_2)x} e^{Kr/(r_1+r_2)} \exp\left(-\dfrac{r}{K(r_1 + r_2)}[x - K]^2\right).$

13. (a) One system of SDEs is

$$db(t) = b(t)\left(\frac{k_{max}n(t)}{k_n + n(t)} - D\right)dt + \left(\frac{k_{max}n(t)b(t)}{k_n + n(t)}\right)^{1/2} dW_1(t)$$
$$-\sqrt{Db(t)}dW_2(t)$$
$$dn(t) = \left[D(n_0 - n(t)) - \beta\frac{k_{max}n(t)b(t)}{k_n + n(t)}\right]dt$$
$$-\beta\left(\frac{k_{max}n(t)b(t)}{k_n + n(t)}\right)^{1/2} dW_1(t)$$
$$+ (D[n_0 + n(t)])^{1/2}dW_3(t).$$

16. The drift and diffusion coefficients are $\tilde{\alpha}(x) = \beta x(1 - x/N) - (b + \gamma)x$ and $\tilde{\beta}(x) = \beta x(1 - x/N) + (b + \gamma)x$. For $\beta = 2$, $b + \gamma = 1$, and $N = 100$, the quasistationary p.d.f. is $p(x) = cx^{-1}e^{2x}(150 - x)^{199}$. The constant c is found by letting $\int_1^{100} p(x)\,dx = 1$. The mean is $\mu \approx 48.9$ and standard deviation is $\sigma \approx 7.2$. The solution can be found using a computer algebra system.

Index

T - #0064 - 101024 - C0 - 234/156/27 [29] - CB - 9781439818824 - Gloss Lamination